Ecology and Sustainable Development

Volume I

S.C. Bhatia
B.E. (Chemical), MBA

ATLANTIC
PUBLISHERS & DISTRIBUTORS (P) LTD

Published by

ATLANTIC

PUBLISHERS & DISTRIBUTORS (P) LTD

B-2, Vishal Enclave, Opp. Rajouri Garden,
New Delhi-110027
Phones : +91-11-25413460, 25429987, 25466842

Sales Office
7/22, Ansari Road, Darya Ganj,
New Delhi-110002
Phones : +91-11-23273880, 23275880, 23280451
Fax : +91-11-23285873
Web : www.atlanticbooks.com
E-mail : info@atlanticbooks.com

Branch Office
5, Nallathambi Street, Wallajah Road,
Chennai-600002
Phones : +91-44-64611085, 32413319
E-mail : chennai@atlanticbooks.com

ISBN 978-81-269-0960-5 (Vol. I)
ISBN 978-81-269-0962-9 (Set)

Printed in India
at Nice Printing Press, Delhi

Preface

Ecology is the study of organisms in relation to the surroundings in which they live. These surroundings are called the environment of the organism. This environment is made up of different components, including other living organisms and their effects and purely physical features such as the type of climate and soil. Ecology can be considered on a wide scale, moving from an individual molecule to the entire global system.

Ecologists are those who study ecology and are always aiming to understand how an organism fits into its environment. Ecology is like an enormous jigsaw puzzle; each organism has certain requirements for life which interlock with those of the many other individuals in the area.

Human activities have been drastically changing the ecological balance of practically every component of the environment, namely atmosphere, hydrosphere, lithosphere and biosphere. There have been efforts on global level for environmental management, identifying environmental concerns and planning strategies suitable for a particular nation.

This comprehensive textbook presents the significant concepts of modern ecology in an intelligible and easy-to-understand manner. It exposes the reader to different frontiers of ecology and evolution through a critical analysis of different eco-compartments such as community ecology, population ecology, biodiversity, threats to ecology, conservation and restorations. The book is designed for students with some knowledge of biology and chemistry; difficult terminology has been minimised and elementary mathematics has been used for quantitative aspects.

The first volume of the textbook *Ecology and Sustainable Development* deals with topics such as ecosystems, community and ecology, population and biodiversity.

The volume is divided into four sections. Section I deals with basic concepts of ecosystems. Ecosystem can be defined as functional unit of ecology where the biotic and abiotic factors interact with each other. Ecosystem or biocoenose is composed of a community and physical environment it occupies.

Chapter 1 presents introduction and discusses the scope of ecology, its relation to other sciences and the community. Chapter 2 focuses on climate factors. Various ecological and climate factors, which are complex, have been grouped and discussed in detail. Adaptation is the result of natural selection acting on inheritable differences in fitness. Non-inheritable traits such as the age of an individual, may also influence survival and reproduction, but such phenomena have no effect on evolution. Considering this, chapter 3 deals with ecological adaptation. Ecological adaptation in xerophytes, halophytes, mesophytes, hydrophytes, land, plants and animals has also been discussed.

Section II focuses on community and ecology. Communities are integrated groups of species but together in a common environment or habitat. In community, organisms share the same habitat growing in a uniform environment and by definition community must include only living (biotic) entities of the area. In nature, different kinds of organisms grow in association with each other. A group of several species (plants/animals) living in a natural area are known as a community. Keeping this in mind, chapter 4 studies community ecology. Chapter 5 focuses on community dynamics/ecology succession. Communities are never stable; they are dynamic, changing more or less regularly over time and space. Chapter 6 highlights ecosystem ecology. An ecosystem or biocoenose is composed of a community and the physical environment it occupies. Various types of ecosystems, their structure, function and productivity are discussed in detail. The chapter also discusses energy flow in ecosystem and various atmospheric and nutrient cycles. Chapter 7 elucidates biotic diversity and community stability. Climate is the major determinant of vegetation. Plants in turn exert some degree of influence on climate. Both climate and vegetation profoundly affect soil development and the animals that live in an area. Considering this, chapter 8 studies interface between climate and vegetation. Chapter 9 throws light on resource acquisition and allocation.

Section III deals with population. Although ecologists are often interested in the complex interaction between species, it is worthwhile remembering that it is individual organisms that are the products of natural selection. Chapter 10 discusses the biology of individuals. Individuals are the fundamental units of population, community, ecosystems and biomes. Chapter 11 focuses on population ecology and population dynamics. The chapter highlights the basic concepts of population ecology, its characteristics, theory of population growth and regulation of population density. Chapter 12 is devoted to interactions between population. Various aspects such as direction interactions, complex population interactions and indirect interactions are briefly discussed. Chapter 13 deals with population growth. Various factors which regulate population size and pattern in population are discussed in detail. Chapter 14 concentrates on habitats and niches. Methods of determining niches and habitats along with Gause's competitive exclusion principles and species coexistence are the highlights of this chapter. A convenient way to analyse the feeding relationship of a community is to analyse its trophic levels. Keeping this in mind, chapter 15 focuses on trophic levels.

Section IV deals with biodiversity. Biodiversity is an abbreviation of biological diversity and is a term encompassing a variety of organisms at all levels—from genetic variants belonging to the same species—to species diversity including the variety of ecosystems. The conservation of biodiversity includes the preservation of genetic variation, the diversity of species and population and also life support properties of ecosystem such as climatic and drainage effects. Chapter 16 is an in-depth study of biological diversity. The biological diversity at its most basic level includes the full range of species on earth such as bacteria, viruses and protists, and multi-cellular kingdom of plants, animals and fungi.

Patterns of biodiversity are created by ecological and evolutionary processes. Considering this, chapter 17 deals with creation of biodiversity. Chapter 18 focuses on types of biodiversity. Various types of biodiversities along with methods of measuring and global biodiversity patterns are briefly discussed. Soil is one of the most important ecological factors. Plants depend for their nutrients, water supply and anchorage upon the soil. Keeping this in mind, chapter 19 concentrates on soil. The chapter describes various factors affecting soil formation, soil erosion and soil conservation. Chapter 20 is an in-depth study of succession which is defined as a continuous, undirectional, sequential change in the species composition of a natural community. A mid-successional community is characterised by high biomass, high levels of organic nutrients and high species diversity. Chapter 21 emphasises on biomes which represent the major community types of the world, consisting of distinctive combinations of plant and animal species characterised by an approximately uniform life form of vegetation such as grass or deciduous trees. Biomes provide a convenient shorthand for describing the world's flora and fauna.

The text in this book is throughout supplemented with diagrams and tables. All the topics have been treated in a cogent, lucid style so as to enable the readers to grasp the information quickly and easily. The References and Index which are given at the end will prove useful study-aids to readers.

This textbook will be most useful to the students of undergraduate, graduate and postgraduate courses in Ecology, Environmental Biology, Environmental Sciences, Life Sciences, Botany, Zoology, Forestry, Agriculture, Toxicology and to those working in these fields.

While utmost care has been taken to make this book useful, exhaustive and error-free, the author would welcome any constructive suggestions from students, teachers and professionals in these fields.

S.C. Bhatia

Contents at a Glance

Contents

SECTION IV

SECTION I

Introduction to Ecosystems

1

Introduction

INTRODUCTION

The two components of nature, organisms and their environment are not only much complex and dynamic but also interdependent, mutually reactive and interrelated. Ecology, relatively a new science, deals with the various principles which govern such relationships between organisms and their environment. The term ecology was coined by combining two Greek words, *oikos* (meaning 'house' or 'dwelling place') and *logos* (meaning 'the study of') to denote such relationships between the organisms and their environment. Thus, literally, ecology is the study of organisms 'at home'.

The recent development in study of ecology has been the recognition of the fact that the biotic (living) and abiotic (non-living) components of nature are not only interrelated but both these components function in an orderly manner as a definite system. Thus, structure and function should be studied together for fuller understanding of this vast nature.

Biotic factors are the other organisms encountered, whether of the same or different species. Abiotic factors are the physical and chemical conditions such as temperature, moisture, respiratory gases and substrate.

BASIC CONCEPTS OF ECOLOGY

Structural Basis

Like other sciences ecology too has its own principles and basic concepts, which are as follows:

1. All living organisms and their environment are mutually reactive, affecting each other in various ways. Animal population, flora and vegetation are interdependent through the environment and are mutually reactive.
2. Environment, which is actually a complex of several inter-related factors and is much dynamic (i.e., varying with time and space), works as a sieve selecting organisms for growth from so many forms, as its one or the other factor becomes critical at critical stages of the life cycle of the species.
3. The species puts each effort to maintain its uniformity in structure, function, reproduction, growth and development by preservation of its genetic pool. However, species is also plastic and reacts to the varying environment to get itself adjusted structurally and physiologically in the changing environment. This is achieved by the degree of plasticity set by the genetic constitution of the species. The various forms of a species, in order to meet the challenge of changed environment,

may arise by virtue of somatic plasticity, the ecads or by the reorganisation of their genes during sexual reproduction, the ecotypes. Thus species may increase their capacity of tolerance towards changing environment by developing ecads and ecotypes.

4. It is not only the environment which influences the life of organisms, but organisms too modify their environment as a result of their growth, dispersal, reproduction, death, decay etc. Thus, the environment is caused to change due to organism's activities. The dynamic environment and organisms make ways for the development of different kinds of organisms through a process known as succession. The process continues till the development of a community which is now more or less stable and is now able to keep itself adjusted in equilibrium with the environment. This final stage of community is called a climax.
5. Clements and Shelford, however, put forth a concept of biome wherein all plants and animals are related to each other by their coaction and reaction on the environment. According to their view, under similar climatic conditions, there may simultaneously develop more than one communities, some reaching to climax stage, others under different stages of succession. This complex of several communities in any area, represented by an assemblage of different kinds of plants, animals etc., sharing a common climate, is called a biome.

Functional and Structural Basis

In the above account, basic concepts of ecology have been explained mainly upon structural basis. However, with the introduction of ecosystem concept in ecology, functional aspects along with the structural ones are also to be strongly emphasised. Tansley thus emphasised the role of environment, with its various factors interacting with each other in his comprehensive term ecosystem which involves all the non-living and living factors working in a complex. With this new concept in modern ecology, following are the basic concepts:

1. When both, biotic and abiotic components are considered, the basic structural and functional units of nature are ecosystems. Discrete biological units consist of population and communities, including biomes. Each population occupies a specific niche, a unique functional position with respect to other organisms with which it interacts.
2. There exist varying degrees of +, – or even neutral interactions among organisms, at both, inter and intraspecific levels, which determine along with abiotic parameters, the degree of success a particular population has within a given habitat. Population ecologists study interactions at population as well as community levels. They study competition, usually between population from the same trophic level (such as herbivores competing for the same grass i.e., population ecology involving individuals of same species) and prey-predator interactions between members of adjacent trophic levels (i.e., population ecology involving interactions between individuals of different species, at community level).
3. Also, there are involved energetics of ecosystem, as energy is the driving force of this system. The radiant energy is trapped by the autotrophic organisms (producers) and is transferred as organic molecules to the heterotrophic organisms (consumers). This energy flow is uni-directional or non-cyclic.
4. The chemical components of the ecosystem move in defined cycles—biogeochemical cycles. Within the ecosphere, biological systems frequently regulate the rate of movement of cycling of the chemicals. Role of water as the universal solvent for biological systems is much relevant here.

5. Successful growth of the organism is governed by limiting factors. For success in growth and reproduction with a particular habitat, an organism requires various essential factors from its environment The success of an organism is limited not only by deficiencies in substances or conditions but also by excesses. The minimal and maximal levels of tolerance for all ecological factors of a species vary seasonally, geographically and according to the age of the population.
6. Under natural conditions, different kinds of population undergo succession. Ecosystems undergo an orderly process of change with time, passing from a less complex to a more complex state. This process involves not only changes in species composition but also changes in the physical environment of a community. The terminal or stabilised state is known as the climax.

According to Evans, the ecosystem involves the circulation, transformation and accumulation of energy and matter through the medium of living things and their activities. Thus, the dynamic abiotic components of the environment and the assemblage of plants and animals there, as a result of interactions between themselves keep modifying and changing each other and this leads to the development of ecosystem.

7. Then come the probabilities of disruption and exploitation of ecosphere. As a result of natural condition or activities of man, species diversity of an ecosystem is reduced. It leads to a set back to the state of development and reduction in the stability of the ecosystem. Man's exploitation of ecosystems is directed toward channeling productivity to his needs. Applied ecology or human ecology is the use of ecological concepts to describe human activities and the determination of ways in which people can best obtain their needs from ecosystems. Ecosystems which are substantially altered by human activities are called *managed*, whereas those free from such disturbances are referred to as *natural*.

SCOPE OF ECOLOGY

Ecology is the science that needs minimum time and labour for its introduction to a layman. Present day problems of varied nature in human life are directly or indirectly very much related to ecology, as their solution needs an ecological knowledge. These days ecology has been contributing very much to socio-economic, political and other similar policies of the world. There are interdependencies not only between ecology and other areas of plant sciences, but also between ecology and physical as well as social sciences. Ecology indeed plays an important role in human welfare. This is primarily a field subject and modern ecology is concerned with the functional interdependencies between living things and their environment.

Ecology plays an important role in agriculture (crop rotation, weed control, etc.), management of grasslands (range management), forestry, biological surveys, pest control, fishery biology and in the conservation of soil, wildlife, forest, water supplies, etc. The international problem of environmental pollution also needs ecological assistance.

ECOLOGY—SCIENCE OR ART: ITS RELATION TO OTHER SCIENCES

It is natural to think and very often asked whether ecology is a science or an art. Before we think over the issue, we must attempt first to explain what we mean by science and art. Art is doing, science is understanding. For instance, a painter who paints beautiful picture is an artist. This is doing, without understanding anything of the physics and chemistry of action of the pigments and oils being used in painting. On the other hand, the scientist aims to understand the glycerides in oils, their chemistry, the degree of unsaturation in the molecules etc. He does not paint the picture that the artist does. The information gathered through understanding of various principles that govern the function of nature (the science of ecology) are then applied in the welfare of mankind that becomes the art of ecology. Although, it may appear quite surprising as how can we harmonise in one profession both, a science and an art, but it is true and ecology, like some other science, has both, its science as well as its art. Indeed science and art go together and are studied best together. Advancement in art depends on science, as more we shall learn and understand, newer the art would be. And it would be almost futile to investigate science unless it is applied in the development of art. Moreover, it would also be not incorrect to say that art inspires science to investigate more. For instance, if a scientist looks at any picture, he might think as how it can be made more attractive by an advancement in the chemistry of pigments etc. The art of ecology developed first, which may be traced back to the period of prehistoric man. The science of ecology like all phases of learning, developed gradually. But at present, we know much of the science of ecology and it has helped us in the development of its art, as conservation of nature and natural resources, solution of pollution problems, etc.

As regards the relation of ecology to other life sciences, we should know first various divisions of biology. Biology, 'the science of life' has its basic divisions such as morphology, ecology, genetics, evolution, molecular biology and developmental biology. These are basic divisions because they are concerned with fundamentals common to all forms of life. The various forms of living organisms, are the taxonomic divisions of biology which deal with the morphology, physiology, ecology etc., of specific kinds of organisms. The three large taxonomic divisions of biology are, botany (plants), zoology (animals) and microbiology (micro-organisms). Other taxonomic divisions are smaller as they deal with limited group of organisms. These are mycology (fungi), phycology (algae), entomology (insects), protozoology, etc. Thus, ecology is a basic division of biology and as such, is also an integral part of any and all the taxonomic divisions.

An ecosystem is the whole biotic community in a given area plus its abiotic environment. It therefore includes the physical and chemical nature of the sediments, water and gases as well as all the organisms. Ecosystem ecology emphasises the movements of energy and nutrients (chemical elements) among the biotic and abiotic components of ecosystems. Because the ecosystem is the highest level of biological organisation, all ecological concepts can be set within its framework. The biotic components of any ecosystem are linked as food chains. Food chains are interlinked to form complex food webs. Food webs are the basic units of ecosystem ecology. Thus, ecology begins with populations and culminates in ecosystems.

Food webs are basic units since it is around them that energy and nutrient transfers take place. In addition to the organisms, there are also exchanges with the abiotic environment. Figure 1.1 shows the basic patterns of energy and nutrient transfers in a generalised ecosystem. The patterns of energy and nutrient movements differ significantly in their relationships with the abiotic environment and with the ecosystem boundary. Energy flows through ecosystems, being acquired from outside as light energy from the sun and being intimately lost from the ecosystem as heat dissipated by the respiration of all community members. Nutrients are cycled within ecosystems to a much greater extent.

ECOLOGY AND LIFE SYSTEMS

To consider the biosphere and its ecology in their entirety, it is necessary to look at several levels in which life exists. The unimaginably huge numbers of individual organisms in the biosphere belong to species or kinds of organisms. Groups of organisms living together and occupying a specified area over a particular period of time constitute a population; and that part of Earth on which they dwell is their habitat. In turn, various population coexist in biological communities.

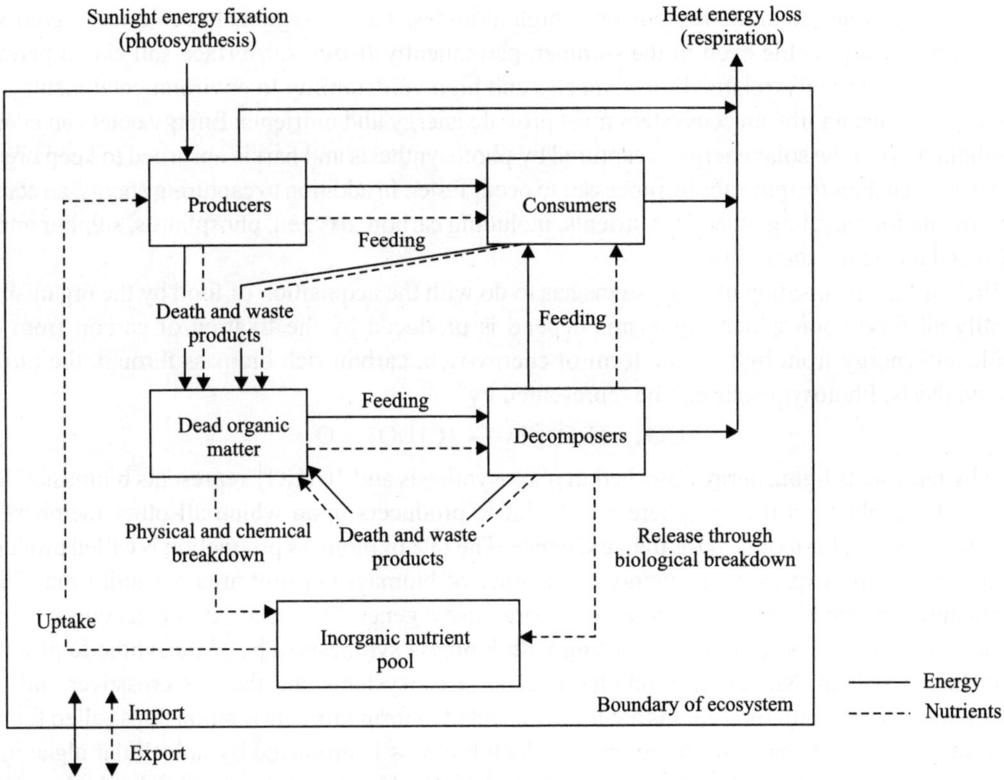

Fig. 1.1. Patterns of energy and nutrient exchanges in a hypothetical ecosystem.

Members of a biological community interact with each other and with their atmospheric, aquatic and terrestrial environments to constitute an ecosystem. An ecosystem describes the complex manner in which energy and matter are taken in, cycled and utilised; the foundation upon which an ecosystem rests is the production of organic matter by photosynthesis. Assemblies of organisms living in generally similar surroundings over a large geographic area constitute a biome. Each biome may contain many ecosystems. The following are examples of important kinds of biomes:

1. Tropical rain forests characterised by warm temperatures throughout the year and having most of their nutrients contained in the bodies of the organisms populating the forest.
2. Warm-climate evergreen forests found in the southeastern United States.
3. Coniferous forests in temperate climates that have distinct summer and winter seasons and that are populated by cone-bearing trees with needles, such as cedar, hemlock and pine.
4. Temperate deciduous forests growing in regions with hot, wet summers and cold winters populated by trees that grow new leaves and shed them annually.
5. Grasslands in which grass is anchored in a tough, dense mass of grass roots and soil called sod.
6. Hot deserts populated by cacti, creosote bush, yucca and other species adapted to high temperatures and sparse moisture.
7. Cold deserts in which tough perennial plants, such as sagebrush and some grasses, survive under cool, dry conditions.

8. Tundra found in arctic regions or at high altitudes. Tundra regions have no trees, cold winters with frost possible even in the summer, permanently frozen subsurface soil called permafrost, low productivity, relatively few species and high vulnerability to environmental insult.

In order to sustain life, an ecosystem must provide energy and nutrients. Energy enters an ecosystem as sunlight. Part of the solar energy is captured by photosynthesis and part is absorbed to keep organisms warm, which enables their metabolic processes to occur faster. In addition to capturing energy, an ecosystem must provide for recycling essential nutrients, including carbon, oxygen, phosphorus, sulphur and trace-level metal nutrients, such as iron.

Much of the organisation of ecosystems has to do with the acquisition of food by the organisms in it. Virtually all food upon which organisms depend is produced by the fixation of carbon from carbon dioxide and energy from light in the form of energy-rich, carbon-rich biomass through the process of photosynthesis. Photosynthesis can be represented by

$$CO_2 + H_2O + h\nu \rightarrow \{CH_2O\} + O_2$$

where hv represents light energy absorbed in photosynthesis and $\{CH_2O\}$ represents biomass. Thus, the photosynthetic plants in the biosphere are the basic producers upon which all other members of the community depend for food and for their existance. The rate of biomass production is called productivity. It is conventionally expressed as energy or quantity of biomass per unit area per unit time. The food manufactured by producers is utilised by other organisms generally classified as consumers.

The sequence of food utilisation, starting with biomass synthesised by photosynthetic producers is called the food chain. Numerous food chains exist in ecosystems and there is crossover and overlap between them. Therefore, food chains are interconnected to form intricate relationships called food webs. An example of a food chain would be one in which biomass is produced by unicellular algae in a lake and consumed by small aquatic organisms (copepods), which are eaten by small fish which are eaten in turn by large fish. Finally, a bald eagle atop the food chain may consume the large fish. As shown by this example, there are several levels of consumption in a food chain called trophic levels. In going up the chain, the first through fourth trophic levels are: (i) producers; (ii) primary consumers (herbivores); (iii) secondary consumers (carnivores); and (iv) tertiary consumers, sometimes called top carnivores. In addition to herbivores and carnivores, there are several other classifications of consumers. Omnivores eat both plant matter and flesh. Parasites draw their nourishment from a living host. Scavengers, such as beetles, flies and vultures, feed on dead animals and plants. Detritovorous, such as crabs, earthworms and some kinds of beetles, feed on detritus composed of fragments of dead organisms and undigested wastes in feces. Ultimately fungi and micro-organisms complete the degradation of food matter to simple inorganic forms that can be recycled through the ecosystem, a process called mineralisation. Food webs can be divided into two main categories broadly based upon whether the food is harvested from living population or from the remains of dead organisms. In a grazing food web food, along with the energy and nutrient minerals that it contains, is transferred from plants to herbivores and on to carnivores. Dead organisms become part of a detritus food web in which various levels of scavengers degrade the organic matter from the dead organisms.

The energy content of food passing through each trophic level may be consumed by respiratory processes to maintain the metabolic activity and movement of the organisms at that level, eventually to be dissipated as heat. Energy incorporated into the bodies of organisms or excreted from them as waste products can become part of the detritus food web. A relatively small, but very important fraction of the energy is passed on to the next trophic level when the organisms are consumed by predators.

It is instructive to consider the flow of energy utilisation through various trophic levels. Typically, based upon 100 per cent of the energy from primary producers, the percentages utilised by various trophic levels are decomposers 25 per cent, herbivores 15 per cent, primary carnivores 1.5 per cent and top carnivorous less than 0.1 per cent.

BIOLOGICAL COMMUNITY

A biological community consists of an assembly of organisms that occupy a defined space in the environment. The nature of such a community depends upon the physical and chemical characteristics that influence the life forms in it and upon the interactions of the organisms in the community. A biological community is the biological component of an ecosystem, which includes the organisms and their physical environment.

The community functions in a manner such that it tends to utilise and convert energy and materials in the most efficient manner that will enable the organisms in it to reproduce and thrive. Therefore, the exchange of matter and energy among the organisms and with their physical environment is a key aspect of an ecosystem. The study of biological communities is called community ecology.

There are many interactions of organisms in a community. Many of these interactions are mutually advantageous. For example, grazing animals derive their food from grass growing on grasslands and return nutrients to the grass as manure, urine and mineralised phosphorus, nitrogen and potassium from their own bodies after death. Fierce competition exists between species in biological communities. Species compete for the same nutrients, food sources, space and sunlight. Coexistence often depends upon species finding their own niches. For example, tall trees utilise the sunlight falling directly on a forest canopy for photosynthesis, whereas smaller plants can exist on the more meager light that filters through from above.

Biological communities are subject to constant change. Some of these changes are relatively short term and cyclical, following daily and seasonal patterns. Some desert plants, for example, may lie largely dormant during prolonged periods of dry weather, only to start growing fiercely and burst into bloom when a rare rainstorm occurs. Other changes occur as a habitat undergoes long-term transitions. These may occur, for example, after a forest fire abruptly kills all the tall trees in a community and the plants and the animals in the community undergo several successions until tall trees are eventually re-established. Many transitions are the result of human activities, such as take place when agricultural land is taken out of production for a number of years. In the past, major transitions in biological communities have occurred with changes in climate, such as happened with the retreat of the glaciers after the last Ice Age. Some evidence suggests that global greenhouse warming will cause long-term changes in many biological communities during the next several decades.

Stable biological communities are characterised by a high degree of order and often complex organisation. The organisms in a community undergo constant opposing and compensating readjustment of their behaviour, feeding and reproduction in response to each other and to their surroundings. Therefore, such communities are in a state of homeostasis. An established, stable biological community as a whole is homeostatic.

PHYSICAL CHARACTERISTICS AND CONDITIONS

Biological communities largely develop in response to their physical environment, which includes the three most critical factors of temperature, moisture and light, as well as other factors, such as quality of soil and nutrient supply. In the case of an aquatic environment, mixing and circulation, which bring nutrients up from sediment deposits are important. Another factor that must be considered is variability of the physical environment. Changes in temperature and moisture level can have particularly strong effects on biological communities, forcing them to change in response to the new conditions. Whereas organisms have adapted well to seasonal differences in their environment, sudden and drastic variations can put a lot of stress on a community. With these factors in mind, it is understandable that some of the most thriving, diverse, productive biological communities are found in the tropics where elevated temperatures favour high metabolic activity, intense light enables a high level of photosynthesis, abundant water favours plant growth and relatively constant climatic conditions make it unnecessary for organisms to waste much of their effort adapting to seasonal change.

Consideration of the crucial factors in maintaining biological communities leads to the following important concepts:

1. Limiting factors determine whether a species can exist in a biological community and, if so, its abundance, growth and distribution.
2. Tolerance limits describe the lower and higher values of limiting factors below and above which a species cannot live.
3. The critical factor is the one that is closest to a tolerance limit.

Consider a species of plant that under normal circumstances might be the primary producer of biomass in a biological community. The most prominent limiting factors that determine its survival are light, temperature and water. Typically, during the late summer months, light is abundant and the temperature remains within a range conducive to plant growth, but the supply of moisture becomes marginal due to drought conditions. Therefore, water is the critical factor and in a very severe drought the water supply may fall below the lower tolerance limit such that plants of the particular species wither and die, thus affecting the whole food chain and supply for the biological community.

Topography can have an important influence on a biological community. The population of a community on a south-facing slope (in the Northern Hemisphere) is influenced by the generally warmer temperatures and greater intensity of sunlight compared to that of a north-facing slope. (Advantage has been taken of this phenomenon in reclaimed land on the sites of brown-coal strip mines in Germany. The soil was replaced in terraces having large, gradual south-facing, facing slopes where crops were planted and short, steep north-facing slopes that were not cultivated. This has extended the growing season by several days in both spring and fall, thereby increasing crop yields.) Marked contrasts in communities may exist on either side of a mountain range over which moisture-laden air flows, as illustrated in Fig. 1.2. As warm moist air flows over the upwind slope, it is forced to rise causing it to cool and release precipitation. On the downwind slope the cool air sinks, becomes warmer and has a drying effect on the vegetation and terrain that it contacts. The absence of rain in this region is called a 'rain shadow'.

EFFECTS OF CLIMATE

Climate refers to weather conditions that exist over the long term, whereas weather describes short-term conditions of temperature, precipitation, humidity, wind and cloud cover. Organisms and their biological communities are largely determined by climate. This section briefly addresses the effects of various aspects of climate on organisms.

The productivity of ecosystems is a function of climate. The most productive ecosystems are those in the tropics. Productivity is lower in the temperate zones and much less near the poles. Seasonal cold has a tremendous effect on plants and hence on productivity. Annual plants are killed by freezing whereas deciduous trees lose their leaves and stop photosynthesis when it freezes. Thus cold weather and freezing temperatures tremendously decrease primary productivity.

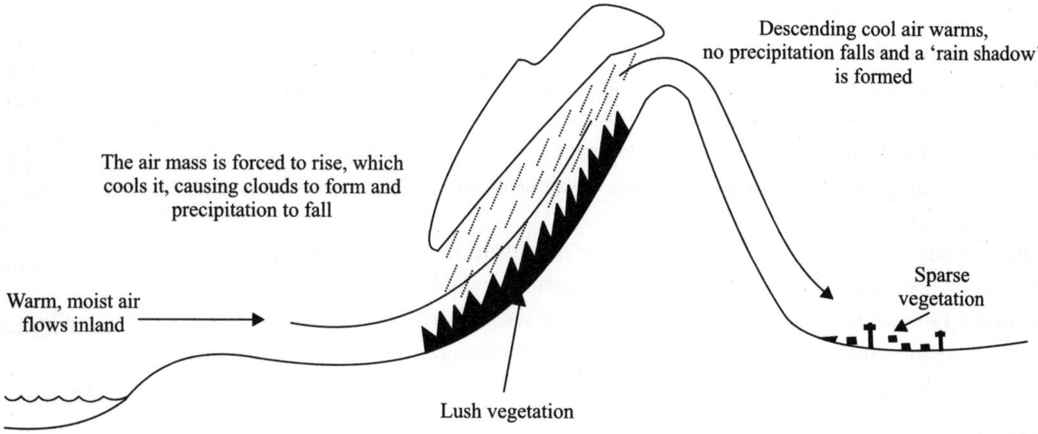

Fig. 1.2. Illustration of the effects of topography on biological communities. A mass of warm moist air forced to rise over a mountain range deposits rain and snow as it does so, enabling a productive community to thrive. On the other side of the range a 'rain shadow' creates drought conditions in which the productivity is low and a much different community exists.

Specific species of plants and animals thrive only within certain temperature ranges. Animals are either poikilothermic, such that their temperatures remain close to those of their surroundings or homeothermic, using intricate physiological processes to maintain their internal temperatures within a narrow range. There are several important ways in which animals maintain desired temperatures. One of the most interesting of these is through circulatory system 'heat exchangers' in which blood flowing in veins from the extremities is warmed by outgoing arterial blood. Fur, feathers, layers of fat and perspiration are all employed to maintain relatively constant temperatures in warm-blooded animals. Some animals hibernate and thus maintain reduced but adequate body temperatures at a low level of metabolism.

The temperatures of plants are largely those of the surroundings. (An interesting exception is the 'warm-blooded' skunk cabbage plant that uses respiratory processes to metabolise energy-producing materials stored in its large root to keep its temperature as much as 30°C higher than the surrounding atmosphere on cold spring days.) However, for most plants temperature extremes cause enzymes to become inactivated and proteins to become denatured. Plant membranes are injured by very high or very low temperatures. Freezing can be very damaging to plants and the first 'killing frost' of autumn signals the demise of many plants that have been thriving during the summer. Ice crystals formed when plants freeze destroy tissue so that it can no longer function. Plants do have some mechanisms to cope with freezing. If the temperature decrease is slow enough, plant tissues can lose water so that they become dehydrated and ice crystals form on the plant surfaces. Resistance to freezing is also accomplished by the presence of sugars, sugar alcohols and amino acids in plant fluids. These substances act as a kind of natural antifreeze. Plants also have mechanisms to resist high temperatures. One of these is for plant leaves to turn at an angle such that they do not receive the full force of mid-day sun.

All life forms require water. Undesirable levels of moisture, usually moisture deficiency, constitute one of the most common environmental stresses faced by organisms. Drought conditions can be very harmful to biological communities. For a terrestrial community a lack of water combined with high temperatures may be devastating to plants and animals. During winter or early spring total levels of water may be adequate, but the water is frozen and not available.

Organisms have various ways to respond to moisture deficiency. Some plants get around the problem by being ephemeral in that their seeds germinate and the plants grow and bloom only when precipitation levels are adequate. Other plant mechanisms for dealing with moisture deficiency include shedding leaves and reducing water loss to the atmosphere by transpiration. This is accomplished by closing stomata and with structures such as thicker leaves or waxy surfaces that slow moisture loss. Succulent plants store large quantities of water in cells, whereas others have extremely deep root systems that reach underground water levels.

Animals have evolved that can tolerate some dehydration. Nocturnal animals avoid the extreme drying conditions of hot desert days. Water generated during metabolic oxidation of food substances is retained in the body. One interesting mechanism of coping with short water supplies is that of producing a very concentrate urine. Such urine, which may have a very pungent odour, is characteristic of the cat family, which is largely adapted to dry conditions.

SPECIES

The kinds of species present in a biological community are largely determined by their environment and in turn, influence their surroundings. The roles played by different species can be categorised in several different ways. Before considering population of organisms, it is instructive to consider the categories and functions of the species that constitute such population.

It should be noted that individuals in species are not identical, but exhibit slight differences resulting in genetic diversity. Living organisms undergo mutations arising from changes in their DNA and through recombinations of their genes that occur with reproduction. Mutations that give rise to favourable traits in individuals give them natural advantages so that they are relatively more likely to survive and reproduce. This natural selection allows species to better adapt to their environment and over many generations, results in permanent changes in genetic composition. This is the process of evolution, through which new species are formed. Natural selection and evolution are very important in determining the occupants of biological communities. Organisms evolve that are best adapted to climatic conditions, particularly those of temperature and moisture and to the presence or deficiency of nutrients. They also evolve in response to competition from other species, diseases caused by bacteria and viruses, parasites and predation by other species.

A diverse biological community cannot exist in which all of the species present are competing for exactly the same limited resources. Therefore, in order to thrive, a species must be specialised and have its own ecological niche, which basically describes where the organism lives and how it functions in its environment. The ecological niche of a species defines the habitat that it occupies, the food and nutrients that it utilises, its predators, its interaction with the other biotic and abiotic components of the ecosystem and the seasons and times that it does various things. In order to coexist with the limited resources available, the members of a biological community must partition resources. Organisms vary in the degree to which they require specialised niches.

Niche specialists occupy very well defined niches, tending to require specific kinds of climatic conditions, habitats and kinds of foods. Specialists are vulnerable to destruction or disturbance of their niches. Such a specialist is the spotted owl, which requires old-growth timber in the US Northwest for its habitat. Tropical rain forests are particularly notable for the wide variety of niche specialists that occupy them. So many niches are possible because of the stratification of the forests and their long-term environmental and biological stability. That is one reason that damage to or destruction of tropical rain forests poses a particular threat of species extinction.

Niche generalists are organisms that are much more adaptable to different environments. Some niche generalists have adapted so well to habitats modified by human that they have become nuisances. An example is the coyote, which coexists with human in some suburban areas, occasionally killing an unwary pet cat. Another example is the Canadian goose, which has taken a strong liking to country club lakes and suburban lawns and may even aggressively object to sharing these niches with the human that they encounter. The human species, itself, is an excellent, resourceful niche generalist.

It is useful to define several other categories of species in discussing ecosystems. Native species are those that are naturally present in an ecosystem, whereas those that have been introduced, usually by human intervention, are called immigrant species or alien species. In some cases immigrant species have been beneficial, whereas in other cases they have displaced native species with catastrophic results. Some species have such important functions in an ecosystem that they are called keystone species. Keystone species include some predators that keep numbers of otherwise destructive species in check, insects that pollinate flowers and species that modify habitat to provide resources and dwelling space for other species. An example of the last of these is the American alligator, which digs depressions in swampland that provide water habitat for numerous species during droughts.

In considering pollution and damage to ecosystems, indicator species are particularly important. These are species whose numbers decline or exhibit symptoms of malaise as a reflection of habitat damage before other major symptoms are observed. The decline of hawks, eagles and other predatory birds at the top of the food chain was indicative of widespread DDT pesticide pollution during the 1950s. The current loss of frogs and other amphibians worldwide is viewed by some experts as an indication of widespread environmental deterioration and pollution.

POPULATION

Recall that population consist of groups of organisms living together and occupying a specified area over a particular period of time. Population have numerous characteristics, including numbers, genetic composition, birth and death rates and age and sex distribution. This section addresses some of the important factors involved with population in biological communities.

All existing species are present on Earth because they have developed exceptional abilities to reproduce and to survive long enough to do so. The potential for reproduction always greatly exceeds the ultimate capacity of an environment to support a population. This capacity is known as the carrying capacity. When members of a population are newly introduced into an area that is amenable to their growth, their numbers tend to increase very rapidly at a rate approaching exponential growth. Eventually numbers reach a level around the carrying capacity at which point rapid growth ceases abruptly because of some limiting factor, such as limited food, nutrients, water, air or shelter or because of stress from crowding. After the very rapid growth phase ceases, the number of organisms tends to fluctuate slightly around the carrying capacity as shown in Fig. 1.3.

In some cases a large growth in population can cause it to temporarily exceed the carrying capacity by a significant margin or the carrying capacity can be suddenly reduced by circumstances such as droughts. This can result in a temporary overpopulation until reduced reproduction rates and increased death rates can adjust the population to accord with the carrying capacity. The result can be an abrupt decrease in population or population crash. If the carrying capacity has been altered by overpopulation, such as destruction of grassland by overgrazing or if it has been reduced by external factors, the new population will stabilise at a figure that reflects the new carrying capacity.

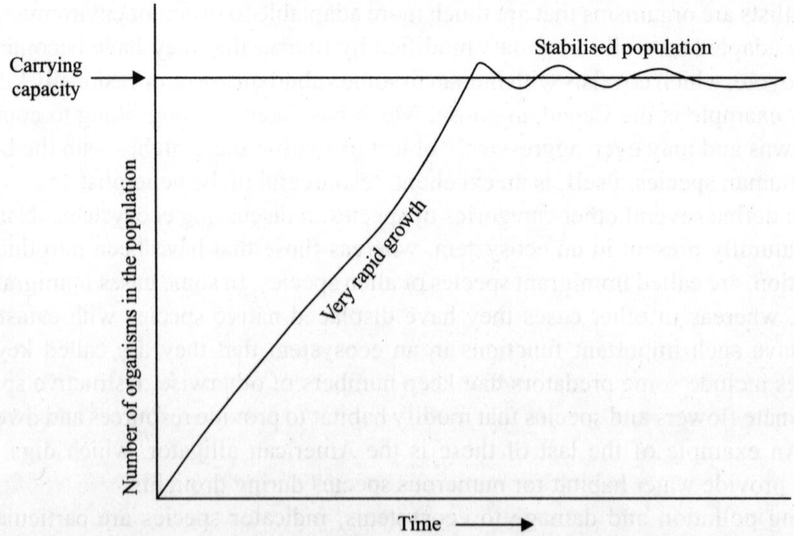

Fig. 1.3. Very rapid growth of a population newly introduced into an environment suitable for its survival, followed by stabilisation of numbers around the carrying capacity. This is an idealised picture subject to perturbation by many factors, such as predation, disruption of habitat or disease.

The physical and chemical conditions of a habitat largely determine the organisms that dwell in it. The biological history of a community is also an important factor. Species may dominate simply because they have been there for quite some time and chance may have played a role in their establishment and survival. The presence and numbers of a species can also be strongly influenced by the other species present. The degree of productivity and the kind of plant biomass produced can largely determine the kinds of organisms in a community. For example, large ruminant animals, such as American Bison, can thrive in areas that are highly productive of prairie grass and squirrels may thrive in wooded areas that produce large quantities of nuts. Species may modify the environment in ways that assist other species. Damming of streams by beavers, which are keystone species in some habitats, can create pools that provide habitat for certain kinds of plants and other kinds of animals. Predation by other species is also important in determining which species exist in a biological community.

Species divide resources for most effective utilisation according to time and space. Some animal species hunt and gather food during the day and others do so at night, thus avoiding needless competition. Some varieties of plants thrive and reproduce in the spring when moisture is abundant, whereas others are adapted to hotter, drier conditions and become dominant during the mid- and late-summer months. One of the major adaptations based upon physical space is that of vertical stratification, which is seen, for example, in forests where tall trees get light for photosynthesis from the top layers of the forests and ferns and mosses use the much less intense, diffuse light at ground level. As shown in Fig. 1.4, vertical stratification is particularly important in bodies of water that are stratified because of the temperature/density behaviour of water.

SURVIVAL OF LIFE SYSTEMS, PRODUCTIVITY, DIVERSITY AND RESILIENCE

Biological communities are subject to various stressful changes that affect their population. These can be divided between the two main categories of catastrophic vs. gradual changes, each of which is divided between those caused by natural factors and those resulting from human activities. Examples of catastrophic natural changes include fire, wind damage, flooding, drought, landslide or volcanic eruption. Catastrophic changes caused by human include deforestation, cultivation, disruption of surface by strip mining and some forms of severe pollution. Gradual natural changes include those due to adaptation, evolution, changes in climate and disease. Gradual changes caused by human include exploitation of natural, soil and wildlife resources; salinisation of soil due to irrigation; addition or elimination of species; and destruction of habitat.

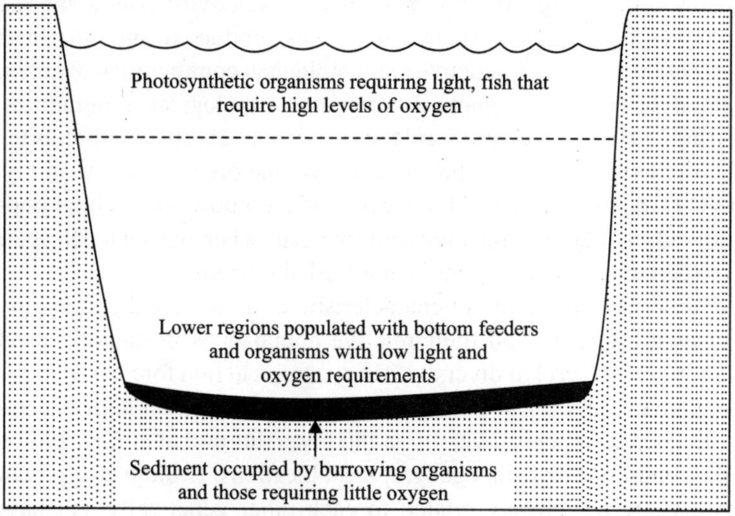

Fig. 1.4. Vertical stratification of a biological community is exhibited in bodies of water, which are often physically stratified because of the temperature/density behaviour of water.

Ecosystems and biological communities that are suffering from stress exhibit warning signs. One of the most common of these is decreased productivity; if the basic food source for a whole community diminishes, the community as a whole must suffer. Commonly associated with decreased productivity is a loss in nutrients. A well-balanced ecosystem recycles and retains nutrients such as nitrogen. Under stress, such nutrients may be permanently lost from the whole system, whereas levels of pollutants may increase as the ecosystem suffers damage. The loss of indicator species is indicative of problems in an ecosystem, as is a decline in the diversity of species present. Increases in the numbers of some species, particularly predatory insects and disease-causing organisms indicate ecosystem damage.

There are several key parameters that are used to describe the ability of biological communities to survive and thrive. These are the following:

1. Productivity, the rate at which a biological community generates biomass, almost always by photosynthesis, required to sustain the organisms in it.
2. Diversity, which describes how many different species are present and their relative abundances.
3. Inertia describes the resistance of a community to being altered, damaged or destroyed.
4. Constancy is the ability to maintain numbers within the optimum limits that can be supported.
5. Resilience, the ability to recover from perturbations, which may be of a catastrophic nature and kill large numbers of organisms in the community.

The productivity of a biological community by the plant producers in it is influenced by many factors. One such factor for land-based biological communities is the nature of the geosphere to which the community is anchored. A key geospheric component is soil; the quality of soil and its nutrient contents largely determine biomass yield. In some communities soil is virtually absent and what little productivity there comes from lichen growing on rock surfaces and hardy plants that anchor their roots between rocks. Topography can influence productivity to an extent. Soil on steep slopes, for example, may be subject to erosion and hence less productive than soil on gently sloping terrain. In general, higher temperatures and higher precipitation favour productivity. Therefore, rain forests tend to be the most productive biological communities and dry, cold deserts have very little productivity.

Overall, ecosystems in the world's oceans have lower productivity than do those on land. However, biological communities along ocean shores are usually very productive and estuaries where fresh water and saltwater merge contain the most productive of all biological communities. Warm seawater in tropical regions provides habitat for the most productive marine-based biological communities. On the high seas, cooler temperatures, limited light penetration and the lack of availability of nutrients tends to keep marine productivity low. Dead biomass and fecal pellets from marine organisms commonly sink to the ocean floor where they are not readily available for recycle of the nutrients, such as phosphorus, that they contain. Areas of particularly high marine productivity occur when nutrient-rich sediments are brought to the surface by convection currents, a phenomenon called upwelling.

Species diversity describes an important characteristic of a biological community based upon how many different species are present and their relative abundances. Relatively productive biological communities often show a high level of diversity. Thus, a tropical rain forest provides habitats for a large variety of plants, mammals, reptiles and lizards; whereas an arctic biological community may have comparatively few. Related to species diversity is the observation that similar biological communities develop under similar conditions in distant geographical locations. Ecological equivalents is a term used to describe species that are not closely related to each other genetically, but have evolved similar characteristics and behaviour patterns that enable them to occupy similar niches in widely different locations. An example is provided by the African jackal and the North American coyote, which, though entirely different species, fill similar ecological niches on their respective continents and exhibit some similarities in their traits and appearance.

High productivity and species diversity can help a biological community to be resilient and stable. A resilient community subjected to stress, such as from a drought or disease affecting one or more of its members, has the ability to deal with and compensate for change; whereas stress applied to a community with only a few members that is marginally productive may be catastrophic for the community as a whole. One that has a large number of species and a cushion of high productivity will be in a much better position to recover from damaging events and circumstances.

RELATIONSHIPS AMONG SPECIES

The interactions of species in a biological community have a strong influence on determining its nature. The crucial aspects of such interactions are the following:

1. Biomass photosynthesised by plants almost always provides the basic food source upon which the rest of the community depends. Most of this food is usually provided by a dominant plant species, such as algae growing in a pond.
2. The physical nature of the environment is modified by the species in it, particularly the dominant plant species. For example, sagebrush on arid lands holds soil in place and provides shelter for small animals.
3. Competition exists between different species for food, sunlight and space.

4. Beneficial and antagonistic relationships may exist between species.
5. Predation occurs when an organism feeds upon another.
6. Parasitism describes the activities of organisms that feed off other live host organisms, usually without causing death to the host. Pathogens are disease-causing organisms, usually microbial bacteria, protozoa, fungi or viruses.

As shown in Fig. 1.5, a dominant plant species anchors the community as its major producer of biomass. In addition to providing most of the food through photosynthesis that the rest of the community uses, the dominant plant species often acts to modify the physical environment of the community in ways that enable the other species to exist in it. For example, the trees in a forest community provide the physical habitat in which birds can nest, relatively safe from predators. In addition the trees provide shade that significantly modifies the habitat at ground level and prevents the growth of most kinds of low-growing plants.

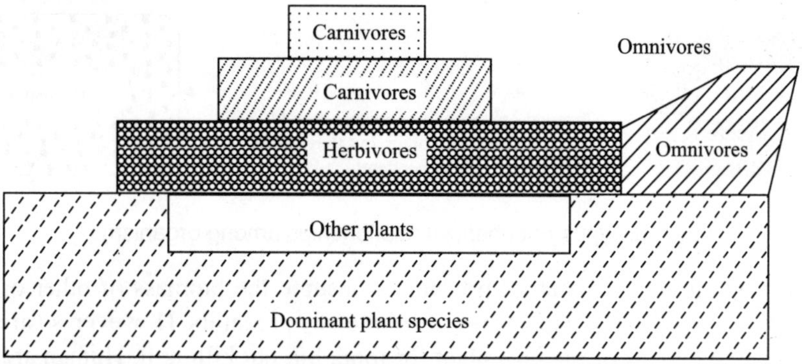

Fig. 1.5. A dominant plant species typically provides most of the food for a biological community. The biomass that it produces is consumed primarily by herbivores, which are fed upon by carnivores, of which there may be more than one level. Omnivores feed on both plants and animals.

Competition exists between organisms for energy, matter and space. Members of the same species may compete, as may members of different species. Adult plants may shade the ground such that seedlings of the same or other plant species cannot get started. Competition can be more overt, such as occurs when plants secrete substances through their leaves or roots that deter the growth of other plants. Animals exhibit territoriality in which they define areas from which they try to exclude competing animals, including those of their own species that are not one of a mating pair or part of a larger family unit. When two or more similar species exist in the same biological community on a steady basis, it is because they have evolved in ways that reduce competition for food and space; that is, they do not occupy exactly the same ecological niche. This phenomenon is called the principle of competitive exclusion.

Symbiotic relationships are those in which species live in close association with each other. Symbiosis is extremely common and is closely related to the other aspects of biological communities, such as the ways in which organisms utilise energy and nutrients. The classic example of symbiosis is that of lichen in which a species of algae exists within the matrix of filaments produced by a fungus.

The fungus provides a hospitable physical environment that is anchored to rock, retaining moisture and extracting nutrients from the rock. The algae fix carbon as biomass that is utilised by the fungi. Other examples include mycorrhizae, in which fungal hyphae associate with plant roots, enabling the roots to take up adequate quantities of water and nutrients and nitrogen-fixing bacteria that exist as nodules on the roots of leguminous plants.

The possible beneficial relationships among organisms can be divided into the three classifications as shown in Fig. 1.6. Mutualism occurs when two species must have each other to exist; commensalism when one species requires the other, but the reverse is not true; and protocooperation when two species benefit from each other, but can exist independently.

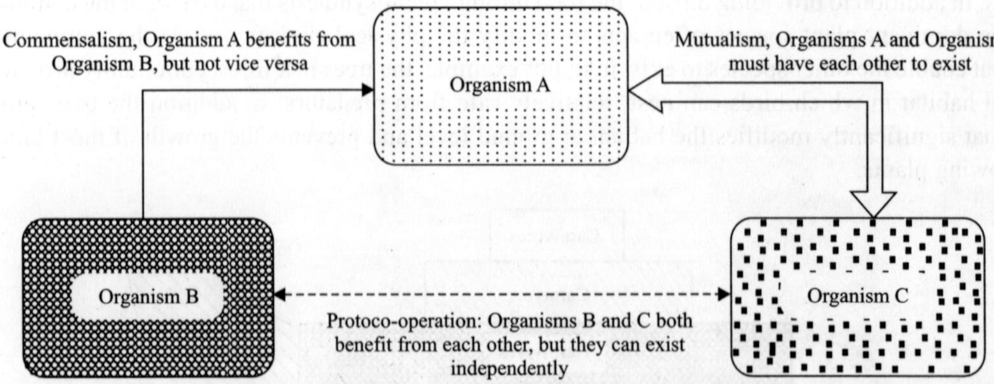

Fig. 1.6. Types of beneficial relationships among organisms.

In the broadest definition of the term, predators consume the biomass of other organisms. Meat-eating carnivores that pursue and kill their prey are obviously predators. However, so are herbivores that graze on grass, consuming part of the host plant for food, but not killing it. Human are omnivores that will eat both meat and vegetable matter.

Predation in a biological community can be quite complex. The predators have highly specialised means for getting food and their potential prey have evolved ways to prevent this from happening. In some cases predation is of benefit to the prey. Through co-evolutionary processes predators and prey have evolved such that both may survive. Stages of the life cycle are very important for both predators and prey. Most very young predators are not sufficiently developed to feed on the prey of adults, who care for them until they can feed themselves. Tadpoles are herbivores, but turn into carnivorous frogs. Very young plants provide excellent food sources for small predators, such as cutworms, whereas the fully grown plants are prey for larger animals, such as ruminant animals.

Different levels of feeding exist in a biological community; these are called trophic levels. Trophic levels are illustrated clearly in some aquatic ecosystems. In such systems the most abundant primary food source constituting the base of the grazing food web may consist of very small, free-floating photosynthesising organisms called phytoplankton. These are fed upon by very small, free-floating, single-celled and invertebrate animals called zooplankton. (The larvae and eggs of many somewhat larger aquatic animals, such as crustaceans, are planktonic.) Small aquatic creatures, such as small fish, feed on the free-floating plankton. Organisms called filter feeders efficiently strain water through special structures to remove plankton for food. Organisms that feed on plankton are in turn consumed by larger creatures. This phenomenon gives rise to several levels of carnivores. It can result in the concentration of some kinds of aquatic pollutants that accumulate in progressively higher levels up the food chain, a phenomenon called biomagnification.

CHANGING COMMUNITIES

The nature of biological communities changes over both space and time. The borderline between communities is often a region of unique character and activity. Communities begin, develop and end, sometimes over very long periods of time, sometimes abruptly. Human intervention is one of the major ways in which biological communities are forced to change.

This section outlines the general way in which a community changes. It should be emphasised, however, that such changes are not as predictable as was once believed. It cannot be known for certain that the succession will proceed in a specific direction or that it will end up with a readily predicted population of organisms. Random chance plays a part in such uncertainty. Furthermore, human effects, which may be indirect and the result of human activities remote from the community add a note of uncertainty. However, human intervention intelligently applied can actually hasten the succession processes in biological communities and direct them in beneficial ways. Such measures include, as examples, replanting of forest trees after clear cutting or deliberate introduction of prairie grass species on worn-out agricultural and that is being restored to a native state.

The most fundamental way in which a biological community can change is when a new community becomes established on a newly created site where life of the sort that eventually populates the community has been essentially absent. The process of establishing life forms in such an area is called primary succession. On land primary succession occurs when a community develops in a location where soil is initially absent. This occurs, for example, with the establishment of a community on a new lava flow or newly exposed rock or gravel removed as overburden in strip mining. The process of colonising such an area is called primary succession. It is one case of the more general process of ecological succession, a term used to describe the changes in species abundance as a result of changes in the environment of a biological community. The initial colonisation of such a site is accomplished by a pioneer species that is hardy enough to exist under the harsh conditions often presented by a new site. Freshly exposed rock is often colonised by colonies of lichen, described above as symbiotic combinations of fungi and algae. The chemical action of lichen dissolves nutrients from the rock and initiates chemical weathering, which is the first step in soil formation. As soil is formed and crevices develop in the rocks, other organisms begin to populate the community and eventually displace the pioneer species. Newly established aquatic ecosystems may eventually fill in with debris and a succession of plants follows that can eventually change the system from an aquatic one to a swamp, then to a grassland and eventually to a forest.

A mature biological community that lasts for very long time periods is called a climax community. (Although climax community is used here, some authorities prefer other terms, such as mature community or relatively stable community, which imply a less permanent state.) Consider the development of a climax community on land. Long before a mature community develops, productivity is low, but goes up gradually as the community acquires increased ability to utilise the resources in it. Soil forms, becomes deeper and increases, in organic content. In general, as the community develops, it becomes richer in organic matter, both living and dead. The number of species usually increases as the total number of organisms rises. The growth of trees and other taller plants enhances the three-dimensional character of the community. The number and variety of habitats become larger. Shading and shelter from larger plants give rise to small microhabitats that are protected from wind, sun and temperature extremes, thus enabling colonisation by small, specialised organisms that thrive in specialised microhabitats. As conditions change, successions of population of various organisms occupy niches in the community. There is a tendency for successive population to be both larger and longer lived, such as is the case with trees in an old-growth forest. Finally, a climax community is achieved that has reached a steady-state condition with its terrestrial, atmospheric and aquatic environment and is in a homeostatic state. It usually has a relatively large amount of organic matter; compare, for example, the high organic content of a rich soil with the negligible amount of organic matter on rock colonised by lichen. A climax community does not exhibit sudden onset of populations of different species or loss of species. The species in the community go through their life cycles in ways that maintain community stability and homeostasis. The following are some important characteristics of a climax community:

1. High diversity of life forms.
2. A high degree of order.
3. Narrow specialisation of species in their appropriate niches with a high degree of stratification.
4. Conservation and recycling of nutrients.

Secondary succession occurs when a less drastic, but still major change has occurred in a biological community. An example of secondary succession would be the establishment of a highly modified community on forest land that has just been subjected to clear-cutting. Annual plants are the first to colonise newly exposed soil. These are followed by small perennials, such as grass and successively larger perennials, such as shrubs, bushes, small trees and finally large trees.

A special kind of community is that which is adapted to periodic and dramatic alteration, especially from fire. Fire-climax communities are populated by species that quickly become re-established following a fire. Some species of trees even require fire to reproduce.

The borders of different biological communities have special characteristics called edge effects. A shoreline where grassland borders on a body of water is such an edge. Such areas provide special habitats for various kinds of plants and animals that can take advantage of both kinds of habitat. Edges often show a high degree of biological diversity. Somewhat similar to edges are patches that are present within a community due to some discontinuity in the physical structure of the community. A rock outcropping in the middle of a hilly grassland is an example of a patch. Patches may show a different distribution of species from those present in the main community.

HUMAN EFFECTS

Human intervention has a large potential effect on biological communities. Sometimes such intervention involves drastic physical alteration of the community, such as by ploughing grasslands or cutting forests. Other types of human intervention may be less subtle, but may still have drastic effects.

An important way in which human may influence biological communities is through the introduction of new species. If it is successful in a community, a new species affects those already there and may significantly modify the physical nature of the habitat. New species may prey upon those already present or serve as prey that attracts predatory species from outside the community. When forests are cut and grasslands established, larger numbers of herbivores and representatives of species not previously present are attracted. These animals in turn attract carnivores that feed on them. Parasites usually accompany newly introduced species that can serve as their hosts.

Some introduced species are particularly destructive to biological communities and habitat. One of the worst of these is the goat, which has a well-earned reputation for indiscriminate consumption of vegetation, destruction of plant life and damage to sod with its hard hooves. Rats introduced onto islands have wreaked havoc with indigenous species. Domestic house cats reverting to a wild state have wiped out whole population of birds. Aggressive bird species, particularly house sparrows and starlings, have displaced more desirable native species.

The human species has become inextricably linked with technology such that in a sense *Homo sapiens* are not 'natural animals'. Much of what is known about the effects of human on biological communities is negative—destruction of habitat, emission of pollutants to the environment, a potential permanent change in climate from greenhouse gases. These kinds of influences are unfortunate and very harmful to biological communities. However, humen are linked to technology irreversibly and it will be necessary for human to adapt themselves and their technologies to the biological communities upon which ultimately humankind depends for its existence.

HUMAN ACTIONS TO PRESERVE AND IMPROVE LIFE ON EARTH

Technology can be harnessed to preserve and improve the condition of life on Earth. A prime example of how that is done is provided by agriculture. With the application of plant genetics, herbicides, fertilisers and advanced cultivation and harvesting techniques, agricultural interests can vastly increase the productivity of a plot of soil. By building terraces and waterways planted to grass that forms a tough, erosion-resistant sod, the productivity of land may be increased while erosion is slowed to a negligible level.

Human intervention can be used to create and enhance habitats that are not maintained for agricultural production. Although not all reservoirs formed by damming streams are desirable, many provide a welcome variety of habitat for species that live in or around bodies of water. Impounding water can cause suspended material to settle from streams, thus improving stream quality below the dams. In a very limited, but encouraging number of cases, human intervention is being applied to reverse damage done to habitats by human activities in the past. Once meandering streams straightened and turned into ugly, erosive ditches by channelisation have been restored in some cases to provide the bending channels that make the stream hospitable to life. Productive wetlands are being restored or even constructed where none existed before, usually as a means to aid waste-water treatment. Badly used, eroded farmland has been converted to forests and grassland. Special structures can be made and sunk in shallow coastal areas to provide shelter and habitat for marine life; even old ship hulls and airplane fuselages have been used for this purpose.

The restoration of ecosystems by human intervention is called restoration ecology. Restoration ecology has become a significant area of human endeavour and it may be hoped that it will increase in importance as technology is used increasingly to benefit the natural environment. The restoration ecologist needs to be familiar with basic ecology, as well as with the kinds of technology used to rebuild ecosystems. A knowledge of related areas, such as geology, hydrology, limnology and soil science is also required. After catastrophic floods along the Missouri and Mississippi Rivers in the US in 1993, the deliberate decision was made to forego reconstruction of some river dikes destroyed by the flood. Restoration ecology was applied to some limited areas to restore wetlands and river bottom lands for wildlife habitat.

Much of the work that has been done to preserve wildlife and to restore ecosystems in which wild species exist has been the result of efforts to maintain and increase numbers of game animals. Enlightened hunting and fishing laws have reduced the harvest of many species to sustainable levels. In some cases these have brought species back from very low numbers or even the brink of extinction. Important examples in the US are American bison, wood ducks, wild turkeys, snowy egrets and white-tailed deer. In addition to hunting and fishing restrictions, habitat restoration has been very important in increasing numbers of game animals. Restoration of wetland breeding areas have enabled significant increases in numbers of waterfowl.

Information is essential in order to understand, preserve and enhance biological systems. The capabilities of technology to gather and process information are enormous. Sophisticated chemical analysis techniques provide detailed profiles of the chemical characteristics of the environment in which organisms live, including both nutrients and pollutants. Sensors for temperature, wind, moisture and sunlight can be used to give a continuous picture of the physical environment. This and other information can be subjected to sophisticated computer analysis to provide a profile of the life system and to direct human intervention in constructive ways.

One of the more useful relatively recent technologies used to study life systems consists of satellite images of Earth, such as those provided by the Landsat satellite. Such images can be gathered by infrared measurements, digitised and processed by computer to provide profiles of geological features, water on Earth's surface and vegetation. By remotely sensing the absorption of electromagnetic radiation at specific wavelengths, instruments mounted on satellites can monitor gases or reactive chemical species in the atmosphere. One example of the latter is chlorine dioxide (ClO_2) a reactive intermediate produced during the photochemical processes that occur as part of ozone depletion from stratospheric chlorofluorocarbons. The levels of greenhouse gases, including carbon dioxide, nitrous oxide and methane can also be monitored. This information can be used to predict the effects of atmospheric species on life forms that may occur from global warming or ozone depletion.

Artificial Habitats and Habitat Restoration

To a limited degree plants can be preserved artificially by seed banks in which seeds are stored for long periods of time under appropriate conditions for their preservation. Botanical gardens and arboreta enable growth of plants under artificial conditions that can prevent at least some species from becoming extinct. The number of animal species that can be maintained in zoos is limited, but this is still of some use for protecting various kinds of animals from extinction. Zoos are being used to a greater degree for wildlife preservation, in some cases with the goal of introducing animal species back into the wild. Captive breeding programmes have been established to salvage individuals of endangered species from the wild, increase their population by breeding in captivity and reintroduce them into the wild state.

The numbers of endangered bird species have been increased by taking eggs from nests of birds in the wild and hatching them in captivity, sometimes with surrogate parents from other bird species. On a much larger scale, fish hatcheries have been in use for many decades to ensure a steady supply of fingerlings, particularly of trout and salmon species. There have been some tentative successes in captive breeding programmes to restore species to the wild. In the US, captive peregrine falcon and blackfooted ferret have been reintroduced to some areas. The Arabian oryx (a large species of antelope) has been restored to some of its former habitats in the Middle East. Golden lion tamarins have been reintroduced to rain forests in Brazil. The widely publicised reintroduction of the California condor, a large carrion-eating bird, from individuals bred in captivity has been difficult because of the deaths of many of the specimens released. A major problem with captive breeding programmes has been the vulnerability of limited numbers of any species population to loss. When only a few individuals remain, the sudden onset of disease can be devastating. Not the least of the problems is the limited genetic diversity of a small population and the adverse effects of inbreeding.

TEN RULES IN ECOLOGY

Rule 1 Ecology is a Science

Ecology is a purely scientific discipline which aims to understand the relationships between organisms and their wider environment. Like any science, the outcomes of ecological studies do not dictate ethical or political actions. It is important to make this distinction because the environmental movement has endowed the word 'ecology' with political connotations. It is right that ecology should inform politics, but as a student of ecology it is imperative to consider ecological research from a rigorous scientific viewpoint.

Rule 2 Ecology is Only Understandable In The Light of Evolution

The huge diversity of organisms and the wealth of variety in their morphologies, physiologies and behaviour are all the result of many millions of years of evolution. This evolutionary history has left an indelible impression on each and every individual. It is only possible to make sense of the patterns we find today in the light of this evolutionary legacy.

Rule 3 Nothing Happens 'For The Good Of The Species'

A common misconception is the notion that patterns of behaviour in organisms which appear to be costly to an individual (for example the dying of a female octopus immediately after giving birth or the defensive suicidal attacks of some soldier ants) are 'for the good of the species'. This argument is absolutely and completely wrong and only advanced by those who failed to grasp the importance of Rule 2. Natural selection will favour those genes which are passed on to the most offspring. If the genes for suicidal behaviour in ants or early death in octopuses were good for the species but bad for the individuals carrying them, evolution would favour their replacement with other genes. Equally fallacious, for the same reason, is the argument that population size is limited via reduced birth rates 'for the good of the species'. Both altruism and population regulation are easily understandable in terms of evolution acting on individuals.

Rule 4 Genes and Environment are Both Important

The environment an organism finds itself in plays an important role in determining the options open to that individual. Environmental conditions will define the birth rate, growth rate and level of mortality of a species. However, the genes which define an organism's makeup are also of fundamental importance. The emergent phenotype of an organism is a joint product of its genetic code and the environmental stimuli that affect it during development.

Rule 5 Understanding Complexity Requires Models

At face value, ecology might appear incomprehensible—millions of species, each with varying numbers and ever-changing behaviours set in the context of a structurally complex and dynamic environment. Clearly, we cannot understand it all at once. The solution is a two-step process, firstly to identify a small and specific question, such as 'why do male blackbirds form territories?' and then to test a specific hypothesis, for example 'male blackbirds with a territory have a better chance of getting a mate'. What we have done is to construct a verbal model of the world, which we then set about testing.

Rule 6 'Story-Telling' is Dangerous

In attempting to explain ecological patterns or relationships, it is easy to slip into a make-believe world where every observation is readily explained by some *ad hoc* assertion— 'story-telling'. The temptation to advance hypotheses as facts should be avoided at all costs.

Rule 7 There are Hierarchies of Explanations

For any observation there is often an immediate cause that can be diagnosed. Often this causal explanation is insufficiently informative and we need to probe deeper to reach a fuller grasp of the situation. Even if a phenomenon is 'explained' there may well be further and deeper explanations which allow us to see the fuller picture.

Rule 8 There are Multiple Constraints on Organisms

Whilst the total diversity of form, function and environmental resilience exhibited by organisms is awe-inspiring, each individual (and to a slightly lesser extent, each species) operates within a relatively narrow range of constraints. Constraints fundamentally take two forms: (i) physical; and (ii) evolutionary. Evolution can never reach 'perfection' because of these constraints and organisms are essentially hotchpotches of numerous compromises.

Rule 9 Chance is Important

Chance events play a critical role in ecology. The opening of a gap in a forest canopy or the breaching of a sand dune after a storm will have a major impact on the ecology of the local fauna and flora, but both are unpredictable in either time or location. The role of chance is also integral to the evolutionary past of organisms. The importance of chance events in ecology does not mean ecological patterns are wholly unpredictable, but it necessarily places boundaries on the potential level of predictive detail.

Rule 10 The Boundaries of Ecology are in The Mind of the Ecologist

Ecology is a broad science, covering both organisms and physical environments and hence excludes little as potentially relevant. Mathematics, chemistry and physics are tools essential to the understanding of ecology.

2

Climate Factors

INTRODUCTION

Before we proceed to discuss the subject matter of ecology at different levels of organisation i.e., at individual, population, community and ecosystem levels, it is worth understanding first in detail, what is environment and how it acts. We would thus analyse the environment into its various components—the ecological factors. Various ecological factors vary with time (temporal variations) and space (spatial variations). These factors are involved at each level of organisation i.e., autecological as well as synecological levels. In this section, an attempt is thus made to understand the possible ways in which these factors might affect the life and govern the distribution of organisms.

The term environment etymologically means surroundings. Thus, environment is a complex of so many things (light, temperature, soil, water, etc.) which surround an organism. Any external force, substance or condition, which surrounds and affects the life of an organism in any way, becomes a factor of its environment. These factors have been variously called as environmental factors, ecological factors or simply factors and may be living (biotic) as well as non-living (abiotic). The sum of all these living and non-living factors makes the environment of an organism. The place, where an organism lives—*habitat*, indeed presents a particular set of environmental conditions—the environmental complex.

Various ecological factors of this environmental complex have been grouped and discussed in different ways. Sometimes all the factors are classified into two groups:

1. Direct factors. These are light, temperature, humidity of air, soil air, soil water, soil nutrients etc.
2. Indirect factors. These include soil structure, soil organisms, altitude, wind, slopes etc.

The environment is complex of variable factors or causes, which includes: (i) substances (soil, water); (ii) conditions (temperature, light); (iii) forces (wind, gravity); (iv) organisms (plants, animals); and (v) time.

Scientists preferred to discuss the environment of plants as: (i) environment of roots (geology and soil); and (ii) environment of shoots (climate, i.e. temperature, light, water, wind etc., and biotic factors).

Daubenmire classified the factors into seven coordinate headings as (i) soil; (ii) water; (iii) temperature; (iv) light; (v) atmosphere; (vi) fire; and (vii) biotic factor.

ECOLOGICAL FACTORS

Most of the ecologists, however, prefer to recognise four categories of ecological factors, which are as follows:

1. Climatic or aerial factors. These include:-
   (a) Light.

(b) Temperature of air (atmospheric temperature).
(c) Rainfall (precipitation).
(d) Humidity of air.
(e) Atmosphere (gases and wind).

2. Topographic or physiographic factors. These are the factors related with physical geography of Earth and include:-
   (a) Altitude.
   (b) Direction of mountain chains and valleys.
   (c) Steepness and exposure of slopes etc.
3. Edaphic factors. These deal with the formation of soil, its physical and chemical properties and details of related aspects.
4. Biotic factors. These are all kinds of interactions between different forms of life, i.e., plants, animals, micro-organisms etc.

INTERACTION OF ECOLOGICAL FACTORS

An important aspect of the effects of environment on the life of an organism is the interaction of ecological factors. Actually under natural conditions, at the same time, the life of organism is affected by the sum of all ecological factors—the environment—and not by any individual factor. All the factors are interrelated. Variations in one may affect the other. For instance, an increase in temperature would certainly bring out a decrease in humidity values. Increased light intensity, in addition to affecting the rate of photosynthesis, would affect the atmospheric humidity. Thus, at a given time effect on the organism's life is an overall effect of these interrelated factors and thus it becomes very difficult indeed sometimes to say as which of the individual factors is actually responsible for the marked effect on organism's life. Thus ecological factors, in affecting an organism's life under natural conditions, operate in conjunction and not in isolation, as we have an interaction between various factors. Figure 2.1 shows possible interactions between five principal components of the environment (energy, soil nutrients, organisms, water and air). Thus, in considering the ecological factors separately for distinguishing and measuring the effect of each, one must remain aware that in nature these factors never act alone. Organism is subjected to many influences at the same time and the effect of one factor usually modified by the other factor(s). The 'real life' of the organism thus involves the simultaneous and continuing impact of all existing ecological factors.

Since various ecological factors affect the life of an organism in a holistic manner (i.e., all the factors operate in conjunction and not in isolation), it becomes difficult to understand this mechanism of the nature of influence of individual factor. To understand this mechanism of environmental influence, it thus becomes essential to study the effect of each factor separately and this is said to be an analytical approach. We shall now discuss in detail each of the above said ecological factors separately to understand how it might influence the life and distribution of organisms. We begin with climatic factors.

Climate of any region is chiefly determinated by such meteorological influences as relative humidity of air; temperature, wind pressures and evaporation rates. Climatic factors characterise, in general, a particular region. These factors, concerned mainly with the aerial environment of organisms, include light, temperature of air, humidity of air, precipitation (rainfall) and gaseous components of atmosphere including wind. We will discuss each of them in detail.

Light Factor

Light is well known for its effects on such basic physiological processes of plants as photosynthesis, transpiration, seed germination, flowering etc., and thus is of much concern in ecology, where it plays important role in the species composition and development of vegetation. Light intensity, reaching the Earth's surface, shows much spatial variations being influenced by some such factors as: (i) atmosphere, chiefly atmospheric gases, nitrogen and oxygen, that absorb and disperse small fractions of the shorter wavelengths; (ii) suspended particles, solid particles dispersed in air, such as dust and smoke or in water, such as clay, silt, plankton, bog colloids etc., have a screening effect; (iii) water layers; (iv) layers of vegetation, chiefly in terrestrial habitats, as forests; and (v) such topographic factors as direction and slope of the land surfaces which cause marked variations in intensity and daily duration of insolation. Radiant energy of sun, reaching the Earth is transformed to other forms of energy, such as thermal, mechanical, electrical etc., which bring about changes in several physical processes in atmosphere as well as on Earth. Light intensities are closely related with atmospheric temperature and relative humidity. Thus light affects organisms directly as well as indirectly.

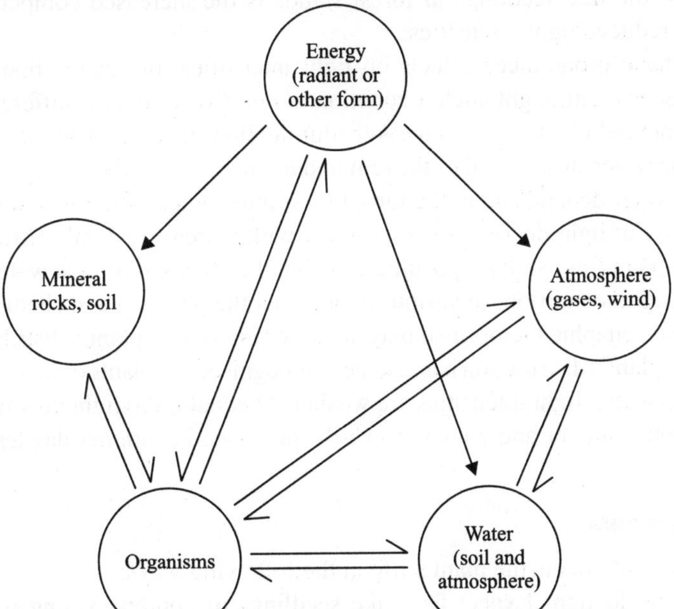

Fig. 2.1. Diagram showing interrelations between principal components of the environment.

The great deal of solar radiation is absorbed in the atmosphere. The amount of light thus is greater at higher altitudes than at sea level. This is due to thinner layers of air at higher altitudes. On a cloudy day, light may be reduced to 4 per cent of the normal intensity. Although red and blue parts of the spectrum are useful in photosynthesis, in ecology relatively less importance is given to the quality of light for its role in plant distribution.

The layers of vegetation bring about variations in light intensities reaching at various heights of mountain. This becomes evident, when we study light conditions in forests, where a major proportion of light intensity is absorbed by tree vegetation and the light reaching to the lower part of the ground vegetation is considerably reduced by 90–98 per cent of that in the exposed areas. The amount of light reaching the forest floor depends upon the height of canopy, crown development of trees, age of trees and phenological characteristics of the constituent species. Thus, in a forest, the mature tallest tree receives full insolation, under-shrubs receive subdued illumination and herbs and especially epigenous cryptogams grow in weaker light conditions. When leaves of a tree are fully expanded, its canopy may reduce light to less than 1 per cent of full solar radiation. Salisbury, as a result of his studies on light conditions in forests, showed that light intensities govern the development of ground flora of a forest. But it is seen that light conditions of forests, with and without ground flora, are not very much different and thus it seems that not only light, but other factors as temperature, moisture, plant growth, mineral nutrition, humus content and root competition are also involved. It is also seen that during shade phase i.e., when full leaves are present on the trees, light conditions are poor, but in nature sun-flecks and movement of leaf canopy of trees play an important role and these compensate for the reduced light, making it available to the ground flora. Oosting and Shirely, too, on the basis of a series of experiments, concluded that for each light intensity growth of forest plants may be increased by reducing the root competition. Generally, the cause of death of the tree seedlings in forest stands is the increased competition in roots for the nutrients rather than reduced light intensities.

Layers of water have pronounced effects on light intensities. In aquatic bodies, plants are found growing in distinct zones, althought such a zonation is not always due to different light intensities at various depths. Submerged plants receive weaker illumination than the exposed plants, as some of the light is reflected at the water surface and of the remainder, much is absorbed by the upper layers of water. Light penetration in water depends upon the turbidity, solute content, motion and planktonic growth of the water. The intensity of light decreases successively with increasing depth. However, in their studies on distribution of plants in English lakes, could show that the zonation in such water bodies is not solely due to differences in light, conditions at various depths, but the physical and chemical characteristics of the rooting medium i.e., edaphic factors also play an important role in plant's distribution. Importance of mud characteristics in plant's distribution has also been recognised for plants of low-lying lands. Generally at dawn, sunset and in water, light intensities are weaker. At equator daylight prevails for about 12 hours out of every 24, in both summer and winter. Towards the poles, in summer day length becomes longer than 12 hours.

Light in relation to plants

Directly or indirectly light affects the plant's life in the following ways:

1. Chlorophyll production: Expect few, like seedlings of conifers young fronds of ferns, some mosses and algae which may become green even without light, most of the plants need light for chlorophyll formation. Thus existence of other life forms, which depend upon plants for food is due to light.
2. Heating action: Exposure of parts of the plants to light raises their temperature, which effect their related processes.
3. Effect on transpiration rate: Indirectly light affects transpiration rates through increase in temperature. Transpiration rates correspondingly affect water absorption also. Thus high light intensities are always associated with dry habitats and high transpiration rates.
4. Stomatal movement: Opening and closing of stomata is regulated by light. This is thus related to transpiration and absorption.
5. Distribution of plants: Light conditions at poles are different from other parts of Earth. Thus total amount of radiation received by the Earth's surface differs with latitude (distance from the equator). This may be one of the reasons for differences in the vegetation at poles and at other parts of the Earth.

6. Overall vegetative development of plant parts: Sometimes plants are classified ecologically on the basis of their relative light requirements and the effects of light on their overall vegetative development into:-
   (a) Heliophytes. Growing best in full sunlight.
   (b) Sciophytes. Which grow best at lower light intensities.

There are some heliophytes, which grow best in Sun, but can grow fairly well under shade. Such plants are called facultative sciophytes. Similarly there are facultative heliophytes which, although grow best at lower light intensities, but can grow well in full sunlight. Some plants, likely to be injured by bright light, develop some characteristics in themselves to minimise such effects. For example, plants of sunny habitats have vertically oriented leaf blades. Some variations in the pigments complex of leaves also take place. In moist climates, vegetation is generally a complex series of superimposed layers of tall trees, low trees, shrubs, herbs and mosses etc., most of which such as maples, beech, spruces, firs and hemlocks are obligate or facultative sciophytes. In dry climates, species of the flora are mostly heliophytes, which are mostly pines, soft maple, willows, tulip tree, birches, larches and junipers.

In comparison to plants grown in the shade, those developing in full sunlight, usually exhibit such morphological and physiological features as: thicker stems with well developed conducting elements and mechanical tissues, shorter internodes, more frequent branching, smaller chloroplasts, well-developed palisade with poorly developed spongy parenchyma, roots longer, more in number and much branched with higher root/shoot ratio, less chlorophyll content, low photosynthetic rate per unit surface, high respiration rate, low water content (dry weight basis), higher concentration of salts and sugar and high osmotic pressure, decrease in acidity of cell sap, low potassium contents, early appearance of flowers and much more resistance to temperature injury, drought and parasitic organisms.

7. Photoperiodism: Total length of the daily light period to which plants are exposed—photoperiod has pronounced effect on the vegetative growth as well as flowering of the plants. On the basis of their photoperiod, the plants may be: (i) short-day plants, that develop normally only when the photoperiod is less than a critical (between 12–14 hours) maximum. Examples are *Salvia splendens*, *Datura stramonimum*, *Cannabis sativa*, *Andropogon virginicus* and *Cosmos bipinnatus*; and (ii) long-day plants, those demanding a photoperiod in excess of certain critical minimum. For example, *Brassica rapa*, *Nigella avroensis*, *Secale cereale*, *Sonchus oleraceum*, *Sorghum vulgare*. Most of the plants are short-day, which require a relatively long period of uninterrupted darkness for flowering. Some plants are indifferent to the above conditions and are called day neutral plants. For example, *Cucumis sativus*, *Gossypium hirsutum*, *Nicotiana tabacum*, *Phaseolu lunatus*, *Poa annua* and *Solanum tuberosum*. Thus, light as photoperiod plays an important role in local distribution of plants, by affecting stem elongation, flowering, fruit development and other processes.
8. Succession: Also, there are some reports in literature indicating the role of light in plant succession, where it is generally mentioned that the light requirements of pioneers are comparatively much more than the climax species. But these could not receive much experimental support. Some such factors as mainly the temperature and soil fertility may, under natural conditions, modify the influences brought about by light. Most of the temperature and light influences are interrelated in their effect on plants.

Thus, light as an ecological factor affects the distribution of plants through its effects on photosynthesis, vegetative shape, tissue differentiation, chlorophyll production, number and position of chloroplasts, leaf structure, stomatal movement, transpiration, growth and development of flowers, seeds and fruits. These effects are well known in the study of physiology of plants.

Light in relation to animals

So much has been said of the role of light in the life of plants; what about animals? Light has far-reaching effects on animals also, by affecting their several types of activities like pigmentation, reproduction, development, growth, locomotion, migration etc., which may be summarised as follows:

1. Metabolism: Light affects metabolic processes of animals through its heating effect on tissues and by ionisation of the protoplasm. This all results into an increase in the enzymic activities, general metabolic rate and degree of solubility of salts and minerals. Animals receiving poor light in caves thus shows slow rate of metabolism. Light also affects photo-oxidation and respiration rates.
2. Reproduction: In animals like birds, light initiates the breeding activities. In some birds gonads become active with increased light intensity during summer. Thus animals are long-day, short-day and indifferent to day-length. Some sheep, deer and goats are short-day animals, which may be brought to sexual act by decreasing the length of exposure to daylight. Spring breeding birds and animals as ferrest, starlings, turkeys and others are long-day animals. Ground squirrels, guinea pigs etc., are indifferent towards the length of exposure to daylight.
3. Development: In *Salmon*, larvae undergo normal development only under sufficient light conditions, in the absence of which there may take place rapid death of larvae. In *Mytilus*, larvae in their earlier stages grow larger in darkness than in light.
4. Eyes: The degree of eye development sometimes depends upon the available light intensities. In animals, living in caves as *Proteus anguinus* and in the deep sea fishes, the eyes are absent or rudimentary. In surface dwelling animals as crustaceans and fishes, the ratio of the eyes to head is called as normal. In the ocean, with increasing depth, size of eyes in animals generally goes on increasing.
5. Vision: Higher animals including man are able to see various objects only in the presence of one or other form of light. Some fishes, as *Lepomis*, also depend on eyesight for location of their food.
6. Pigmentation: Most of the chemical changes that result into the formation of pigments are initiated by light energy. The process of pigmentation in animals is thus influenced by light in a number of ways as: (i) in the development of skin colour. For instances, in some animals living in caves, as amphibians and some fishes, the pigments in skin are lacking due to absence of light; (ii) in protective colouration. Some animals develop colouration in their skin which more or less resembles with the background in which they live. Thus leaf insect *Phyllium*, Australian fish, *Phyllopteryx eques* and butterfly, *Kallima paralecta* etc., have colouration exactly like that of their background and this affords a good means of protection from their enemies. Similarly some birds, mammals and fish display differential shading, with darker pigmentation on back and lighter one below, that helps them in providing protection; (iii) colour changes. Some insects, crustaceans, fishes, amphibians, reptiles and cephalopods are able to change their colour or patterns rapidly according to the environmental conditions where they live. This helps in thermoregulation and in affording protection from enemies.
7. Locomotion: The speed of locomotion in some lower animals is regulated by light. This phenomenon is known as photokinesis. There is evidence, that blind larvae of mussel crab move faster if exposed to increased light intensities. More or less similar responses in terms of locomotion towards different light intensities are known to occur in some aquatic invertebrates. In some animals, light plays a role in the orientation of locomotion and this phenomenon of movement of animals in response to light is called phototaxis and such movements are known as phototactic.

Animals like *Euglena, Rantara* etc. are positively phototactic, moving towards the light source, whereas earthworms, slugs and some zooplanktons like copepods etc., are negatively phototactic. In some animals, like many tubicolous worms and polyps of many coelenterates, only a part of their body shows movement in response to light and this has been termed as phototropism.

8. Photoperiodism: Animals are known to respond to the length of the day, through the effect of total light period on such processes as gonad activities, reproduction, metamorphosis, migration etc. In certain catfishes longer days induce early maturity. Migration of eels, salmons and birds is known to be affected by photoperiodism, as some birds are seen migrating towards north during summer when days are longer and towards south during winter when days are shorter. In some animals development of fur and feathers during winter is shown to develop in response to photoperiodism.

Daily responses of animals to light conditions are known as circadian rhythms, whereas the annual rhythms known as circannual rhythms. Circadian rhythms are internally driven or endogenous, are affected little by temperature changes, are sensitive to a great variety of chemical inhibitors and are innate, not learned from or imprinted upon the organism by the environment. The innate character of circadian rhythm is demonstrated by several animals. When Drosophila are kept under constant conditions from the larval stage on, they will still emerge from pupae with a regular circadian rhythm. Eggs of chicken and lizards kept under constant condition produce animals that later show regular circadian cycles. The circadian rhythms have been observed in zooplanktons polychaete annelids, many insects, most birds and some mammals. Planktons of sea and lakes show diurnal changes in their vertical distribution. Solary day, lunar day, tidal rhythms, monthly and annual rhythms are also of common occurrence among animals. Endogenous annual cycles or circannual rhythms have been known in ground squirrels, warblers and other birds, some crayfishes and slugs.

Temperature Factor

The term 'temperature' here, besides the aerial temperature of terrestrial organisms, also includes the temperature of water for aquatic life. In fact the atmosphere to which aquatic forms of life are exposed, is the water medium. Temperature is well known for its effects upon plants and animals.

Effects of temperature on plants and animals

Temperature affects plants and animals in various ways, which are as follows:

1. Effects on metabolism: All metabolic processes are influenced by temperature. Since temperature regulates the activity of enzymes, all chemical reactions in the body of organism are controlled by temperature. It affects the rates of transpiration, photosynthesis in plants and respiration rates and other metabolic processes in plants as well as animals. In plants it affects seed germination also.
2. Effects on reproduction: Flowering in plants is affected by temperature through thermoperiodism—i.e., the response of plant to rhythmic diurnal fluctuations in temperature. Temperature is an important factor in the phenology of plants (i.e., study of periodical phenomena of plants, as the time of flowering in relation to climate). In animals, the maturation of gonads or sex cells and the liberation of gametes take place at a particular temperature, which varies from species to species. Breeding in some animals remains unaffected by temperature throughout the year, whereas in some, breeding occurs only in summer or in winter. Some species may have two breeding seasons, one in spring, other in fall. In blow-fly *Calliphora sericate*, the number of eggs laid per

female, increases with increase in temperature up to 32.5°C, but the number decreases with further increase in temperature.

Thus temperature also affects fecundity of animals. Fecundity is the reproductive capacity of animals. The female of the insect, acridid *Chrotogonus traychpterus* becomes sexually mature at 30°C and 35°C than at 25°C and maximum eggs per female laid at 30°C. In grasshoppers, *Melanoplus sanguinipes* and *Camnula pellucida* produced 20–30 times as many eggs at 32°C than those at 22°C.

3. Effect on growth and development: Both, extremely low and high temperature have adverse effect on the growth of plants. Low temperatures bring about such cold injuries as desiccation, chilling injury and freezing injury. In desiccation, tissues are dehydrated and injured due to rapid transpiration and slow absorption during winter. Chilling injury is the killing or injury of plants of hot climate, when exposed to low temperatures for sometime. In some plants of temperate climates, if exposed to low temperatures, water is frozen into ice crystals in the intercellular spaces causing injury to cells. This is known as freezing injury. Some perennials can tolerate extremely low temperatures and this ability is called cold resistance. Extremely high temperatures cause stunting and final death of plants, which is due to adverse effects on a number of physiological processes as respiration, transpiration, protein metabolism etc. This is called heat injury.

   In animals also temperature affects growth as well as development. For example in the *Oyster*, *Ostraea virginea*, the length of body increases from 1.5 mm to 10.3 mm with an increase in temperature from 10°C to 20°C. Sea urchin, *Echinus esculentus* shows maximum size of its body in warmer water. Corals also do not flourish well when water temperature drops below 21°C. Effect of temperature on the developmental stages has been studied in detail in insects. Thus, in blow fly, *Calliphora sericate*, incubation period decreases with increase in temperature.
4. Effect on crossing over: In animals like fruit flies, *Drospohila* spp., temperature is shown to affect the crossing over and somatic expressions of gene characters. Development of wings, eyes etc., is affected if larvae or pupae are kept at low or high temperature. Wings tend to be longer at high and shorter at low temperatures.
5. Effect on sex-ratio: In some animals, as rotifers and daphnids, sex-ratio is affected by temperature. Under normal condition daphnids give parthenogenetic eggs that develop into females, whereas with increase in temperature they give sexual eggs which after fertilisation develop either into males or females. In *Macrocyclops albida* there is significant increase is number of males with increase in temperature. Similarly in plague fea., *Xenopsylla cheopes*, males out numbered females on rats on days when the mean temperature remains in between 21–25°C.
6. Effect on colouration: Some insects, birds and mammals in warm humid climates bear darker pigment than the races of same species present in cool and dry climates. This phenomenon is known as Gloger rule. In the frog *Hyla* and the horned toad *Phrynosoma* low temperatures induce darkening.
7. Effect on morphology: Temperature also affect the absolute size of an animal and the relative proportions of various body parts—Bergman's rule. Birds and mammals attain greater body size in cold regions than in warm areas. But poikilotherms are smaller in cold region. The tail, snout, ears and legs of mammals are relatively shorter in colder part than in the warmer areas—Allen's rule. For example, there occurs difference in the size of ears of arctic fox (*Alopex lagopus*), red fox (*Vulpes vulpes*) and the desert fox (*Megalotis zerda*). *Gazella picticanda* of Himalayas has shorter legs, ears and tail than *G. benetli* found in the plains. The races of birds with relativeiy

narrow and more accuminate wings tends to occur in colder regions, while those in warmer areas tend to be broader (Rensch's rule). Temperature also affects the morphology of some fishes and is found to have some relationships with the number of vertebrae – Jordon's rule.

Thus, through its effect on organisms in various ways described above, temperature affects geographical distribution of plants and animals. Together with moisture, temperature determines the climatic condition of a particular region.

Variations in temperature and its effects on distribution of plants

Temperature shows pronounced temporal as well as spatial variations. Especially in a country like India, there are experienced marked seasonal extremes and fluctuations in the temperature values. Soil temperature is the result of heat gained by the absorption of solar heat energy and soil temperatures are greatly influenced by the latitude (distance from the equator) of the particular place. Obviously, the temperature values are maximum at the equator, decreasing gradually towards the poles. The division of Earth's vegetation into different zones, as equatorial, tropical, coniferous, alpine vegetation etc., rests on the marked variations in temperature at different latitudes. Similarly, altitude (height above the sea level) affects temperature values. Besides latitude and altitude, colour and composition of surfaces, plant cover, water content of soil and such physiographic factors as the steepness of slope, exposure of slope and direction of mountain chains etc., affect greatly the temperature conditions. For instance, on mountains, we generally experience a decrease in temperature with increasing altitude and thus vegetation at different altitudes is different, showing distinct zonation. Generally, mean temperature of air decreases about 10°F for every degree of latitude north of equator and for every 300' of altitude. In nature, valleys and lowlands are sometimes much cooler due to sinking in of the heavier cold air than the mountain tops.

Changes in both, the latitude and the altitude, show more or less similar effect upon the type of major vegetation of the world. The various zones of different kinds of vegetation at increasing latitudes from equator towards poles (Fig. 2.2) and at increasing altitudes on mountains (Fig. 2.3) are more or less similar, showing that increase in latitude and altitude brings about more or less similar influence upon vegetation. Thus, we may find distinct latitudinal as well as altitudinal zonations in the vegetation.

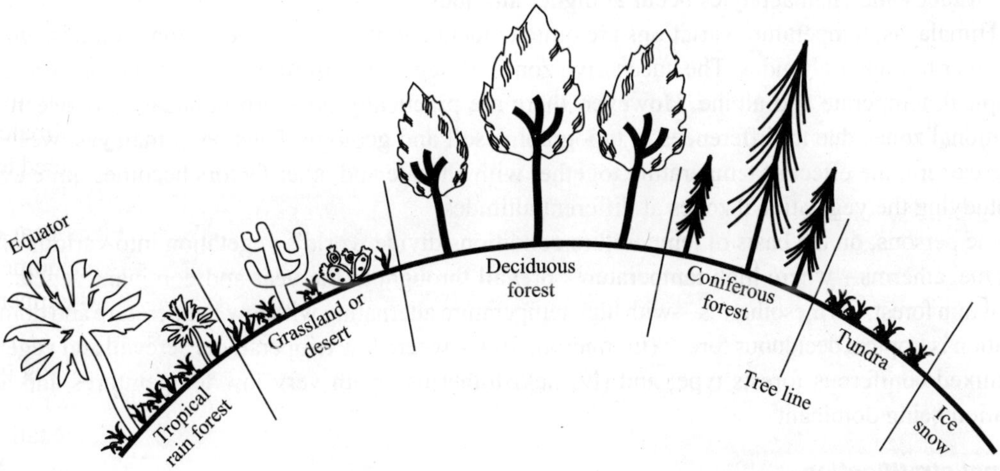

Fig. 2.2. Latitudinal zonation of vegetation. Note the different types of vegetation from equator towards poles (increasing latitude).

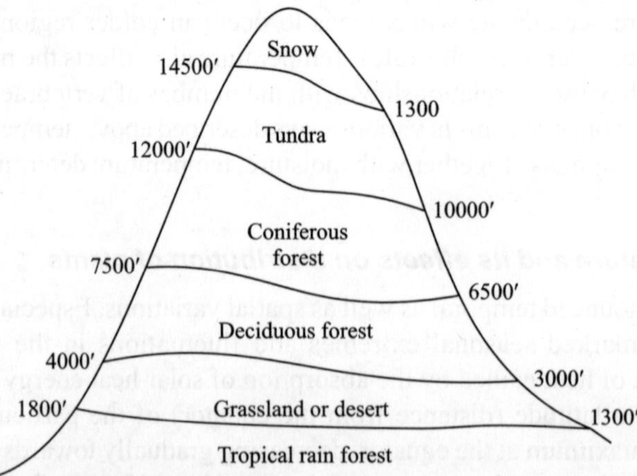

Fig. 2.3. Altitudinal zonation of vegetation on mountains. Note the different types of vegetation with increasing altitude. If compared with Fig. 2.2 the vegetational types are more or less similar in both.

On mountains, temperature and rainfall are commonly said to determine the type of vegetation at different altitudes. But collin has shown that at different altitudes, not only temperature and rainfall but edaphic conditions (soil characteristics) are also important. On mountains with deviations in elevation, in general, southern and western slopes register higher temperatures than do the northern or eastern slopes. The angle of the slope in relation to sun governs the total amount of radiation. These differences are probably due to different solar radiation received, differences in the amount of rainfall, snowfall, relative humidity and wind movement on the two slopes. At higher elevations, generally temperature decreases and rainfall increases and these certainly affect the development of soil and vegetation. Due to low temperature and high rainfall the organic matter content of the soil increases at higher altitudes, with an increase in soil nitrogen and a decrease in its pH values. Generally, xerophytes are more common at lower altitudes and chamaephytes occur at higher altitudes.

In Himalayas, temperature variations are quite evident and there is a general zonation of vegetation from lower to higher altitudes. The successive zones of vegetation from base upwards are tropical and subtropical, temperate and alpine. However, there are practically no sharp boundaries between these vegetational zones due to differences in topography, soil and geology. Thus on Himalayas, western as well as eastern, the effect of temperature together with altitude and other factors becomes quite evident after studying the vegetational zones at different altitudes.

Some persons, on the basis of temperature conditions divide world's vegetation into various classes as: (i) megatherms—where high temperatures prevail throughout the year and dominant vegetation is tropical rain forest; (ii) mesotherms—with high temperature alternating with low temperature and dominant vegetation is tropical deciduous forest; (iii) microtherms—where low temperatures perevail and vegetation is of mixed coniferous forests type; and (iv) hekistotherms—with very low temperatures and alpine vegetation being dominant

Thermal stratification

We have described above the variations in temperature and its influence on plants of terrestrial conditions. Similarly, temperature values may show variations in aquatic habitats also which are considered below.

In aquatic habitats, submerged forms experience lesser fluctuations of temperature in the water medium. In large bodies of water, due to its high latent heat, it does not become so cool in night hours and, moreover, wind effects and water currents mix up and lessen the effects of local cooling and heating. Thus, diurnal variations in temperature are not much pronounced. However, in small ponds such diurnal variations may be pronounced. In deep water of lakes, thermal stratification may be observed during summer. During summer, the atmospheric temperature reaches up to 27°C, which raises the temperature of upper layers of water to 22°C–23°C, whereas that of the bottom layers it is as low as only 5°C. Between this range of 5°C–22°C, there are generally differentiated three different zones of vegetation viz.: (i) epilimnion—vertical gradient of gradually decreasing temperature from the surface, followed by; (ii) thermocline or metalimnion—a short zone of rapidly falling temperature; and (iii) hypolimnion—bottom cold zone where no temperature gradient is evident. At the onset of summer, however, there is mixing up from bottom mud into lake water.

Effect of temperature on distribution of animals

As discussed above, there are much variations in temperature and these affect plants distribution in various ways. It also affects animals. All animals are broadly classified into two group:

1. Homoiothermic or endothermic animals (warm-blooded). These animals like birds and mammals are able to maintain their body temperature at a constant level irrespective of the environmental temperature.
2. Poikilothermic or ectothermic animals (cold-blooded). These are animals as reptiles, fishes, amphibians in which the body temperature fluctuates with changes in the environmental temperature.

Thermoregulation in homoiotherms

Though several factors may operate, the most important role is played by their skin. Almost each of the various structures present in the skin play some part in the temperature regulation. A homoiotherm has different response to cold or heat as follows:

1. Response to cold. When a typical homoiotherm as a mammal is subjected to severe cold, following adaptations help the animal from excessive cooling:—
   (a) The subcutaneous tat serves as insulator and reduces heat loss from body.
   (b) The hair is raised and brought up into a more-or-less vertical position by the contraction of the erector-pili muscles. Thus air gets trapped in the spaces between hairs. This air is warmed by the body and being a poor conductor of heat it serves as an insulatory layer around the animals. In birds, the same function is done by feathers.
   (c) The superficial blood vessels in the skin constrict so that blood is directed from the surface to the deeper layers. This reduces loss of heat from the blood to surrounding atmosphere.
   (d) Extra heat is produced by increase in the metabolic rate particularly of the liver and muscles.

Homoiotherm besides above said structural adaptations also have physiological capability of increasing their heat production.

2. Response to heat. Heat production is cut down and heat loss encouraged as follows:-
   (a) Animals of hot climates have relatively little subcutaneous fat. The fat deposits are localised. Thus camel's fat is stored only in hump and in buffalo and bison on top of the neck.
   (b) The hair is lowered by relaxation of the erector-pili muscles so that it lies flat against the body surface. There is no space between the hair and no air can be trapped against the skin.

(c) The superficial blood vessels are dilated so that blood is brought up near the surface from which it can lose heat to atmosphere.

(d) Sweating or panting occurs. Sweating by sweat glands of skin and evaporation of sweat from the body surface cools the skin and the blood flowing through it.

(e) Metabolic rate falls in hot conditions so that less heat is generated by the body.

Thus, in mammals the hypothalmic centre functions as a thermostat. It is sensitive to temperature changes of blood flowing through it and responds by sending nerve impulses through different nerves to the appropriate effectors. If the temperature of the blood is slightly higher than it should be, the thermo-regulatory centre of hypothalamus of brain detects this and sets into motion process that collectively encourage heat loss. On the other hand, if the temperature falls below the normal the centre initials processes that produce and conserve heat.

Thermoregulation in poikilotherms

Some poikilotherms avoid both heat and cold by underlying dormant in period of environmental stress. Many insects, some crustaceans, mites and snails, enter diapause, a state of dormancy and arrested growth. Eggs, embryonic larvae and pupae are involved. Some poikilotherms become dormant in temperature extreme conditions. Amphibians and turtles bury them in the pond mud, snakes and lizards seek burrows.

A few exceptional poikilotherms, especially insects, certain amphibians and reptiles, exercise some degree of thermoregulation by physiological or behavioural mechanism. For example, hawk-moths can raise the temperature of their flight muscle to 32°C–36°C by vibrating the wings before take off and gregarious butter fly larvae may raise their temperature 1.5°C–2°C by clustering together. Locust and grasshoppers may increase temperature 10°C by basking sideways in the sun. Ants move their larvae to warm or cool places.

When temperature drops, lizards bask in the sun to achieve more temperature. Once this temperature is attained, they will divide their time between sun and shade to maintain it. The frogs and reptiles lower their body temperature slightly by evaporating cooling through the skin or via the respiratory tract by panting.

Temperature affects the distribution of animals also, as different species have different ranges of their temperature requirements. For instance, coral reefs require a minimum temperature of 21°C for their suitable growth, hence are completely absent in colder regions, where prevailing temperatures are below 21°C. Speckled trout, *Salvelinus fontinalis*, prefers 14°C to 19°C and thus cannot tolerate a temperature above 25°C, being thus completely absent from streams where the temperature exceeds 24°C, for some time. Distribution of intertidal barnacles is also affected by temperature. For example, an intertidal barnacle, *Pectin groenlandicus*, is hardly known to occur in areas with 0°C temperature, whereas another one, *Balanus balanoides*, occurs at a temperature below 7.2°C.

Adaptations developing in response to extremes of temperature

Both in plants as well as animals there has been the development of a number of morphological and physiological characteristics during the course of evolution, in order to meet the extremes of temperature conditions. These may be briefly summarised as follows:

1. Production of spores, cysts, eggs, pupae, seeds: Plants and animals, to resist the extremes (both minimal as well as maximal) of temperature, produce thick-walled structures as spores, cysts, eggs, pupae, seeds etc. Thus amoeba in encysted state may tolerate a temperature below 0°C.

2. Increase in osmotic concentration: In some plants, there take place such physiological changes as increase in osmotic concentration that prevents freezing. Increase in osmotic values enhances the amount of bound water in colloidal form. Thus, in winter rye at 0°C, leaves and stems are flexible and not brittle.
3. Removal of water from tissues: Dried seeds, spores and cysts like structures avoid freezing because no liquid remains in them to freeze. Dry seeds are able to germinate, even after their exposure to –190°C for a period of three weeks.
4. Dormancy: This includes both, hibernation as well as aestivation. The former is overwintering in dormant state. This is common in poikilothermic as well as homoiothermic animals, such as bats, hedgehogs, ground squirrels and jumping mice. During this period, metabolic rate and heartbeat rate are reduced. Aestivation occurs during summer to resist higher extremes of temperature, which is very common among insects, some plants, lungfishes, amphibians and some mammals. Lungfish, *Protopterus*, during summer, burrows into the mud, secretes a cocoon of slime around it and lies dormant. In insects aestivation is achieved by the formation of diapause, a stage during which morphological growth and development remain suspended. Some insects undergo hibernation and aestivation during different months of the year, depending upon the environmental temperatures. One very well known example is the insect, *Coccinella septempunctata*, which has different hibernation and aestivation periods in different countries i.e., London, Moscow and Formosa, with no hibernation in the last country.
5. Homoiothermy: Birds and mammals are able to maintain their body temperature at a constant value and thus are not affected at all by variations in the environmental temperature extremes. This is achieved through evaporation of water from bodies during summer and through insulating action of fur, feathers, fats, etc., during cold weather of winter months.
6. Thermal migrations: These are found only in animals. The journeys taken by animals that enable them to escape from extremely hot or cold situations are referred to as thermal migrations. This is very common in such locomotory forms as desert animals like amphibians, burrowing animals, deer, bear, insects, birds, mammals, etc. Fishes are also known to migrate.
7. Phenotypic changes: Temperature is known to affect the phenotype of some animals. For instance, in *Drosophila melanogaster* temperature affects the eye size and number of legs. Moreover, in some animals like *Daphina*, the body form undergoes different modifications in relation to seasonal changes in temperature and this phenomenon is known as cyclomorphosis. During winter, the head of the animal is round, in spring there develops a helmet-like projection on the head, which attains its maximum size during summer and during next winter the head becomes round again. Thus degree of warmth of water and size of the helmet are in one way or the other correlated.

Precipitation (Rainfall)

Precipitation is the chief source of soil water. The water available to plants and animals from soil comes as a result of rainfall. There occurs an interchange of water between the Earth's surface and the atmosphere, forming the water or hydrologic cycle. The two important events which are involved in this cycle are precipitation and evapo-transpiration.

Infact, water during rainfall is directly of little or no use to the organisms. It is used only after reaching to the soil. Precipitation is the result of gravitational pull on the vapour in atmosphere. Precipitation occurs in various forms. These are, drizzle, rain, snow, dew and frost, sleet and hail. Drizzle involves minute drops appearing as to float in air. Rain is the drops of liquid water, which are larger than drizzle and also heavier. Snow is the moisture as solid state and dew and frost are formed due to condensation of moisture directly on the surfaces of objects, plants, animals, soil, etc. Sleet is in the form of small grains or pellets of ice, whereas hail consists of balls or lumps of ice. Snow is injurious to plants, breaking tender branches, flowers and fruits. Hail and sleet also cause similar damage. Some sedges grow in snow patches.

Of all the above forms of precipitation, the rain is the most important. It is the source of soil water and also affects humidity of atmosphere. Rains in India are caused by monsoons. About 45 per cent of the water available during annual precipitation flows into rivers, 20 per cent percolates into the ground and the remaining 35 per cent is lost by evaporation. The amount of annual rainfall greatly influence the vegetation as well as animal population of a particular region. Rainfall distribution in different seasons of the year also may further be important in regional distribution of vegetation. On high mountains, the regions are correlated with the distribution of rainfall. And, there are distinguished three regions: (i) lower region with scanty rainfall; (ii) middle region or the cloudy region with lot of rain; and (iii) upper dry region, above the clouds covered with snow and ice.

Annual rainfall determines the type of vegetation in any region. For example, in tropical areas with heavy rainfall throughout the year, main vegetation are evergreen forests. In countries with heavy rainfall during winter and low during summer, there are present sclerophyllous forests. The plants are shrubs stunted in height, with leathery, thick, evergreen leaves. The areas with heavy rainfall during summer and low during winter are characterised by the presence of grasslands. The regions where rainfall is scanty, are seen with deserts and Xerophytic vegetation.

In India, due to differences in the amount of annual rainfall and its distribution in different seasons of the year, we find that vegetation types in different parts of the country are much different from each other. Similarly, with changes in vegetation we also find different types of animals in different geographic regions.

Humidity of Air

It is one of the different forms of water in nature. Atmospheric moisture in the form of invisible vapour is known as humidity. The humidity of air is expressed in terms of relative humidity values. It is 'the amount of moisture in air as percentage of the amount which the air can hold at saturation at the existing temperature'. Humidity is greatly influenced by intensity of solar radiation, temperature, altitude, wind, exposure, cover and water status of soil. High temperature increases the capacity of the air to retain moisture and cause lower relative humidity. Low temperatures cause higher relative humidity by decreasing the capacity of air for moisture. Relative humidity is measured by the instrument called psychrometer or by paper strip hygrometer or a thermo-hydrograph. Daily variations in relative humidity values depend upon the type of habitat conditions. In plains and deserts, it may show variations during day, whereas in oceanic islands there is little variation, being same throughout the year.

Of the other factors of aerial environment, humidity plays an important role in life of plants and animals. Processes as transpiration, absorption of water etc., are much influenced by atmospheric humidity. The saturation deficit, temperature and wind velocity modify evapo-transpiration rates. Saturation deficit (as mm Hg) is the difference between the pressure of water vapour in the atmosphere at a given time and the maximum vapour pressure that it could contain at the same temperature. A decrease or increase in the saturation deficit causes a fall or a rise in evaporation rate respectively. Similarly temperature variations also influence evaporation. Dry winds decrease the amount of air moisture by removing moist air about plants and mixing it with dry air which decreases the humidity and thus increases transpiration.

Thus humidity affects the life of plants in various ways. Effects of moist air on plants are more or less similar to those of reduced light intensity. Some plants as orchids, lichens, mosses etc., make direct use of atmospheric moisture. In fungi and other microbes, it plays an important role in germination of spores and subsequent stages in life cycle.

Atmosphere – Gases and Wind

The term 'atmosphere' here includes only the thick gaseous mantle surrounding the Earth. The soil too has its own atmosphere. There are pore spaces in between soil particles as well as between aggregates and such spaces make the atmosphere of soil.

Atmospheric gases

Up to the height of about 300 km above the Earth's surface, there is present some sort of a thick gaseous mantle. With increasing height above sea level, the atmospheric pressure gradually decreases. In atmosphere, about 95 per cent of the total air is present up to the height of about 20 km above Earth's surface and the remaining 5 per cent in the rest about 280 km height. In the gaseous mantle, there is found a mixture of different gases in different proportions. Of these various gases, nitrogen, oxygen and carbon dioxide are the major components. Relative proportions of different gases, by volume, are shown in Table 2.1.

Table 2.1. Composition of atmosphere.

Components	*Quantity in %*
Major components	Nitrogen (78.09)
	Oxygen (20.94)
	Water vapour (0.1 – 5)
Minor components	Argon (9.34×10^{-1})
	Carbon dioxide (3.25×10^{-2})
Trace components	Neon (1.82×10^{-3})
	Helium (5.24×10^{-4})
	Methane (2×10^{-4})
	Krypton (1.14×10^{-4})
	Nitrous oxide (2.5×10^{-5})
	Hydrogen (5×10^{-5})
	Xenon (8.7×10^{-8})
	Sulphur dioxide (2×10^{-8})
	Ozone (trace)
	Ammonia (1×10^{-6})
	Carbon monoxide (1.2×10^{-5})
	Nitrogen dioxide (1×10^{-5})
	Iodine (trace)

Besides gases, some other constituents are water vapours, industrial gases, dust and smoke particles in suspended state, micro-organisms, pollen grains, fungal spores etc. Cycling of gases as carbon dioxide, nitrogen etc., between the atmosphere, soil and organism is the regular feature of nature.

The detailed structure of atmosphere is shown in Fig. 2.4.

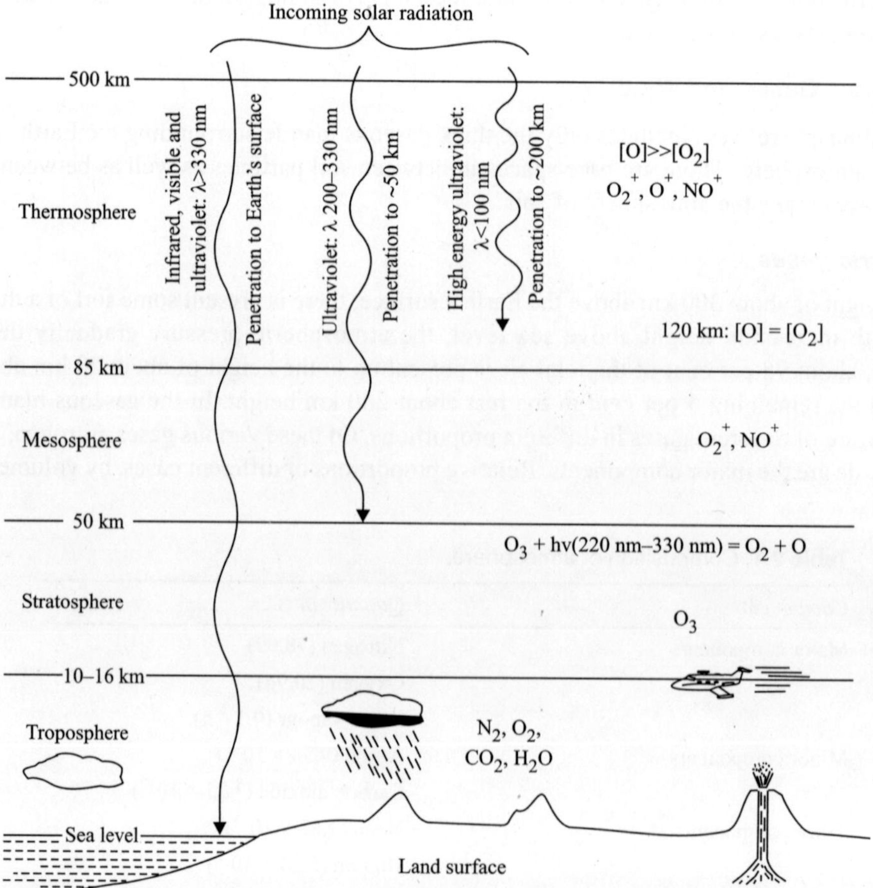

Fig. 2.4. Diagrammatic sketch showing the principal zones of atmosphere along with variations in temperature.

Atmosphere can be roughly divided into four zones:

1. Troposphere: The basal part of atmosphere which is about upto 20 km above the Earth's surface (on poles only upto about 8 km) is known as troposphere. For organisms this is the most important zone of the atmosphere. The important events, such as cloud formation, lightning, thundering, thunder storm formation etc., all take place in troposphere. Air temperature in this zone gradually decreases with height at the rate of about 6.5°C per km. Towards the upper layers of troposphere, the temperature might decrease upto –60°C and this limit of troposphere, which gradually merges into the next zone, is known as tropopause.
2. Stratosphere: Next to troposphere, the second zone of about 30 km height, is called stratosphere, where the temperature values show an increase up to 90°C. This increase in temperature is due to ozone formation under the influence of ultra-violet component of sunlight. Upper layers of stratosphere form stratopause.

3. Mesosphere: About 40 km in height, the zone, next to stratosphere is called mesosphere. In this zone, temperature shows again a decrease up to –80°C. Upper layers of this zone are mesopause.
4. Ionosphere: The rest above the mesosphere, upto the height of above 300 km above Earth's surface is ionosphere. Most of the gaseous components which become ionised under the influence of radiant energy, remain as ions. Due to their much less density, they are rarely present in molecular form.

Thus, radiant energy brings about considerable changes in the atmosphere. Water vapour in the atmosphere affects the temperature and light intensities of atmosphere.

Wind Factor

Air in motion is called wind. It is an important ecological factor of the atmosphere, as it affects plants' life mainly on flat plains, along seacoasts and at high altitudes in mountains. Wind is directly involved in transpiration, in causing several types of mechanical damage and in dissemination of pollen, seeds and fruits. It also modifies the water relations and light conditions of a particular area. The velocity of wind is affected by such factors as geographic situation, topography and vegetation masses and position with respect to seashores. Effects of winds are much pronounced in plants along the sea coasts and at high altitudes on mountains. Air moves from a region of high to that of low pressure. The pressure differences are mainly due to differential heating of atmosphere. The equatorial regions receive more heat than north or south regions, thus low pressure occurs at lower latitude. The air generally moves from the poles towards equator.

Wind brings about a number of physical, anatomical and physiological effects on plants, which are as follows:

Physical effects

1. Breakage and uprooting: A wind of much high velocity may cause the breaking of living branches of trees and sometimes even their complete uprooting. However, in forests where the canopies of individual trees at different heights reduce the velocity of wind to about 80 per cent, such effects are uncommon. Thus, such vegetation in forests serve as natural windbreaks (shelter belts). Sometimes special trees and shrubs are planted to provide protection against wind for fields, orchards, buildings or livestock. A windbreak is a densely planted strip of tall vegetation, usually between 50 and 200 ft. wide, oriented at right angles to the direction of prevailing winds, in order to reduce their velocity near the ground. Windbreaks afford such benefits to plants, as reduced rates of evaporation and transpiration, reduction of damage by breakage, lodging and abrasion and reduction in the movement of soil by wind. Usually violent winds may cause breakage of soft woods of such plants as cottonwoods and river maple. Single trees or groups of trees which are uprooted in forest are often known as wind throws or windfalls.
2. Deformation: Strong winds from a constant direction sometimes cause permanent alteration in the form and position of the shoots. Deformation is very common in trees growing on ridges and along coasts.
3. Lodging: It is a type of wind injury common in grasses, as wheat, maize, oat, sugarcane etc., where violent winds cause the flattening of these herbaceous plants against the ground.
4. Abrasion: Particles of soil or ice carried by wind may act as strong abrasive force, by which the buds and other parts of plants may be eroded away. Crops grown on sandy soils usually suffer with such damage.

5. Erosion and deposition: The roots of plants whose covers are very thin may be exposed, as the soil is generally eroded out by strong winds. Moreover, the areas where these additional soils are deposited become unsuitable for plants growing there.
6. Salt spray: Along seacoasts, the salts of the water are carried along by strong winds in the vicinity of the ocean. Such salts have injurious effects on some plants growing there.

Anatomical and physiological effects

1. Compression wood: In some tree boles, as a result of wind deformation, there develops a dense, reddish type of xylem called the compression wood, on the compressed side. In some herbaceous plants there may be stimulated the formation of more collenchyma in the wind-deformed organs.
2. Desiccation: Strong winds cause an increase in the rates of evaporation and transpiration. Under such circumstances, plants fail to maintain an internal water balance and thus suffer from desiccation. Other plants with crowded branches may escape such wind damage. They are prevalent in windy habitats and are known as cushion plants. Their root systems attain a cushion form.
3. Dwarfing: Plants growing under the influence of drying winds generally suffer from dehydration and consequent loss of turgidity. Under these conditions, their organs become dwarfed. This is common in trees on seacoasts, arctic or alpine timberlines.

Fire Factor

Except those fires caused by lightning or volcanic activity, which are rare and confined only to certain specific areas, most of the fires very common in most of the regions are of biological origion. These are mostly man-caused and sometimes, chiefly in forest, develop due to mutual friction between tree (bamboos etc.) surfaces.

Fires are generally classified as: (i) ground fires—which develop in such conditions where organic matter accumulates richly as heaps and they catch fire which generally smoulder for longer periods. These fires are flameless and subterranean and kill almost all plants rooted in the burning material except some woody species; (ii) surface fires—which sweep over the ground surface rapidly and their flames consume the litter, living herbaceous vegetation and shrubs and also scorching the tree bases if come in contact; and (iii) crown fires—which are most destructive, burning the forest canopy, surface debris and bringing about vegetational destruction at larger scale.

These develop in dense woody vegetations. Temperatures that prevail when fire is in progress may reach up to 1300°F.

Effects of fire

In addition to its direct lethal (killing) effects on plants, fire also affects them indirectly in the following ways:

1. Some trees, as a result of injury by fire, have large scars on their stems. Such scars may serve as suitable avenues of entry of parasitic fungi and insects.
2. Fire brings about marked alteration of such environmental factors as light, rainfall, nutrient cycles, fertility of soil, litter and humus contents of soil, pH and soil fauna. Sometimes the habitat needs thousands of years in the return of normal conditions. The nature of such alterations depends mainly on the nature of fires.
3. Fire plays an important role in the removal of competition for surviving species. Such species that fortunately survive fires (fire-tolerants) generally increase in abundance at the expense of

those killed by fire (fire-sensitive) as a result of considerable reduction in competition and possibly due to alteration in other conditions.

4. Some plants as *Populus tremuloides* and *Epilobium angustifolium* are stimulated to growth by fires. A number of such grasses as *Aristida stricta*, *Cynodon dactylon*, *Paspalum notatum* etc., are stimulated by fire to produce large quantities of seeds. In some grasses and legumes the seeds would germinate only after these get fire treatment. Their seeds have much degree of tolerance to temperatures due to fires. Common examples studied are species of *Themeda*, *Heteropogon*, *Andropogon*, *Rheus*, *Tephrosia* and *Stipa*.
5. Some fungi, mainly some ascomycetes, grow in soils of burnt areas. Such fungi are known as pyrophilous, as for example *Pyronema confluens*.

Adaptations to fire

In areas, where fires are very common, the plants have developed some features that favour their survival under such conditions. These adaptive features are as follows:

1. Some shrubs belonging to the genera *Rheus*, *Ceanothus*, *Aractostaphylos* produce seeds in large numbers and whose seed-coats are very hard. The breakage of hard seed-coats depends on fire, after which seedlings appear in large numbers. It is experimentally proved that their seeds need a temperature of about 80°C for maximum germination.
2. Some woody plants that develop fire-resistant seeds or fruits have unique property of completing their life cycles within a short period by virtue of their rapid rate of growth and development. Within this short period, there never accumulates organic debris and thereby chances of surface fires reduced. Seedlings of *Pinus palustris* grow at rapid rate and in *P. muricata* cone development starts when plants are only one meter tall.
3. In some plants leaves are fire-resistant due to poor contents of such compounds as resin or oil and thus may check surface fires. Similarly, *Quercus macrocarpa*, *Larix occidentalis*, *Pinus palustris* and *P. ponderosa* have thick bark which often escape fire injury.
4. Some plants as *Pinus rigida* and species of *Eucalyptus* have adventitious or latent axillary buds which may develop into new branches. Similarly, *Betula papyrifera* and *Vaccinium* spp. may develop new shoots after fire kills the older ones.
5. In chaparral vegetation of Europe, North America and Australia, many shrubs, as species, of *Eucalyptus* and *Aractostaphylos* have evolved special turnip-shaped swellings of their axes remaining entirely below the ground surface, which have latent buds. These structures are known as lignotubers which may produce new shoots.
6. A number of conifers, as *Pinus contorta* var. *latifolia*, *P. clausa*, *P. leiophylla*, *Picea mariana* produce serotinous cones i.e., cones remaining on trees with viable seeds for several years sometimes, up to 75 years. Fires cause the prompt opening of these cones.

3

Ecological Adaptations

INTRODUCTION

A physiological background makes it quite evident that water is an important ecological factor in the life of organisms. Water is universal solvent in which practically all the minerals, present in soil, may be dissolved. It is the medium of solutes, raw material in photosynthesis and is essential for maintenance of turgidity of cells and sometimes plays an important role in fertilisation, pollination and dissemination. Water relations of plants bring about in them many physiological responses by affecting seed germination, water uptake, transpiration, photosynthesis and growth, respiration, translocation and some biochemical processes. Water shortage, in particular, which has been the main concern of human race, leads to many physiological effects in plants which in turn have many ecological effects as their counterparts. Drought, thus has been a major physiological and ecological problem to be tackled by the scientific community of the world.

FORMS OF WATER IN NATURE

In nature, water is found in three principal forms–atmospheric moisture, precipitation and soil water, of which precipitation is the chief source. Atmospheric moisture is present as invisible vapour, known as the humidity and also as visible vapour, known as fog or cloud. The invisible vapour or humidity of the air is usually expressed as relative humidity. Warm air can hold more water vapour than the cold air. Generally this capacity of air of holding vapour increases with an increase in temperature. Relative humidity values of air show daily rhythm and these values are measured by such instruments as Psychrometer (a dry and wet bulb thermometers system) and Hygrometer. The visible vapours, cloud and fog, are actually water droplets or tiny ice crystals formed by cooling of the air to a temperature below its dewpoint. Clouds are formed due to the movement of air upward from land surface towards cooler sides of atmosphere. They are always separated from the earth. Fog is caused due to cooling of air at or near the land surface and is generally continuous from the surface upward. Atmospheric vapours affect plants through their effects on the intensity of solar radiation and on evaporation and transpiration. These are also sometimes sources of soil moisture and are also of direct use for some such plants as desiccated mosses and lichens and some tropical orchids, that absorb moisture from humid atmosphere without its condensation.

Precipitation (rainfall) is actually the chief source of soil moisture. From the soil, plants meet their water requirements. Directly, rains are of little importance to plants. Precipitation affects plants directly in several such forms as snow, hail and sleet. Snow is injurious, as well as beneficial to plants. It breaks off tree branches, flowers and fruits and may cause bending of slender stems. Some sedges prefer to grow in snow patches. Deep snow may press seedlings down to the ground and there are chances of their being parasitised by fungi and other pathogens. Sometimes melting snow facilitates water supply to soil. Hail and sleet cause damage to aerial parts of the plants.

Soil moisture is the chief source of water for plants. Its chief source is rainfall. All the water which falls during precipitation is never available to soil as soil moisture. Some of the precipitated water, before reaching the ground, evaporates in the atmosphere, affecting its relative humidity values. During long period of rainfall, relative humidity may increase up to 100 per cent due to sufficient evaporation of precipitating water to the atmosphere. Some water is lost as run off i.e., the water that runs off the land directly flowing back as streams to the source of precipitation-seas. Characteristics of soil and nature of the topography govern the amount of loss of water as run off. These include the moisture content, temperature and porosity of soil and smoothness and steepness of the slopes.

Since at some length the various forms of water in soil and the different terms, generally used to specify the water status of a soil including those that indicate soil water-plant relationships. Of the various forms, capillary water is mostly available to plants. It has also been pointed out earlier that the amount of water held up by a soil is, besides other factors, greatly influenced by its texture and structure. Thus coarse-textured soils as sand hold water only loosely due to larger pore specs, whereas clay and silt soils are able to hold much water and are poorly drained. Loam, a mixture of sand and silt and/or clay are supposed best for plants' growth as they can hold much water as well as minerals and are also properly aerated.

SOIL-PLANT-ATMOSPHERE-CONTINUUM (SPAC)

Importance of water in the life of plants is evident by the strong relationship between the water status of soil, plant and the atmosphere. The rooting zone of the soil (zone of soil in which the water absorbing organs, roots, root hairs are present), the plant body and the lower layer of atmosphere behave as a continuum in relation to water transfer and must be considered as a whole in any complete analysis.

Movement of water in soil takes place in the liquid phase by capillary flow and in the gaseous phase by molecular diffusion or mass flow. In moist soil, capillary flow predominates, whereas vapour transfer is generally important in rather dry soils. Water transport in the plant occurs by liquid-phase viscous flow in channels or by diffusive transfer through membranes. Final escape to the atmosphere is by vapour-phase molecular diffusion across the intercellular spaces of the mesophyll to the stomatal pores. From here, turbulent diffusion in the leaf and canopy boundary layers completes the transfer to the atmospheric sink. Thus flow-path of water through the SPAC is essentially complex, a sequence of 'series-parallel' resistances and capacitances complicating the theoretical analysis of the system. Solar radiation is the primary energy source for the water transport process in the SPAC, the sink being the latent heat change in the evaporation of water at mesophyll cell surface and at the soil surface.

ADAPTATIONS OF LAND PLANTS—WATER-BALANCE PROBLEMS

The above account of SPAC shows that the water deficit conditions pose some sort of water balance problems in plants, particularly for those of dry habitats, where such 'deficit-days' in the year are of quite a high frequency.

Aquatic plants have hardly any problem of water, since such phenomena as transpiration, wilting, etc., are generally not much pronounced or may be even absent but in land plants there is always a water balance problem. Due to the limited amount or available water during the year, they are to evolve some characteristics that may check more loss of water from their exposed organs.

They have to maintain always some sort of balance between the loss and absorption of water. Due to these water-balance conditions, the structure of plants is greatly influenced. In land plants, which grow under unfavourable water conditions, there develop some characteristics, which are absent in plants growing under optimum moisture conditions. These characteristics are morphological as well as physiological. The former type includes the reduced size of shoot, elaborate root system, smaller blades with smaller stomata and smaller veinlets and more hairs on leaf surfaces, thicker cuticular layers and cell walls, well developed palisade tissue, poorly developed intercellular spaces and spongy parenchyma and xylem cells smaller and much lignified. The latter i.e., physiological ones include a more rapid rate of transpiration as well as photosynthesis per unit area, higher osmotic pressure with an increase in protoplasmic permeability and decrease in its viscosity, grater resistance to wilting and early flowering and fruiting.

ECOLOGICAL GROUPS OF PLANT AND ECOLOGICAL ADAPTATIONS

The ecological groups of plants are indeed much broader groups without any airtight boundaries between them. This is a classification or rather a very broad grouping of plants based on their water relationships. Warming classified plant communities on the basis of plants' dependence upon and relation to water. Water as an ecological factor occupied the foremost position in distribution of vegetation and its structure. On the basis of their water requirements, Warming primarily recognised three major groups of plants:

1. Aquatic plants (hydrophytes), living in abundance of water with their lower parts i.e., roots, rhizomes etc. and leaves immersed in water.
2. Land or terrestrial plants, living in normal water conditions with their assimilatory organs adapted to existence in air. Land plants exhibit many grades of adaptation to their mode of life. Thus, those which encounter greatest difficulties in securing water are the xerophytes, while others as mesophytes, which in some respects occupy an intermediate position between the two extremes, the hydrophytes and the xerophytes.
3. Helophytes or marshy plants, an intermediate group between the acquatic and land plants.

An account of these ecological groups of plants, i.e., hydrophytes, mesophytes, xerophytes and halophytes will be presented here.

Hydrophytes

They grow on extremely wet soil where water is available to plants in abundance. According to the way in which they develop in water, they are further subdivided into the following five categories:

Free-floating hydrophytes

They remain in contact with water and air, but not soil. They float freely on the water surface. Leaves in some are very minute, while in others quite large. Some of the free-floating hydrophytes, as *Wolffia, Lemna, Spirodella, Azolla, Eichhornia, Salvinia* and *Pistia.*

Rooted hydrophytes with floating leaves

Their roots are fixed in mud, but leaves have long petioles which keep them floating on the water surface. The remaining plant, except leaves, remains in water. Some of the rooted hydrophytes with floating leaves, as *Trapa, Nelumbo, Nymphaea* and *Marsilea.*

Submerged floating hydrophytes

These are the plants in contact with only water, being completely submerged and not rooted in the mud. Their stems are long and leaves generally small. Some of the examples are *Ceratophyllum*, *Utricularia*, *Najas*, etc. In *Ceratophyllum,* roots are lacking even in embryo stage, sometimes leafy branches being modified into 'rhizoids'.

Rooted submerged hydrophytes

These are hydrophytes like *Hydrilla, Potamogeton, Isoetes* and *Vallisneria* that remain completely submerged in water and rooted in soil. In *Hydrilla*, *Potamogeton* and *Chara*, the stem is long, bearing small leaves at the nodes. Hydrilla is a slender weed with fibrous roots. In *Isoetes* and *Vallisneria* stem is tuberous (corm-like) with long leaves, which are narrow, ribbon shaped.

Rooted emergent hydrophytes

They grow in shallow waters. These are such hydrophilous forms which, although require excess of water, but their shoots (assimilatory organs) are partly or completely exposed to air. The root system is completely under water, fixed in soil. In some, as *Sagittaria* and *Ranunculus*, shoots are partly in water and partly emerging i.e., exposed to air, whereas in others such as *Scirpus* and *Cyperus* also known as marshy plants, the shoots are completely exposed to air.

Warming realised the importance of soil also in plants' life and on the basis of the nature of soil and water requirements, he divided plant communities, into a number of classes as-Hydrophilous: (i) hydrophytes (in water); (ii) helophytes (in marsh); Xerophilous; (iii) oxylophytes (on acid soils); (iv) psychrophytes (on cold soils); (v) halophytes (on saline soils); (vi) lithophytes (on rocks); (vii) psammophytes (on sand and gravel); (viii) chersophytes (on wastelands); (ix) eremophytes (deserts and steppe); (x) psilophytes (savannah); (xi) sclerophyllous formations (bush and forest); (xii) coniferous formations (forest) and Mesophilous; and (xiii) mesophytes.

Ecological Adaptations in Hydrophytes

Hydrophytes may differ from each other in some aspects, but most of the characteristics, morphological as well as anatomical, are common to all of them. Due to these similarities, they belong to the same ecological group–the hydrophytes. We shall now deal with all such features of these plants which enable them to become adapted to such hydrophytic conditions.

Morphological features

1. Roots: Due to availability of water in plenty roots, the principal organs of water absorption, in such plants become of secondary importance and less significant. Their overall development is usually very poor and insignificant in most of the hydrophytes:-
   (a) Roots may be entirely absent as in *Wolffia, Salvinia* and *Ceratophyllum* or poorly developed as in *Hydrilla*. In *Salvinia* submerged leaves compensate for roots. However in emergent forms, as *Ranunculus* which grow in mud, roots are well developed with distinct root caps.
   (b) Root hairs are absent or poorly developed.
   (c) Root caps are usually absent. In some cases, as *Eichhornia* root caps are replaced by 'root pockets'.
   (d) Roots, if present, are generally fibrous, adventitious, reduced in length, unbranched or poorly branched. In *Lemna* roots act simply as balancing and anchoring organs.

(e) In some plants, as *Jussiaea repens*, there develop some floating aerial roots also in addition to normal adventitious roots.

2. Stems:-
   (a) In submerged forms as *Hydrilia* and *Potamogeton* stem is long, slender, spongy and flexible. In free-floating forms it may be slender, floating horizontally on water surface as in *Azolla* or thick, short, stoloniferous and spongy as in *Eichhornia*. In forms which are rooted with floating leaves it is a rhizome, as in *Nymphaea* and *Nelumbo*.
   (b) Vegetative propagation by runners, stolons, stem and root tubers, dormant apices, offsets etc., is most extensive and common method of reproduction. Thus most of them are perennials.
3. Leaves:-
   (a) In submerged forms, leaves are thin and are either long and ribbon-shaped as in *Vallisneria* or long and linear as in *Polamogeton* or finely dissected as in *Ceralophyllum* Floating leaves are large, flat and entire as in *Nymphaea* and *Nelumbo* with their upper surfaces coated with wax; their petioles are long, flexible and often covered with mucilage. In *Eichhornia* and *Trapa* petioles become swollen and spongy.
   (b) Emergent forms as *Ranunculus* and *Sagittaria* show heterophylly with submerged, floating and aerial leaves.
   (c) Submerged leaves are generally translucent.
4. Flowers and seeds: These are less common in submerged forms. Where flowers develop, seeds are rarely formed.

Anatomical features

1. Roots: The distribution of various tissues in roots of hydrophytes, in general, becomes clear from showing transverse sections of roots of *Polamogeton* and *Eichhornia* respectively. It may be concluded from these figures that:-
   (a) Cuticle is either completely absent or, if present it is thin and poorly developed.
   (b) Epidermis is usually single-layered made up of thin-walled parenchymatous cells.
   (c) Cortex is well developed, thin-walled and parenchymatous, major portion of which is occupied by well developed prominent air cavities the 'aerenchyma' which offers resistance to bending stress, increases buoyancy and allows a rapid gaseous exchange.
   (d) Vascular tissues are poorly developed and least differentiated in submerged forms, as *Potamogeton* with thin-walled elements. In xylem, vessels are less common, tracheids being generally present. In floating forms as *Eichhornia* are comparatively differentiated to some extent. However, in emergent forms as *Ranunculus* and *Typha* etc., vascular elements are comparatively much distinct and well developed.
   (e) Mechanical tissues are generally absent except in some emergent forms, as *Typha* where pith cells are sclerenchymatous.
2. Stems: Distribution of various tissues in stems of hydrophytes becomes clear from transverse section of stem of *Hydrilla*. It may be seen that:-
   (a) Cuticle is either absent or poorly developed and thin.
   (b) Epidermis is usually single-layered made up of thin-walled parenchymatous cells. However, rhizomes of *Nymphaea* and *Nelumbo* show well-developed epidermis. In emergent forms as Typha, cuticle as well as epidermis are generally well developed.

(c) Hypodermis is completely absent in submerged forms as *Hydrilla* and *Potamogeton*. However, in floating and emergent forms, it may be present as thin-walled parenchyma or collenchyma.

(d) Cortex is well developed, thin-walled and parenchymatous, extensively traversed by air cavities as in roots. Cells of cortex generally possess chloroplasts and are thus photosynthetic. In some, as *Nymphaea*, there are found large number of vascular bundles scattered in the cortex.

(e) Endodermis is generally distinct; especially in rhizomes and similar organs.

(f) Vascular bundles generally lack bundle sheaths. Vascular elements are thin-walled, lignified elements being absent. However, in emergent forms as Typha, vascular elements are comparatively well differentiated and developed.

(g) Mechanical tissues are usually absent.

3. Leaves: In their internal structure, leaves in different hydrophytes show variations. However, some of the anatomical features are common to most of the leaves. Distribution of different tissues in leaves are *Anacharis, Potamogeton* and *Nymphaea* respectively. From these we may conclude that in leaves:-

(a) Cuticle is usually absent in submerged forms as *Anacharis* and *Potamogeton*. In floating forms as *Nymphaea* it is poorly developed confined only to upper side and is thin. In emergent forms also it is thin.

(b) Epidermis is single-layered, made up of thin-walled cells with abundance of chloroplasts.

(c) Stomata are completely absent in submerged leaves as *Anacharis* and *Potamogeton*. In floating forms as *Nymphaea* stomata are confined only to the upper surface of leaf, whereas in emergent forms they are generally found on both surfaces of leaves.

(d) Mesophyll is undifferentiated in submerged leaves, where it is generally single-layered as in *Potamogeton*. In *Anacharis* also it is thin and undifferentiated. In floating leaves as *Nymphaea*, it is, however, differentiated into palisade and spongy parenchyma with well-developed air cavities. In emergent leaves, mesophyll is well differentiated with air cavities.

(e) Vascular tissues are very much reduced and sometimes difficult to be differentiated into xylem and phloem as in leaves of submerged forms like *Anacharis*. Wherever differentiated as in *Nymphaea*, xylem elements are thin-walled, phloem being well developed. However, in aerial leaves, vascular elements are comparatively well differentiated with vessels in xylem.

(f) Mechanical tissues are absent.

(g) The petioles, wherever well developed, as in *Nymphaea* also possess internally the various tissues characteristic of a typical hydrophyte, i.e., an abundance of aerenchyma, thin-walled cells, lack of differentiation in vascular tissues and absence of any lignified mechanical tissues.

Mesophytes

They are very extensive on the land surface. They are such land plants which grow in moist habitats and need well-aerated soils. They prefer soil and air of moderate humidity and avoid soil with standing water or containing a great abundance of salts. In some respects they stand in between the hydrophytes and xerophytes. Broad-leaved trees growing in wet depressions, along lakes and rivers, are mesophytes. They generally lack the special structural and physiological adaptations found in hydrophytes and xerophytes.

The generally morpho-anatomical features of mesophyte are as follows:

1. Root system is well developed. Roots are generally fairly branched, with root caps and root hairs.
2. Stems are generally aerial, solid and freely branched.
3. Leaves are generally large, broad, thin and varied in shapes; generally oriented horizontally, green, without hair or waxy coatings.
4. Cuticle in all aerial parts moderately developed.
5. Epidermis well developed, without any hair or waxy coatings and cells without chloroplasts.
6. Stomata generally present on both surfaces of leaves. Guard cells show frequent movements.
7. Mesophyll in leaves is differentiated into palisade and spongy parenchyma, with many intercellular spaces.
8. Vascular and mechanical tissues fairly developed and well differentiated.
9. They may show temporary wilting during noon-hours.

Xerophytes

The term 'xerophyte' has been defined and interpreted variously. Morphological features and rates of transpiration have almost failed to explain the true nature of this group of plants. They are sometimes loosely defined as 'plants of dry habitats'. But if taken in this sense, most of the mesophytes should be called xerophytes. Daubenmire in his discussion on the group defined xerophytes as 'plants which grow on substrata that usually become depleted of growth water to a depth of at least two decimeters during a normal season'. Thus, in arid zones, all plants not confined to the margins of streams or lakes have been considered as xerophytes, whereas in regions of heavy rainfall the class would be represented only by shallow-rooted plants of sandy soils, by plants of dry ridgetops and by algae, mosses and lichens which grow on tree barks or rock surfaces etc. The true nature of xerophytism is not clearly understood. For example, it is difficult to decide whether a xerophyte is really xerophilous and occurs only in dry habitats and deserts or is merely drought-resistant. There are instances, for example, that, if desert plants are grown in soils with moderate amount of water, they show better growth as in *Artemisia fasciculata*; *Zygophyllum fabago* and *Centaurea solstitialis*, the yield of plants increased to 1.5–2 times by increasing water-holding capacity of soil from 40 per cent to 60 per cent. Thus, such xerophytes are not xerophilous at least so far as the soil moisture is concerned. Oppenheimer, in a literature survey, favours a much broader interpretation including all the morphological, anatomical and physiological modifications which may assist the plant to cope with environmental water deficit.

On the basis of their morphology, physiology and life cycle pattern, xerophytes are generally classified into the following three categories.

Ephemeral annuals

They are also called as 'drought evaders' or 'drought escapers'. They are mostly found in arid zones. They are annuals's, which complete their life cycles within a very short period of 6–8 weeks or so. With their small size and large shoots in relation to roots, they are well adapted to such dry habitats. They actually avoid and not withstand dry seasons and thus escape dryness in external and internal environments. Some do not prefer to call them true xerophytes. Examples *Argemone mexicana*, *Solanum xanthocarpum*, *Cassia tora*, *Tephrosia purpurea*, etc.

Succulents

They are the plants that suffer from dryness in external environment only. Their succulent, fleshy organs (stems, leaves, roots) serve as water-storage organs which accumulate large amount of water during brief rainy seasons. In cacti moreover, the root systems also become shallow. Their root system is shallow, stem swollen and leaves thick, leathery and succulent. Some of the examples are *Aloe, Euphorbia* and *Opuntia* and various cacti, *Agave* and *Ceiba parviflora*. As the succulents avoid drought, some prefer to exclude them from true xerophytes. This is indeed a unique mode of adaptation.

In some of them, stems become succulent, which are also called the 'fleshy xerophytes' as in Opuntia and Euphorbia. In such xerophytes, cuticle is thick and well developed hypodermis is two-to-three layered, collenchymatous; cortex is also thick-walled with chlorenchymatous cells, below which there is present a prominent 'water-storage region', whose cells are thin-walled with a few intercellular spaces.

Those succulent xerophytes in which leaves become fleshy; are also known as 'malacophyllous xerophytes', such as *Aloe Begonia, Salsola, Bryophyllum, Agave, Yucca, Tradescantia* etc. Their leaves possess turgescent parenchymatous cells with thin walls. *Asparagus* and *Ceiba parvifolia* store water in their roots.

Non-succulent perennials

These are actually the true xerophytes or drought-resistants, because they possess a number of morphological, anatomical and physiological characteristics which enable them to withstand critical dry conditions. They are the plants that suffer from dryness both in their internal as well as external environments. Their morpho-physiologic features include rapid elongation and extensive root system, high osmotic pressure and endurance of desiccation, ability to reduce transpiration to an extremely low level during permanent wilting and reduction in size of leaf blades and cells. Non-succulent perennials like *Calotropis procera, Acacia nelotica, Zizyphus jujuba* and *Capparis aphylla*. Other examples are *Prosopis, Casuarina, Nerium, Alhagi, Saccharum* and *Salvadora*.

The general characteristic features of non-succulent xerophytes are as follows:

1. Root system is very extensive. For example, in *Prosopis*, *Calotropis* and *Alfalfa*, roots may be more than 30 meters long.
2. Following characteristics help minimising the rate of transpiration:-
   (a) Dying back of leaves as in many grasses.
   (b) Rolling and folding of leaves as in many grasses like *Ammophila arenaria*.
   (c) Delicate leaves which are shed under conditions of less supply of water.
   (d) Heavy cuticular and epidermal layers.
   (e) Waxy coatings on leaves.
   (f) Sunken type of stomata on leaves.
   (g) Very small, narrow, sometimes scaly leaves.
   (h) Ridged stems.
3. Aerial organs may become vigrously modified according to the prevailing climatic conditions.
4. Experimentally, the reduction of water supply has been shown to reduce the shoot/root ratio in some grasses.

Ecological Adaptations in Xerophytes

We have discussed above three different types of xerophytic plants and their chief characteristic features. These three major groups of xerophytes may differ in several ways from each other but at the same time they all possess certain features in common. On the basis of these similarities, they are grouped into the same ecological group–the xerophytes. These features, which are more or less common to all, and which enable them to survive under the prevailing xeric conditions, which are as follows:

Morphological features

1. Roots: In contrast with hydrophytes which develop in conditions with plenty of water, xerophytes develop under water-deficient conditions. Thus, in order to secure water, which is present in less amount and, moreover, in deeper layers of soil, roots in xerophytes become the principal organs of primary importance. The root system is thus very well developed, with the following characteristics:-
   (a) It is very extensive, which in some cases is several times longer than the shoot. Roots are long, tap roots, with extensive branching spread over wide areas.
   (b) Root hairs and root caps are very well developed.
2. Stems:-
   (a) Mostly they are stunted, woody, dry, hard, ridged and covered with thick bark.
   (b) In some as *Saccharum*, stem becomes underground, whereas in *Opuntia* it becomes fleshy, green, leaf-like (phylloclade) covered with spines. In *Euphorbia* also it becomes fleshy and green.
   (c) On stems and leaves there are generally hairs and/or waxy coatings.
3. Leaves:-
   (a) Leaves are very much reduced, scale-like appearing only for a brief period, sometimes modified into spines. Lamina may be long, narrow or needle-like as in *Pinus* or divided into many leaflets as in *Acacia*.
   (b) Foliage leaves, wherever present, may become thick, fleshy and succulent or tough and leathery in texture.
   (c) Leaf surfaces are generally shiny and glazed to reflect light and heat.
   (d) In some monocots as *Ammophila, Poa* and *Agropyron*, leaves become folded and rolled in such a manner that the sunken stomata become bidden and thus rate of transpiration is considerably minimised.
   (e) In some of them as *Euphorbia, Acacia nelotica, Zizyphus jujuba* and *Capparis aphylla* stipules become modified into spines.

Anatomical features

1. Roots:-
   (a) Root hairs and root caps are well developed. In *Opuntia*, root hairs develop even at the root tips.
   (b) Roots may become fleshy to store water as in *Asparagus*.
   (c) In Pinus *edulis* and *Calotropis*, roots possess rigid and thickened walls.
2. Stems:-
   (a) In succulent xerophytes, stems possess a water-storage region.
   (b) In stems of most of the non-succulent xerophytes, such as *Casuarina*.
3. Leaves:-
   (a) In succulent leaves of malacophyllous xerophytes, such as *Peperomia*, epidermal cells of leaves serve as water-storage organs.

Similarly, succulent leaves of *Aloe* and *Salsola* have prominent water-storage regions in their mesophyll. Moreover, in such leaves cuticle is thick and outer walls of the epidermal cells are heavily deposited with cutin and cellulose.

(b) Leaves of non-succulent xerophytes, such as *Nerium* and *Pinus* etc.

To sum up this detailed descriptive account of various features for the purpose of quick revision, the same has been presented in a comparative way in Table 3.1.

Table 3.1. Generalised morphological, anatomical and physiological features of three ecological groups of plants.

Hydrophytes	*Mesophytes*	*Xerophytes*
Morphological		
Excepting the emergent forms, roots generally greatly reduced, even absent: if present, unbranched without root hairs	Root system well developed, tap or fibrous roots with root haris	Root system generally very deep, extensive, reaching deeper layers of soil; several times larger than shoot, hard, woody
Stem very much reduced, in some modified into rhizomes, thin, delicate	Stem is rigid and stout	Stem much dwarfed, branched, sometimes under ground
Leaves thin, narrow, linear, some with long petioles and large lamina, covered with wax or hairs, in some dissected into segments	Leaves large, thin, generally without waxy surfaces	Leaves small, sometimes much reduced to scales or modified into spines, in some leathery and thick with shining surface and waxy coating, covered with hairs
Anatomical		
Aerenchyma extensive almost in all vegetative organs	Aerenchyma lacking.	Aerenchyma lacking
Cuticle generally absent	Cuticle developed.	Cuticle thick, well developed
Stomata either absent or if present only at upper surface or even non-functional	Stomata on one or both the surfaces	Stomata less in number, generally confined to lower surfaces of leaves, sunken
Mesophyll undifferentiated into palisade and spongy parenchyma	Palisade well developed and differentiated	Palisade generally on both sides of leaves, cells and vacuoles small
Chlorophyll in addition to leaves also in other parts of plant	Chlorophyll mostly in leaves	Chlorophyll mostly in stem and leaves
Epidermal cells thin-walled	Epidermal cells thick	Epidermal cells conspicuously thick-walled
Lignified mechanical tissues lacking, conducting elements very few, non-lignified	Mechanical and vascular tissues well developed	Mechanical and vascular tissues very well developed
Physiological		
Habitats mostly deficient in oxygen content, thus having an ability to respire anaerobically or have low oxygen requirements. Special aerating organs are present	Normal physiological processes, with temporary wilting at room temperature	Most of the features directed to reduce the transpiration rates; ephemerals complete life cycle in short period. High osmotic pressure (as much as 3100 atms. in some halophytes) and endurance of desiccation

Halophytes

They are a special type of xerophilous plants which grow on saline soils with high concentration of salts like NaCl, $MgCl_2$ and $MgSO_4$. Soils are physiologically dry. Their osmotic pressure values are very high. Most of them are succulents, their leaves are evergreen, thin, small and leathery with water storage tissues, thick cuticle and well developed palisade tissues. Most of them produce special type of branched, negatively geotropic roots that come out of the mud surface to encourage the entry of oxygen gas. These roots are called pneumatophores which possess breathing pores for gaseous exchange. Some of the halophytes show vivipary i.e., germination of seeds before they are shed from parent plant. The universal feature of these plants is selective absorption and tolerance of sodium chloride and the ability to grow in soils so saline that they are either physiologically toxic to normal plants or impose so serious an osmotic water stress that water absorption is impeded. Plant cells are bathed in a medium of high osmotic potential and thus should maintain a higher vacuole osmotic potential in order to avoid loss of water.

Halophytic communities have been divided into: (i) lithophilous; (ii) psammophilous; (iii) peldphilous; and (iv) helophilous types, growing on rock and stones, sand, mud and swamp respectively. Helophilous halophytes have been further sub-divided into: (i) salt-swamp and salt-desert; and (ii) littoral swamp-forest (mangrove) types, of which the latter are most extensive, which occur in all tropical seas, especially on flat, muddy shores, where the water is relatively calm, as in lagoons, inlets, estuaries etc. Soil is flooded with water either permanently or at high tide. These littoral swamp-forests form a characteristic vegetation, known as the mangroves or mangrove formation in tropical and subtropical areas.

In India, mangroves are very much prevalent on seashores of Bombay and Kerala and in Andaman and Nicobar islands. They grow on water-logged soil, where habitat is characterised by sandy, swampy and saline soils, high precipitation, high atmospheric humidity and almost no fluctuation in temperature throughout the year. The two common halophytes of this type are *Rhizophora mucronata* and *Sonneratia*. They produce stilt roots, pneumatophores and show vivipary. They show xeromorphic characters as evident from the anatomical characters of vegetative organs of *Rhizophora mucronata* as given below.

Roots

Roots are of two types-aerial and subterranean. The subterranean root in transverse section shows:

1. Several-layered cork.
2. Cortex made up of star-shaped or stellate cells, connected with each other by lateral arms. Cortical cells have peculiar thickening ridges which are lignified. Some cells of cortex are filled with oil and tannin.
3. Pith cells are thick-walled with pitted thickenings of lignin and contain tannin and oil.

Stem

Young stem in transverse section shows that:

1. There is very thick cuticle even at young stage.
2. Epidermal cells heavily thickened, almost squarish, filled with oil and tannin.
3. Hypodermis several-layered, of heavily thick-walled cells.
4. Primary cortex possesses a number of lacunae, cells of which thickened, filled with tannin and oil. Some cells also possess calcium oxalate crystals. H-shaped heavily thickened spicules present in this region provide mechanical strength to cortex.
5. Inner cortex possesses a number of branched, thick-walled cells-sclereids that provide mechanical strength.

6. Pericycle several-layered, sclerenchymatous.
7. Pith also possesses lacunae and H-shaped spicules, cells of pith also contain tannin.
8. Vascular tissues are well developed.

Leaf

Transverse section of the leaf shows that:

1. Cuticle is very well developed.
2. Epidermal cells are heavily thickened and may contain calcium oxalate crystals.
3. Stomata, confined only to lower surface and sunken.
4. Below the upper epidermis there are present several layers of thin-walled cells that remain filled with water, the outer layers of this layer filled with oil and tannin. Cells of lowermost layer are mucilaginous that remain embedded into palisade tissue.
5. Mesophyll is well differentiated.
6. There are present corky areas on the ventral surface.

Psammophilous halophytes may be annuals or perennials, the former living and growing during rainy season. The perennials show much pronounced xerophytic characters. Common examples of this group of halophytes are *Suaeda fruticosa* and *Salsola foetida.*

Ecological Adaptations in Animals

So far we have discussed the role of water in the life of plants and its ecological significance in their distribution. It has been made clear, as how plants get themselves adapted in conditions with abundance of water, with normal supply of water and those with scarcity of water. But what about animals? Like plants, animals too have to face the similar problem in different kinds of habitats in respect of the available water. Animals too are subjected to problems of scarcity as well as abundance of water. For their survival under these conditions, animals also have developed in their bodies a number of devices to meet the challenge of the environment where they live. It would be clear from the following account, where we shall discuss the water problems faced by animals in different habitats.

Aquatic conditions-hydrocoles

The plenty of water present in different water bodies as rivers, lakes, ponds, pools, puddles, seas, oceans etc., poses no problem as such from the quantity point of view for the animals living there. But the water contains varying amounts of materials dissolved in it. The concentrations of these substances in water and those inside the fluid present in animals' bodies are, however, different. Thus, there exist difference in the values of osmotic concentration of the body fluid of animal and that of the external water medium. Due to difference in osmotic concentration values, there may exist a problem for animals, that is to be overcome in one way or the other. In marine waters, whose osmotic concentration is greater than the body fluid of fishes living there, there is possibility of loss of water from the body of fish. Thus for their survival, the fishes have to conserve more water. In fresh water habitats as rivers, lakes, ponds etc., the osmotic concentration of water medium is generally less than that of the body fluid of the animals. Thus, water generally enters inside the body of animals. In the freshwater fishes the continuous entrance of water is very common. It becomes a problem and the water in one way or the other should be passed out to avoid bursting of the body. Thus the fishes may start moving to other areas.

Such hydrocoles possess a number of primary aquatic adaptations. These adaptations thus occur only in those animals as fish, which had never lived on land.

Their ancestors too were perhaps aquatic. They have the following main adaptations:

1. Head body and tail are compressed to give an elongated stream-lined body. There is no any protuberance over the spindle-shaped body that would interfere in the swift passage of the animal in water.
2. Presence of paired (pectoral and pelvic) and the unpaired (median, dorsal anal and caudal) fins. The pectoral fins and dorsal fin act as stabilisers or balancers and the caudal fin provides a forward push to body.
3. Arrangement of body muscles in the form of bundles separated by myocommata.
4. Respiration by gills thus making use of gases dissolved in water.
5. Presence of air-bladders filled with air, that act as an accessory respiratory as well as hydrostatic organs in many teleost fishes. By use of these bladders fishes are able to maintain themselves at a desired depth of water.
6. Presence of lateral-line system in fishes that contains neuromast organs which function as rheoreceptors, helpful in echolocating the objects in water.
7. Integument rich in mucous glands or protected with scales.

Amphibious conditions–secondary hydrocoles

During the course of evolution, there had been migration of life from aquatic to terrestrial conditions. During this migration, only those that were able to retain sufficient water in their bodies for adverse dry conditions could succeed. Thus some birds, mammals and most reptiles could succeed in developing abilities to respire in air. However, others such as some amphibious animals as well as plants, could come only half the way out of water. These, amphibians are said to be the first to live on land. They are able to live both in air as well as in water. Such animals live in water during reproductive phase of their life and the remaining part may be successfully passed on land. When conditions become dry, they migrate to water or damp places. These animals have developed in themselves a number of devices for living in water as well as on land. For instance, some fishes may live in wet tuft of grasses for a period of about 64 hours. Frogs burrow in the mud where they remain dormant during dry periods of the life. African lungfish, *Protopterus*, constructs special mud cocoon, remaining there during the drought period. Moreover, it also secretes an impervious sheath around its body that prevents the loss of water from body surface. Some lower animals as protozoans, sponges, crustaceans etc., develop thick-walled eggs or cysts during adverse dry conditions. Tidal zones of the seashores, subjected to fluctuating drought and flood conditions are inhabited by such life forms as seaweeds, barnacles, mussels, claims etc.

Such adaptations are infact secondary which occur in those animals that are lung breathers. Their ancestors were perhaps terrestrial but due to harsh environmental conditions with scarcity of food or severe competition etc., these were forced to adopt an aquatic mode of life. Such secondary aquatic adaptations occur in amphibians like frog, mud puppy (*Necturus)*, *Amphiuma* and *Siren*; reptiles as turtles, tortoise, snakes (*Natrix*, *Hydrophis*), crocodiles, gavials and alligators; birds as albatrosses, petrals, penguins, ducks; and mammals as whales, dolphins and porpoises.

The main adaptations are as follows:

1. Stream-lined form of body, neck constriction absent and tail enlarges (as in Cetacea, Sirenia, Pinnipedia and Ichthyosauria).
2. Body-size becomes larger (whales).
3. Shortened, dorsoventrally flattened and more wider cranium and the facial portion becoming an elongated, slender, snout or rostrum as in Porpoises and Ichthyosaurs.

4. Neck becomes short and immobile.
5. Bones become light and spongy.
6. External ear disappears. Its presence could have hindered in locomation in water.
7. Nostrils and eyes shift on the apex of face.
8. Limbs become fleshy and fin-like expansions or paddles for swimming.
9. In aquatic mammals, the power of mastication of food is lost, therefore the cornoid process of mandible is reduced.
10. In aquatic mammals, skin becomes smooth and naked.

Terrestrial conditions

In terrestrial conditions, there are found both the mesocoles as well as the xerocoles. Water problem in such conditions indeed becomes very much critical. Due to deficiency of water, animals have somehow or the other retain sufficient water in their bodies. For this, animals develop in themselves a number of adaptive features, such as selective food habits, impervious sheathy structures over their bodies, internal lungs or tracheal system, dry excretion, aestivation, humidity control, burrowing and nocturnal habits, migration and emigration etc.

Goats on mountains derive water from their food; carnivorous get water from blood of their prey; desert animals live upon their water of metabolism. In birds and mammals, the impervious skin cover, in insects the chitinous exosakeleton and in molluscs the mucous secretions, all are meant to avoid excessive loss of water from the body. Internal lungs and internal tracheal systems of pulmonate snails, land isopids, spiders and higher vertebrates, all contribute to water conservation. Some insects, reptiles and birds dispose off their concentrated nitrogenous wastes as solid uric acids. Some animals burrow into deeper layers of soil and some are active only during night, living in other protective spatial niches during day when temperatures are high. Many birds and mammals of arid areas may be seen migrating to other areas when water conditions are in scarcity.

Adaptations in the mesocoles

Earthworms, amphibians and land planarians secrete a mucus coating to maintain a moist situation. They also develop some behavioural adaptations such as burrowing, coiling or forming aggregates.

Arthropods have an external covering over the respiratory surfaces. Insects and myriopods have a well-developed covering over the respiratory surface and tracheal system. Vertebrates' skin has many cellular layers besides the well-protected respiratory surface. There are many structural modifications in different animals that help in preventing loss of water. For example, there is operculum in gastropod molluscs. In insects a wax layer (in epicuticle) makes the integument very much impermeable to water. The integument of terrestrial crustacea is less permeable to water than that of predominantly littoral crustacea. Some animals (amphibians) are able to regulate the rate of loss of body water. These are able to slow the rate of water loss if exposed to dry air.

In the crab, *Carcinus maenas* the rate of water loss is reduced by increased calcification of the exoskeleton. Some animals obtain their water requirements from food as partial replacement of water lost through excretion.

Adaptations in the xerocoles (desert animals)

Like plants, the desert animals also face water-shortage problems.

They meet these critical conditions as follows:

Drought evaders

These xerocoles, like ephemeral annual plants, adopt either a short annual life cycle or undergo aestivation or other dormant state. The eggs of some insects and other invertebrates and insect pupae may be dormant for several months to year. On the onset of rains, a variety of animal may be seen in desert. These include crickets, grasshoppers, ants, wasps, bees, moths, butterflies, bettles etc. Amphibians like spadefoot toad, *Scaphiopus* may undergo aestivation for about 8–10 months. The birds make nest during the rainy reason when there is abundant food. During drought in the breeding season, birds rarely reproduce.

Drought resistants

Such xerocoles are active and carry their normal function throughout the year. They are able to circumvent aridity and heat through morpho-physiological adaptations or by modifying their feeding and activity patterns (i.e. behavioural adaptations). Some animals merely avoid heat by adopting nocturnal habits and remaining underground or in the shade during the day time. Some xerocole rodents that are active in the day, periodically seek burrows and passively lose heat through conduction by pressing their bodies against the burrow walls. Some birds as *Phalaenoptilus nuttallii* and *humming* birds and bats go into a daily torpor.

The animals like jackrabbits (*Lepus*) and kit fox (*Vulpes velox*) have large ears that reduce the need of water evaporation to regulate the body temperature. Their ears function as efficient radiators to the cooler desert sky, which on clear days may have a radiation temperature of 25°C below that of the animal. By seeking shade and sitting in depressions, *Lepus* could radiate 5 Kcal/day through its two large ears (400 cm^2).

The Kangaroo rat, which seals its burrow by day to keep its chamber moist, can live throughout year without drinking water. Similar adaptations are present in the jerboas and gerbils of Africa and the Middle East and the marsupial kangaroo mice of Australia. They feed on dry seeds and dry plants even when succulent green plants are available.

The Kangaroo rat obtains its water from its own metabolic processes and from hygroscopic water in its food. To conserve water the animal remains in the borrow in day, it has no sweat glands, its urine is highly concentrated and its faeces is dry.

The large xerocoles like camel are able to use water effectively for evaporative cooling through the skin and respiratory system. Their low surface area-to-body-size ratio and lower internal heat production result in slower accumulation of heat. The camel not only excretes highly concentrated urine but can also withstand dehydration upto 25 per cent of its body weight. It loses water from body tissues rather than from blood. The body temperature of camel drops to 33.8°C over night and rising to 40.6°C by day when the animal begins to sweat. The camel accumulates fat in the hump, which speeds heat flow away from the body and its thick coat prevents the flow of heat inward toward the body.

One very good example are some African antelopes such as *Oryx*, in which extreme adaptations to desert condition are found. Many African ungulates migrate to escape the heat and dryness but the *Oryx* does not. Its water requirements are low as it stores heat in its body during the day. This results into a much rise in body temperature. It also reduces daytime evaporative loss by suppressing heating. It reduces its metabolic rate and conserves water by decreasing the rate of internal production of calories, thus reducing the amount of evaporating cooling. These are daytime adaptations. By night, *Oryx* decreases its non-sweating cutaneous evaporation and its metabolic rate by 60 per cent. Its respiratory rates are proportional to its respiratory efficiency and inversely proportional to body temperature. A cool animal breaths more slowly than a warm one. With a lowered night-time body temperature, the saturation level for water vapour in the exhaled air is lower. The animal normally does not drink water; it uses metabolic water and also feeds on grasses and shrubs, mainly succulents.

Some desert birds, like marine birds utilise a salt gland to help in the maintenance of water balance. The desert bird like black-throated sparrow (*Amphispiza bilineata*) of North America, budgerigar and zebra finch of Australia and grey-backed finch lark of Africa feed on dry seeds. When they have some access to green vegetation, they do not drink water. These birds can survive indefinitely without drinking by reducing the water content of their excreta from about 81 to 57 per cent.

SECTION II

Community and Ecology

4

Community Ecology

INTRODUCTION

The population ecology is concerned with the individual population (a group of individuals of single species) in relation to environment. However, community ecology deals with the groups of different kinds of population in the area.

In nature, different kinds of organisms grow in association with each other. A group of several species (plants and/or animals) living together with mutual tolerance (adjustment) and beneficial interactions in a natural area is known as a community. In a community, organisms share the same habitat growing in an uniform environment. A forest, a grassland, a desert or a pond are natural communities. By definition, a community must include only living entities of the area. If non-living (abiotic) factors together with the living (biotic) entities are also considered, then we would be dealing with an ecosystem rather than a community.

Thus, community is an assemblage of species population that occur together in the same place at the same time. It is the biological part of an ecosystem, as distinct from its physical environment. The community has interesting and complex properties arising from the interaction between species. Competition, predation, parasitism and the mutualism occurring among individuals appear to lie behind many patterns in community organisation. At the other extreme, the community can be viewed from the wider perspective in terms of species diversity and distribution, food-webs, energy flow and the interactions among guilds of similar species.

Early this century the community was perceived as a 'superorganism' whose member species were somehow bound together such that the community was born, lived, died and evolved as a whole. This view has been largely rejected in favour of the 'individualistic' concept which focuses on the community as a collection of individual species where community patterns can be explained by processes at the level of the individual.

The concept of community is by no means so recent and it may be traced back to the time of Theophrastus (370–250 BC) who had recognised the existence of plant communities or associations of species in different environmental areas. Many centuries later, Grisebach recognised the plant formation as an important unit of vegetation. Thereafter Edward Forbes, while studying the molluscs of Aegean Sea, noticed for the first time that in different areas (i.e., at different depths), there might be distinguished different types of assemblages of species. It was indeed in the later part of the 19th century that authentic studies on communities began, when Carl Mobius coined the term biocoenosis for the associations of organisms in an oyster-bed community. In 1902, Schroter and Kirchner introduced the term *synecology* referring to plant species on the ocean bottom as an association or community. Later developments in this particular field were the methods of studying such associations and their developmental trends (successions) in varying habitat conditions, etc. Further studies in community ecology led to the development of following five lines of traditional thoughts of classifications:

1. In USA, Cowles (1899) and Clements (1905) emphasised the process of succession.
2. In Britain, Tansley (1904), also emphasised the idea of succession in community classifications.
3. In North Europe, at Uppsala (Sweden), Hult (1881) and Du Rietz (1930) emphasised phytosociological, whereas Raunkiaer (1934) physiognomic methods of community studies.
4. In Southern Europe, Zurich-Montpellier school, led by Braun-Blanquet (1932), who also developed phytosociological studies.
5. In addition to above, with a different view point, not very clear, there developed also a Russian school of thought.

A community has its own structure, developmental history and behaviour. All these characteristics in correlation with the environmental factors (synecology), just as population of same species (population ecology) and individual species (autecology) are correlated.

CHARACTERISTICS OF A COMMUNITY

Like a population, a community too has its own characteristics, which are not shown by its individual component species. These characteristics which have meaning only with reference to community level of organisation, are as follows:

1. Species diversity: Each community is made up of much different organisms-plants, animals, microbes, which differ taxonomically from each other. The number of species and population abundance in community also vary greatly. There are two levels of species diversity: (i) regional diversity of whole nations or parts of continents within which many different communities exist; and (ii) local diversity in a given nation where different communities exist at different latitudes.
2. Growth form and structure: Community is described in terms of major growth forms as trees, shrubs, herbs, mosses etc. In each growth form as in trees, there may be different kinds of plants as broad-leaved trees, evergreen trees etc. These different growth forms determines the structural pattern of a community. According to the mode of arrangement of the various growth forms, the communities exhibit: (i) zonation-(horizontal layering) as well as; and (ii) stratification (vertical layering). The details of these structural patterns will be given later.
3. Dominance: In each community all the species are not equally important. There are relatively only a few of these, which determine the nature of the community. These few species exert a major controlling influence on the community. Such species are known as dominants.
4. Succession: Each community has its own developmental history. It develops as a result of a directional change in it with time.
5. Trophic structure (self-sufficiency): Nutritionally, each community, a group of autotrophic plants as well as heterotrophic animals, exists as a self-sufficient, perfectly balanced assemblage of organisms.

COMPOSITION, STRUCTURE, ORIGIN AND DEVELOPMENT OF A COMMUNITY

Each community has its own composition, structure and developmental history.

Composition

Communities may be large or small. Larger ones extend over areas of several thousands of square kilometers, as forests, others such as deserts etc., are comparatively smaller with dimensions in hundreds of kilometers and still others such as meadows, rivers, ponds, rocky plateaus etc., occupying a more restricted area. Very small-sized communities are the groups of micro-organisms in such microhabitats as leaf surface, fallen log, litter, soil etc. The number of species and population abundance in communities vary greatly.

In each community there are diverse species. All these species are not equally important but these are only a few overtopping species which by their bulk and growth modify the habitat and control the growth of other species of the community, thus forming a sort of characteristic nucleus in the community. These species are called the dominants. Generally in most of the communities, only a single species, due to being particularly conspicuous, is dominant and in such case the community is called by the name of dominant species as for example, spruce forest community. In other communities, there may be more than one dominants as in oak-hickory forest community.

Structure of Communities

Besides composition and dominance, the communities exhibit a structure or recognisable pattern in the spatial arrangement of their members. Thus structurally, a community may be divided horizontally into 'subcommunities', which are units of homogeneous life-form and ecological relation. This horizontal division constitutes the zonation in the community. In shallow ponds, zonation is very little. However, in deep ponds and lakes, there may be recognised three zones (Fig. 4.1), viz., littoral zone, limnetic zone and profundal zone. In each zone, organisms differ from each other. Another aspect of structure that is more common, is stratification which involves vertical rather than horizontal changes within the community. In each horizontal zone, there may be recognised distinct vertical storeys. Sometimes the stratification is very complex where community possesses a number of vertical layers of species, each made up of a characteristic growth form. In grassland communities, there is a subterranean floor, containing basal portions of the vegetation such as rhizomes of grass covered by litter and debris of plants as well as animals and herbaceous substratum consisting of upper parts of the grasses and herbs with a characteristic fauna. However, stratification in a forest community (Fig. 4.2) is most complicated where as many as five vertical subdivisions may be recognised. These vertical subdivisions are: (i) subterranean subdivision; (ii) forest floor; (iii) herbaceous vegetation; (iv) shrubs; and (v) trees. In some tropical rain forests, there may be as many as eight vertical strata. Thus, based upon the light and relative humidity requirements, we find stratification in above ground parts. Similar stratification may also be found in the underground parts, roots, rhizome or other structures below the soil.

The key points about communities are that they are collections of species which occur together in some common environment or habitat and that the organisms making up the community are somehow integrated or interact as a society.

Of course, communities are not constructed only of plants. Some communities are mainly animals, like the association of fish and invertebrates which make up coral reef communities. Most communities are composed of a mixture of plants, animals, fungi, prokaryotes and protoctists.

Communities are usually recognised in two ways. One way a community can be identified is from the form of the environment or habitat in which it occurs. Some communities take their names from physical features; this is the case for rock pools, lakes and sand dunes. Other communities are recognised by the dominant species in the association. These are usually the largest or most abundant plant species present, hence deciduous oak wood, cypress swamp, grassland and sphagnum bog communities. There is no fixed size for a community. They can range from the very small and constrained, like the association of micro-organisms found in the mammalian gut, through to huge and variable expanses of grassland or forest.

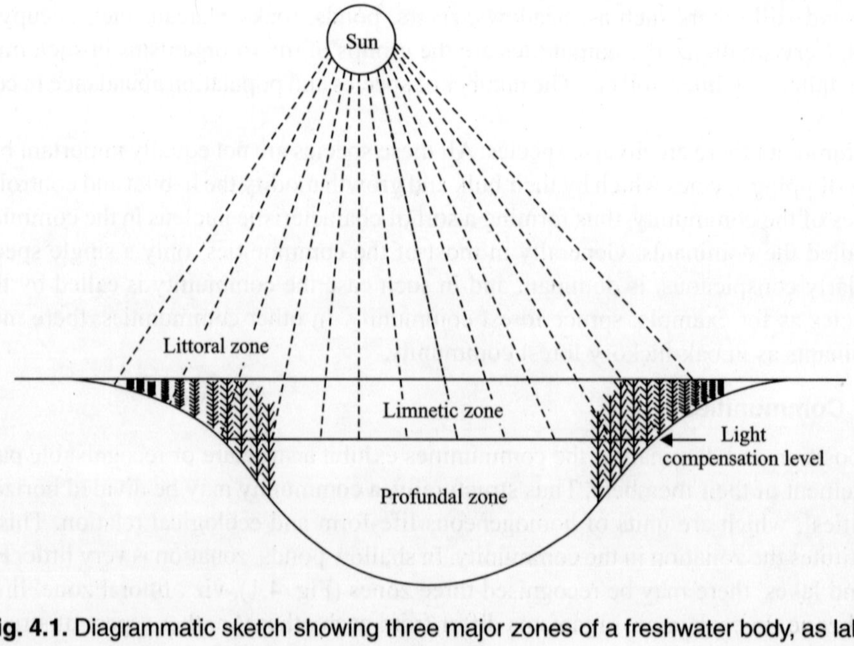

Fig. 4.1. Diagrammatic sketch showing three major zones of a freshwater body, as lake.

But why are the same associations of species found again and again in similar habitats. This could be explained if the habitat is a collection of niches. If the same niches occur in similar habitats, then they will be filled with the same species. These could include niches for particular plants, grazing animals, parasites, predators, decomposers and so on. So the community is built up of species with other dependent species in recognisable combinations. The communities found in a particular habitat type will not be identical. Minor differences in the environment or the effects of chance or history of the site, may mean that some of the species usually found together are missing and other more unusual species are present. The species present in a community will vary depending on where in the world that community is to be found. Hence species growing in pools on rocky shores in Australia will be very different from the species in similar pools in Britain. Even on one coastline the species may be different at either end if there is a cline in environmental conditions like sea temperature or changes in rock type. However, wherever the rock pools are to be found, they will contain similar relationships between species occupying the same collection of niches.

Investigation of communities

Communities are almost like living organisms. Within any community is a whole series of complex interactions between individuals of different species. The whole collection of population then fit together into a working unit. Often a glance at a community does not reveal much about the complexity of the system. More time and observation is needed to study communities.

To understand this complex mesh of interactions, an ecologist has to unravel the associations and events which are important in the community.

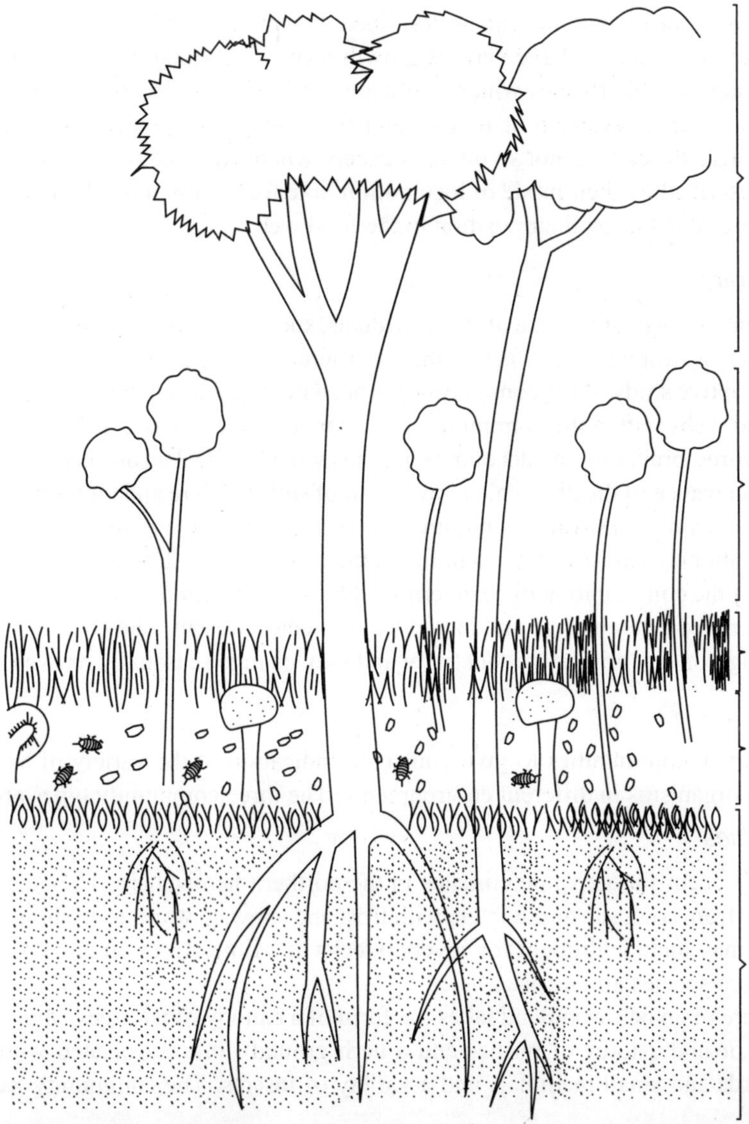

Fig. 4.2. Diagrammatic representation of a complex deciduous forest community showing stratification, where five vertical sub-divisions of different vegetational types (subterranean, forest floor, herbaceous vegetation, shrubs and trees) may be easily distinguished.

Easiest ways of investigation of community

First, the abiotic features where the community lives can be noted. For example, the water situation: whether the physical environment is marine, freshwater, marsh, drained soil or desert will be important. So will the mineral nutrients, geology and topography. Initially, this may be a brief study, but as knowledge of a community develops, the role of the abiotic factors may become more important and require more detailed investigation.

Second, the overall form of the community can be described. In terrestrial communities this is usually determined by the vegetation. Plants can have various growth forms, including: (i) Trees which create forest or woodland communities; (ii) Bushes which dominate scrub; (iii) Herbs and grasses which are the major types in grasslands like savannah or prairie; and (iv) Wet mosses which build up bogs. In aquatic habitats, this physical structure is not so obvious except where calcareous reefs occur. Aquatic organisms can often be described by their mode of locomotion, like free swimmers, planktonic drifters, bottom dwellers and sessile adults or by their position in the food web.

Species in the community

A species may be dominant because of the size of its individuals, such as trees or because it is abundant. Dominant species may have a strong biotic effect on the environment of all the other species.

Once this mainly descriptive study of the community has been made, a more functional approach can be taken. The trophic food webs within the community can be investigated to assess the importance of primary producers, herbivores, predators and decomposers. This will provide data on energy and nutrient cycles. At this point, the relevance of the physical environment of soil structure and nutrient availability, mineral input from erosion, toxins and so on may require re-investigation. The way organisms regenerate and how this affects population dynamics and the stability of the community will also need investigating.

The overall structure of the community will be determined by a combination of several features such as the physical environment, community size and longevity of species present. The community may be stable or unstable, with high or low primary productivity and may change seasonally or even daily.

Different communities

There are three very different communities to give you some indication of the variety of associations which can occur between organisms in different environments. The three community types are:

Oak woodland communities

Deciduous oak woodlands occur in temperate climates in the northern hemisphere. They experience a high degree of seasonality from cold, subzero winters with frost and snow, to warm mild summers which are often wet, but which may include drought for several weeks at a time. Trees are the dominant life form and the most abundant species are the oaks.

The vegetation structure of an oak woodland takes the form of a canopy of tall trees, about 12–20 m high, with a lower layer or understorey of bushes and a herb ground flora. Associated with this is a diversity of woodland birds and small mammals and a variety of insects. The community has a fairly high diversity.

Oak woodland is the major natural vegetation type throughout most of England and is common in Wales, western Ireland and southern Scotland. The community has probably existed in Britain for about 7500 years. Different tree species dominated in different areas including oak, limes (*Tilia* spp.) and beech (*Fagus sylvatica*) but most woodland consisted of mixture of many tree and shrub species. Most of this mixed, broad-leaved deciduous forest has been destroyed by a succession of human cultures from Neolithic times onwards.

Although oak trees predominate in oak woodland communities in Britain, they occur with a number of other tree species. These include ash (*Fraxinus excelsior*), maple (*Acer campestre*), hornbeam (*Carpinus betulus*) and elms (*Ulmus* spp.), together with some lime (*Tilia cordata*) which is most common where oak woodland grades into the other main southern English woodland community of deciduous lime woodland. There are also a number of shorter trees, which do not usually grow tall enough to reach the full canopy, including holly (*Ilex aquefolium*) and the rowan (*Sorbus aucuparia*). The shrub layer under this tree canopy includes hazel (*Corylus avellana*), roses (*Rosa* spp.) and the woodland or midland hawthorn (*Crataegus laevigata*).

The simplest way to investigate the functioning of the community is to follow the changes which occur in the wood throughout the year.

To maintain the population of adult oaks does not require many acorns to grow into trees. Many oaks live for 200 to 400 years. Only one of the thousands of acorns a tree produces during its lifetime has to survive to maturity for the population to be stable. By producing so many acorns, oaks provide a food source for many of the community animals, including jays, squirrels, voles and even some of the insects such as knopper gall wasps (Cynipidae), a recent invader from the continent and weevils (Curculioniidae) whose larvae develop inside the growing acorn.

Oaks have to suffer the attacks of many insect species, including several other cynipids which create galls on twigs and leaves or in buds. The other most notable insect pest is the oak leaf roller moth (*Tortrix viridana*). The adult moth lays eggs on the twigs of oaks. In spring the eggs hatch and tiny caterpillars creep up to the oak buds where they crawl between the bud scales as these open up. They are then in an ideal position to begin eating the new protein-rich leaves, which are low in protective tannins. In some years the moths can totally defoliate an oak tree. This does not kill the tree, but it does mean that the tree has to grow a new set of leaves which requires an extra input of energy. Extensive defoliation like this means the tree gets a late start to its productive photosynthesis. This probably results in a lower production of acorns that year, but bearing in mind the number of acorns which are eaten anyway, this may not seriously damage the tree's chances of leaving progeny.

The best chance an oak has of avoiding defoliation is having its buds still tightly closed when the caterpillars hatch. If the caterpillars cannot get between the bud scales into the edible bud, they will either die of starvation or be forced to migrate to another tree. The caterpillars are very small and by spinning a silken thread, they can drop off the oak twigs and float to other trees and bushes. The oak does open its buds later than most of the woodland species and this may be an evolutionary response to such caterpillar activity. However, the later the buds open, the more photosynthetic opportunity is lost. Trees which put out their leaves earlier than the oak will thus have more time to develop their leaves fully and invade the oak's leaf space. The oak has to balance the costs and benefits of bud opening times to maximise its opportunities in the community.

Marine rock pools

Marine rock pools can occur on any rocky coastline and are found in both temperate and tropical climates. The shoreline is an extremely variable place as it is the area between extreme high and low tides. The tides advance and retreat in a cycle lasting on average 12 hours 26 minutes, so high and low tide slowly advance around the clock. This means that not only is the degree of exposure changing on the shore as the tides move in and out, but the time at which the exposure occurs also changes. So sometimes areas will be exposed at night, when it is wet and cold and sometimes during the heat of a midday sun. The tide is at its lowest and most of the shore is exposed to the air, except where the rock pools hold some water. You can see what a bare and desolate place it looks, but it is a fascinating habitat with hundreds of species, each with its own pattern of distribution on the shore.

When the tide goes out, the sea level falls and the shore becomes exposed. Seawater is trapped and held in the rock pools until the next incoming tide covers the shore. If the pools are exposed in the daytime, the water will be warmed up by the Sun. Water will evaporate making the pool more saline (salty). If it is raining heavily, the pool may become less saline, although usually the less dense rainwater floats in a layer on top of the salty pool water, so that only the top of the pool is affected.

Vascular plants (pteridophytes, gymnosperms and angiosperms) and mammals are totally absent. The primary fixers of solar energy are photosynthetic multicellular algae or seaweeds. Most of the animals are invertebrates, with the exception of a few fish and the occasional visit from a predatory bird such as the oystercatcher (*Haematopus ostralegus*) or purple sandpiper (*Calidris maritima*). No organism on the rocky shore is large and obvious like the oak tree. This is because the energy of the waves crashing on the rocks would probably break up a large rigid organism and wash it away. As no organism dominates in size and few are very abundant in the pools, the distinctive nature of the habitat is more obvious than the organisms in it. Thus the rock pool community is named after the environment rather than any of the organisms present.

During the night respiration of the whole community tends to lower the oxygen level, sometimes to the point where it is in short supply. The seaweeds cannot photosynthesise in the dark, so they do not take in carbon dioxide. Because they are not using up the carbon dioxide it increases in concentration and the pool water becomes acidic. This switch in pH, from alkaline during the day to acidic at night, is another changeable aspect of pool life which is much less variable in the open seas. Most organisms of the open seas cannot survive the changes which occur in pools every day; the pool-dwelling organism is a very tolerant one.

In winter, violent storms may occur which can destroy some of the algae and sweep organisms out of the pool. In summer new algae may establish themselves. The pools have a very variable fauna; which species occur in anyone pool and how abundant each one is, seems to be largely a matter of chance. Summer is also a time when predators, such as crabs and prawns, invade the higher shore with its pools. A cross-section through a typical pool, showing the mixed community present, is given in diagrammatic form in Fig. 4.3.

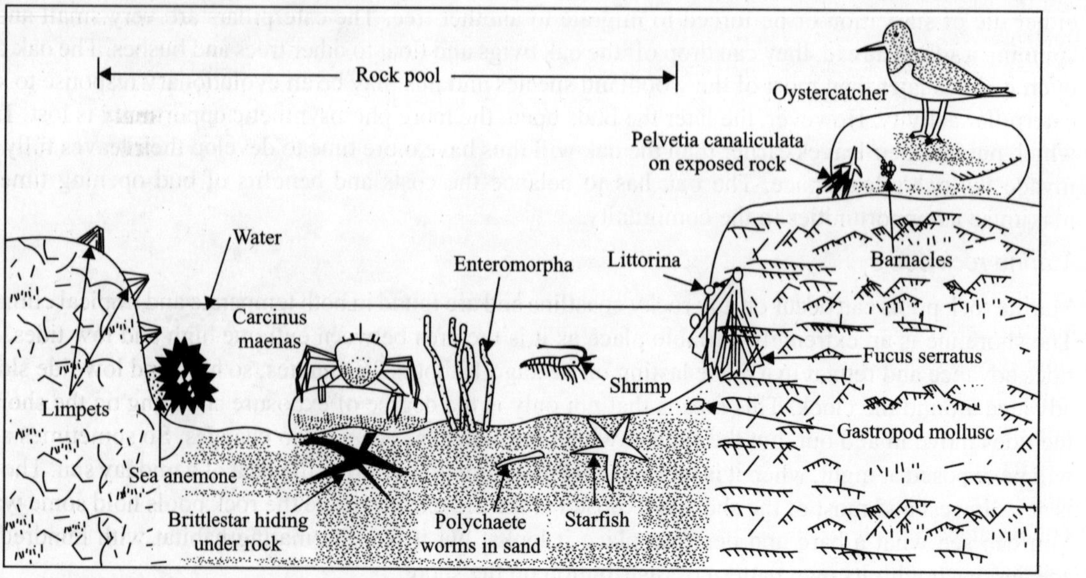

Fig. 4.3. A cross-section of a typical rock pool showing the mixed nature of the community and some of the other organisms which inhabit the more open, rocky shore.

The species of algae which occur vary depending on where on the shore the rock pool is situated. Seaweeds tend to survive a little higher up the shore when growing in a pool than they do when attached directly to exposed rocks.

Many of the animal species found in rock pools are also members of the less sheltered, open rocky shore community. These include gastropod molluscs, like limpets and shore snails (winkles) in the genus *Littorina*. Arthropods such as the green shore crab (*Carcinus maenas*) also roam the whole shore including the pools. Species which are normally only found in the sheltered rock-pool community include a wide variety of animals. Among them are gastropod molluscs, starfish (Asteroidea) and brittle stars (Ophiuroidea), sea anemones (Anthozoa) and some solitary corals, bristle worms (Polychaeta), shore-lice (Isopoda), sandhoppers (Amphipoda), shrimps (Decapoda) and some fish. The typical pool community consists of a variety of species of these groups, with no species dominant. Each pool has a selection of species which must to some extent depend on pool size, depth, position on the shore, the proximity of other similar pools and chance.

Reproduction in such a variable and turbulent environment, with the sea crashing in and retreating twice a day, is obviously going to be a hazardous procedure. So is the establishment of the young so produced. As rock pools are distributed across the whole of the rocky shore, it might be expected that pool organisms would reproduce by releasing large numbers of young into the coastal water so that a few might find their way into some new pools. This is the strategy which many weedy plant species, use when distributing numerous small seeds on land. Many marine organisms, such as corals and sea urchins on reefs and jellyfish (Scyphozoa), use the sea to disperse thousands of eggs and sperm. Fertilisation takes place in the open sea and millions of larval young are carried off suspended in the ocean currents.

It has been mentioned that pools differ from one another in many ways, including temperature, salinity and pH range. An organism surviving well in one of these pools will be suited to that environment, but may not be so well able to survive in a different pool elsewhere on the shore. Producing a small number of well cared for, well adapted offspring which will not travel far beyond their own pool may be the best strategy in this situation.

Mammalian gut communities

The mammalian gut is a long tube which stretches from the mouth of a mammal, where food is ingested, to the anus where the waste products of digestion are egested. In between the mouth and anus, the gut can be divided into various regions where food is broken down by enzyme activity and other regions where food is absorbed by the organism. These regions are summarised in the diagram in Fig. 4.4. The gut environment is obviously an unusual one: constant in temperature at the blood heat of the mammal, but severe in terms of the anaerobic conditions, varying pH and the presence of lysozymes. The gut communities are also unusual in that there are no autotrophs or holozoic feeders. All the organisms present either break down food eaten by the mammal or absorb the breakdown products of the mammal's own digestive system. The main variability in the environment is the supply and type of food being eaten by the mammal and in some systems, the problem of the host organism hibernating and therefore not eating for weeks on end. Little is known about what happens to gut communities under such circumstances.

The organisms present in the mammalian gut communities fall into two main categories: symbionts and parasites. The symbionts include bacteria and protoctists. These are present in the guts of all mammals, but are especially important in mammals which are herbivores and eat a large amount of cellulose in their diet. In ruminants, such as cattle, sheep and deer, huge population of bacteria and protoctists can be found in an enlarged pouch in the foregut called the rumen (Fig. 4.4). An adult ruminant has many billions of bacteria and millions of protoctists in its rumen. Over 100 protoctist species and about 1000 strains of bacteria are known to occur in rumen communities, but anyone community will have only a selection of this variety.

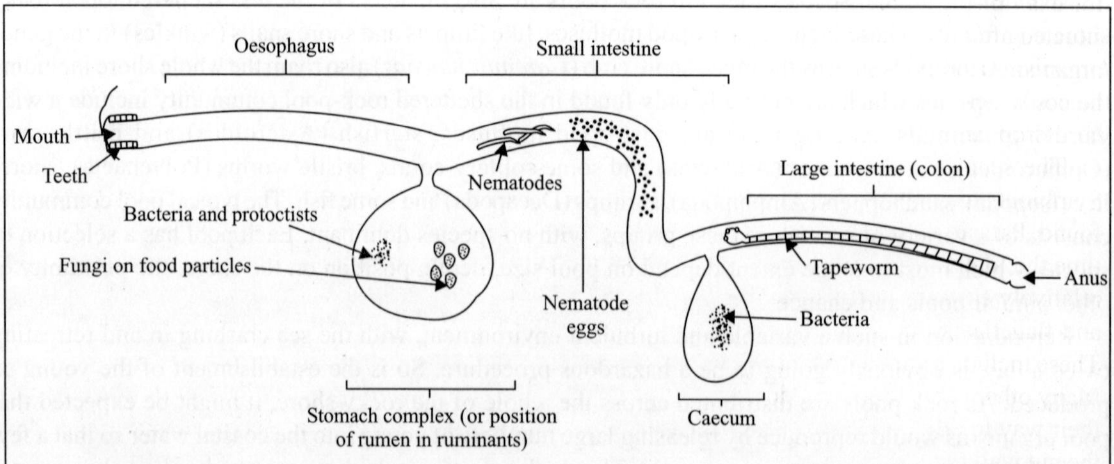

Fig. 4.4. The general structure of the mammalian gut. The different sections of the gut, with their functions, are labelled and organisms from the communities present are shown in diagrammatic form.

The symbiotic bacteria and protoctists are important to the mammal because they produce the enzyme cellulase which breaks down cellulose. Ruminants cannot produce their own cellulase so without the symbionts they are unable to digest the cellulose plant cell walls in their food. For leaf eaters, cellulose is such a high proportion of the plant matter they ingest that without cellulase they cannot obtain enough energy from their food to support themselves. A ruminant which is fed material that harms the organisms in the rumen so that it cannot break down its food quickly loses its appetite, because its rumen remains full of cellulose and eventually dies.

The rumen organisms are provided with an anaerobic, slightly acidic environment, which they require and a constant supply of food. The bacteria also seem able to use nitrogen which has been secreted or ingested into the rumen in the form of urea. They use this to synthesise proteins. The relationship is, perhaps, more complex than the term symbionts suggests as many members of the community are swept, with the digested food, into the small intestine to be digested and absorbed by the mammal, which obtains a source of essential amino acids in this way.

It is only recently that another group of organisms, fungi, have been recognised in the symbiotic gut community. The uniflagellate spores of these fungi had been noted but were mistaken for protoctists. These mobile spores move around the rumen until they come into contact with a piece of swallowed vegetable matter. They then grow long projections, called rhizoids, into the food. The rhizoids produce enzymes which digest the cell walls of the food and it disintegrates into pulp. Eventually a sporangium develops and more flagellate spores are released. Spore production appears to be triggered (in as little as 18 minutes) when a fresh meal of food is eaten by the host mammal. Such fungi appear to be widespread and have now been found in red deer, reindeer, sheep, cattle, kangaroos and elephants.

Most leaf eaters have an enlarged area of the gut, like the rumen in ruminants, where a community of bacteria can aid them in cellulose breakdown. Many leaf-eating monkeys have enlarged stomach areas which function like the rumen. Tree sloths also have huge complex stomachs as do some marsupials, including the tree kangaroos. In other species the stomach seems to be mainly a store for the food and the communities of symbiotic bacteria inhabit a huge elongated hind-gut. In the rabbit (*Oryctolagus cuniculus*), for example, the caecum is greatly enlarged to about half the total gut. As the main gut community is situated after the major digestive and absorptive areas, the rabbit cannot readily utilise the bodies of gut organisms as they are swept along the hind-gut. The rabbit has overcome this problem by producing soft faeces which it eats. After passing for a second time through the gut, the faeces are discarded as small hard droppings.

The second major category of organisms in the mammalian gut communities are the parasites. Symbionts are often concentrated in the fore-gut where the largest quantities of undigested matter are found. Parasites, though, tend to be concentrated in the hind-gut in the large intestine and caecum. By the time the food reaches these areas it has been at least partially digested. It is therefore in a form which is relatively easy for the parasite to utilise. Gut parasites include various protoctists including amoeboids and flagellates; but although the parasites have to be small to fit into the gut, many are multicellular. These include a variety of small invertebrates such as nematodes and rather larger tapeworms (cestodes). Many other parasites which do not normally live in the gut community pass through at least part of it on their way to other organs of the host. Such parasites often enter through the mouth but burrow through the gut wall to reach the bloodstream or other organs of the body. Gut-dwelling parasites live in a severe but fairly constant environment. They do not have to hunt for food or hide from predators. In fact the two main things the parasite has to do in the mammalian gut are to tolerate the environment and reproduce. The main structures and behaviour of the parasites are suited for just these two problems.

The environment a gut parasite faces is harsh because of the function of the gut as a place where organic matter is broken down into absorbable molecules. Any organism in the gut runs the risk of being digested, unless it can protect itself from the gut enzymes. Gut parasites also have to contend with the immune reactions of the host to parasite infection. Within mammal population, some hosts seem to be genetically susceptible to parasite infection. Other hosts develop immune responses which may eliminate the parasites present in the gut. In many cases, parasite numbers remain low in the gut community, below the level of infection which would trigger immune responses from the host. This seems to be the case, for example, in mice.

Many of the smaller multicellular parasites are very mobile in the gut. The larger ones, such as tapeworms, are at risk of being ejected by gut and food movements. They tend to have special hooks and suckers which they use to attach themselves to the host gut wall.

The size of the parasite also influences its mode of reproduction. The whipworm (*Ascaris lumbricoides*) is a small nematode which parasitises humans. There are separate males and females and sexual reproduction takes place. The female lays large numbers of tough, knobbly-walled eggs which leave the gut community in the faeces. In the soil, a larva develops within each egg case. It can survive there for many years, until eaten with food. The larval stage temporarily rejoins the fore-gut community, where the eggshell dissolves. It then burrows through the small intestine wall and travels to the lungs. After further development it breaks through into the lung cavity, moves up the trachea (the air passage to the lungs) and back into the fore-gut down the oesophagus. Once in the hind-gut it matures into an adult and the cycle is complete.

Larger parasites, like the human tapeworm (*Taenia*), have a problem in that their population size in the gut is usually very small. Often only one tapeworm is present there. To rely on outcrossing of separate sexes is risky as individuals might end up alone in the gut or only with members of the same sex. The adult tapeworms are hermaphrodite. Each mature segment or proglottid, of the tapeworm contains both male and female sex organs. Self-fertilisation occurs and the ripe proglottids, full of eggs, pass out of the intestines in the faeces. Infection of the secondary host, the pig, then occurs.

Many gut parasites have extremely complex life cycles, infecting one or more other species. With the human tapeworm, both host species are mammals, but in other parasites, hosts such as insects, gastropods and birds may be involved. Our understanding of how these complex methods of reproduction and dispersal evolved is poor as internal parasites have almost no chance of being fossilised. Some of the earliest *in situ* gut communities of which we know are only about 2–3000 years old.

ORIGIN AND DEVELOPMENT

A community with its particular environment constitutes an entity which has its origin and development. However, a brief outline of the process is being given here. In a barren area there reach the seeds and propagules of the species. This is known as migration. The seeds or propagules after germination develop into the seedlings which then develop into adults. But only a few of these survive as by this time the new environment does not suit to all of them and thus only a few are capable of successful growth and this process of seedling establishment and successful growth is called *ecesis*. As a result of migration and subsequent ecesis, species colonise the new areas—colonisation. By this time with the changing environment due to plants' growth, several other species of both plants and animals start colonising the area and sooner or later the area is colonised by a definite community. This community constitutes the biotic environment of that particular area. For quite sometime, species live together in a community under a particular set of environmental conditions. But time and place affect the living organism very much, as a result of which there are chiefly two major types of relationships between the organisms of a community and their environment. These are as follows:

1. Interrelations between organisms themselves: Such relations are chiefly in respect of food and space. These include:-
   (a) Competition: In the process of rapid colonisation, individuals become aggregated at a place and there develop inter-as well as intraspecific competition for space and food. As a result of these actions and interactions on the habitat, the pioneer species modify their own environment. Those, which are unable to adapt themselves to the changing environmental conditions, disappear from the area. This changed habitat is now colonised by newer species. This community may lead to another, entirely different type of community. This orderly change, which leads to the development of vegetation, is called succession. Sooner or later this change comes to a stop, when a particular community comes to stay and both, community and the habitat have reached a stabilisation. This final community is called the climax community.
   (b) Stratification: In the vertical strata in a community for example, in forests we find several storeys or layers of plant species and in each layer plant species differ in size, form, behaviour etc., Various plants in each stratum are adapted to that particular set of local conditions and moreover, the plants of one layer modify the environment, which is suitable for species in another layer of the community. Stratification thus is the result of inter-dependencies of species, as for example, lianas and epiphytes grow on other plants.
   (c) Cohabitation: There are reports of a number of chemicals secreted by vegetative organs of plants of a community which modify the edaphic conditions. Similarly, chemicals secreted by fungi, bacteria and actinomycetes in soil affect the root system as well as shoots of higher plants.

2. Relationships between species and the environment: Organisms and the environment interact with each other and bring about several modifications. This is one of the chief reasons that a particular community changes to another type sooner or later in a particular area.

CHARACTERS USED IN COMMUNITY STRUCTURE

It is pointed out above that each community is characterised by its species diversity, growth forms and structure, dominance, successional trend etc. To study the details of these aspects of any community there are taken into consideration a number of characters (parameters). These are then used to express the characteristics of a community. For instance, if we say that this particular species is dominant in the community, what may be the possible characters on the basis of which such a conclusion can be drawn.

These various characters used for such purposes are broadly classified into two major categories:

1. Analytical characters: These are generally expressed in terms of 5-point scale.
2. Synthetic characters: These are actually computed from analytical characters.

Analytical Characters

These are of two types: (i) quantitative, which are expressed in quantitative terms; and (ii) qualitative, which are expressed only in qualitative way although finally grouped under 5-point scale.

Quantitative characters

These include such characters as frequency, density, abundance, cover and basal area etc.

Frequency

Various species of the community are recorded by different phytosociological methods, by taking any sampling unit like *quadrat*, *transect* and *point*. Details of these methods will be described later. Frequency is the number of sampling units (as %) in which a particular species occurs. Thus, frequency of each species is calculated as follows:

$$\text{Frequency (\%)} = \frac{\text{Number of sampling units in which the species occurred}}{\text{Total number of sampling units studied}} \times 100$$

After determining the percentage frequency of each species, various species are distributed among Raunkiaer's five frequency classes depending upon their frequency values as follows:

Frequency %	Frequency class
0–20	A
21–40	B
41–60	C
61–80	D
81–100	E

Density

Frequency does not give correct idea of the distribution of any species, unless it is correlated with other characters, such as density etc. Density represents the numerical strength of a species in the community. The number of individuals of the species in any unit area is its density. Density gives an idea of degree of competition.

It is calculated as follows:

$$\text{Density} = \frac{\text{Total number of individuals of the species in all the sampling units}}{\text{Total number of sampling units studied}}$$

The value thus obtained is then expressed as number of individuals per unit area.

Abundance

This is the number of individuals of any species per sampling unit of occurrence. It is calculated as follows:

$$\text{Abundance} = \frac{\text{Total number of individuals of the species in all the sampling units}}{\text{Number of sampling units in which the species occurred}}$$

But, abundance thus obtained in quantitative terms gives little idea of the distribution of the species.

Cover and basal area

Quantitative values of frequency, density etc., do not give correct information on the distribution of a species. Since, different species differ in their growth forms, such values are to be supplemented with other characters like cover and basal area.

Cover, although sometimes used in general sense for the area occupied by a plant, (which may be the herbage cover or the cover or basal area), it is generally used for aboveground parts. Thus cover or herbage cover signifies primarily the area of ground occupied by the aboveground, parts of plants, such as leaves, stems and inflorescence as viewed from above. It is a good measure of the herbage availability. It is estimated by chart-quadrat, line intercept or point frame.

Basal area refers to the ground actually penetrated by the stems and is readily seen when the leaves and stems are clipped at the ground surface. It is one of the chief characteristics to determine dominance. It is measured either 2.5 cm above ground or actually on the ground level. It is measured by callipers, line-interception or point-centered quadrat method.

Since plants differ in their growth form the relationship between herbage cover and basal area differs in different types of plants as shown in Fig. 4.5.

Based on their cover values, species are grouped into six groups as follows: Group 1 (less than 5 per cent), 2(5–25 per cent), 3(25–50 per cent), 4(50–75 per cent), 5(75–95 per cent) and 6(95–100 per cent).

Dominance

Infact, dominance is a synthetic character. However, Daubenmire considered it as a analytical character and expressed its six types. Infact the number of organisms sometimes may not give correct idea of the species. If we base our conclusion on number, a single or few trees in a grassland or few grasses in a forest should be of little value. But if we consider the species on the basis of area occupied or weight (biomass), the situation may be different. Thus in dominance, cover is included as important character. The single tree in grassland may occupy fairly a large area and may have much mass.

Qualitative characters

These include physiognomy, phenology, stratification, abundance, sociability or gregariousness, vitality and vigour, life form (growth form) etc., most of which are described and may be grouped in point scales.

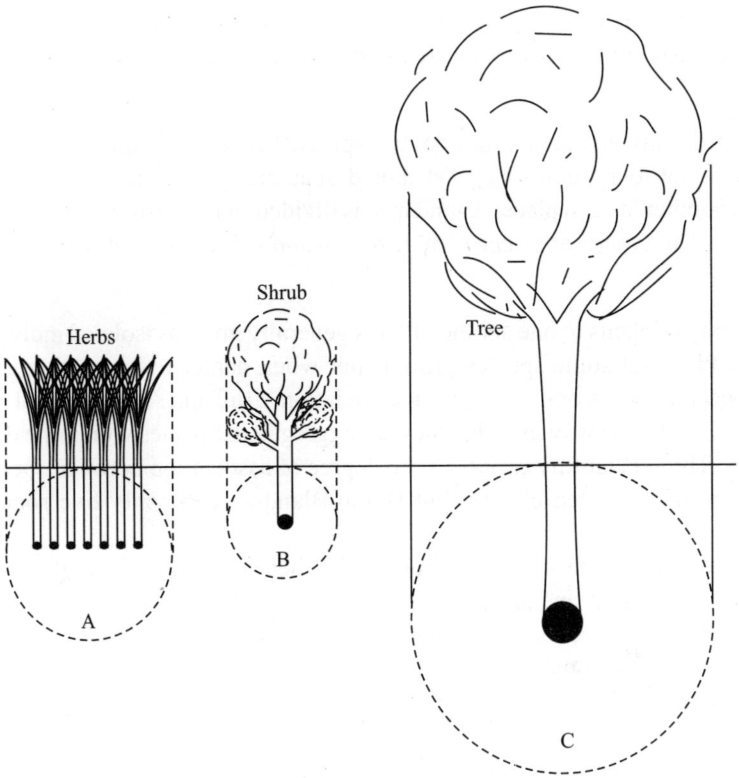

Fig. 4.5. Diagrammatic sketch showing relationship between herbage cover and basal area in different types of growth forms (herbs, shrubs and trees) of plants.

Physiognomy

This is the general appearance of vegetation as determined by the growth form of dominant species. Such a characteristic appearance can be expressed by single term. For example, a simple look to such a community of plants where larger trees are dominant with some shrubs, would tell that it is a forest. Similarly on the basis of appearance it may be a grassland, desert etc.

Phenology

It also includes periodicity, aspection etc. Phenology is the scientific study of seasonal change i.e., the periodic phenomena of organisms in relation to their climate. Different species have different periods of seed germination, vegetative growth, flowering and fruiting, leaf fall, seed and fruit dispersal etc. Such data for individual species are recorded. A study of the date and time of these events is phenology. In other words, phenology is the calendar of events in the life history of the plant. Thus, phenology of different species may differ from each other and in a community, we find species with different phenology, changing the composition of community.

Stratification

Stratification is the way in which plants of different species are arranged in different vertical layers in order to make full use of the available physicial and physiological requirements. Data on stratification are obtained by using a bisect. Vertical projection is constructed and stature of the species may be plotted to find out distribution pattern of stem, roots etc., of different species.

Abundance

Although related with density, abundance may not be expressed generally in quantitative terms. Organisms, particularly plants are not found uniformly distributed in an area. They are found in smaller patches or groups, differing in number at each place. Abundance is divided in five arbitrary groups, depending upon the number of plants. The groups are, *very rare*, *rare*, *common*, *frequent* and *very much frequent*.

Sociability

It denotes the proximity of plants to one another. Plants generally grow as isolated individuals, in patches, colonies or groups. Plants of some species grow better when nearer to each other and produce thick population. Others become weak or die in such an association. Fruit and seed dispersal etc., and nature of reproduction are affected by the way in which they are aggregated. Species with same density, may differ in sociability values. Thus sociability expresses the degree of association between species. Some divide the sociability into as many as ten classes, but Braun-Blanquet used only five sociability groups as follows:

S_1—Plants (stems) found quite separately from each other, thus growing singly.
S_2—A group of 4–6 plants at one place.
S_3—Many smaller scattered groups at one place.
S_4—Several bigger groups of many plants at one place.
S_5—A large group occupying larger area.

Vitality

This is the capacity of normal growth and reproduction, which are important for successful survival of a species. The vitality depends upon weight of species. In plants, stem height, root length, leaf area, leaf number, number and weight of flowers, fruits, seeds etc. determine the vitality. On this basis plants are grouped into different vitality groups. Smith and Allen gave three groups as: (i) well developed plants completing their life cycles regularly, producing viable seeds; (ii) plants reproducing vegetatively; and (iii) ephemerals, short lived. Braun-Blanquet gave four groups. Daubenmire gave vitality groups as follows:

V_1—Plants whose seedlings die.
V_2—Seedlings grow, but unable to reproduce.
V_3—Reproduce only vegetatively.
V_4—Reproduce sexually, but uncommon.
V_5—Reproduce sexually and grow regularly.

Life-form (growth form)

In past, ecologists used the general appearance of plant communities (as determined by the stature of the plant species, their spread and characters of life-form), as the basis of the study of communities. These methods are known as physiognomic methods. Of these, Raunkiaer's life-form method has been commonly employed. A life-form is 'the sum of the adaptation of the plant to climate'. He considered that the way in which different species overcome the adverse environmental conditions determines their limits of distribution. Thus the plants' climate can be expressed by the statistical distribution of life-forms in the flora of a particular region. On the basis of the position of perennating buds on plants and the degree of their protection during adverse conditions, Raunkiaer classified plants into five broad life-form categories which are shown in Fig. 4.6. These are as follows:

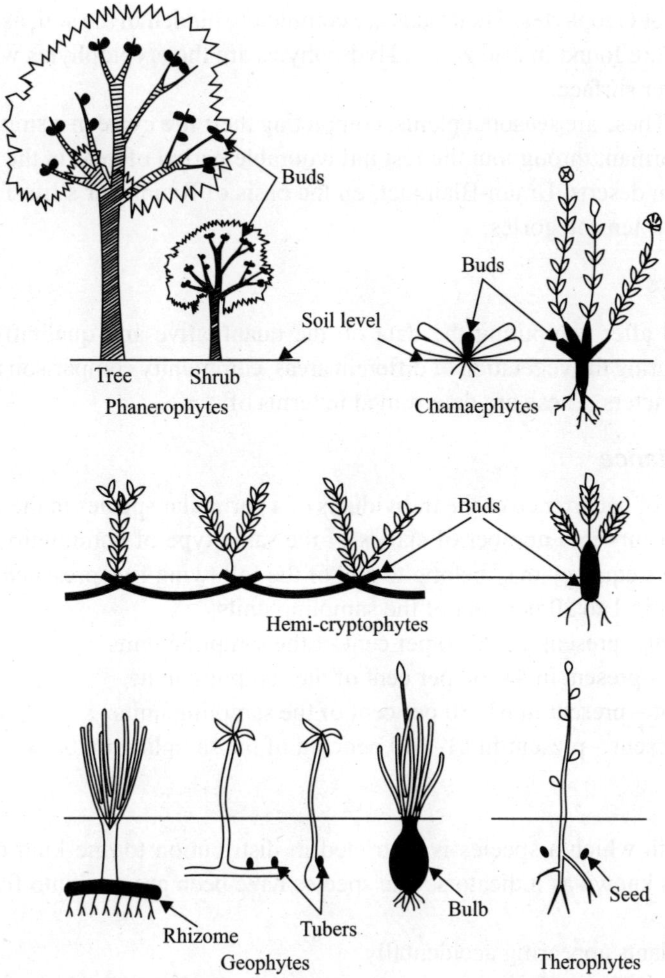

Fig. 4.6. Diagrammatic representation of Raunkiaer's life-forms. The barred region in phanerophytes and solid black regions in rest of the life-forms, survive(s) during adverse periods of the growing season. Note the successive increasing degree of protection of renewal buds, organs or seeds (shown as black solid bodies) from phanerophytes to therophytes.

1. Phanerophytes. Their buds are naked or covered with scale and are situated high up on the plant. These life-forms include trees, shrubs and climbers generally common in tropical climates. Depending upon the size, they may be further subdivided into megaphanerophytes (over 30 meters high), mesophanerophytes (8–30 meters high), micro-phanerophytes (2–8 meters high) and nano-phanerophytes (under 2 meters). Epiphytes are either included in phanerophytes or sometimes included under a separate life-form.
2. Chamaephytes. Their buds are situated close to the ground surface. They are common at high altitudes. Example—*Trifolium repens*.
3. Hemicryptophytes. These are mostly found in cold temperate zone. Their buds are hidden under soil surface, protected by the soil itself. Their shoots generally die each year. Examples—most of the biennial and perennial herbs.

4. Cryptophytes or Geophytes. Their buds are completely hidden in the soil, as bulbs and rhizomes. Most of them are found in arid zones. Hydrophytes are the cryptophytes whose buds are found below the water surface.
5. Therophytes. These are seasonal plants, completing their life cycle in a single favourable season and remain dormant throughout the rest unfavourable period of year in the form of seeds. They are common in deserts. Braun-Blanquet, on the basis of Raunkiaer's life-forms, divided plants into as many as ten categories.

Synthetic Characters

These are determined after computing the data on the quantitative and qualitative characters of the community. For comparing the vegetation of different areas, community comparison needs the calculation of their synthetic characters. These are determined in terms of:

Presence and constance

It expresses the extent of occurrence of the individuals of a particular species in the community i.e., how uniformly a species occurs in a number of stands of the same type of community. The species on the basis of its percentage frequency may belong to any of the following five presence classes:

1. Rare—present in 1 to 20 per cent of the sampling units.
2. Seldom present—present in 21–40 per cent of the sampling units.
3. Often present—present in 41–60 per cent of the sampling units.
4. Mostly present—present in 61–80 per cent of the sampling units.
5. Constantly present—present in 81–100 per cent of the sampling units.

Fidelity

This is the degree with which a species is restricted in distribution to one kind of community. Such species are sometimes known as indicators. The species have been grouped into five fidelity classes as follows:

Fidelity 1 – Plants appearing accidentally.
Fidelity 2 – Indifferent plants, may occur in any community.
Fidelity 3 – Occur in many communities, but predominant in one (preferants).
Fidelity 4 – Specially present in one community but may occasionally occur in other communities as well (Selectives).
Fidelity 5 – Occur only in one particular community and not in others (Exclusives).

Dominance

It has been described under quantitative characters. Here, the dominance is expressed in synthetic form. On the basis of density, frequency and dominance (cover) values, there has been proposed the idea of importance value index (IVI). IVI of a species in the community gives the idea of its relative importance. For IVI, values of relative density, relative frequency and relative dominance (cover basis) are obtained as follows:

$$\text{Relative density} = \frac{\text{Density of the species} \times 100}{\text{Total density of all the species}}$$

$$\text{Relative frequency} = \frac{\text{Frequency of the species} \times 100}{\text{Total frequency of all the species}}$$

$$\text{Relative dominance (cover)} = \frac{\text{Dominance (cover) of the species} \times 100}{\text{Total dominance (cover) of all the species}}$$

Now for IVI, three values are added. IVI values of different species are then arranged in decreasing order.

Other synthetic characters

Besides the above mentioned ones, there have also been proposed some other characters. These have been quite useful in comparative studies on communities. Such characters include, interspecific association and association index, index of similarity, dominance index, species diversity and diversity index etc.

METHODS OF STUDY OF COMMUNITIES

The various characters used to characterise a particular community are described in this section. To obtain these characters, several methods have been used from time to time. The field of community ecology has been investigated in detail particularly for plant communities and thus methods described are mainly related with plants.

Various methods of study of plant communities are broadly grouped into three major categories:

Floristic Methods

These are rather obsolete methods. Here the flora is studied by listing various genera and species present in the community. Thus, vegetation is described in terms of the flora. These are actually descriptive methods. Such methods, however, give little or no idea of the composition, structure, growth forms etc., of the community.

Physiognomic Methods

The various species of the community are studied chiefly in terms of their life-forms (growth forms), general stature, spread etc. Out of a number of physiognomic methods. Raunkiaer's life-form method became most popular. We will describe this method in detail here.

Raunkiaer's life-form method

This method has two parts:

1. Record of different life-forms: The various species of the community are recorded. These are then distributed among the different life-forms—i.e., phanerophytes, chamaephytes, hemicryptophytes, cryptophytes and therophytes, on the basis of the position of renewal bud or organs in the species.
2. Biological (Phyto-climatic) spectrum. Now the biological spectrum of the area is prepared. Biological spectrum represents the percentage (percentage of the total species in the community) distribution of species among the various (five here) life-forms. Thus, through calculation of percentage value of each life-form, biological spectrum of the area is obtained. Such a biological spectrum for a desert is shown in Fig. 4.8. This is now compared with the normal biological spectrum of phanerogamic flora of the world (as shown in Fig. 4.7). Raunkiaer, as a result of

extensive study of the world flora, prepared a normal biological spectrum, where the per cent values (percentage of the total species) of each life-form is as follows: (i) Phanerophytes–46; (ii) chamaephytes–9; (iii) hemicryptophytes–26; (iv) Cryptophytes–6; and (v) Therophytes–13.

Now, a comparison of the biological spectrum of the area studied (Fig. 4.8) with the normal biological spectrum (Fig. 4.7) shows that the value of therophytes is higher in the spectrum of the area studied than the values of this life form in normal spectrum. It indicates the deserty nature of area, since abundance of therophytes is the characteristic of desert climates, with long dry season.

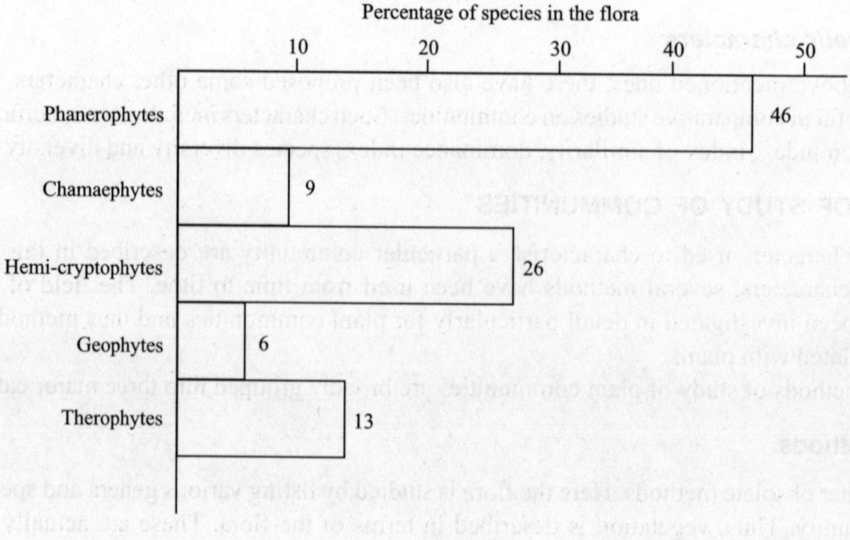

Fig. 4.7. Raunkiaer's normal biological spectrum for the world's phanerogamic flora.

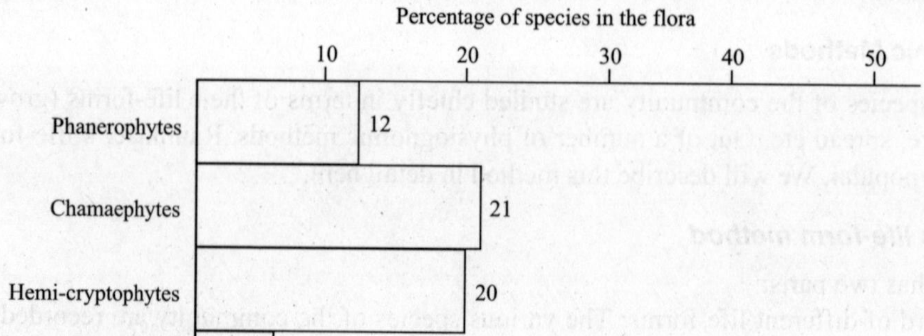

Fig. 4.8. Biological spectrum of a desert. Note the higher value of therophytes (than their percentage value in Fig. 4.7) showing the climate of the area as deserty.

However, the usefulness of the biological spectrum as an indicator of the climate is limited. It does not take into consideration such ecological factors as the biotic disturbances. Since the climate of Indo-Gangetic plains in India is suitable for forest development, values of phanerophytes must come higher, if the biological spectrum of this area is prepared. However, when it is done, we find the dominance of therophytes. This is due to biotic influence of man during agricultural practices, grazing effects, scraping etc., which alter the biological spectrum of the area.

Phytosociological Methods

None of the two above-said methods provides detailed information on the composition, structure, species diversity, growth, trends of succession and other characteristics of the community. Methods, based upon some such characteristics–phytosociological methods were developed simultaneously in USA, Britain and Southern and Northern Europe.

Sampling units

In phytosociological methods, which are many there are generally three forms of sampling units: *area*, *line* and *point*.

In area and line as sampling units, there is a definite area (suitable size) for sampling. However, in such communities as thick forests with long grassy and shrubby belts, it becomes difficult to have a definite area. For such analysis, point is taken as a sampling unit.

On the basis of the nature of sampling units, there are three popular methods of study of communities:

1. Quadrat method, where the sampling unit is definite area which is taken in the form of a square, a rectangle or a circle.
2. Transect method, where the sampling unit is line of suitable length. In fact, a segment of suitable length along the line may be considered as a quadrat.
3. Point method, where the sampling unit is a point.

Quadrat method

Sampling unit is a quadrat which is an area of a definite size. In shape, it may be square, rectangle or a circle. Depending upon the type of vegetation and purpose of study a quadrat may be:

1. List quadrat: The species present in the area are simply listed.
2. List-count quadrat: Besides listing the various species, mumerical counts of individuals of each species are also made.
3. Chart quadrat: Here a detailed to-scale growth and distribution of each species in space is recorded. The individuals are recorded on a miniature quadrat on graph-paper by the use of pantograph at intervals of months or years.
4. Experimental permanent quadrat: The chart quadrat is left undisturbed and the area is studied for vegetational changes over a long period, to study periodic changes in community. Such quadrats are useful in successional studies.

Study of community by quadrat method involves three steps: (i) To determine the minimum size of quadrat by Species-Area-Curve method; (ii) to determine the minimum number of quadrats to be laid down; and (iii) record of the species—their listing and counting of the individuals of each species.

After the record of various species, the values of frequency, density, abundance are determined for each species of the community. Methods of calculating these values are given earlier, while dealing with quantitative analytical characters of the community. (See under the heading of characters used in community).

After determining the frequency (per cent) of each species, these are then distributed among Raunkiaer's five frequency classes—A, B, C, D, E, depending upon their frequency (%) values as follows:

Frequency %	Frequency class
0–20	A
21–40	B
41–60	C
61–80	D
81–100	E

Now find out the values (as per cent of the total species) of each of the five frequency classes to prepare the frequency diagram.

As a result of extensive study of Raunkiaer, he also proposed a normal frequency diagram (Fig. 4.9) in which the percentage values of different frequency classes are—A–53, B–14, C–9, D–8 and E–16. The frequency diagram prepared for the area (a grassland) studied (Fig. 4.10) is then compared with normal frequency diagram. A comparison makes it clear that values of frequency classes B, C, D in Fig. 4.10 are comparatively higher than their respective values in normal diagram (Fig. 4.9). Thus, community studied is a heterogeneous one. Higher values of E indicate the homogenous nature of the community.

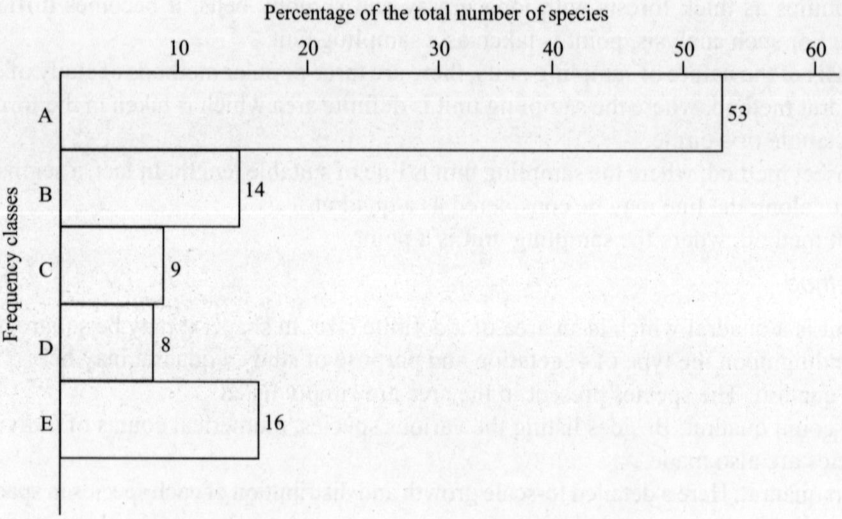

Fig. 4.9. Raunkiaer's normal frequency diagram.

Transect method

A transect is a sampling strip extending across a stand or several stands of vegetation where line is the sampling unit. It may be a:

1. Line-transect: Line-transect is a thin line, which is generally used for grasslands.
2. Belt-transect: Belt-transect is a belt of suitable breadth, used for forests.

The transect is divided into segments of suitable length. This method is useful to analyse the vegetation changing in its composition through an ecotone. Each segment may be taken as a quadrat. Frequency, density values of each species are calculated.

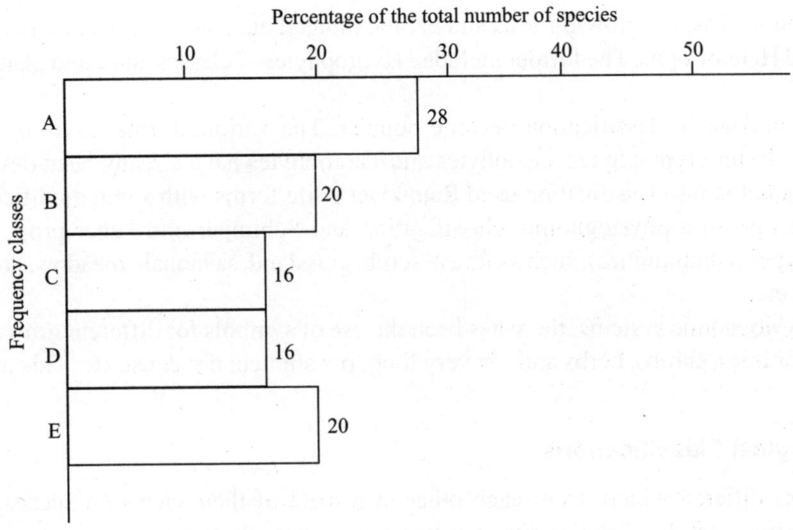

Fig. 4.10. Frequency diagram of a grassland community. A comparison with Fig. 4.7 makes it evident that the values of frequency classes B, C and D are relatively higher than their values in Fig. 4.7, thus, the community is heterogeneous in nature.

Point method

The sampling unit is a point. For convenience, a number (usually 10) of movable pins (about 50 cm long) are inserted in a wooden frame, known as point-frame. The pins are laid down at random in the area. Plants hit by pins are recorded. Thus frequency (percentage) of each species is calculated. One hit of pin taken by a species is taken as its occurrence in the sampling unit. For density values, there are two alternative procedures: (i) random pair method; and (ii) point centered quarter method.

For each of the above methods, frequency diagrams may be prepared which are then compared with Raunkiaer's normal frequency diagram to draw possible conclusions.

UNITS OF VEGETATION – CLASSIFICATION OF COMMUNITIES

Vegetation is the continuous, sum total growth of various forms of plant population in the area. To describe vegetation one has to handle the communities. Communities are the units of vegetation. The classification of communities has since long been the subject of much controversy. Some ecologists are of the view that communities are organised basic units of different species and thus may be classified according to the principles of systematics.

Without going into details, we shall present a brief account of this aspect here. A number of classifications of communities proposed from time to time, may be broadly put under two main categories:

1. Classifications based on habitat, growth-forms i.e., physiognomy.
2. Classifications based on species dominance, succession etc. i.e., phytosociology.

Physiognomic Classifications

These include the classifications of Humboldt, dividing communities into 19 groups and that of Griesbach, who included seven main groups, dividing them into as many as 60 types. The main groups were as woody (Ligniden), herbaceous (Herbiden), grasses and so on.

Warming, on the basis of growth forms and habitat divided the communities into two main classes—Autotrophs and Heterotrophs. The former included Hydrophytes–7 classes and Land plants (terrestrial)—many classes.

Of these Raunkiaer's classification became popular. The various forms used i.e., Phanerophytes, Chamaephytes, Hemi-cryptophytes, Geophytes and Therophytes have already been described.

Braun-Blanquet in his classification used Raunkiaer's life forms with some modifications.

Schimper proposed a physiognomic classification and Schimper and Faber proposed as many as 15 formation-types (communities), such as forest, scrub, grassland, savannah, meadow, groveland, steppe, marsh, swamp etc.

In some physiognomic systems, there has been the use of symbols for different growth forms such as A, B, C, etc., for trees, shrubs, herbs and v = very long; p = stunted; d = dense etc. This practice began in England.

Phytosociological Classifications

Since ecologists differed widely from each other in respect of their views on successional trends in vegetation and final stabilised state (climax community), there developed as many as five continental schools of thought on community classification. Due to such different views, there has been much controversy in the boundary limits of a community. It led to the introduction of a number of terms in synecology. There appeared some sort of terminology struggle between ecologists.

CLEMENTSIAN UNITS OF VEGETATION

According to Clements, on the basis of different units of vegetation, the communities may be classified as follows:

Plant Formation

The plant formation is the major unit of vegetation, the climax community. Whenever the vegetation reaches the highest development becoming more or less stable for more or less definite period, under the existing climate, it is called a plant formation. Thus, a plant formation is a geographically widespread climax vegetation unit growing in uniform climatic conditions or we may say that the formation occupies a climatic unit area or a region where the climate is more or less similar. The dominant members of a plant formation have similar growth-forms. For example, in a deciduous forest formation, all the dominant species are broad-leaved trees. In desert formation the dominant species are usually shrubs. Sometimes, there may be present some areas in a climax formation where local variations in vegetation due to successional trends may occur. The principal formations of the world are rain forests, deciduous forests, coniferous forests, grasslands, deserts, mangroves, scrubs etc.

Associations

In a plant formation, there may be present several associations. As pointed out, in plant formation occupying a wide extensive area, there may be present several dominant species. Thus, a plant formation may be divided into several associations. In each association, there is present a community which has two or more dominant species. Each smaller community (association) of a plant formation with its two or more dominant species is known as an association. The number of associations in any formation is determined by the kinds of sub-climates within the general climate of the formation, in that area. Thus, an association is more uniform and similar than the formation in its physiognomy, structure and floristic composition.

The developmental or seral communities of an association are known as *associes*. However, sometimes due to local variations in some other factors (but under the general climate like that of association), there may develop two sub-divisions in an association. These are simply local variants of an association. These are:

Faciation

It is actually a local variant of an association which is related within a general climate to small differences in moisture relations and temperature. It is not very much distinct from an association within a formation. Similar to association, in faciation too, there are present two or more dominant species. Although, general climate is like that of association, it differs from it in having specific precipitation, evaporation and temperature. But still it is related with the general sub-climate of the association. The seral communities are accordingly known as facies.

Lociation

This is also a localised variant .of an association, differing from it in having different types of some main sub-dominant and chief secondary species. Seral communities are accordingly known as locies. Thus, in a formation, besides several associations, there may also exist faciations and lociations which are simply local variants of association differing from it in minor respects.

Consociation

Further, due to local variations in edaphic factors, temperature, precipitation etc., there may develop several plant communities within a plant association. Each community in the association is dominated by only a single species. Each such community, with a single dominant species, is known as a consociation. Thus in one association, there may be present many consociations, each with a single dominant species. The developmental or seral communities of a climax consociation are known as consocies. In a consociation, consocies are controlled by facies.

Society

It is a community characterised by one or more subdominant species. In areas with dominance of consociation or faciation, many other species are found growing in abundance. These species are of indeed local importance, in influencing the communities present in the area and are thus called the subdominants. We may think of society as a dominance within the dominance. The seral communities are known as socies. Societies are of two general types:

Aspect society

These are found in places with marked seasonal variations. Different species appear that differ in periods of their vegetative growth, flowering etc. A society offers a conspicuous aspect to the association in a particular season. Another society would similarly give another aspect to the association in another particular season. Thus each species can form its own society. Aspect societies are common in desert and grasslands.

Layer society

These societies are common in forest formation that show complex stratification. Under the influence of dominant species, the trees of the forest formation, there develop one or more layers of shrubs followed by layers of herbaceous communities. Besides them, there is also present a ground community or society of mosses, lichens etc., in the forests. The various units of vegetation constituting a particular 'formation' are diagrammatically represented in Fig. 4.11.

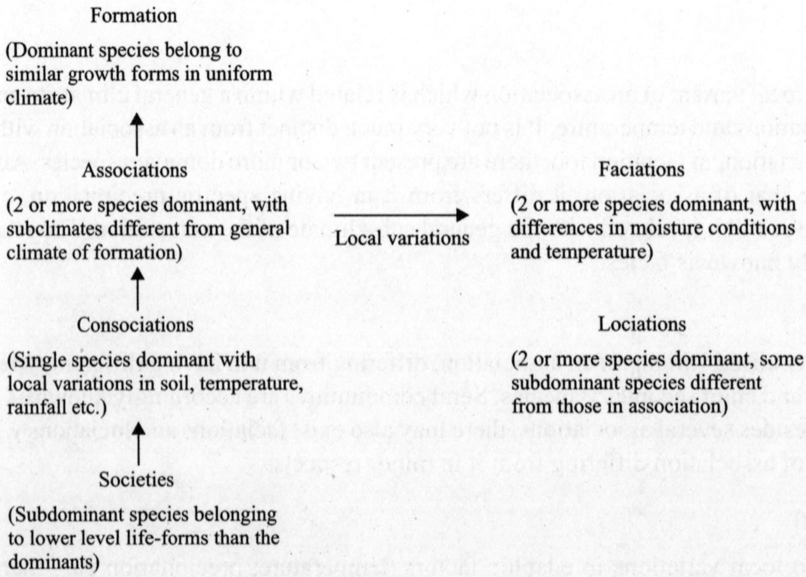

Fig. 4.11. Diagrammatic sketch to show different units of vegetation according to Clements' scheme.

MAIN CONCEPTS IN CLASSIFICATION OF COMMUNITIES

It was indicated in the beginning of the subject of community classification that two main concepts prevailed in this subject. One, that communities are organised basic units and may by classified and another that a community represents simply an assemblage of species with similar environmental requirements and thus can not be classified as discrete units. Accordingly there developed two main concepts in community classification.

Individualistic Concept

We have seen above in the five different classificatory traditions of community classification, that although differing in respect of terminology and choice of the terms, these all recognise one fact. According to all of them, basic unit of community may be recognised at one or other level of organisation. And are to be classified in one way or the other. In all these theories, communities have been recognised as equivalent to a species or an organism and the existence of a particular species in each community is recognised. Thus each community is characterised with particular species—dominant, constant etc.

The above said traditions of community classification were called 'association unit theory' or 'community unit theory' by Whittaker.

However, there were several objections to the attempts for classifying communities. Of several authors, H.A. Gleason in America put his views against classification of communities. His views were recognised as the individualistic concept according to which each individual species tends to be distributed independently of others, occurring where it can disperse and survive. Furthermore conditions favouring each species differ. Consequently the organisms of a given piece of ground are simply a gathering of those species finding it within their tolerances.

Thus, according to Gleason, each species has a distribution within its total range of tolerance to environmental factors, which depends upon the limits of its own population dynamics, physiology and genetic adaptability. This is referred to as the principle of species individuality.

Vegetational Continuum Concept

The 'individualistic concept' was badly criticised by many persons who believed in the inter relations between species of a community. According to them, the individualistic concept, as usually expressed, gives too little weight to coactions and especially their co-evolutionary development. According to supporters of this concept, communities are supposed to vary continuously in space, with each point of the continuum being equally probable. The concept of vegetational continuum was developed by Whittaker and Curtis and their associates. The continuum concept has been supported on the basis of environmental gradient through gradient analysis or ordination. Together with environmental gradient, there exists also community gradient (coenocline or vegetational gradient) in nature. This may be better seen on mountains, where with changing altitude, there is gradual change in climatic and edaphic conditions associated with gradual changes in vegetation.

Thus, under natural conditions, there exists no clear cut lines of separation between vegetation as well as environmental conditions. This is better explained by calculating continuum index. Accordingly, by elaborating the continuum concept through gradient analysis, communities are expressed in the form of landscape pattern. Parallel to the environmental gradient, there is found vegetational gradient or ecocline. Whittaker recognised five kinds of ecoclines correlated with climatic conditions in different geographic areas. Also, all living communities are grouped into 20 biomes.

ECOTONE

Under natural conditions, climatic and physical factors become more adverse, as we approach the outer limits of a terrestrial community. Finally, there occurs a point at which a different groups of plants and animals representing another community becomes apparent. In nature, however, there is never distinguishable a point or sharp line indicating the beginning of one community and the end of another; instead, there is a zone of transition or tension in which the conditions for each of the adjacent communities become more adverse and there is usually an intermixing of species from both the communities. Such a region presenting a situation of special ecological interest is known as an ecotone or a tension zone. The extent of ecotones is variable. The border between forest and grassland, the bank of a stream running through a meadow etc., are good examples of ecotones. A general characteristic feature of the ecotone is that it has sufficiently greater number of species and moreover, the density of most of the species is higher than that in the neighbouring communities. This unique general feature of an ectone is known as principle of edges. Some ecologists have introduced the continuum concept in ecology, which means that there are no distinct communities with well-defined boundaries, but there is a gradual change in space and time along some sort of gradient which may be in terms of moisture, temperature, soil type, amount of insolation, altitudinal or other abiotic factors etc. There are no sharp borders or changes in species-composition in areas where a continuum can be demonstrated.

5

Community Dynamics

INTRODUCTION

Although a typical community maintains itself more or less in equilibrium with the prevailing conditions of the environment, in nature this is hardly true. Communities are never stable, but dynamic, changing more or less regularly over time and space. They are never found permanently in complete balance with their component species or with the physical environment. Environment is always kept on changing over a period of time due to: (i) variations in climatic and physiographic factors; and (ii) the activities of the species of the communities themselves. These influences bring about marked changes in the dominants of the existing community, which is thus sooner or later replaced by another community at the same place.

This process continues and successive communities develop one after another over the same area, until the terminal final community again becomes more or less stable for a period of time. This occurrence of relatively definite sequence of communities over a period of time in the same area is known as ecological succession. Thus ecological succession is the change in species composition and community structure and function over time. Succession is usually defined as a continuous, unidirectional, sequential change in the species composition of a natural community. Odum preferred to call this orderly process as ecosystem development rather than the more often known ecological succession. He made an elaborate statement to define this process and in his own words 'ecosystem development' or what is more often known as ecological succession, may be defined in terms of the following three parameters:

1. It is an orderly process of community development that involves changes in species structure and community processes with time, it is reasonably directional and therefore, predictable.
2. It results from modification of the physical environment by the community, that is, succession is community controlled even though the physical environment determines the pattern, the rate of change and often sets limits as to how far development can go.
3. It culminates in a stabilised ecosystem in which maximum biomass (or high information content) and symbiotic function between organisms are maintained per unit of 'available energy flow'.

CAUSES OF SUCCESSION

Since succession is a series of complex processess, it is natural that there may not be a single cause for this. Generally, there are three types of causes:

1. Initial or initiating causes: These are climatic as well as biotic. The former includes factors, such as erosion and deposits, wind, fire etc., caused by lightning or volcanic activity and the latter includes the various activities of organisms. These causes produce the bare areas or destroy the existing populations in an area.

2. Ecesis or continuing causes: These are the processes as migration, ecesis, aggregation, competition, reaction etc., which cause successive waves of populations as a result of changes, chiefly in the edaphic features of the area.
3. Stabilising causes: These cause the stabilisation of the community. According to Clements, climate of the area is the chief cause of stabilisation, other factors are of secondary value.

TRENDS OF SUCCESSION

An ecological succession proceeds along the following four lines:

1. A continuous change in the kinds of plants and animals.
2. A tending increase in the diversity of species.
3. An increase in the organic matter and biomass supported by the available energy flow (but in heterotrophic succession reverse is true).
4. Decrease in net community production or annual yield.

Basic Types of Succession

The various types of succession have been grouped in different ways on the basis of different aspects. Some basic types of succession are, however, as follows:

In primary succession initial conditions are often severe. At first, the only organic matter will be wind-blown and no soil is present. On exposed boulder clay till deposited by retreating glaciers in Glacier Bay, Alaska, the nutrient poor clay is first colonised by mosses and shallow-rooted herbaceous species such as mountain avens *(Dryas sp.)*. These early colonists alter the soil conditions allowing new species to colonise the habitat. Mountain avens has nitrogen fixing symbionts that increase the nutrient content of the substrate. Litter accumulation also leads to soil development, allowing colonisation by shrub and eventually tree species such as cottonwood and hemlock. Alder (also a nitrogen fixer) has a strong acidifying effect, reducing the pH of the surface soil and creating conditions suitable for invasion by sitka spruce. The alder is eventually displaced by the spruce leading to a mature sitka spruce–hemlock forest.

Primary succession to woodland is a slow process, taking in the region of a hundred years or more. More rapid primary succession can be observed on intertidal boulders overturned by wave action. Bare substrate is rapidly colonised by the pioneer green alga *Ulva*, followed by several species of red algae. The red algae are slower to establish and can only invade where the *Ulva* has died, been grazed off or removed by some other disturbance. Selective grazing by a species of crab particularly partial to *Ulva* accelerates succession to the tougher, longer-lived red or brown algae.

In any of the basic environments (terrestrial, fresh water, marine), one type of succession is primary succession which starts from the primitive substratum, where there was no previously any sort of living matter. The first group of organisms estabilishing there are known as the pioneers, primary community or primary colonisers.

Secondary succession

Old field succession is an example of secondary succession. Here the pioneer species are opportunistic annuals that can respond quickly to the appearance of a new habitat. The germination of *Ambrosia artemisiifolia* is triggered by disturbance, unfiltered light, fluctuating temperature and reduced CO_2

concentration, all of which are associated with a newly cleared site. Summer annuals such as Ambrosia are replaced by winter annuals which establish earlier in the season, giving them a headstart in the competition for space, light and other resources. Late successional species are shade-tolerant and slow colonisers. Secondary succession which starts from previously built up substrata with already existing living matter. The action of any external force, as a sudden change in climatic factors, biotic intervention, fire etc., causes the existing community to disappear. Thus, area becomes devoid of living matter but its substratum, instead of primitive, is built up. Such successions are comparatively more rapid.

Autogenic succession

After the succession has begun, in most of the cases, it is the community itself which, as a result of its reactions with the environment, modifies its own environment and thus causing its own replacement by new communities. This course of succession is known as autogenic succession.

Allogenic succession

Serial replacement of species can result from external environmental factors, such as geophysico-chemical changes. This is in contrast to autogenic succession (above) which is dependent on the biological action of the organisms on their environment.

Allogenic succession has occurred over most of North America and northern Europe in response to climate warming, following the retreat of the last Pleistocene ice sheet about 10000 years ago. The ice sheet advanced and retreated three times in Europe causing a similar advance and retreat in the biota, each advance consisting of a somewhat different mix of species. As the climate ameliorated, easily dispersed, light-demanding trees such as pine and birch advanced northward to be replaced by the slower dispersing, shade-tolerant species such as oak and ash. The next glacial period favoured spruce, fir and eventually treeless tundra vegetation.

These changes have taken place over thousands of years and are evidenced by pollen cores taken from lake deposits. Allogenic transition over shorter time scales occurs where sediment is accreting (e.g., on sand dunes or estuaries). Silt is accreting in the Fal estuary in Cornwall, England causing the salt marsh to extend seawards and the valley woodland to invade the landward limits of the marshland. The recent spread of the salt marsh grass, *Spartina anglica*, is also contributing to the extension of salt marsh. Species are able to colonise at a particular height above sea level according to their tolerance of inundation by brackish water at high tides. Pioneer grasses and sedges are the only higher plant species tolerant of the brackish mudflats, while slightly higher up tidal scrub and then tidal woodland have developed. This geographic transition mirrors the changes apparent from the woodland soil profile and hence the historic origins of the woodland community. In some cases, however, the replacement of the existing community is caused largely by any other external condition and not by the existing organisms. Such a course is referred to as allogenic succession.

On the basis of successive changes in nutritional and energy contents, successions are sometimes classified as:

Autotrophic succession

It is characterised by early and continued dominance of autotrophic organisms like green plants. It begins in a predominantly inorganic environment and the energy flow is maintained indefinitely. There is gradual increase in the organic matter content supported by energy flow.

Heterotrophic succession

It is characterised by early dominance of heterotrophs, such as bacteria, actinomycetes, fungi and animals. It begins in a predominantly organic environment and there is a progressive decline in the energy content. In ecological literature, there are mentioned still so many other kinds of succession, depending mainly upon the nature of the environment (primarily based upon moisture relations), where the process has begun and thus it may be a hydrosere or hydrarch–starting in regions where water is in plenty, as ponds, lakes, streams, swamp, bog, etc; a mesarch–where adequate moisture conditions are present; and a xerosere or xerarch–where moisture is present in minimal amounts, such as dry deserts, rocks etc. Sometimes, there are further distinguished, the lithosere– initiating on rocks, psammosere–on sand and halosere–in saline water or soil.

Successional Processes

The examples above and studies involving models of succession indicate that it is strongly influenced by three processes:

1. Facilitation: Changes in the abiotic environment that are imposed by the developing community. Pioneer species 'prepare the ground' allowing other species to invade, e.g., mountain avens and alder on boulder clay till.
2. Inhibition: Species of one seral stage resist invasion by later successional species such that invasion is only possible following disturbance or by replacing individuals as they die, e.g., green algae (*Ulva*) on intertidal boulders.
3. Tolerance: Late successional species invade because they are able to tolerate lower resource levels and can outcompete early successional species for limited light and space, e.g., old field succession.

The tolerance model suggests a predictable sequence of species replacement based on species' strategies for exploiting resources. Tilman, in his resource ratio hypothesis, argues that species dominance is determined by the relative availability of two resources, light and nutrients. During succession, nutrient availability increases with litter accumulation and soil formation, while light levels decrease as a result of shading. Figure 5.1 shows how species A, with the highest requirement for light and the lowest requirement for nutrients is the first to colonise, followed by a predictable sequence of invaders until eventually the community is dominated by nutrient hungry, shade-tolerant species E.

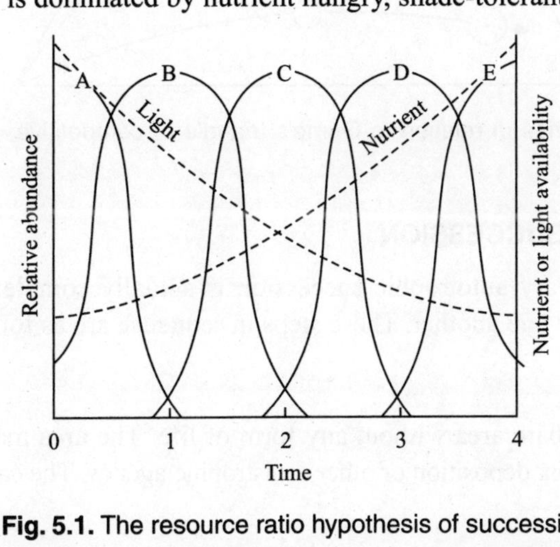

Fig. 5.1. The resource ratio hypothesis of succession.

Nobel and Slatyer also stress the importance of individual species properties in determining their place in succession. These properties are their method of recovery after disturbance and their ability to reproduce in the face of competition. Species unable to reproduce in competitive environments and able to recover rapidly after disturbance (e.g., by means of a seedling pulse from a seed-bank) occur early in the succession. Competitive species only able to invade slowly through seed dispersal into the habitat dominate later. Typically, early successional species are r-selected, having developed efficient colonisation mechanisms to escape from competition. Late successional species have evolved characteristics enabling them to persist longer in competitive situations and are thus responding to K-selection. In general, good colonisers arc poor competitors and *vice versa,* i.e., there is a trade-off between competitive strength and colonisation ability.

The three primary strategies proposed by Grime–competitive (C), stresstolerant (S) and ruderal R–allow any species of plant to adopt a combination of strategies, trading off performance in one type of environment with performance in another. Using this model, succession in a temperate forest clearing involves a transition from ruderal (pioneer) species to fast growing trees and shrubs which have a competitive strategy (Fig. 5.2). At later stages in the course of succession the proportion of stress-tolerant species increases. Secondary succession on sites with low soil fertility has a more shallow parabola, as productivity is constrained.

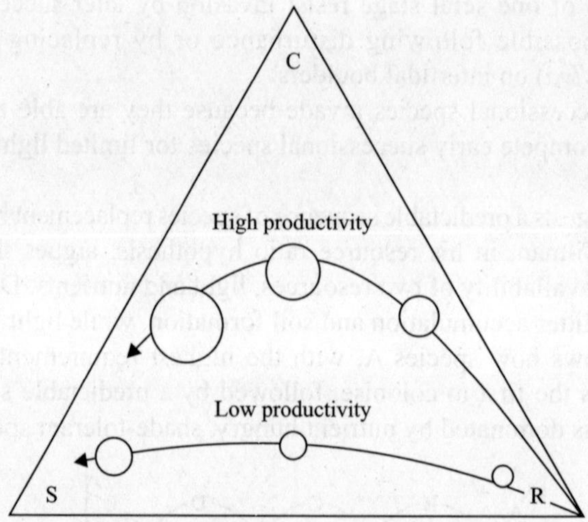

Fig. 5.2. Successional trajectiories in relation to Grime's triangle of competitive, stress-tolerant and ruderal (CSR) strategies.

GENERAL PROCESS OF SUCCESSION

The whole process of a primary autotrophic succession is actually completed through a number of sequential steps, which follow one another. These steps in sequence are as follows:

Nudation

This is the development of a bare area without any form of life. The area may develop due to several causes such as landslide, erosion, deposition or other catastrophic agency. The cause of nudation may be:

Topographic

Due to soil erosion by gravity, water or wind, the existing community may disappear. Other causes may be deposition of sand etc., landslide, volcanic activity and other factors.

Climatic

Glaciers, dry period, hails and storm, frost, fire etc., may also destroy the community.

Biotic

Man is most important, responsible for destruction of forests, grasslands for industry, agriculture, housing etc. Other factors are disease epidemics due to fungi, viruses etc., which destroy the whole population.

Invasion

This is the successful establishment of a species in a bare area. The species actually reaches this new site from any other area. This whole process is completed in following three successive stages.

Migration (dispersal)

The seeds, spores or other propagules of the species reach the bare area. This process, known as migration, is generally brought about by air, water, etc.

Ecesis (establishment)

After reaching to new area, the process of successful establishment of the species, as a result of adjustment with the conditions prevailing there, is known as ecesis. In plants, after migration, seeds or propagules germinate, seedlings grow and adults start to reproduce. Only a few of them are capable of doing this under primitive harsh conditions and thus most of them disappear. Thus as a result of ecesis, the individuals of species become established in the area.

Aggregation

After ecesis, as a result of reproduction, the individuals of the species increase in number and they come close to each other. This process is known as aggregation.

Competition and Coaction

After aggregation of a large number of individuals of the species at the limited place, there develops competition (inter-as well as intraspecific) mainly for space and nutrition. Individuals of a species affect each other's life in various ways and this is called coaction. The species, if unable to compete with other species, if present, would be discarded. To withstand competition, reproductive capacity, wide ecological amplitude etc., are of much help to the species.

Reaction

This is the most important stage in succession. The mechanism of the modification of the environment through the influence of living organisms on it, is known as reaction. As a result of reactions, changes take place in soil, water, light conditions, temperature etc., of the environment. Due to all these the environment is modified, becoming unsuitable for the existing community which sooner or later is replaced by another community (seral community). The whole sequence of communities that replaces one another in the given area is called a sere and various communities constituting the sere, as seral communities, seral stages or developmental stages. The pioneers are likely to have low-nutrient requirements, more dynamic and able to take minerals in comparatively more complex forms. They are small-sized and make less demand from environment.

Stabilisation (Climax)

Finally, there occurs a stage in the process, when the final terminal community becomes more or less stabilised for a longer period of time and it can maintain itself in equilibrium with the climate of the area. This final community is not replaced and is known as climax community and the stage as climax stage.

Generalised process of succession and different plant communities appearing in the process with developing environmental complex are shown in Fig. 5.3 taking hydrosere as a model. General process of succession as outlined above will show that the whole process involves several stages, each stage having characteristic organisms together with their environment. Each such developmental stage is called a seral stage. These stages are infact continuous with each other and the whole sequence from beginning till the climax stage is known as a sere. The species which colonise the bare area in the beginning of succession are called pioneers.

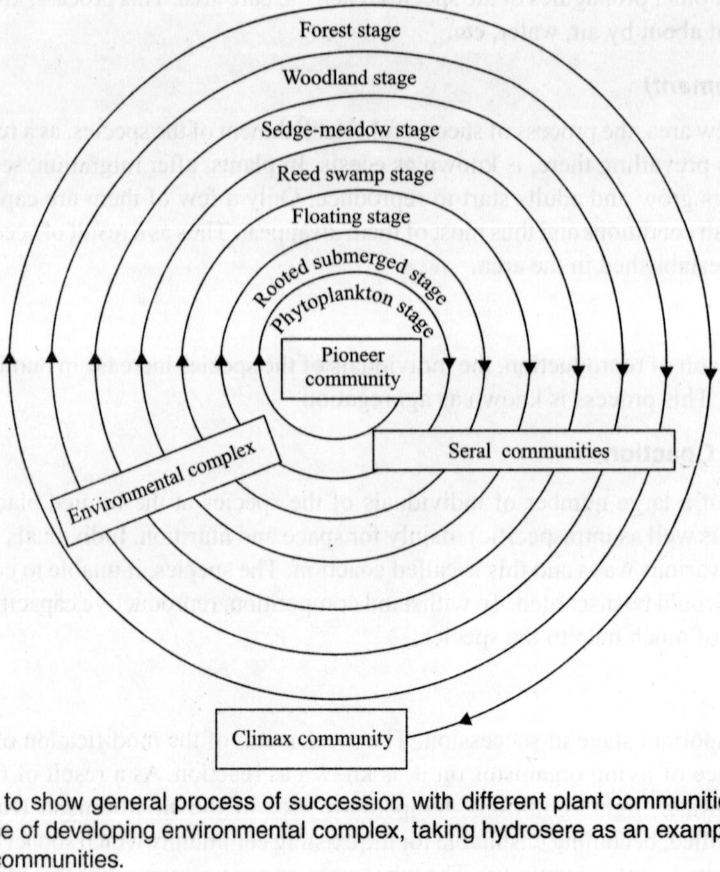

Fig. 5.3. Diagram to show general process of succession with different plant communities appearing therein under the influence of developing environmental complex, taking hydrosere as an example. Note the pioneer, seral and climax communities.

Fluctuations and the Climax Community

Studies have shown that the model of the single climax community is inaccurate. The climax community will vary depending on local variations in climatic and edaphic conditions and in response to other factors such as the disturbance regime and the influence of animals, particularly herbivores. There is no single climax but a continuum of climax types varying along environmental gradients.

The stable climax community also proves elusive in practice. No community is stable for long because of natural disturbances such as storms and fire. A forest climax may take 300 years to develop but the probability of disturbance may be so high that succession never reaches this stage. Non-successional, short term, reversible changes in the floristic composition of a community or fluctuations, are also common. These changes are the result of environmental stochasticity and range from changes in species dominance to more subtle variations such as changes in the age structure of a species.

Succession was considered a predictable unidirectional process, but it is now known that this is not always the case. For example, cyclic succession occurs on a small scale in most communities. Where a forest tree dies and falls, pioneer species germinate in the gap to be replaced by mid- and late-successional species until the gap is again filled by trees. Thus, even old, seemingly stable communities are in a state of flux and are a shifting mosaic rather than a steady state. In disturbed habitats, cyclic succession will prevent the development of a climax community and arrest seral development at an early stage. Although succession is undoubtedly an important and widespread process, the physical and biological structure of many communities may be dominated by other influences.

Some ecologists, have talked of retrogressive succession in which continuous biotic influences have some degenerating influence on the process. Due to destructive effects of organisms, sometimes the development of disturbed communities does not occur and the process of succession instead of progressive becomes retrogressive. As for example forest may change to shrubby or grassland community. This is called retrogressive succession.

Sometimes due to changes in local conditions as soil characteristics or microclimate, the process of succession becomes deflected in a different direction than that presumed under climatic condition of the area. Thus the climax communities are likely to be different from the presumed climatic climax community. This type of succession is called deflected succession. In India, with a monsoon type of climate, in some habitats like temporary ponds, pools etc., it is common to observe each year the development of different kinds of communities in different seasons of the year. Dudgeon called them as seasonal succession. However, such changes are simply recurrent and not developmental and thus should not be designated as succession.

The above-mentioned general process of succession would become more clear by studying in detail the following different kinds of seres in different habitat conditions.

Hydrosere or Hydrarch

The various stages in a hydrosere are well studied in ponds, pools or lakes and we shall study for the present purpose this process of succession occurring in a pond. Hydrosere, orignating in a pond, starts with the colonisation of some phytoplanktons which form the pioneer plant community and finally terminates into a forest, which is a climax community together with their chief components of vegetation.

Just like other primary autotrophic successions, in a hydrosere too, successive changes take place in plants as well as animals' life. However, as the changes are more obvious in plants than animals, it looks as it is a succession of plants only. But this is due to the fact, that changes in plants are so obvious, that we designate the different stages of succession, based upon the dominant forms of plants in that community, as phytoplankton stage, rooted submerged stage, etc. and so on. These various stages together with their chief components of plant species of a hydrosere as shown in Fig. 5.4, are as follows:

Phytoplankton stage

They constitute the pioneer community. Some blue-green algae, green algae, diatoms and bacteria etc. are the first organisms to colonise the primitive medium of the pond. The soils are very much reduced with a pH value of not more than 5.00. They multiply and grow for some time.

Rooted submerged stage

As a result of death and decomposition of phytoplanktons and their mixing with the silt, brought from the surrounding land by rain water and by wave action of pond water, there develops a soft mud at the bottom of pond. This new habitat which tends to be a bit shallower and where light penetration may now occur easily becomes now suitable for the growth of rooted submerged hydrophytes like *Myriophyllum, Elodea, Hydrilla, Potamogeton, Vallisneria, Utricularia* etc. These plants bring about further build up of the substratum as a result of their death and decay. The water level also decreases making the pond more shallower. This new habitat now replaces these plants giving way to another type of plants which are of floating leave type.

Rooted floating stage

By now the water depth is almost 2–5 feet. These plants colonise the habitat with their rhizomes. They all are rooted hydrophytes with their large leaves floating on the water surface. These are species of *Nelumbo, Nymphaea, Limnanthemum, Aponogeton, Trapa, Monochoria* etc. Some free floating species as *Azolla, Lemna, Wolffia, Pistia, Spirodella, Salvinia* etc. also become associated with the rooted plants, due to availability of salts and other minerals in abundance. The water level by now becomes very much decreased, making the pond more shallower. The decomposing organic matter formed due to death of these plants brings about further build up of the substratum. Thus floating species sooner or later disappear from the area.

Reed-swamp stage

This stage is also known as amphibious stage as the plants of community are rooted but most parts of their shoots (assimilatory organs) remain exposed to air. Species of *Scirpus, Typha, Sagittaria* and *Phragmites* etc., are the cheif plants of this stage. They have well-developed rhizomes and form a very dense vegetation. The water level is by now very much reduced and finally becomes unsuitable for the growth of these amphibious species.

Sedge-meadow stage

Due to successive decrease in water level and further changes in the substratum, species of some Cyperaceae and Gramineae, such as *Carex*, *Juncus*, *Cyperus* and *Eleocharis* colonise the area. They form a mat-like vegetation towards the centre of the pond with the help of their much branched rhizomatous systems. As a result of high rate of transpiration, there is much rapid loss of water and sooner or later the mud is exposed to air as a result of which nutrients like ammonia, sulphides etc., become oxidised to nitrates and sulphates. Thus mesic conditions approach the area and marshy vegetation disappears gradually.

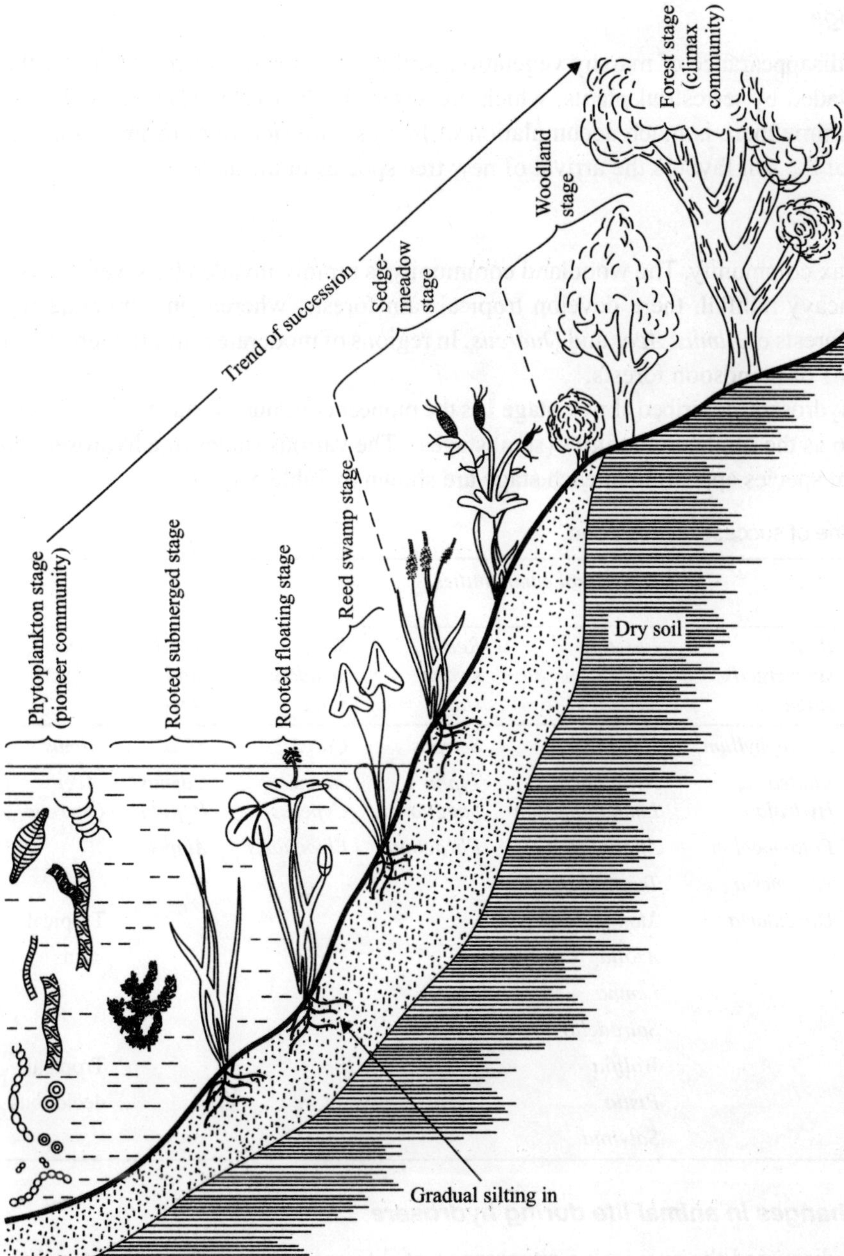

Fig. 5.4. Diagram showing different plant communities appearing at different stages of a hydrosere originating in a pond. Note the gradual decrease in water level and successive development of the soil in area to give finally a terrestrial habitat for forest climax community.

Woodland stage

By the time of disappearance of marshy vegetation, soil becomes drier for most time of the year. This area is now invaded by terrestrial plants, which are some shrubs (*Salix, Cornus*) and trees (*Populus, Almus*). By this time there is much accumulation of humus with rich flora of micro-organisms. Thus, mineralisation of the soil favours the arrival of new tree species in the area.

Forest stage

This is the climax community. The woodland community is rapidly invaded by several trees. In tropical climates with heavy rainfall, there develop tropical rain forests, whereas in temperate regions, there develop mixed forests of *Almus*, *Acer* and *Quercus*. In regions of moderate rainfall, there develop tropical deciduous forests or monosoon forests.

Thus in the hydrosere, described above, stage 1 is the pioneer community, stage 7 the climax community and stages 2 to 6 as the seral communities (seral stages). The various stages of a hydrosere and the chief component plant species appearing at each stage are shown in Table 5.1.

Table 5.1. Scheme of succession in a pond.

Pioneer community	*Seral communities*						*Climax community*
Phytoplankton stage	*Rooted submerged stage*	*Floating stage*	*Reed-swamp stage*	*Sedge-meadow stage*	*Wood-land stage*	*Forest stage*	
Blue-green algae,	*Myriophyllum*	*Nelumbo*	*Scirpus*	*Carex*	*Salix*	*Ulmus*	
Green algae,	*Elodea*	*Nymphaea*	*Typha*	*Juncus*	*Cornus*	*Acer*	Temperate mixed
Diatoms,	*Hydrilla*	*Limnanthemum*	*Saggitaria*	*Cyperus*	*Populus*	*Quercus*	
Bacteria etc.	*Potamogelon*	*Aponogeton*	*Phragmites*	*Eleocharis*	*Almus*		
	Vallisneria	*Trapa*					
	Utricularia	*Monochoria*				Tropical	
		Azolla				rain	
		Lemna					
		Spirodella					
		Wolffia				Tropical	
		Pistia				deciduous	
		Salvinia					

Successive changes in animal life during hydrosere

So far we have discussed the successive appearance of differnt communities of plants that develop at various stages of succession progressing in a pond-the hydrosere. What about the animal life? Is there any change in this form of life also together with plants? Certainly, there take place changes in animal life also although these are not as obvious as those in the plant communities.

Thus, with the ageing of a pond and the development of marshes, the animal life also undergoes changes.

These are as follows:

The protozoans like *Paramecium, Amoeba, Euglena* etc. are the pioneers. But, if the planktonic growth forms are very rich, other animal life as blue gill fish, sun fish, large mouth bass etc., start appearing. Some caddisflies are also found. In the next, submerged stage the caddisflies are replaced by other animals that may creep over the submerged vegetation. Thus, dragon flies, mayflies and some crustaceans as *Asellas, Gammarus, Daphnia, Cypris, Cyclops*, etc., inhabit the pond at this stage.

At the floating stage, the animal life is chiefly represented by *Hydra* spp., gill breathing snails, frogs, salamanders, diving beetles, whirligig beetles and other insects. There also appear some turtles and snakes.

At the reed-swamp stage, the pond becomes shallower and the bottom starts becoming exposed. The floating animals are replaced by different species of mayflies and dragon flies, whose nymphs remain attached to submerged parts of the vegetation and adults present on the surfaces of emergent parts. Gill breathing snails are replaced by lung breathers as *Lymnea, Physa* and *Gyraulus*. Among insects, water scorpion, giant water bug, scavenger beetles etc., are present at this stage. The bottom of the pond is now inhabited by some annelids, mud pickrel and bull heads. Red winged black birds, king fisher, great blue heron, swamp sparrow, ducks, musk rats, beavers etc., become common in the area.

At the sedge-meadow stage, the animals like anails as *Anodonta, Psidium* etc., become common. Finally, at the woodland stage, under terrestrial conditions most of the terrestrial forms of animal life appear in the area.

Functional Changes In Hydrosere

To sum up, we may conclude that during succession in a pond (hydrosere) the major changes from a structural standpoint i.e., changes in composition, are: (i) a change in species compostion; and (ii) a change in variety or diversity. Besides them there take place marked changes in number and kinds of both, autotrophs and heterotrophs i.e., functional changes.

Two major functional changes include: (i) a progressive increase in amount of both, living (biomass) and dead organic matter; and (ii) a shift in community metabolism from young ponds, which are autotrophic ($P/R>1$, where P-rates of production; R-rates of respiration), to those which are stabilised ($P/R = 1$) and finally to heterotrophic ($P/R<1$) condition.

The increasing heterotrophy of these aquatic system applies to that phase of their succession in which they achieve a terrestrial condition. These four major structural and functional attributes of ecological succession (increase in species diversity, increase in structural complexity, increase in organic matter and tendency towards metabolic stability) are, in a sense, both causes and effects of the very processes of change and the eventual stability that characterise ecosystems in general.

Lithosere—a Xerosere on Rock

This is a type of xerosere orginating on bare rock surfaces. The orginal substratum is deficient in water and lacks any organic matter, having only minerals in disintegrated unweathered state. The pioneers to colonise this primitive substratum are crustose type of lichens and through a series of successive, seral stages the succession finally terminates into a forest which constitutes the climax community.

As pointed out in hydrosere, in a lithosere also successive changes take place in both plants as well as animals. But here also, as in most of the primary autotrophic successions, changes in plants life are more obvious than those in animals. The changes in plants are obvious to the extent that it looks as a succession of plants only and various stages are named on the basis of particular stage dominated by a particular plant species.

The various stages and their component plant species of a lithosere appearing on a rock are shown in Fig. 5.5. These are as follows:

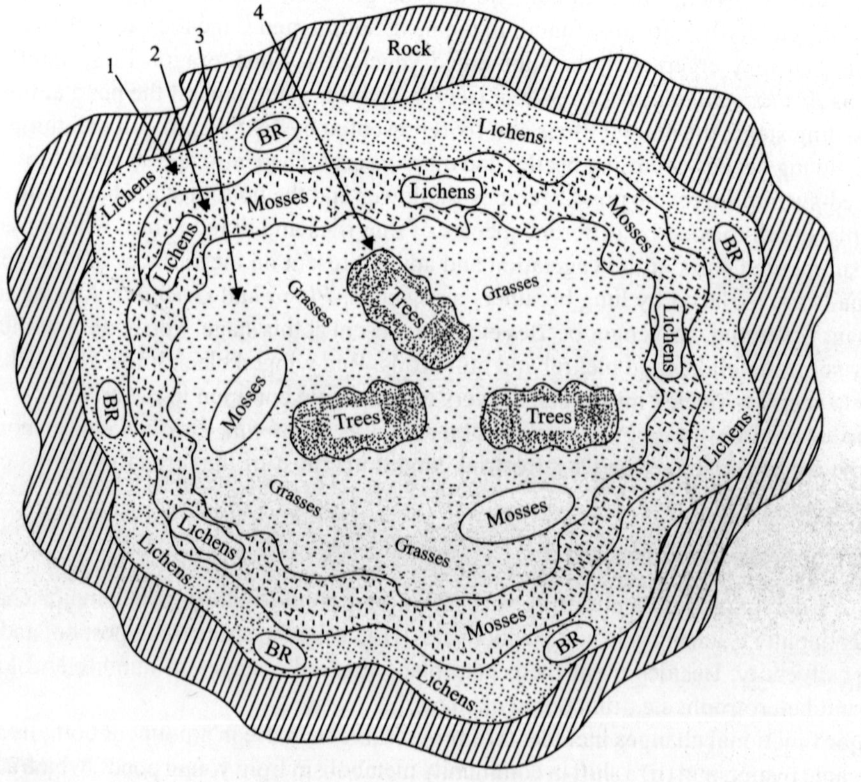

Fig. 5.5. Diagrammatic representation of different plant communities of a lithosphere appearing on a rock. Note the vegetational zonation showing the pioneer community of lichens around the outer edge and more advanced stages of trees located in the centre. The various zones from outside towards the centre of rock are (1) lichens (pioneers); (2) ring of mosses; (3) grasses (broad zone); and (4) trees (seedlings) scattered. BR-bare rock.

Custose lichens stage

As mentioned earlier, the substratum colonised by these pioneers is very poor in moisture and organic matter, subjected with extremes of temperature. The lichens of this stage are species of *Rhizocarpon, Rinodina* and *Lecanora.* They produce some acids which bring about weathering of rocks. The dead organic matter of lichens becomes mixed with the small particles of rocks. However, this process is very slow. These lichens are then replaced by foliose type of lichens.

Foliose lichens stage

They appear on the substratum partially built up by the crustose lichens. This community includes species of *Parmelia, Dermatocarpon* etc., which have large leaf-like thalli. They can absorb and retain more water and are able to accumulate dust particles which help in the further build up of the substratum. Thus some humus becomes accumulated. The weathering of rocks and its mixing with humus results into the development of a fine thin soil layer on rock surface and thus there is a change in the habitat.

Moss stage

The development of thin soil layer on rock surface, especially in the crevices, favours the growth of some such xerophytic mosses as species of *Polytrichum, Tortula* and *Grimmia*. At their successful growth, they compete with the lichens. Due to their death and decay there is further addition of organic matter in the soil. The thickness of the soil layer now increases.

Herbs stage

Due to more extensive growth of mosses there accumulates more soil and there are added more minerals to it due to leaching out from the overlying vegetation. This changed habitat favours the growth of some herbaceous weeds which are chiefly the annuals, in turn being followed by some biennials and perennials. Due to their growth and death there is much more accumulation of humus in soil together with further weathering of rock. Thus, habitat changes with decreasing xeric conditions. This stage is constituted by such shallow rooted grasses as *Aristida, Festuca, Poa, Solidago,* etc., which in turn are replaced by shrubs.

Shrub stage

Due to much accumulation of soil, the habitat becomes suitable for shrubs which start migrating in the area. These are species of *Rhus, Phytocarpus* etc. They overshadow the herbaceous vegetation. The soil is further enriched by this dense shrub by growth. These in turn are finally replaced by trees which make up the climax community.

Forest stage

Some xerophytic tree species invade the area. Further weathering of rocks and increasing humus content of the soil favour the arrival of more trees and vegetation finally becomes mesophytic. Thus, there develops finally a forest community.

The various stages of a lithosere and the chief component plant species appearing at each stage are shown in Table 5.2.

Table 5.2. The trend of succession on a rock.

Pioneer community		*Seral communities*			*Climax community*
Crustose lichens stage	*Foliose lichens stage*	*Moss stage*	*Herbs stage*	*Shrub stage*	*Forest stage*
Rhizocarpon	*Parmelia*	*Polytrichum*	*Artistida*	*Rhus*	Mesophytic trees
Rinodina	*Damatocarpon*	*Tortula*	*Festuca*	*Phytocarpus*	
Lecanora		*Grimmia*	*Poa*		
		General trend of succession			

Changes in animal life during lithosere

As in a hydrosere, there occur successive changes in animal life during the lithosere also. Associated with the lichens, the pioneers, are a few mites. Fauna is sparse in terms of species composition. There are a few ants and few spiders present in the cracks and crevices of rock. These are the pioneer animal species which are exposed to harsh environment—the thermal extremes etc. The mites become more varied in terms of species and small spiders, springtails as well as tradigrades become associated with the secondary community of mosses. At later stage of succession and when grasses start developing, the fauna undergoes a quantitative and qualitative increase. Nematodes and larval insects, collembola, ants, spiders and mites appear in this new environment. With the development of forest climax community, there develop a rich fauna consisting of invertebrates as well as vertebrates. These include slugs, snails, wire worms, millipedes, centipedes, ants, sow bugs, springtails, mites, squirrels, shrews, mammals like fox, chipmunk mouse and mole, birds like grouse and flycatcher, reptiles like skinks, turtles, snakes etc. and amphibians such as salamanders and frogs.

HETEROTROPHIC (MICROBIAL) SUCCESSION

Succession of micro-organisms like fungi, bacteria, actinomycetes etc., that occurs within a microhabitat (microenvironment or microclimate), has been vigrously called as *microsuccession*, *serule* or *microsere*. This type of succession, which is heterotrophic in nature, begins on a habitat, which in contrast with that of autotrophic succession, is rich in organic matter content. It follows the course of an autogenic succession, i.e., determined in part by the organisms themselves. However, most of such successions involve the opposite progression, from an initial condition of high energy content to a final stage of nil energy. The substratum here, instead of being progressively complex (as happens in autotrophic successions) becomes gradually and gradually depleted in terms of organic matter content. Such successions of micro-organisms play an important role in ecosystems, where the decomposers are an important biotic component. These bring about mineralisation of dead organic matter making the mineral elements again available to soil. Succession of fungi on dead plant remains in the soil has been extensively studied by Garrell and such a schema is outlined in Table 5.3. In the succession of fungi on plant remains in the soil (roots etc., dying through natural senescence), the primary colonisers are likely to be weak parasites and/or saprophytes of sugar fungi group. On dead tissues, the ascomycetes and their imperfect forms (cellulose decomposers) are the secondary colonisers. At this stage, some Mucorales may also appear as secondary saprophytic sugar fungi. Finally, the basidiomycetes (lignin decomposers) appear in the succession. The chemical make-up of the substrate determines the qualitative characteristics of fungal flora appearing at a particular stage of decomposition.

Table 5.3. General scheme for fungal succession on plant remains in the soil.

Senescent tissue		*Dead tissue*	
Primary colonisers		*Secondary colonisers*	
Stage l a	*Stage 1*	*Stage 2*	*Stage 3*
Weak parasites	Primary saprophytic sugar fungi, living on sugars and carbon compounds (mucoraceous phycomycetes)	Cellulose decomposers and associated secondary saprophytic sugar fungi, sharing products of cellulose decomposition (ascomycetes and some mucorales)	Lignin decomposers and associated fungi (basidiomycetes and others)

Fungal succession on dungs and similar excreta of herbivores has also been studied by several workers, the schema for which is outlined in Fig. 5.6. On herbivores' dung the pioneer colonisers are phycomycetes, followed by secondary colonisers, the ascomycetes, in turn being succeeded finally by the basidiomycetes. Deuteromycetes also develop simultaneously alongwith the other groups of colonisers.

Microbial successions on a variety of other substrates as senescent and decaying herbaceous plant parts, deciduous and evergreen tree litters, fallen logs and branches etc., lying above the soil have been investigated by many workers. These are now well established successions in the field of microbial ecology. In a tropical country like India, successions of fungi on decaying plant remains from their early senescence onwards have been studied by various scientists. These studies have shown that on senescent and dead plant parts, various groups of fungi appear in an orderly sequence, which may be divided into pioneer and secondary colonisers.

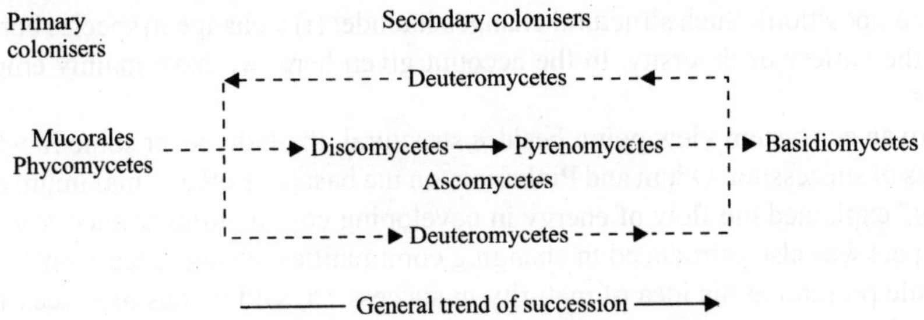

Fig. 5.6. The schema of fungal succession on herbivores' dung.

The available information on the successions of fungi on a variety of plant remains above the soil has been reviewed by Hudson and general schema of such fungal successions is outlined in Fig. 5.7.

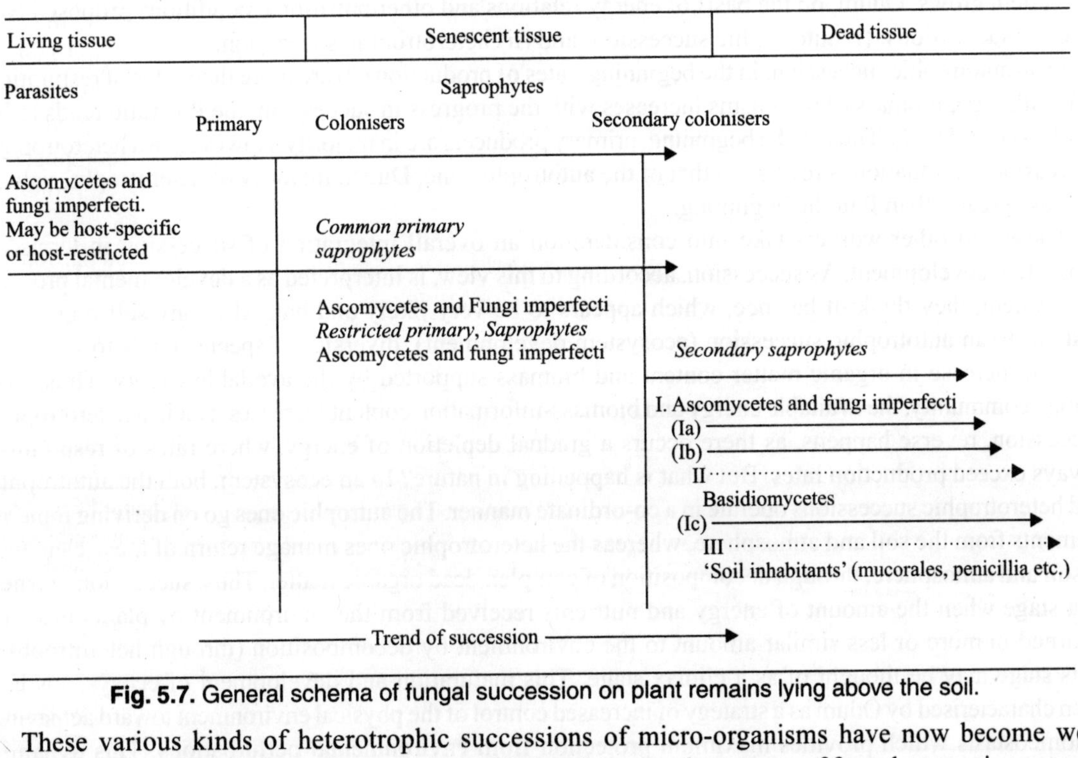

Fig. 5.7. General schema of fungal succession on plant remains lying above the soil.

These various kinds of heterotrophic successions of micro-organisms have now become well established in the field of microbial ecology. On these substrates, the process of fungal succession appears to be governed by such factors as humidity and temperature of air, moisture content of the substrate, host resistance and competition between colonisers themselves.

ECOSYSTEM DEVELOPMENT

It was pointed out in the beginning of the chapter that with the development of ecosystem concept in ecology, ecologists like Odum, preferred to call the process of succession as ecosystem development. The account of two autotrophic successions, a hydrosere and a lithosere and a heterotrophic succession makes it clear that there occur major changes in each succession from a structural point of view (i.e., changes in composition). Such structural changes include: (i) a change in species composition; and (ii) a change in the variety or diversity. In the account given here, we have mainly emphasised these structural aspects.

However from an ecosystem view point, besides structural, there do occur some functional changes also in the process of succession. Odum and Pinkerton, on the basis of Lotka's 'maximum energy content in living systems,' explained the flow of energy in developing communities in succession. In this way, the functional aspect was also introduced in changing communities during succession.

Margalef, while presenting his idea of maturity in succession, said 'status of succession in ecology may be thought of as equivalent to development in biology.' According to him, with the progressive changes the succession attains maturity. Odum could elaborate the idea of maturity and modifications put forth his conept of ecosystem strategy. According to this concept, in any ecosystem, there is a tendency of homeostasis while under the influence of unfavourable external environment. But in a balanced state, the system grows. Odum, on the basis of energy relations and other nutritional conditions proposed two types of successions: (i) autotrophic succession; and (ii) heterotrophic succession.

In an autotrophic succession, in the beginning, rates of production (P) are more than rates of respiration (R). Although, biomass of organisms increases with the progress in succession, the P/R ratio tends to be similar (i.e., $P/R = 1$). Thus, in the beginning, primary producers are in majority. However, in a heterotrophic succession the situation is reverse to that of the autotrophic one. Due to majority of heterotrophs, values of R are greater than P in the beginning.

These and other workers take into consideration an overall integration of succession in terms of ecosystem development. As succession, according to this view, is interpreted as a developmental process in a system, they think of balance, which appears to be very likely and natural in any self regulatory system. In an autotrophic succession (ecosystem development), diversity of species tends to increase, with an increase in organic matter content and biomass supported by the available energy. Thus, in a climax community, the available energy and biomass-information content increases. But in a heterotrophic succession, reverse happens, as there occurs a gradual depletion of energy, where rates of respiration always exceed production rates. But what is happening in nature? In an ecosystem, both the autotrophic and heterotrophic successions operate in a co-ordinate manner. The autrophic ones go on deriving mineral elements from the soil and atmosphere, whereas the heterotrophic ones manage return of these elements to soil and atmosphere, through decomposition of complex dead organic matter. Thus, succession reaches to a stage when the amount of energy and nutrients received from the environment by plants is again returned in more or less similar amount to the environment by decomposition (through heterotrophs). This stage may be thought of as a climax stage. This maturation and development of ecosystems has been characterised by Odum as a strategy of increased control of the physical environment toward achieving a homeostasis which provides maximum protection from environmental perturbations. This dynamic and largely predictable process of community-controlled modification of the environment sooner or later results into a stabilised ecosystem. Odum has identified a number of trends to be expected in the development of ecosystems, which are shown in Table 5.4.

Table 5.4. Trends of changes in ecosystem development.

Ecosystem attributes	*Developmental stages*	*Mature stages*
	Community energetics	
Gross production/community respitation (P/R ratio)	Greater or less than 1	Approaches 1
Gross production/standing crop biomass (P/B ratio)	High	Low
Biomas supported/unit energy flow (B/E ratio)	Low	High
Net community production (yield)	High	Low
Food chians	Linear, predominantly grazing	Weblike, predominantly detritus
	Community structure	
Total organic matter	Small	Large
Inorganic nutrients	Extrabiotic	Intrabiotic
Species diversity-variety component	Low	High
Species diversity-equitability component	Low	High
Biochemical diversity	Low	High
Stratification and spatial heterogeneity (pattern diversity)	Poorly organised	Well-organised
	Life history	
Niche specialisation	Broad	Narrow
Size of organism	Small	Large
Life cycles	Short, simple	Long, complex
	Nutrient cycling	
Mineral cycles	Open	Closed
Nutrient exchange rate between organisms and environment	Rapid	Slow
Role of detritus in nutrient regeneration	Unimportant	Important
	Seletion pressure	
Growth form	For rapid growth	For feedback control
Production	Quantity	Quality
	Overall homeostasis	
Internal symbiosis	Undeveloped	Developed
Nutrient conservation	Poor	Good
Stability (resistance to external perturbations)	Poor	Good
Entropy	High	Low
Information	Low	High

CLIMAX CONCEPT IN SUCCESSION

The final, terminal and more or less stabilised community in succession that is able to establish some sort of equilibrium with the environmental conditions of that area, was termed climax by Clements. The subject of climax has been much controversial and is reviewed from time to time by many ecologists. According to Clements, climax has the following three principal characteristics:

1. Unity: Climax is a unit and index of the climate of area. Life or growth forms of plants indicate the climate type. Unless all the species are not taken as an organised unit, climax would not indicate the climate.
2. Stability: The form of climax community is more or less stable with the climate. This climax community cannot be replaced through competition by any group of species. In other words, according to Clements, in a particular climate area there may develop climax communities only with a few characteristic dominant species.
3. Origin and phylogenetic relations: Climax community is to be treated equivalent to an organism, like which it takes birth, grows and develops and becomes mature. In this organismic-concept of climax community, Clements went to the extent of designating it a super organism. In any climatic region, the developmental stages of the climax community have their own characteristics, which reflect the type of climate. Similar to the development of organism with changing age, climax communities have also simultaneously undergone changes with changing climate. Thus phylogenetic relations may be estalished between different climax communities of the world.

The three popular theories about the climax concept in ecology are discussed below:

Monoclimax Theory

As evident from the above account, Clements emphasised the importance only climate in the stabilisation of the climax community. According to the monoclimax theory, within a given region all land surfaces eventually tend to be occupied by a single kind of community which is climax. The climax is determined by the regional climate. Given a stable climate the climax community is stable indefinitely. However, his views were strongly criticised by many ecologists. Cowles, disagreeing with the idea of stability, stated that 'equilibrium state is never reached and succession is in fact a variable approaching a variable rather than a constant.' According to Cooper, all progressive as well as retrogressive changes in communities reflect succession. Thus he considered climax state as stage of minimum change, rather than finally changed state of succession. Climax may not be taken equivalent to an organism, from which it differs in several respects. Therefore, community may not be considered as a unit.

It is common to observe that in the same climate, climax communities are different. Because, depending upon their primary stages and habitat characteristics, their sucessional stages are also different. Also, in same climate, a lithosphere and hydrosphere beginning with different pioneer communities and proceeding through different successive developmental stages, might develop into similar climax communities. Similarly, it may not always be true that successions with similar pioneer communities and seral communities would develop into similar climax communities.

The most objectionable point and subject of controversy about the climax (with Clements view point) has been its intimate relationship with the climate. In many conditions, in an area of uniform climate, it is common to observe different types of climax communities according to the soil, topography and other factors. Under such situations it would not be correct to consider only the climate as determining factor in climax. For example, Hult in Finland, long before the Clements concept of climax, described as many as seven different habitat types under similar climatic conditions. Tansley too disagreeing with Clements put forth his concept that climax communities are infact controlled by more than one factor and not solely by climate. In this way, the climatic-climax concept of Clements became popularly known as monoclimax theory, although Clement did not use this term.

Polyclimax Theory

It is indicated above, how the climatic-climax concept or Clements was, strongly opposed by ecologists including Tansley, who believed that climax is controlled by many (and not one-climate) factors. Accordingly Tansley's concept became popularly known as polyclimax theory. Clements, although agreed with the possible control of factors other than climate on climax, but he thought that these communities would sooner or later develop into climatic-climax types. Consequently, with an attempt to accommodate these stages in his own hypothesis, Clements introduced in literature a number of new terms. These are as follows:

The stage in succession just preceding the climatic climax community was called a sub-climax. The community, which became stabilised at any of the seral stages of succession due to microclimate or effects of factors as soil, fire etc., was called a sereclimax.

A disclimax (disturbed climax) was called the community which became more or less stabilised due to recurrent disturbance by man or other biotic factors, which prevent the establishment of climatic climax community in the area. At the same place, under similar climatic conditions, at smaller distances, due to differences in water and soil characters, there may develop different climax communities. Under such circumstances, Clements coined two more terms: (i) preclimax. It is the community with life forms lower than those in the expected climatic climax, which might develop in a habitat drier than that of the expected climatic climax; and (ii) postclimax, a strip of community with life-forms higher than those in the expected climatic climax and which develops under more moist or colder habitats.

As a result, followers of Clements coined a number of new terms. For instance, coclimax, superclimax, quasiclimax, paraclimax, anti-climax, conclimax, peniclimax, metaclimax, pseudoclimax etc. The only way to come out of the jargoons of Clements terminology was the polyclimax theory. There is evidence that even under primeval conditions it was difficult to find large areas of uniform vegetation. According to this theory the climax stage may be controlled by any factor of the environment and not only by climate. Accordingly, the climax stage is to be named, depending upon the nature of the factor in stabilisation. Thus, in addition to climatic climax, controlled by climate, there may develop:

1. Edaphic climax: On an underdeveloped soil, it develops due to edaphic effects.
2. Biotic climax: Developed due to biotic disturbances. Some ecologists make further; categorisation into anthropogenic climax (due to man), grazing climax (due to grazing effects) and zootic climax (due to animils).
3. Topographic climax: Due to differences in topographic factors at mountains, hills, mounds etc.
4. Fire climax: Due to repeated effects of fire.

According to Daubenmire, if one accepts monoclimax theory, it means to admit that other factors of the environment are of secondary value.

Climax-pattern Hypothesis

This concept about climax was put by Whittaker who rejected the classification approach. On the basis of community gradient analysis, it was postulated that the communities developing at a particular place are in accordance with all the factors of environment. In this area, the vegetation may not be divided into smaller basic units. In fact different climax-types are orderly organised parallel to the environmental gradient. According to this hypothesis, there is, in a sense, only one big community that changes according to soil, slope and other habitat factors.

6

Ecosystem Ecology

INTRODUCTION

At present, ecological studies are made at ecosystem level and it has been one of the most recent developments in ecology, which is generally referred to as the bioenergetic approach. At ecosystem level, the units of study are comparatively very large. And there are indeed no practical units, if the nature is conceived as a single, giant ecosystem—the biosphere. The overall view of this type of approach is that living organisms and their non-living environment are inseparably interrelated and interact with each other. The term ecosystem is defined as 'the system resulting from the integration of all the living and non-living factors of the environment'. Thus, the ecosystem is regarded as including not only the organism-complex but also the whole complex of physical factors forming the environment. However, the term ecosystem is most preferred, where 'eco' implies the environment and 'system' implies an interacting, interdependent complex.

Thus any unit that includes all the organisms i.e., the communities in a given area, interact with the physical environment so that a flow of energy leads to clearly defined trophic structure, biotic diversity and material cycle (i.e., exchange of materials between living and non-living components) within the system, known as an ecological system or ecosystem. Figure 6.1 shows a generalised scheme of nutritional relationships (food links) amongst the living organisms of such a system. Keeping this in view, we may think of the Earth, we live upon as a giant ecosystem where abiotic and biotic components are constantly acting and reacting upon each other bringing forth structural and functional changes in it. This vast ecosystem—biosphere (Fig. 6.2) is, however, difficult to handle and thus for convenience we generally study nature by making its artificial subdivisions into units of smaller ecosystems (*Terrestrial —forest, desert, grassland*; man—engineered as a cropland; *Aquatic—freshwater, marine* etc.) of different sizes. An ecosystem may thus be as small as a pond, a cropland or as large as an ocean, desert or forest. It must be remembered, however, that these unit ecosystems are simply separated from each other with time and space, but functionally they all are indeed linked with each other, forming as an integrated whole. There exist practically no functional boundaries between them.

The recent development in ecological studies has been to undertake besides structure, the similarities and differences in food and energy relationships among living components of ecosystem, that is generally referred to as the bioenergetic approach in modern ecology. Modern ecology is thus, broadly defined as the study of ecosystems.

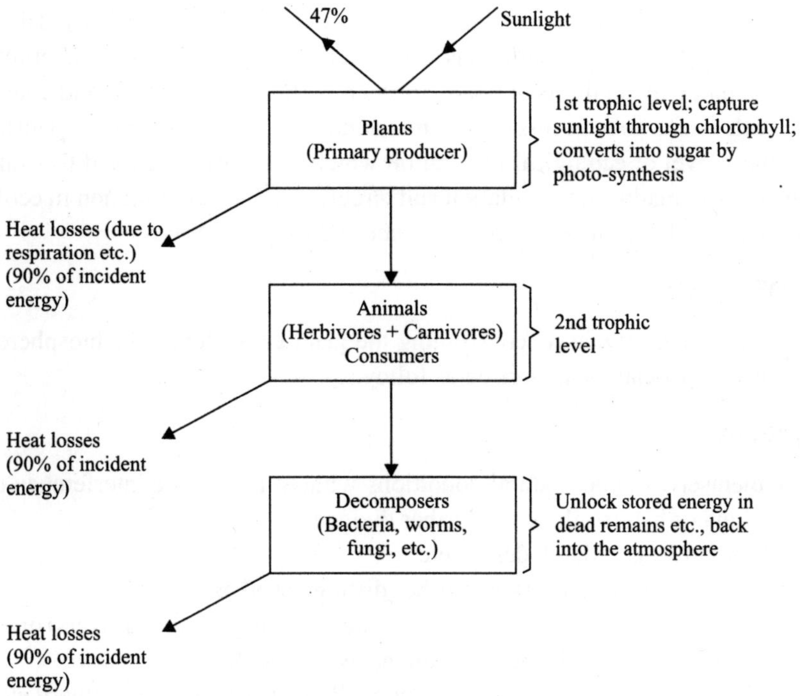

Fig. 6.1. A generalised scheme of nutritional relationships among different biotic components of an ecosystem.

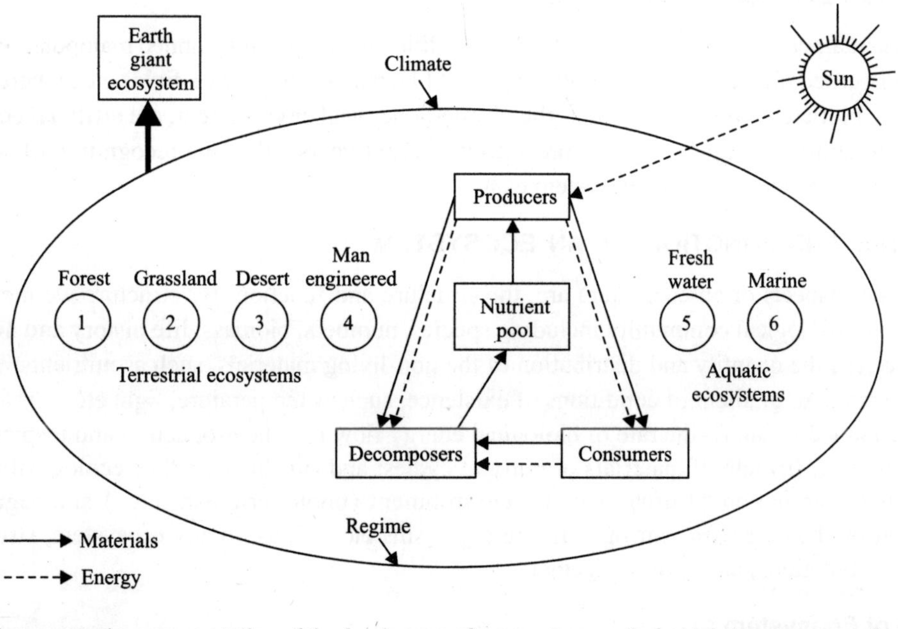

Fig. 6.2. Diagrammatic representation of the basic type of ecosystems, all of which together constitute the giant ecosystem—the biosphere. Note, in the center, the generalised scheme of the structure and function of any unit ecosystem of the biosphere.

An ecosystem is an overall integration of whole mosaics of interacting organisms and their environment. It is normally an open system with a continuous, but variable, influx and loss of materials and energy. It is a basic, functional unit with no limits of boundaries, consisting of both biotic and abiotic components interacting with each other, both necessary for maintenance of life upon Earth. Thus, an ecosystem represents the highest level of ecological integration which is energy-based and this functional unit is capable of energy transformation, accumulation and circulation. Its main function in ecological sense is to emphasise obligatory relationships, interdependence and causal relations.

KINDS OF ECOSYSTEM

Different types of ecosystem of nature, constituting the giant ecosystem—the biosphere, are shown in Fig. 6.2. These may be artificially categorised as follows:

Natural Ecosystems

These operate by themselves under natural conditions without any major interference by man. Based upon the particular kind of habitat, these are further divided as:

1. Terrestrial, as forest, grassland, desert, etc.
2. Aquatic (water-open), which may be further distinguished as:-
   (a) Freshwater, which may be *lotic* (running-water as spring, stream or rivers) or *lentic* (standing-water as lake, pond, pools, puddles, ditch, swamp etc.).
   (b) Marine, such deep bodies as an ocean or shallow ones as a sea or estuary etc.

Artificial (Man-Engineered) Ecosystems

These are maintained artificially by man where, by addition of energy and planned manipulations, natural balance is disturbed regularly. For example croplands like maize, wheat, rice-fields etc., where man tries to control the biotic community as well as the physico-chemical environment, are artificial ecosystems. In addition to above the rapid progress made during recent years led to the recognition of some other such types of ecosystem as space ecosystem etc.

STRUCTURE AND FUNCTION OF AN ECOSYSTEM

The two major aspects of an ecosystem are, the structure and function. By structure we mean: (i) the composition of biological community including species, numbers, biomass, life history and distribution in space etc.; (ii) the quantity and distribution of the non-living materials, such as nutrients, water etc.; and (iii) the range or gradient of conditions of existence, such as temperature, light etc.

By function we mean: (i) the rate of biological energy flow i.e., the production and respiration rates of the community, (ii) rate of materials or nutrient cycles; and (iii) biological or ecological regulation including both regulation of organisms by environment (photoperiodism etc.,) and regulation of environment by the organism, (nitrogen fixing organisms etc.). Thus, in any ecosystem, structure and function (rate functions) are studied together.

Structure of Ecosystem

As shown in Fig. 6.6, an ecosystem has two major components:

Abiotic (non-living) component

It includes: (i) the amount of inorganic substances as P, S, C, N, H, etc., involved in material cycles. The amount of these inorganic substances, present at any given time in an ecosystem, is designated as the standing state or standing quality; (ii) amount and distribution of inorganic chemicals, such as chlorophylls etc., and of organic materials, as proteins, carbohydrates, lipids etc., present either in the biomass or in the environment i.e., biochemical structure that link the biotic and abiotic components of the ecosystem; and (iii) the climate of the given region.

Biotic (living) component

This is indeed the trophic structure of any ecosystem, where living organisms are distinguished on the basis of their nutritional relationships. From this trophic (nutritional) standpoint, an ecosystem has two components:

1. Autotrophic component: In which fixation of light energy, use of simple inorganic substances and build up of complex substances predominate. The component is constituted mainly by green plants, including photosynthetic bacteria. To some lesser extent, chemosynthetic microbes also contribute to the build up of organic matter. Members of the autotrophic component are known as producers.
2. Heterotrophic component: In which utilisation, rearrangement and decomposition of complex materials predominate. The organisms involved are known as consumers, as they consume the matter built up by the producers (autotrophs). The consumers are further categorised as:-
   (a) Macroconsumers: These are the consumers, which in an order as they occur in a food chain are, herbivores, carnivores (or omnivores). Herbivores are also known as primary consumers. Secondary and tertiary consumers, if present, are carnivores or omnivores. They all are phagotrophs which include chiefly animals that ingest other organic and particulate organic matter.
   (b) Microconsumers: These are popularly known as decomposers. They are saprotrophs (=osmotrophs) and include chiefly bacteria, actinomycetes and fungi. They breakdown complex compounds of dead or living protoplasm, absorb some of the decomposition or breakdown products and release inorganic nutrients in environment, making them available again to autotrophs.

The biotic component of any ecosystem may be thought of as the functional kingdom of nature, since they are based on the type of nutrition and the energy source used. The trophic structure of an ecosystem is one kind of producer consumer arrangement, where each 'food' level is known as trophic level. The amount of living material in different trophic levels or in a component population is known as the standing crop, a term applicable to both, plants as well as animals. The standing crop may be expressed in terms of: (i) number of organisms per unit area; and (ii) biomass i.e., organism mass in unit area, which can be measured as living weight, dry weight, ash-free dry weight or carbon weight or calories or any other convenient unit suitable for comparative purposes.

Ecological pyramids

Trophic structure, i.e., the interaction of food chain and the size metabolism relationship between the linearly arranged various biotic components of an ecosystem is characteristic of each type of ecosystem.

The trophic structure and function at successive trophic levels, i.e. producers→herbivores→carnivores, may be shown graphically by means of ecological pyramids where the first or producer level constitutes the base of the pyramid and the successive levels, the tiers making the apex. Ecological pyramids are of three general types: (i) pyramid of numbers, showing the number of individual organisms at each level; (ii) pyramid of biomass, showing the total dry weight and other suitable measure of the total amount of living matter; and (iii) pyramid of energy, showing the rate of energy flow and/or productivity at successive trophic levels. The pyramids of numbers and biomass may be upright or inverted depending upon the nature of the food chain in the particular ecosystem, whereas pyramids of energy are always upright.

Pyramids of numbers

They show the relationship between producers, herbivores and carnivores at successive trophic levels in terms of their number. The pyramids of numbers in three different kinds of ecosystems are shown in Fig. 6.3 (A-C). In a grassland (Fig. 6.3a) the producers, which are mainly grasses, are always maximum in number. This number then shows a decrease towards apex, as the primary consumers (herbivores) like rabbits, mice etc., are lesser in number than the grasses; the secondary consumers, snakes and lizards are lesser in number than the rabbits and mice. Finally, the top (tertiary) consumers hawks or other birds, are least in number. Thus, the pyramid becomes upright. Similarly, in a pond ecosystem (Fig. 6.3b) the pyramid is upright. Here the producers, which are mainly the phyto-planktons as algae, bacteria etc. are maximum in number; the herbivoes, which are smaller fish; rotifers etc., are lesser in number than the producers; and the secondary consumers (carnivores), such as small fish eating each other, water beetles etc., are lesser in number than the herbivores. Finally, the top (tertiary) consumers, the bigger fish are least in number. In a forest ecosystem (Fig. 6.3c), however, the pyramid of numbers is somewhat different in shape. The producers, which are mainly large-sized trees, are lesser in number and form the base of the pyramid. The herbivores, which are the fruit-eating birds, elephants, deers etc., are more in number than the producers. Then there is a gradual decrease in the number of successive carnivores, thus making the pyramid again upright. However, in a parasitic food chain (Fig. 6.3d) the pyramids are always inverted. This is due to the fact that a single plant may support the growth of many herbivores and each herbivore in turn may provide nutrition to several parasites, which support many hyperparasites. Thus, from the producer towards consumers, there is a reverse position, i.e. the number of organisms gradually shows an increase, making the pyramid inverted in shape.

Actually the pyramids of numbers do not give a true picture of the food chain as they are not very functional. They do not indicate the relative effects of the 'geometric', 'food chain' and 'size' factors of the organisms. They generally vary with different communities with different types of food chains in the same environment. It becomes sometimes very difficult to represent the whole community on the same numerical scale (as in forests).

Pyramids of biomass

They are comparatively more fundamental, as they, instead of geometric factor, show the quantitative relationships of the standing crops. The pyramids of biomass in different types of ecosystem are shown in Fig. 6.4(a-c). In grassland and forest (Fig. 6.4 a, b), there is generally a gradual decrease in biomass of organisms at successive levels from the producers to the top carnivores. Thus pyramids are upright. However, in a pond (Fig. 6.4c) as the producers are small organisms, their biomass is least and this value gradually shows an increase towards the apex of the pyramid, thus making the pyramid inverted in shape.

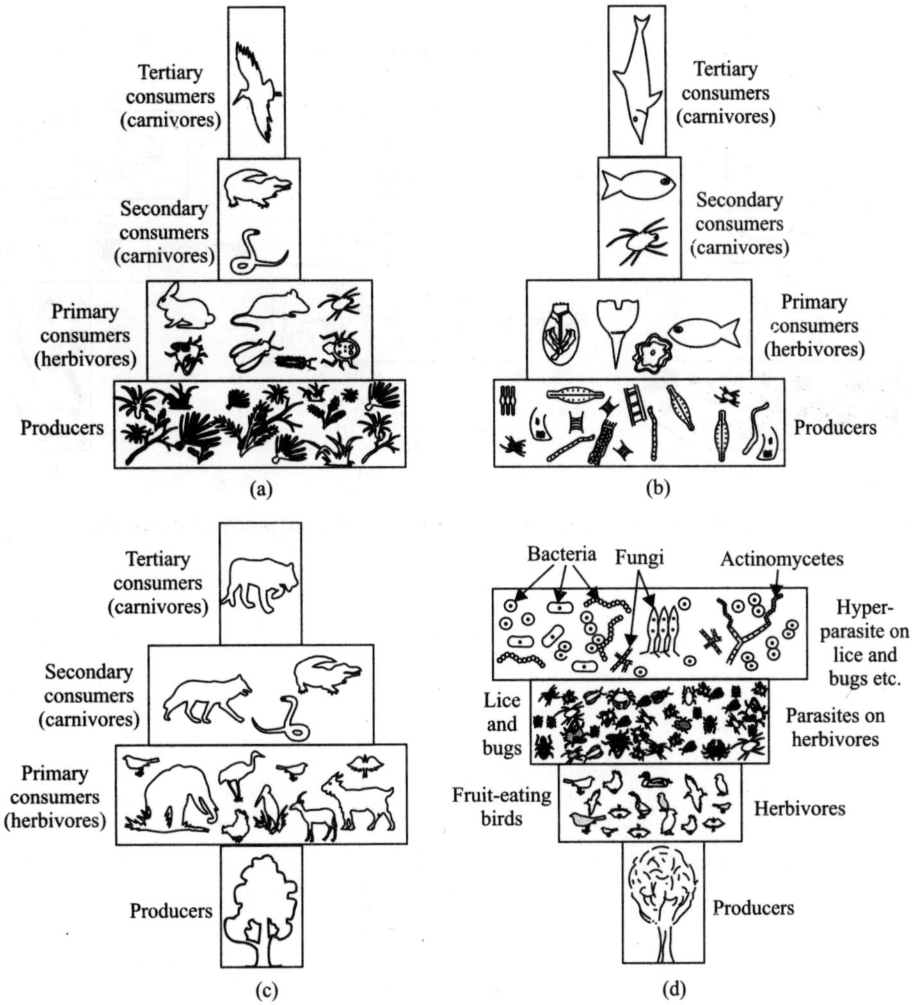

Fig. 6.3. Pyramids of numbers (individuals per unit area) in different kinds of ecosystems/food chains: (a)—grassland ecosystem, (b)—pond ecosystem, (c)—forest ecosystem. In (a)—(c) parasitic micro-organisms and soil animals are not included, (d)—parasitic food chain.

Pyramid of energy

Of the three types of ecological pyramids, the energy pyramids give the best picture of overall nature of the ecosystem. Here, number and weight of organisms at any level depends not on the amount of fixed energy present at anyone time in the level just below but rather on the rate at which food is being produced. In contrast with the pyramids of numbers and biomass, which are pictures of the standing situations (organisms present at any moment), the pyramid of energy (Fig. 6.5) is a picture of the rates of passage of food mass through the food chain. In shape it is always upright, as in most of the cases there is always a gradual decrease in the energy content at successive trophic levels from the producers to various consumers.

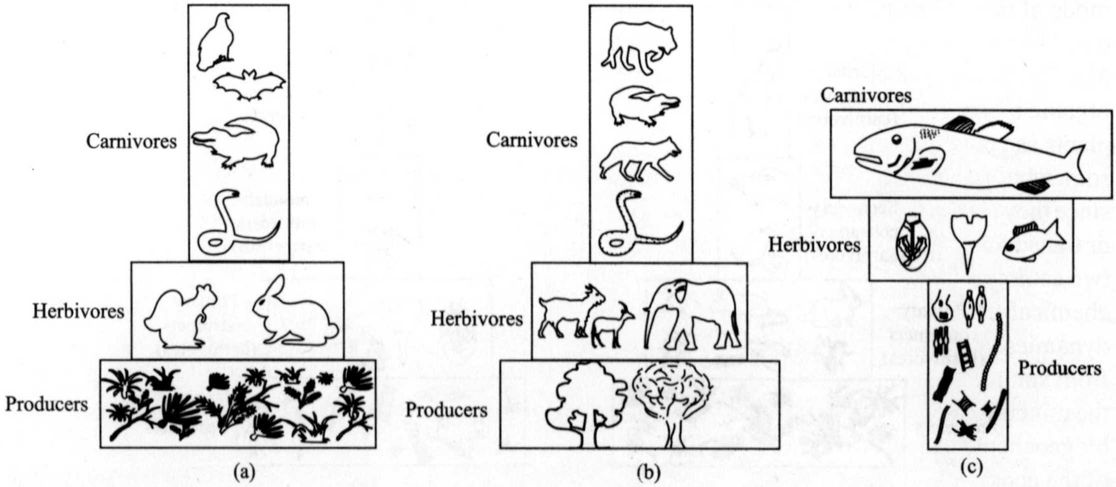

Fig. 6.4(a-c). Pyramids of biomass (g dry wt. per unit area) in different kinds of ecosystems. A— grassland, B—forest, C—pond.

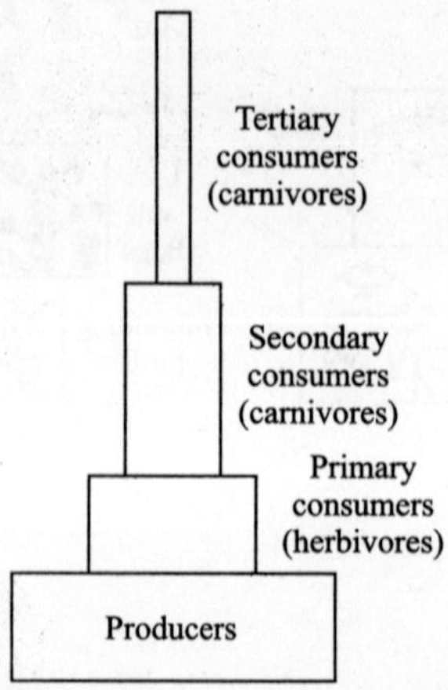

Fig. 6.5. Pyramid of energy (K cal per unit area within unit time, season or years) in any ecosystem.

The species structure includes not only the number and kinds of species but also the diversity of species i.e., the relationship between species and number of individuals or biomass; and the dispersion (spatial arrangement) of individuals of each species present in the community.

Functional Aspects of an Ecosystem

The foregoing account presents the gross structure of an ecosystem. But for a fuller understanding of nature, functions (particularly rate functions) must also be investigated. Both, structure and function, are best studied together and now we shall consider a brief outline of gross function of an ecosystem i.e., how an ecosystem works or operates under natural conditions?

From the operational viewpoint the living and non-living components of ecosystem are so interwoven into the fabric of nature that their separation from each other becomes practically very such difficult. The mode of movement of materials and energy in an ecosystem is shown in simple model presented in Fig. 6.6. The producers, green plants, fix radiant energy and with the help of minerals (C, H, O, N, P, L, Ca, Mg, Zn, Fe etc.) taken from their soil and aerial environment (nutrient pool) they build up complex organic matter (carbohydrates, fats, proteins, nucleic acids etc.). Some ecologists prefer to call the green plants as converters or transducers, since they feel that the term 'producer' from an energy viewpoint is somewhat misleading. Their viewpoint is that green plants produce carbohydrates and not energy and since they convert or transduce radiant energy into chemical form, they must be better called the converters or transducers. However, the term 'producer' is so widely used that it is preferred to retain it as such. The two ecological processes of energy flow and mineral cycling, involving interaction between the physico-chemical environment and the biotic communities, may be thought as the 'heart' of the ecosystem dynamics. According to the model shown in Fig. 6.6, energy flows in non-cyclic manner (unidirectional) from sun to the decomposers via producers and macroconsumers (herbivores and carnivores), whereas the minerals keep on moving in a cyclic manner. The cycling of the minerals is accomplished by different biogeochemical cycles super imposed upon the unidirectional energy flow through the biotic component of the ecosystem. It would become, however, clear later that energy not only flows unidirectionally but also lost from the system in several ways and that minerals too similarly show a net loss in several ways.

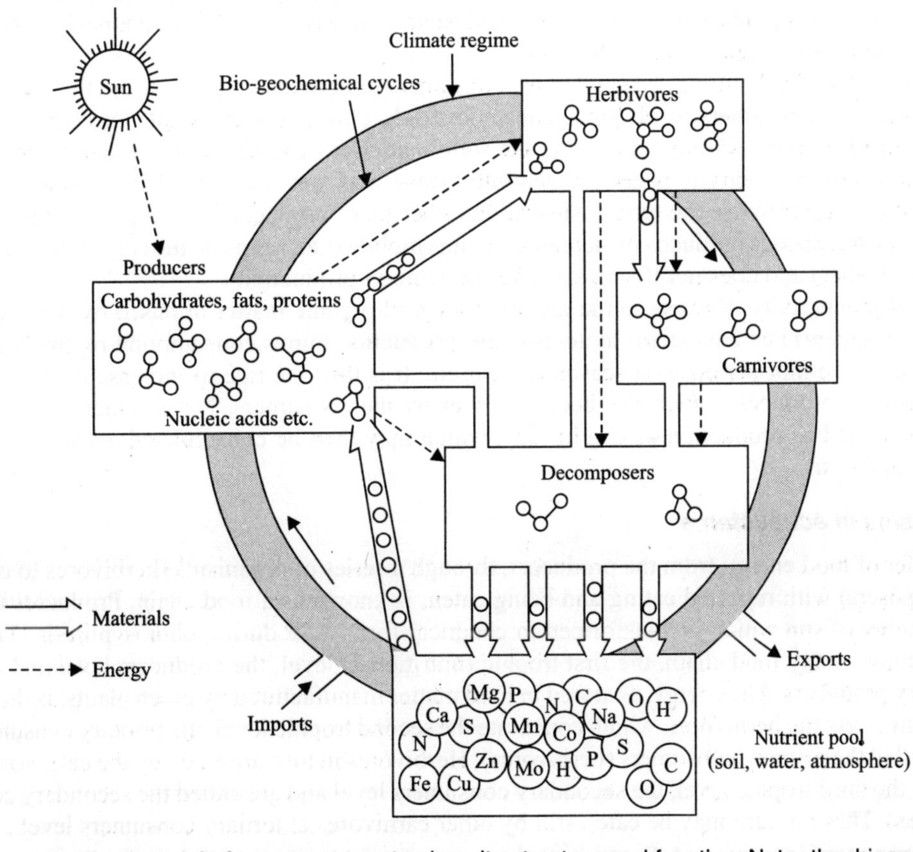

Fig. 6.6. A generalised model of an ecosystem to show its structure and function. Note, the, biogeochemical cycles superimposed on the various components of the ecosystem; movement of materials in cyclic manner and that of energy in non-cyclic (unidirectional) manner.

Productivity of ecosystem

The productivity of an ecosystem refers to the rate of production i.e., the amount of organic matter accumulated in any unit time. Productivity is of the following types:

1. Primary productivity: It is associated with the producers which are autotrophic, most of which are photosynthetic and to a much lesser extent the chemosynthetic micro-organisms. These are the green plants, higher macrophytes as well as lower forms, the phytoplanktons and some photosynthetic bacteria. Primary productivity is defined as 'the rate at which radiant energy is stored by photosynthetic and chemosynthetic activity of producers. Primary productivity is further distinguished as:-
   (a) Gross primary productivity: It is the total rate of photosynthesis including the organic matter used up in respiration during the measurement period. This is also sometimes referred to as total (gross) photosynthesis or total assimilation. It depends on the chlorophyll content. The rate of primary productivity are estimated in terms of either chlorophyll content as, Chl/g dry weight/unit area or photosynthetic number i.e., amount of CO_2 fixed/g Chl/hour.
   (b) Net primary productivity: It is the rate of storage of organic matter in plant tissues in excess of the respiratory utilisation by plants during the measurement period. This is thus the 'rate' of increase of biomass and is also known as apparent photosynthesis or net assimilation. Thus, net primary productivity refers to balance between gross photosynthesis and respiration and other plant losses as death etc.
2. Secondary productivity: It refers to the consumers or heterotrophs. These are the rates of energy storage at consumers level. Since consumers only utilise food materials (already produced) in their respiration, simply converting the food matter to different tissues by an overall process, secondary productivity is not divided into 'gross' and 'net' amounts. Thus, some ecologists as Odum, prefer to use the term assimilation rather than 'production' at this level, the consumers level. Secondary productivity actually remains mobile (i.e., keeps on moving from one organism to another) and does not live *in situ* like the primary productivity.
3. Net productivity: It refers to the rate of storage of organic matter not used by the heterotrophs (consumers) i.e., equivalent to net primary production minus consumption by the heterotrophs during the unit period, as a season or year etc. It is thus the rate of increase of biomass of the primary producers which has been left over by the consumers. Net productivity is generally expressed as production of C g/m^2/day, which may then be consolidated on month, season or year basis.

Food chains in ecosystems

The transfer of food energy from the producers, through a series of organisms (herbivores to carnivores to decomposers) with repeated eating and being eaten, is known as a food chain. Producers utilise the radiant energy of sun which is transformed to chemical form, ATP during photosynthesis. Thus green plants occupy, in any food chain, the first trophic (nutritional) level, the producers level and are called the primary producers. The energy, as stored in food matter manufactured by green plants, is then utilised by the plant eaters the herbivores, which constitute the second trophic level, the primary consumers level and are called the primary consumers (herbivores). Herbivores in turn are eaten by the carnivores, which constitute the third trophic level, the secondary consumers level and are called the secondary consumers (carnivores). These in turn may be eaten still by other carnivores at tertiary consumers level i.e., by the tertiary consumers (carnivores). Some organisms are omnivores eating the producers as well as the carnivores at their lower level in the food chain. Such organisms thus may occupy more than one trophic levels in the food chain. This classification of all the living organisms of any ecosystem is one of their functions and not of species. Species that are taxonomically widely different from each other may occupy the same trophic level as they all have the similar function in the food chain. *Typha*, *Nymphaea*, *Chara*, *Volvox*, *Nostoc*, photosynthetic bacteria, although taxonomically much different but all belong to the same trophic level, the producers level, as all have a common function i.e., the fixation of radiant energy into chemical form. In any food chain, energy flows from primary producers to primary consumers (herbivores), from primary consumers to secondary consumers (carnivores) and from secondary consumers to tertiary consumers (carnivores/omnivores) and so on. This simple chain of eating and being eaten away is known as food chain. A food chain in grassland ecosystem starts with grasses and forbs and goes through grasshoppers, the frogs, the snake, the hawk in an orderly sequential arrangement based on the food habits, whereas in a pond the order would start with phytoplanktons, going through water fleas, smaller fish, bigger fish, birds, larger animal and so on.

In nature, we generally distinguish two general types of food chains:

1. Grazing food chain: This type of food chain starts from the living green plants, goes to grazing herbivores (that feed on living plant materials with their predators) and on to carnivores (animal eaters). Ecosystems with such type of food chain are directly dependent on an influx of solar radiation. This type of chain thus depends on autotrophic energy capture and the movement of this captured energy to herbivores. Most of the ecosystems in nature follow this type of food chain. From energy standpoint, these chains are very important. The phytoplanktons—zooplanktons—fish sequence or the grasses rabbit—fox sequence are the examples of grazing food chain.
2. Detritus food chain: This type of food chain goes from dead organic matter into micro-organisms and then to organisms feeding on detritus (detritivores) and their predators. Such ecosystems are thus less dependent on direct solar energy. These depend chiefy on the influx of organic matter produced in another system. For example, such type of food chain operates in the decomposing accumulated litter in a temperate forest. A good example of a detritus food chain is based on mangrove leaves. In the brackish zone of Southern Florida, leaves of the red mangrove–*Rhizophore mangle* fall into the warm, shallow waters. Only 5 per cent of the leaf material was removed by grazing insects before leaf fall. As shown in Fig. 6.7a the fallen leaf fragments (acted on by such saprotrophs as fungi, bacteria, protozoa etc., and colonised mainly by phytoplanktonic and benthic algae) are eaten and re-eaten (coprophagy) by a key group of small animals. These animals include crabs, copepods, insect larvae, grass shrimps, mysids, nematodes, amphipods, bivalve molluscs etc. All these animals are detritus consumers. These detritivores are the key group of small animals, comprising only a few species but very large number of individuals. They ingest large amounts of the vascular plant detritus. These animals are in turn eaten by some minnows and small game fish etc., i.e., the small carnivores, which in turn serve as the main food for larger game fish and fish eating birds which are the large (top) cornivores. The mangroves considered generally as of less economic value make a substantial contribution to the food chain that supports the fisheries, an important economy in that region. Similarly, detritus from seagrasses, saltmarsh grasses and seaweeds support fisheries in many estuarine areas.

Thus, we see that the detritus food chain ends up in a manner similar to the grazing food chain (big fish eat little fish), but the way in which the two chains begin is quite different. As shown in Fig. 6.7b, in detritus chain, the detritus consumers, in contrast to grazing herbivores, are a mixed group in terms of trophic levels. These include herbivores, omnivores and primary carnivores. As a group, the detritus feeders obtain some of their energy directly from plant material, most of it secondarily from micro-organisms and some tertiarily through carnivores (for example by eating protozoa or other small invertebrates that have fed on bacteria that have digested plant material).

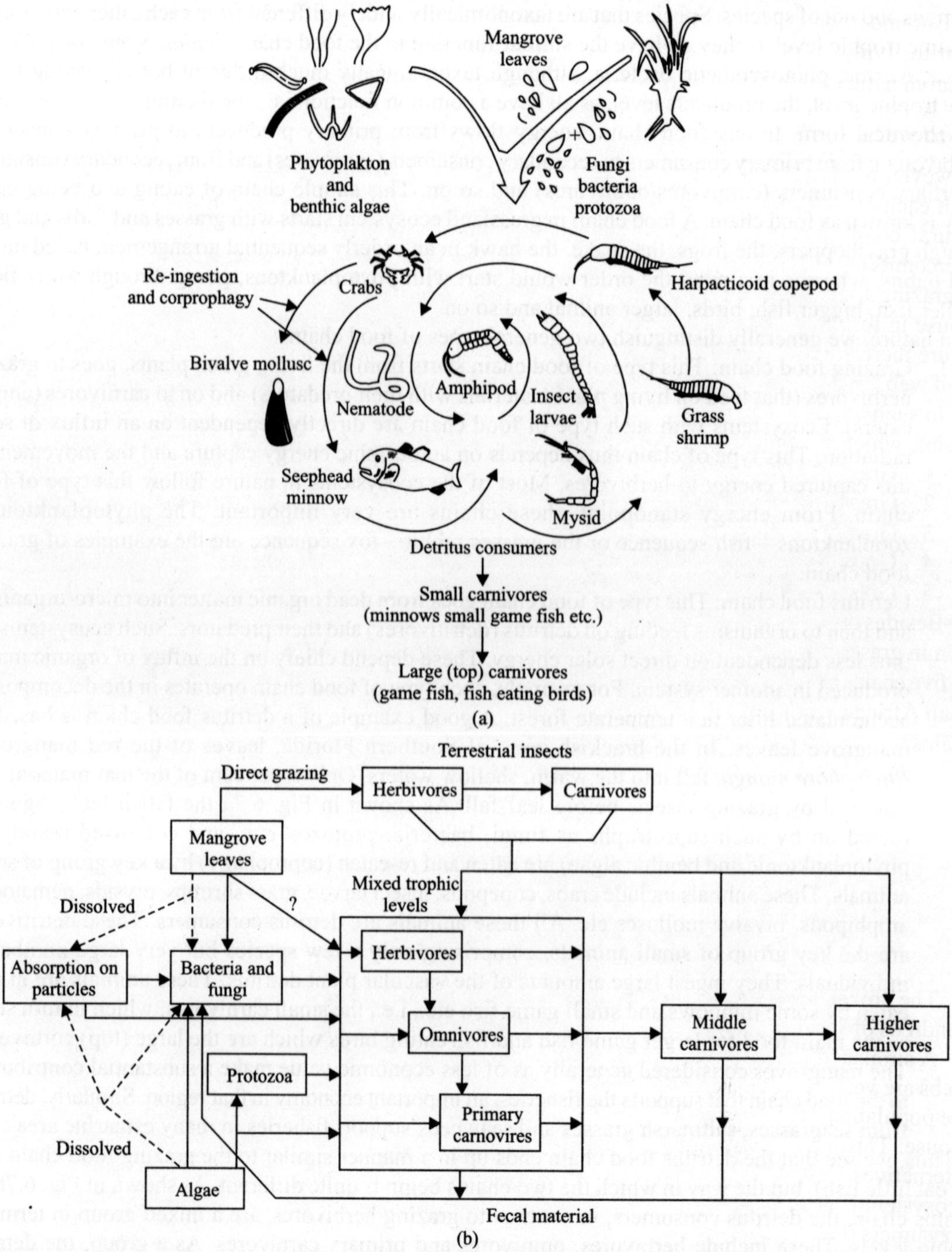

Fig. 6.7. A detritus food chain based on mangroves leaves falling into shallow estuary waters. Leaf fragments acted on by the saprotrophs and colonised by algae are eaten and re-eaten (coprophagy) by a key group of small detritivores which in turn, provide the main food for game fish, herons, storks and ibis. A-a 'pictorial model' of the food chain, B-a 'compartmental model'.

But under natural situations, a system must always be self sufficient. In fact this type of food chain (detritus type) is simply a sub-component of another ecosystem. And the above said two types of food chain in nature are indeed linked together belonging to the same ecosystem.

Food webs—interlocking pattern of organisms

However, food chains in natural conditions never operate as isolated sequences, but are interconnected with each other forming some sort of interlocking pattern, which is referred to as a food web. Under natural conditions, the linear arrangement of food chains, hardly occurs and these remain indeed interconnected with each other through different types of organisms at different trophic levels. For example, in grazing food chain of a grassland, in the absence of rabbit, grass may also be eaten by mouse. The mouse in turn may be eaten directly by hawk or by snake first which is then eaten by hawk. Thus, in nature there are found alternatives, which all together constitute some sort of interlocking pattern—the food web.

In such a food web in grassland, as shown in Fig. 6.8, there may be seen as many as five linear food chains, which in sequences are:

1. Grass → Grasshopper → Hawk.
2. Grass → Grasshopper → Lizard → Hawk.
3. Grass → Rabbit → Hawk (or vulture or fox or even man, if present).
4. Grass → Mouse → Hawk.
5. Grass → Mouse → Snake → Hawk.

Besides those shown in Fig. 6.8, there may also be present some other consumers as vultures, fox and man in grasslands and if so, the food web may be even more complex than shown here. However, these all five chains are interlinked with each other at different points, forming food web. Similarly the population in an African grassland may be characterised by their feeding relationships in the following two food chains:

(1) Live grass (producer) ⟶ Wildebeest (herbivore) ⟶ Lion (primary cornivore in chain 1)

(2) Dead grass (dead organic matter) ⟶ Termite (decomposer) ⟶ Aardvark (primary cornivore)

(broken arrows: Live grass ⇢ Dead grass; Aardvark ⇢ Lion)

The broken arrows show how the two chains are linked into a food web. Real food webs usually have hundreds of species interlinked by their feeding habits. Food webs are basic units of ecosystem ecology.

A similar food web in a pond, with different interlinked food chains is shown in Fig. 6.9. The food webs are very important in maintaining the stability of an ecosystem in nature. For example, decrease in the population of rabbit would naturally cause an increase in the population of alternative herbivore, the mouse. This may decrease the population of the consumer (carnivore) that prefers to eat rabbit. Thus alternatives (substitutes) serve for maintenance of stability of the ecosystem, Moreover, a balanced ecosystem is essential for the survival of all living organisms of the system. For instance, had primary consumers (herbivores) not been in nature, the producers would have perished due to overcrowding and competition. Similarly, the survival of primary consumers is linked with the secondary consumers (carnivores) and so on. Thus, each species of any ecosystem is indeed kept under some sort of a natural check so that the system may remain balanced.

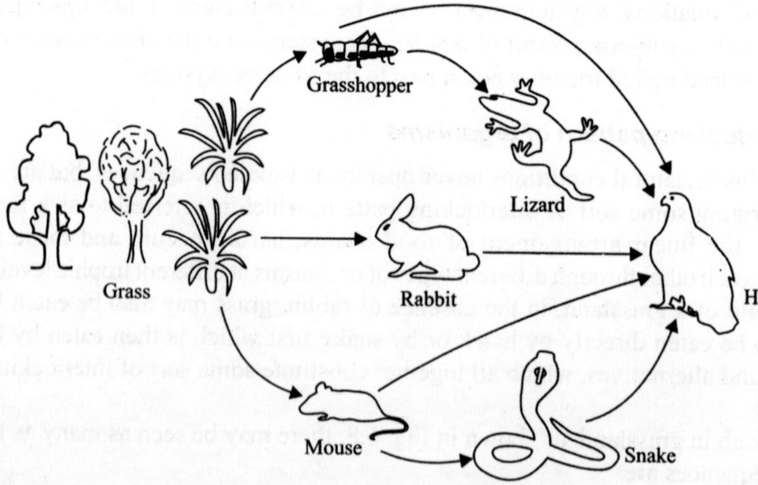

Fig. 6.8. Diagrammatic sketch showing a food web in a grassland ecosystem. There may be seen five possible food chains interlocked together making the food web.

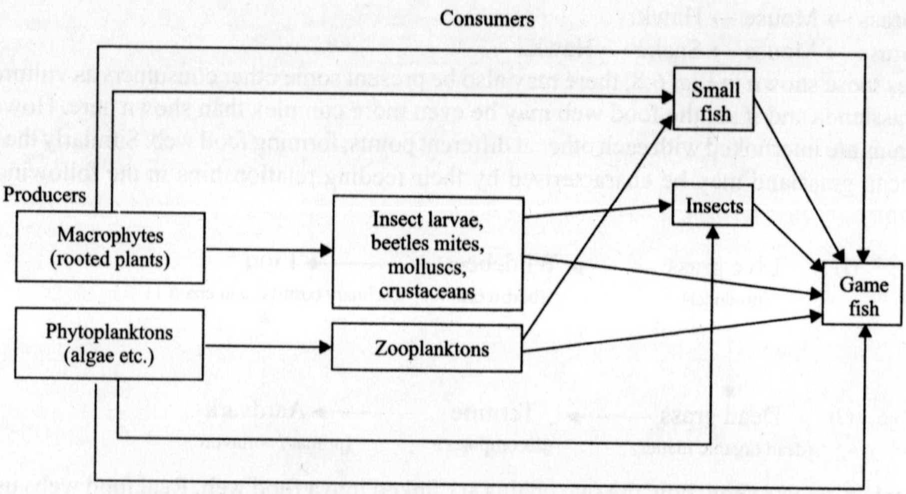

Fig. 6.9. Diagrammatic sketch showing the food web in a pond.

The complexity of any food web depends upon the diversity of organisms in the system. It would accordingly depend upon two main points: (i) length of the food chain. Diversity in the organisms based upon their food habits would determine the length of food chain. More diverse the organisms in food habits, more longer would be food chain; (ii) alternatives at different points of consumers in the chain. More the alternatives, more would be the interlocking pattern. In deep oceans, seas etc. where we find a variety of organisms, the food webs are much complex.

ECOLOGICAL ENERGETICS

The energy used for all plant life processes is derived from solar radiations. A fraction i.e., about 1/50 million of the total solar radiation reaches the Earth's atmosphere. Solar radiation travels through the space in the form of waves, wavelength ranging from 0.03Å to several km. While most radiations are lost in space, those ranging from 300 mμ to 10μ and above 1 cm (radiowaves) enter the Earth's outer atmosphere (which is about 18 miles or 28 km altitude). The energy reaching the Earth's surface consists largely, of visible light (390–760 mμ) and infrared component. On a clear day radiant energy reaching the Earth's surface is about 10 per cent UV, 45 per cent visible and 45 per cent infra-red. Plants absorb strongly the blue and red light (400–50 mμ and 600–700 mμ respectively).

In ecological energetics, we study: (i) quantity of solar energy reaching an ecosystem; (ii) quantity of energy used by green plants for photosynthesis; and (iii) the quantity and path of energy flow from producers to consumers.

About 34 per cent of the sunlight reaching the Earth's atmosphere is reflected back into its atmosphere, 10 per cent is held by ozone layer, water vapour and other atmospheric gases. The rest, 56 per cent reaches the Earth's surface.

Only a fraction of this energy reaching the Earth's surface (1 to 5 per cent) is used by green plants for photosynthesis and the rest is absorbed as heat by ground vegetation or water. In fact, only about 0.02 per cent of the sunlight reaching the atmosphere is used in photosynthesis. Nevertheless, it is this small fraction on which all the organisms of the ecosystem depend.

Energy Flow in Ecosystem

The behaviour of energy in ecosystem can be termed energy flow due to unidirectional flow of energy. From energetics point of view it is essential to understand for an ecosystem: (i) the efficiency of the producers in absorption and conversion of solar energy; (ii) the use of this converted chemical form of energy by the consumers; (iii) the total input of energy in form of food and its efficiency of assimilation; (iv) the loss through respiration, heat, excretion etc.; and (v) the gross net production.

Single-channel energy models

The principle of food chains and the working of the two laws of thermodynamics can be better made clear by means of energy flow diagrams shown in Figs. 6.10 and 6.11.

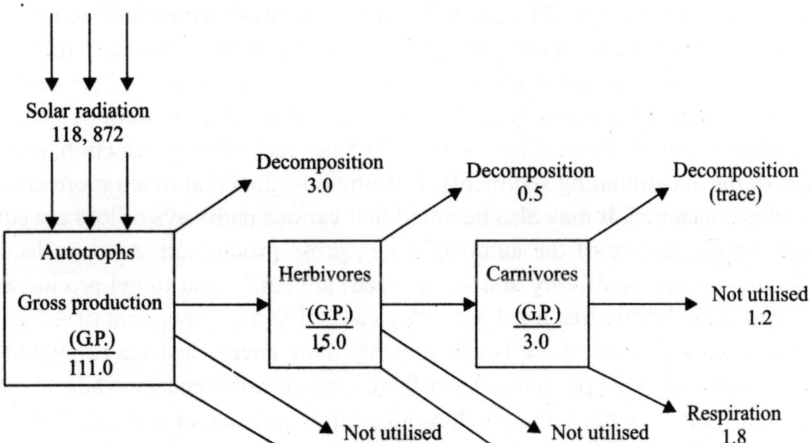

Fig. 6.10. Energy flow diagram for a lake (freshwater ecosystem) in g cal/cm^2/year.

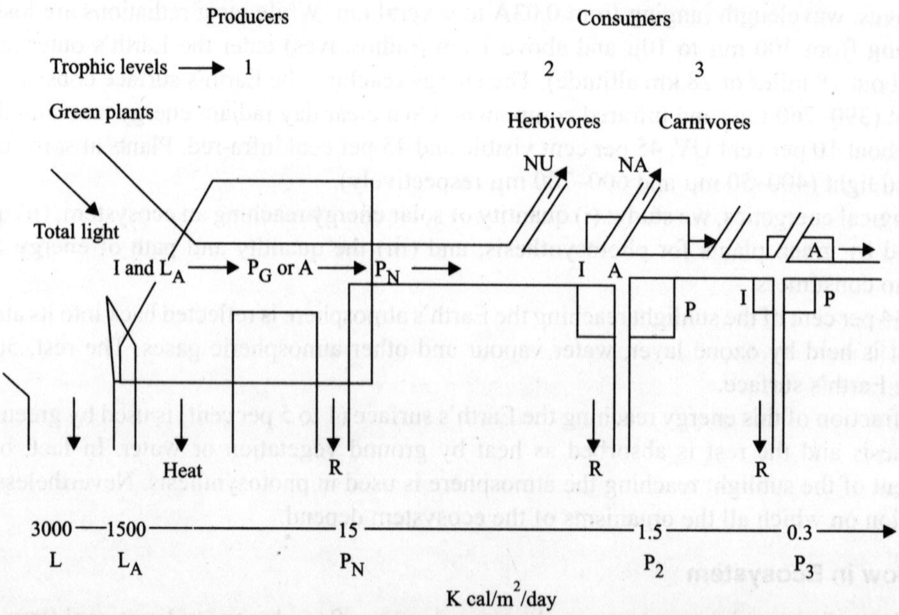

Fig. 6.11. A simplified energy flow diagram depicting three trophic levels (boxes numbered 1,2,3) in a linear food chain. I—total energy input; LA—light absorbed by plant cover. PG—gross primary production; A—total assimilation; PN—net primary production; P—secondary (consumer) production; NU—energy not used (stored or exported); NA—energy not assimilated by consumers (egested); R—respiration. Bottom line in the diagram shows the order of the magnitude of energy losses expected at major transfer points, starting with a solar input of 3000 K cal per square meter per day.

As shown in Fig. 6.10 out of the total incoming solar radiation (1,18,872 g cal/cm^2/year), 1,18,761 g cal/cm^2/year remain unutilised and thus gross production (net production plus respiration) by autotrophs is 111 gcal/cm^2/year with an efficiency of energy capture of 0.10 per cent. It may also be noted that 21 per cent of this energy or 23 g cal/cm^2/year is consumed in metabolic reactions of autotrophs for their growth, development, maintenance and reproduction. It may be seen further that 15 g cal/cm^2/year are consumed by herbivores that graze or feed on autotrophs—this amounts to 17 per cent of net autotroph production. Decomposition (3 gcal/cm^2/year) accounts for about 3.4 per cent of net production. The remainder of the plant material, 70 g cal/cm^2/year or 79.5 per cent of net production, is not utilised at all but becomes part of the accumulating sediments. It is obvious, then that much more energy is available for herbivory than is consumed. It may also be noted that various pathways of loss are equivalent to and account for total energy capture of the autotrophs i.e., gross production. Also, collectively the three upper 'fates' (decomposition, herbivory and not utilised) are equivalent to net production. Of the total energy incorporated at the herbivores level, i.e., 15 g cal/cm^2/year, 30 per cent or 4.5 g cal/cm^2/year is used in metabolic reactions. Thus, there is considerably more energy lost via respiration by herbivores (30 per cent) than by autotrophs (21 per cent). Again there is considerable energy available for the carnivores, namely 10.5 g cal/cm^2/year or 70 per cent, which is not entirely utilised; in fact only 3.0 g cal/cm^2/year or 28.6 per cent of net production passes to the carnivores. This is more efficient utilisation of resources than occurs at autotroph→herbivore transfer level. At the carnivore level about 60 per cent of the carnivores' energy intake is consumed in metabolic activity and the remainder becomes part of the not utilised sediments; only an insignificant amount is subject to decomposition yearly. This high respiratory loss compares with 30 per cent by herbivores and 21 per cent by autotrophs in this ecosystem.

From the energy flow diagram shown in Fig. 6.10, two things become clear. Firstly, there is one-way street along which energy moves (unidirectional flow of energy). The energy that is captured by the autotrophs does not revert back to solar input; that which passes to the herbivores does not pass back to the autotrophs. As it moves progressively through the various trophic levels it is no longer available to the previous level. Thus due to one-way flow of energy, the system would collapse if the primary source, the Sun, were cut off. Secondly, there occurs a progressive decrease in energy level at each trophic level. This is accounted largely by the energy dissipated as heat in metabolic activities and measured here as respiration coupled with unutilised energy.

In Fig. 6.11 the 'boxes' represent the trophic levels and the 'pipes' depict the energy flow in and out of each level. Energy inflows balance outflows as required by the first law of thermodynamics and energy transfer is accompanied by dispersion of energy into unavailable heat (i.e., respiration) as required by the second law. Figure 6.11 presents a very simplified energy flow model of three trophic levels, from which it becomes evident that the energy flow is greatly reduced at each successive trophic level from producers to herbivores and then to carnivores. Thus at each transfer of energy from one level to another, major part of energy is lost as heat or other form. There is a successive reduction in energy flow whether we consider it in terms of total flow (i.e., total energy input and total assimilation) or secondary production and respiration components. Thus, of the 3000 Kcal of total light falling upon the green plants, approximately 50 per cent (1500 Kcal) is absorbed, of which only 1 per cent (15 Kcal) is converted at first trophic level. Thus net primary production is merely 15 Kcal. Secondary productivity (P_2 and P_3 in the diagram) tends to be about 10 per cent at successive consumer trophic levels i.e., herbivores and the carnivores, although efficiency may be sometimes higher, as 20 per cent, at the carnivore level as shown (or $P_3 = 0.3$ Kcal) in the diagram.

It becomes evident from Figs. 6.10 and 6.11 that there is a successive reduction in energy flow at successive trophic levels. Thus shorter the food chain, greater would be the available food energy as with an increase in the length of food chain there is a corresponding more loss of energy.

As shown in Fig. 6.11, it is natural to argue that with a reduction in energy flow (shown as 'pipes' in the diagram) at each successive trophic level, there is also a corresponding decrease in standing crop or biomass (shown as 'boxes' in the diagram). But it never means that there exists any correlation between the biomass and energy. Indeed energy as taken here represents rate functions, say production rates. The relationships between biomass and energy content may differ according to the situations. For example, one gram of an alga (lesser biomass) may be equal to many grams (more biomass) of a forest tree leaves, due to the fact that rate of production (metabolism rates i.e., rates at which radiant energy is fixed to form carbohydrates) of the alga is higher than that of tree leaves. Thus standing crop or biomass (structure) and productivity (rate functions) should not be confused with each other and the two are not necessarily correlated. Organism with higher biomass does not necessarily mean it would be more productive too.

Y-shaped energy flow models

Figures 6.10 and 6.11 describe simple, single-channel energy flow diagrams. There are similar energy flow models for different kinds of ecosystems. For instance Teal prepared an energy flow diagram of Root Spring in USA. Similarly Odum prepared energy flow model for Silver Springs, Florida, USA. In Silver Springs, there is 4,10,000 Kcal/m^2/year effective solar radiation falling on green plants for photosynthesis. Of this 3,89,190 Kcal/m^2/year was lost as heat; only 20810 Kcal/m^2/year could be used in gross primary production. Again, out of this 11977 Kcal/m^2/year was lost by way of respiration and only 8833 Kcal/m^2/year remained for net production. Thus, energy followed again as in Figs. 6.10 and 6.11 three main paths—transfer to next trophic level (producer to consumer); heat loss as respiration; and through death to decomposers. In these two diagrams, there were basic differences. In model given by Teal for Root Spring, most of the energy rich material eaten by heterotrophs entered the system as plant debris (detritus), whereas in that given by Odum for Silver Springs, most of the heterotroph's food in food chain was produced by green autotrophic plants. Odum noted this basic feature that to begin with in some systems heterotrophs consume living plants while in others they feed on dead plant parts (detritus). Thus in this case in Root Springs, the chain began with dead plant parts, whereas in Silver Springs with live plant parts. Odum pointed out thus in nature that there are present two basic food chains in any system. We have already considered these two kinds of food chains: (i) the grazing food chain beginning with green plant base going to herbivores and then to carnivores; and (ii) the detritus food chain beginning with dead organic matter acted by microbes, then passing to detritivores and their consumers (predators). Figure 6.12 presents one of the first published energy flow models as pioneered by Odum. There is shown a common boundary and, in addition to light and heat flows, the import, export and storage of organic matter are also included. Decomposers are placed in a separate box as a means of partially separating the grazing and detritus food chains. Decomposers are in fact a mixed group in terms of energy levels.

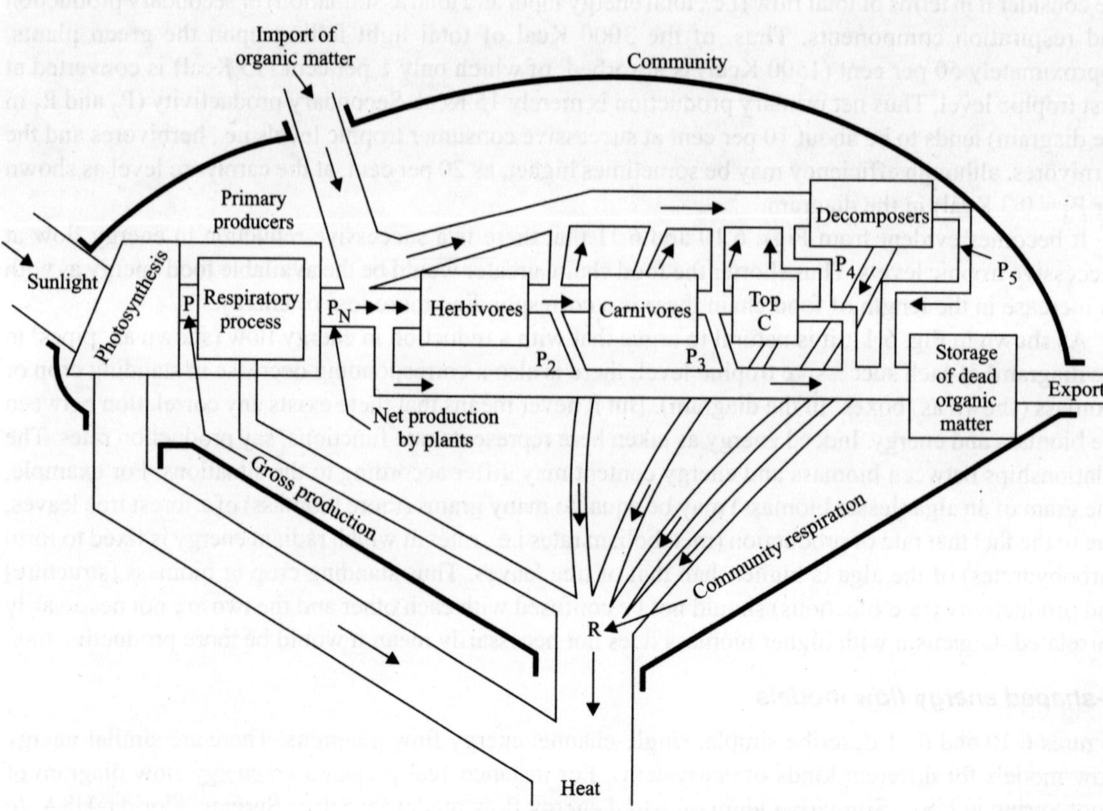

Fig. 6.12. The relationship between flow of energy through the grazing food chain and detritus pathway.

Figure 6.13 shows two Y-shaped or 2-channel energy flow models. In each Y-shaped model one arm represents the herbivore food chain and the other, the decomposer (detritus) food chain. The two arms differ fundamentally in the way in which they can influence primary producers. In each model the grazing and detritus food chains are sharply separated.

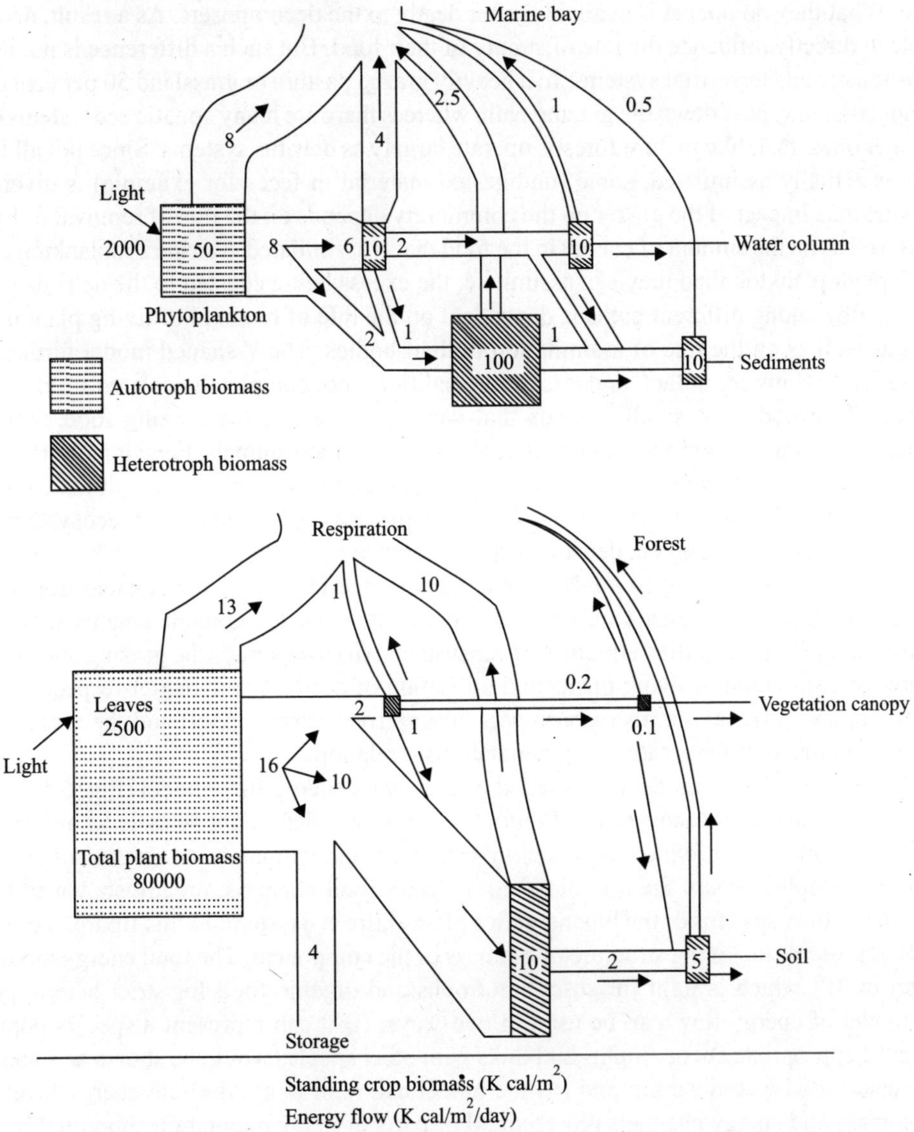

Fig. 6.13. A Y-shaped or 2-channel energy flow diagram that separates a grazing food chain (water column or vegetation canopy) from a detritus food chain (sediments and in soil). Estimates for standing crops (shaded boxes) and energy flows compare a hypothetical coastal marine ecosystem (upper diagram) with a hypothetical forest (lower diagram).

This figure (Fig. 6.13) contrasts the biomass-energy flow relationships in the sea and the forest. In the marine bay the energy flow via the grazing food chain is shown to be larger than via the detritus pathway (forest), whereas the reverse is shown for the forest in which 90 per cent or more of the net primary production is normally utilised in detritus food chain. Thus in marine ecosystem the grazing food chain is the major pathway of energy flow whereas in the forest ecosystem the detritus food chain is more important. In grazing chain herbivores feed on living plants and therefore, directly affect the plant population. What they do not eat is available, after death, to the decomposers. As a result, decomposers are not able to directly influence the rate of supply of their food. But such a difference is not necessarily inherent in aquatic and terrestrial systems. In a heavily grazed pasture or grassland 50 per cent or more of the net production may pass down the grazing path, whereas there are many aquatic ecosystems especially shallow water ones, that, like mature forests, operate largely as detritus systems. Since not all food eaten by grazers is actually assimilated, some (undigested material in feces for example) is diverted to the detritus route; thus impact of the grazer on the community depends on the rate of removal of living plant material as well as on the amount of energy in the food that is assimilated. Marine zooplankton commonly graze more phytoplankton than they can assimilate, the excess being egested to the detritus food chain. Thus, energy flow along different paths is dependent on the rate of removal of living plant material by herbivores as well as on the rate of assimilation in their bodies. The Y-shaped model further indicates that the two food chains are in fact, under natural conditions, not completely isolated from one another. For instance, dead bodies of small animals that were once part of the grazing food chain become incorporated in the detritus food chain as do the feces of grazing food animals. Functionally, the distinction between the two is of time lag between the direct consumption of living plants and ultimate utilisation of dead organic matter. The importance of the two food chains may differ in different ecosystems, in some grazing is more important, in others detritus is major pathway.

The important point in Y-shaped model is that the two food chains are not isolated from each other. This Y-shaped model is more realistic and practical working model than the single-channel model because: (i) it confirms to the basic stratified structure of ecosystems; (ii) it separates the grazing and detritus food chains (direct consumption of living plants and utilisation of dead organic matter respectively) in both time and space; and (iii) that the microconsumers (absorptive bacteria, fungi) and the macroconsumers (phagotrophic animals) differ greatly in size-metabolism relations.

Odum gave a generalised model of Y-shaped or 2-channel energy flow model (Fig. 6.14) applicable to both terrestrial and aquatic ecosystems. Figure 6.15 presents what might be called a universal model, one that is applicable to any living component, whether a plant, animal, micro-organism or individual, population or a trophic group. Such a model may depict food chain as already shown in single and Y-shaped energy flow systems or the bioenergetics of an entire ecosystem. In this figure, the shaded box labelled 'B' represents the living structure or biomass of the component. The total energy input or intake is indicated by 'I', which is light for strict autotrophs and organic food for strict heterotrophs. This universal model of energy flow can be used in two ways: (i) it can represent a species population in which case the appropriate energy inputs and links with other species would be shown as a conventional species-oriented food-web diagram; and (ii) the model can represent a discrete energy level in which case the biomass and energy channels represent all or parts of many population supported by the same energy source. Foxes, for example, usually obtain part of their food by eating plants (fruit etc.) and part by eating herbivores (rabbit, field mice etc.). A single box diagram could be used to represent the whole population of foxes if our objective is to stress intra-population energetics. On the other hand, two or more boxes (as shown on the right of Fig. 6.15) would be employed if we wish to separate the metabolism of a population into two or more trophic levels in accordance with the proportion of plant and animal consumed.

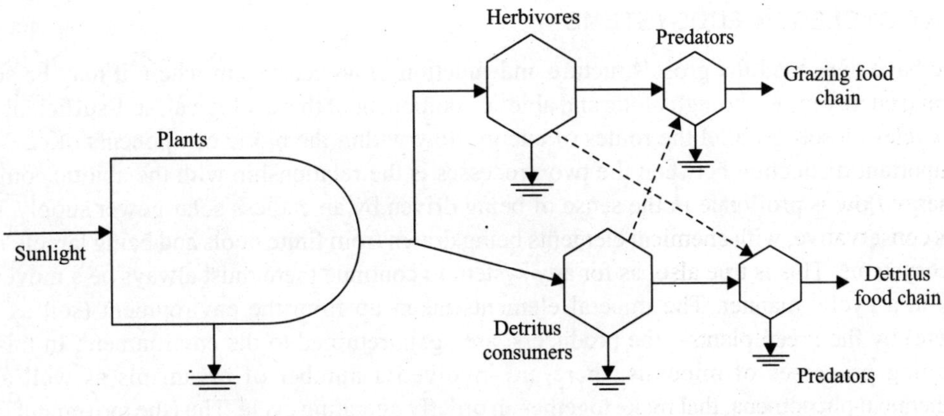

Fig. 6.14. The Y-shaped energy flow model showing linkage between the grazing and detritus food chains.

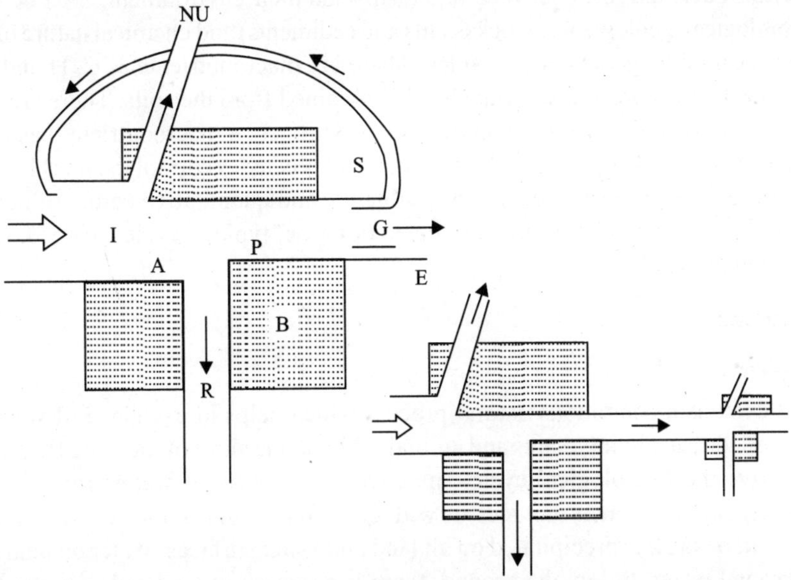

Fig. 6.15. Components for a 'universal' model of energy flow. I = input or ingested energy; NU = not used; A = assimilated energy; P = production; R = respiration; B = biomass; G = growth; S = stored energy; E = excreted energy.

It must however, be remembered that these models depict the basic pattern of energy flow in ecosystem. In practice, under natural conditions, the organisms are interrelated in a way that several food chains become interlocked and this results into a complex food web. We have already referred to food webs in a grassland and in pond ecosystems. The complexity of food web depends on the length of the food chains. Thus in nature there operate multi-channel energy flows, but in these the channels belong to either of the two basic food chains i.e., will be either a grazing or a detritus food chain. Interlocking pattern of such several chains in food web of an ecosystem would lead to a multi-channel flow of energy. Thus in practice, under field conditions, we might face difficulties in measuring the energetics of ecosystem.

NUTRIENT CYCLES IN ECOSYSTEMS

So far we have discussed the gross structure and function of an ecosystem where it may be seen that minerals and energy move through biotic and abiotic component of this ecological, self-sufficient system. Nutrient cycles closely parallel the routes of energy flow within the biotic components of ecosystems, but an important distinction between the two processes is the relationship with the abiotic component. While energy flow is profligate in the sense of being driven by an endless solar power supply, nutrient cycling is conservative, with chemical elements being drawn from finite pools and being largely retained within ecosystems. This is true also, as for any system to continue there must always be a movement of materials in a cyclic manner. The mineral elements taken up from the environment (soil as well as atmosphere) by the green plants—the producers, are again returned to the environment. In this taking and returning processes of minerals, there are involved a number of organisms as well as some physicochemical phenomena, that make together an orderly operating cycle. Thus the movement (imports and exports) of minerals is accomplished by the operation of different chemical cycles that keep on passing the materials back and forth between organisms and their environment.

To stress the biological, geological (in rocks, soils and sediments) and chemical nature of the processes, they are sometimes called biogeochemical cycles. There are macronutrients as C, H and O which have cycles with an atmospheric store and some as P and K obtained from the soils. There are micronutrients as Cu, Fe, Co, which have soil-based or edaphic cycles. Nitrogen, a macronutrient, has inorganic pools in both the atmosphere and soil. A variety of biogeochemical processes control the links between available and unavailable forms of edaphic nutrients. The qualitative and quantitative nature of nutrient cycles in natural ecosystems is very diverse. It is not possible to describe 'typical' cycles. Indeed, very few cycles have been elaborated in detail.

Atmospheric Cycles

Hydrological cycle

The hydrological cycle is a continuous natural process which helps in exchange of water between the atmosphere, the land, the sea, living plants and animals. About one-third of the solar flux absorbed by the earth is used to drive the hydrological cycle—massive evaporation of water from the oceans, cloud formation and precipitation—which provides us with our supply and reserves of fresh water.

Water as rain, hail or snow is precipitated on all land and water surfaces. Water on land surfaces seeps into the soil as ground water. Below the ground, there is a natural water level or water table. The soil below the water table is sustained by the underlying clay and rock strata. Ground water does not remain stationary, but moves in various directions, water moves up above the water table by capillary action and thereby maintains a continuous supply of water to the surface layer of soil, where it is absorbed by plant roots in absence of rain. Another important underground water resource is the aquifers. The latter exist above impermeable rock strata—water percolates through the porous rocks and forms underground lakes or reservoirs. From these aquifers water can be extracted by sinking wells, tube wells and pumping it to the surface.

In the event of good rainfall, all the rainwater precipitated on land does not percolate into the soil. Surface water or run-off flows into streams, rivers, lakes and catchment areas or reservoirs. The land surface and all water surfaces on the earth lose water by evaporation by solar energy. Normal evaporation from the oceans exceeds precipitation by rain into the seas by about 10 per cent. This excess 10 per cent ultimately moves as water vapour over land surface and thereby balances the hydrological cycle and meets our requirements for additional water. Plants absorb capillary ground water, but give off excess water through leaves by the process of transpiration. This is an important process which ensures conduction of water and dissolved mineral salts throughout the plant.

Hence the hydrological cycle consists of a balanced continuous process of evaporation, transpiration, precipitation, surface run-off and ground water movements. A rough idea of the quantum of annual rainfall and evaporation may be given. Each year approx. 50 Tm (1 Tm = 1 Terra Cubic metre = 10^{12} i.e., one million cubic metres; 1 cubic metre or 1 m^3 = 1000 liters) water is evaporated and the same quantity of water is precipitated over the earth's surface. From the land about 44.5 Tm^3 water flows to the sea per year and this is available for man's water supplies.

The *hydrologic cycle* consists of five parts which provide the global water supply (Fig. 6.16). About one-third of the solar flux absorbed by the earth's surface is used to drive the hydrologic cycle—massive evaporation of water from the oceans, cloud formation and precipitation—which provides us with our supply and reserves of fresh water.

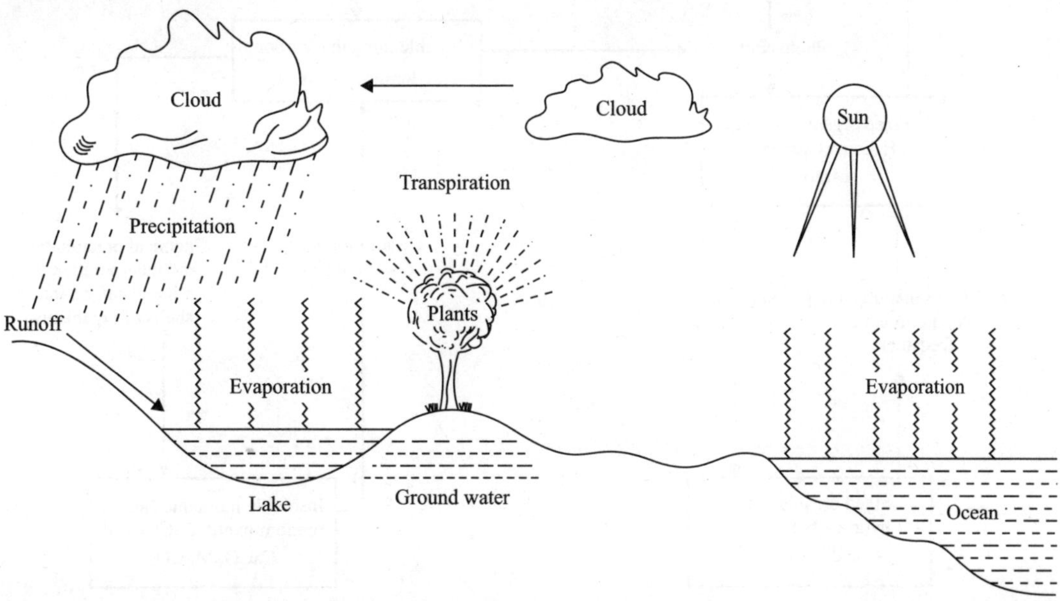

Fig. 6.16. Global hydrological cycle. Most of the storage is in the oceans, but most of the flux is to and from the atmosphere. Units: stores, 10^{18}g: fluxes, 10^{18}g a^{-1}.

Carbon cycle

The simplest of nutrient cycles is the carbon cycle where general components are well recognised. Carbon is returned to the atmosphere as fast as it is removed. Carbon moves in the gaseous phase (CO_2) from the atmosphere to producers. Here, it enters the food chain and is respired out at each trophic level and finally released during decomposition, thus returning to the atmosphere. More than the amount of CO_2 getting locked up in the biomass, is now being thrown into the atmosphere by the processes of combustion of fossil fuels, burning of wood and volcanic activities. The vast ocean helps greatly in maintaining the level of CO_2 in the atmosphere by absorbing and releasing the gas when its levels are high and low, respectively. The atmospheric concentration of carbon dioxide is 0.03–0.04 per cent while oceanic water can retain up to fifty times this level and hence is capable of regulating carbon dioxide in the atmosphere.

Carbon dioxide from the atmosphere is converted into organic compounds in plants through photosynthesis. Organic carbon from plants may go into animals. There it goes through various stages of digestion and assimilation. It may re-enter the atmosphere from plants or animals as CO_2 by oxidation or decomposition of dead organic matter. In animals, carbon may get tied up in hard parts such as shells and hence remain in the form of inorganic carbonates for long. Marine deposits of animal carbonates and inorganic precipitation of carbonates in water may result in limestone. Limestone can then return carbonates very slowly to the living carbon cycle through erosion and dissolution. Carbon may also get locked into organic deposits of coal and petroleum, where it may remain for million of years unless released by combustion (Fig. 6.17).

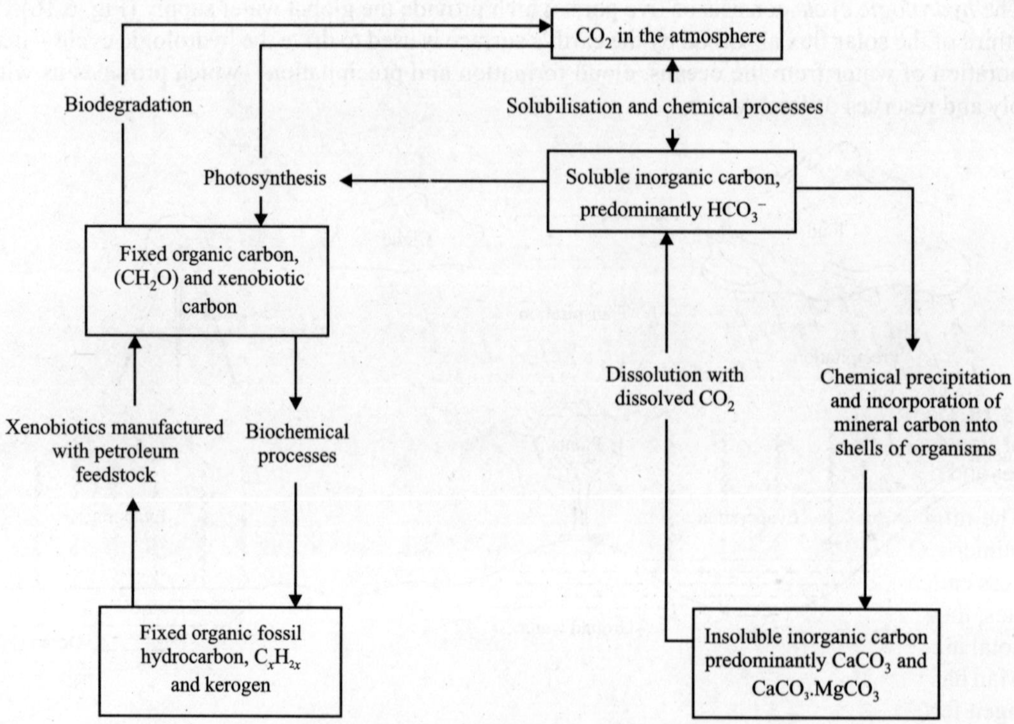

Fig. 6.17. Global carbon cycle.

Edaphic Nutrient Cycles

Terrestrial plants obtain nutrients, other than carbon, in solution from the soil. 'Macro-nutrients' are N, P, K, Ca, Mg and S. Trace elements or micronutrients are Cu, Zn, Mn, Fe, Bo, Mo and Co.

Nitrogen cycle

Nitrogen in its gaseous form constitutes 79 per cent of the atmosphere. However, it cannot be used directly by most forms of life. It must first be 'fixed' before it can be utilised by plants and animals. By fixation, nitrogen is converted into its chemical compounds, largely nitrates (NO_2) and ammonia (NH_3). The fixation of nitrogen takes place through both physiochemical and biological means, although the latter is by far the much bigger contributor. The biological fixation is limited to a few, but abundant, organisms like the free living bacteria *Azetobacter* and *Clostridium*, nodule bacteria on leguminous plants like *Rhizobium* and some blue-green *Algae*. These are the keys to the movement of the nitrogen from the atmospheric reservoir into the cycle shown in Fig. 6.18.

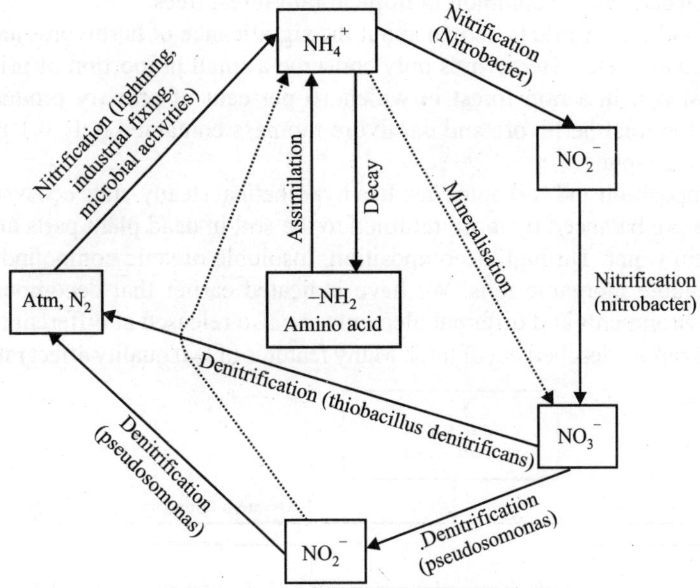

Fig. 6.18. Components of the nitrogen cycle in a terrestrial ecosystem. Solid lines are major and dotted lines minor transfers and processes. The very complex processes whereby the intractions between dead organic matter and decomposers mineralise nitrogen are not shown.

The nitrates are assimilated to form amino acids, urea and other organic residues in the producer, consumer and decomposer cycles. The amino acids and urea are then converted to ammonia through a process called 'ammonification'. To complete the cycle, denitrifying bacteria convert the ammonia into nitrites, then into nitrates and then back into gaseous nitrogen. In this way, under normal circumstances, the total amount of nitrogen fixed equals the total amount returned to the atmosphere as gas.

Man has interfered with this natural cycle by industrially fixing nitrogen. This includes production of nitrogen fertilisers and oxidation of nitrogen during fossil fuel combustion. Most of the excess nitrogen is carried off into rivers and lakes and ultimately reaches the ocean. This increased runoff has greatly increased the productivity in many aquatic environments and has contributed to the process of eutrophication.

Cycles of mineral elements

Strictly speaking, mineral nutrients are those derived in the long-term from geological sources. However, in most ecosystems, mineral ions are obtained from the soil as a result of breakdown of dead organic matter. Nitrogen is not really a mineral element in terms of its origin, but its cycling has many features in common with other edaphic nutrients. Figure 6.19 is a generalised scheme for edaphic cycles, various aspect of which are considered below:

1. Uptake and accumulation in vegetation: Symbiotic fungi are often intimately associated with plant roots and they play an important role in uptake of nutrients from the soil solution, receiving in return photosynthetic products from the plant. These fungi are called mycorrhizae and are of two types: (i) sheathing (ectotrophic) mycorrhizae are mainly Basidiomycotina and are typically

associated with temperate trees and are less common with tropical trees, although dipterocarps of south-east Asia have these; (ii) vesicular–arbscular (endotrophic) mycorrhizae are phycomycetes of Endogonaceae and are common in tropical rain forest trees.

2. Herbivore food web: Little is known about the significance of herbivores and their predators in tropical nutrient cycles. Herbivores only consume a small proportion of primary production in most ecosystems. In a rain forest in which 10 per cent of primary production was eaten by herbivores, the total herbivore and carnivore biomass contained only 0.1 per cent of the total minerals in the biota.
3. Litter decomposition and soil nutrients: In a hypothetical steady-state ecosystem, nutrients taken up by plants are balanced by those returned to the soil in dead plant parts and washed from the plants by rain water. Through decomposition, insoluble organic compounds are broken down, releasing soluble inorganic ions. We have indicated earlier that decomposition rates vary in different environments and different elements are also released at different rates to one another. Potassium is rapidly leached out of litter. Many features of leaf quality affect rate of decomposition.

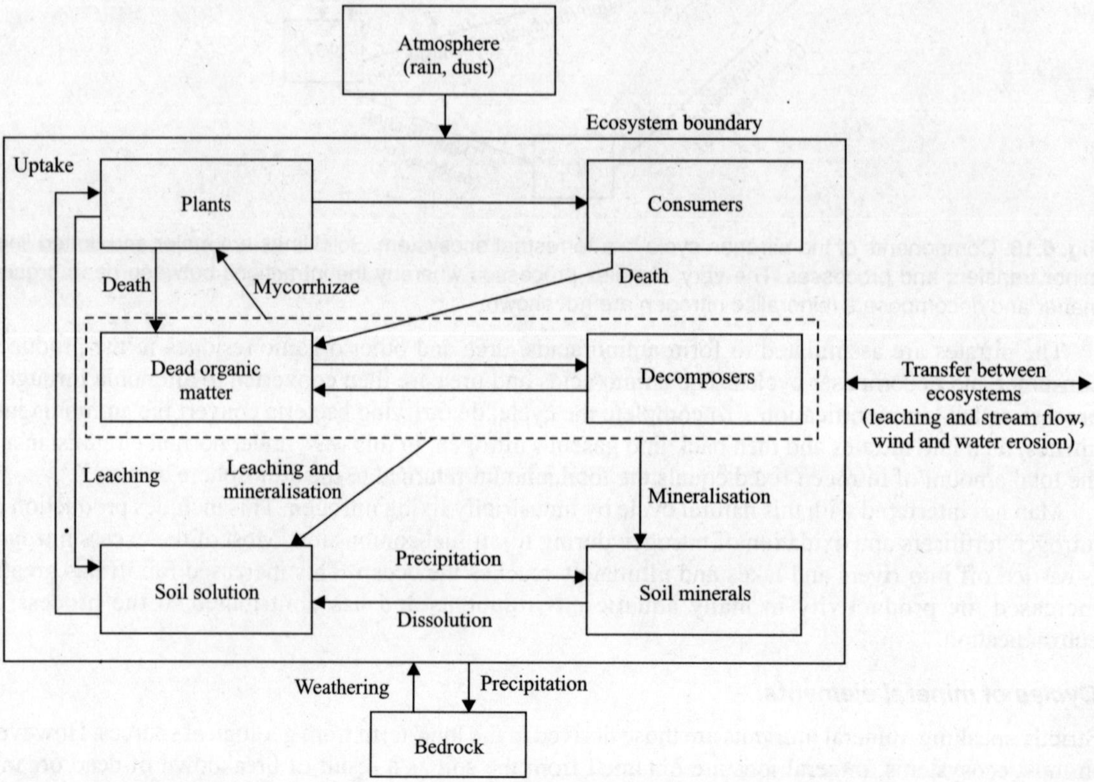

Fig. 6.19. Generalised scheme of mineral cycles in terrestrial ecosystems. Most transfers are within and between the biotic and inorganic soil components. Additional inputs come to the ecosystem as rainfall (or dust) and by weathering of parent rock and losses occur through leaching from and loss of soil and precipitation of soil minerals as unavailable forms.

To become available to plants, the various elements have to be mineralised to organic forms-in particular those that enter the soil solution. Otherwise the elements remain locked up in the undecomposed dead organic matter or immobilised in the tissues of living decomposer organisms. Fire is an important decomposer and mineraliser in many savanna. Some proportion of the elements enters the atmosphere as gases (N, S, P) or small particles (Ca, Mg, K), although these may be precipitated later by gravity or rainfall. However, much of the inorganic material is deposited in a nutrient-rich ash, which subsequently becomes incorporated into soils and new plant growth. The maintenance of soil nutrients in ionic forms available to plants result from the complex interactions of soil chemistry.

Decomposition is not the only source of soluble nutrients to the soil. Rainfall contains significant quantities of mineral elements. Rain also leaches further nutrients from the plant canopy to give the total throughfall reaching the ground. In arid areas, additional inputs or outputs may occur due to wind erosion, that, carries silica, aluminium, organic matter, Ca, Mg and K.

4. Rate of nutrient cycling: In forests, the dominant plants are trees which live much longer than their decomposers or consumers. But not all the organs of trees are long-lived. Thus flowers, fruits and leaves are short-lived. In terms of quantitative significance, leaves and stems are the most important and they represent short and long-nutrient retention respectively. The proportion of total nutrient capital that is cycled rapidly through leaves varies with the element and the type of forest. The proportion is usually 5–25 per cent of the total for a particular element in the vegetation with calcium at the lower end of this range. In low land forests, leaf litter decomposition is rapid and thus there is no slow down nutrient acquisition by plants.

The turnover rate of nutrients in the vegetation can be measured as uptake relative to the quantity contained in the living biomass. The elements are 'turned over' more quickly in the tropical humid forest than in a temperate deciduous forest (Table 6.1). However, there is much variation and some overlap when different forests and different elements are compared. This table also reflects relative mobility of elements; K is the most mobile (circulating most quickly), followed by Mg. These both elements also are generally the most mobile in throughfall and release from litter. Calcium is usually the least mobile element.

Vitousek has assessed the efficiency of nutrient cycles in terms of the amount of organic matter produced per unit of nutrient uptake. His analysis of litter fall in 62 tropical forests showed a wide range of efficiencies, both in comparisons of a single element in different forests and when different elements are compared. Lowland tropical forests do not cycle nitrogen as efficiently as montane tropical forests or temperate forests because leaf litter has a high concentration of this element.

Table 6.1. Turnover time (years) of elements in two forests. Turnover time estimated as the standing crop of an element in the total forest biomass divided by the annual uptake by plants.

Type and location	*Turnover time*				
	N	*p*	*K*	*Ca*	*Mg*
Lowland, humid, tropical forest (Kade, Ghana)	7	7	3	7	5
Temperate, deciduous forest (oak ridge, Tennessee, USA)	8	15	10	12	5

Cycles of micronutrients

Little is known of these in tropical ecosystems. In a tropical rain forest on average, more than 80 per cent of total store of these elements is found in the soil, with progressively less in vegetation, litter and the fauna. However, the concentration of these trace elements increases significantly along food chains in the sequences soil – plants – herbivores – carnivores and litter – detritivores. For example concentrations of iron in a rain forest were 6 ppm in soil, 218 ppm in vegetation, 3054 ppm in herbivores and 3387 ppm in the carnivores. Many of these micronutrients are toxic to plants in high concentrations.

MAJOR ECOSYSTEMS

So far there has been presented a detailed account of gross structure and function of an ecosystem in general. We find various types of ecosystem operating as self-sufficient interacting systems in the biosphere. For example, a pond, a lake, a river, a stream, a spring, an estuary, the sea, a desert, a grassland, a forest, a coral reef, a cropland (field of maize, wheat, sugarcane, paddy) etc. operate as individual ecosystems of nature.

Although, all these ecosystems have a more or less similar fundamental plan of their gross structure and function, however, they differ in respect of their species composition and rates in production etc. Thus we shall now consider in brief individual organisational pattern of each of these ecosystems and then, to some extent, would assess the relative value of each in respect of the total primary production of biosphere. Some of the ecosystems are discussed below.

Pond Ecosystem

A pond as a whole serves a good example of a freshwater ecosystem. A pond indeed exhibits a self-sufficient, self-regulating system. Not only is the pond a place where plants and animals (living organisms) live, but plants and animals make the pond what it is (physico-chemical environment). This would become clear if you examine a bottle full of pond water or a scoop full of bottom mud, which shall show the living organisms (plants as well as animals) and a mixture of inorganic and organic compounds. Some larger forms of life are also present in pond. Thus, whole system becomes much, complex indeed. However, we may study the pond as an ecosystem by making its convenient division into some basic components, as shown in Fig. 6.20. These components are as follows:

Abiotic component

The chief substances are heat, ligh, pH value of water and the basic inorganic and organic compounds, such as water itself, carbon dioxide gas, oxygen gas, calcium, nitrogen, phosphates, amino acids, humic acid etc. Some proportions of nutrients are in solution state but most of them are present as stored in particulate matter as well as in living organisms. The amount of the minerals present at any time in the physical environment of the pond—'standing state' may be estimated by appropriate methods. The light intensity is measured by a Lux-photometer. Turbidity index of water at different depths is obtained by a *Sechhi disc*. Rates of evapo-transpiration are calculated. The pH of water and mud is determined by an electric pH meter. Dissolved oxygen content, carbon dioxide content, solute contents including colloidal suspensions, phosphate and nitrogen contents of water and plant and animal matter are estimated by appropriate methods. Amounts of various organic compounds (carbo-hydrates, proteins, lipid etc.) are also estimated for biomass determination.

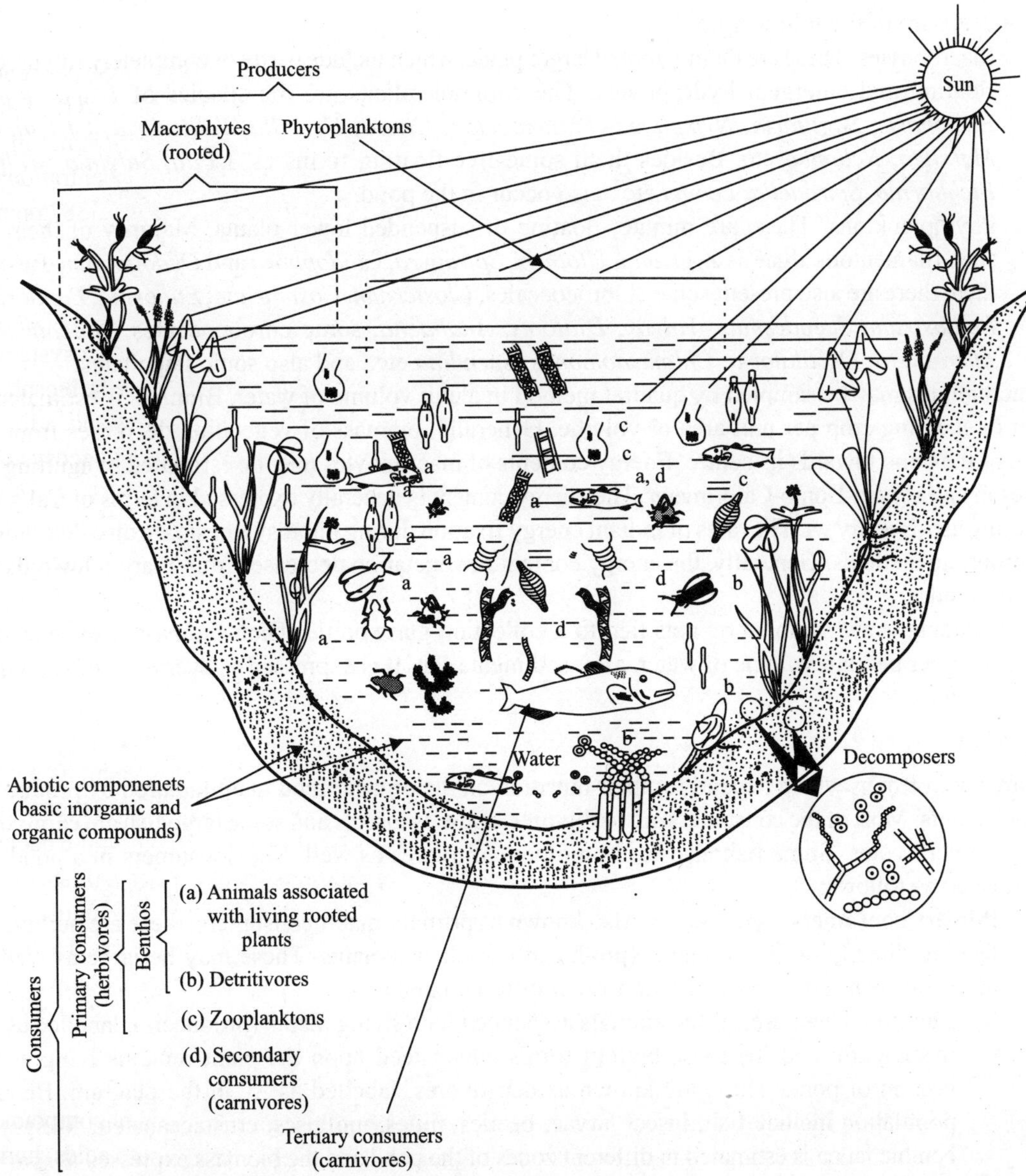

Fig. 6.20. Diagram of the pond ecosystem, showing its basic structural units—the abiotic (inorganic and organic compounds) and biotic (producers, consumers and decomposers) components.

Biotic component

The various organisms that constitute the biotic component are as follows:

Producers

These are autotrophic, green plants and some photosynthetic bacteria. The producers fix radiant energy and with the help of minerals derived from the water and mud, they manufacture complex organic substances as carbohydrates, proteins, lipids etc.

Producers are of the following types:

1. Macrophytes: These are mainly rooted larger plants which include partly or completely submerged, floating and emergent hydrophytes. The common plants are the species of *Trapa*, *Typha*, *Eleocharis*, *Sagittaria*, *Nymphaea*, *Potamogeton*, *Chara*, *Hydrilla*, *Vallisneria*, *Utricularia*, *Marsilea*, *Nelumbo* etc. Besides them some free-floating forms as *Azolla*, *Salvinia*, *Wolffia*, *Eichhornia*, *Spirodella*, *Lemna* etc., also occur in the pond.
2. Phytoplanktons: These are minute, floating or suspended lower plants. Majority of them are such filamentous algae as *Zygnema*, *Ulothrix*, *Spirogyra*, *Cladophora* and *Oedogonium*. Besides them there are also present some chlorococcales, *Closterium*, *Cosmarium*, *Eudorina*, *Pandorina*, *Pediastrum*, *Scendesmus*, *Volvox*, *Diatoms*, *Anabaena*, some chroococcales, *Gloeotrichia*, *Microcystis*, *Oscillatoria*, *Chlamydomonas*, *Spriulina* etc., and also some flagellates.

Macrophytes may be sampled by quadrat method in a unit volume of water. Biomass is estimated as weight of standing crop per unit area or volume. Generally, biomass of vegetation decreases from the margin of the pond towards its centre. Energy contents of macrophytes may be estimated by igniting the samples in an Oxygen-Bomb-Calorimeter. The energy content is generally expressed in terms of Cal/g dry wt or Cal/g ash-free dry wt. The rates of radiant energy fixation during photosynthesis are also determined by appropriate methods. Generally, the energy content of vegetation decreases from margin towards the centre of pond.

Phytoplanktons are sampled by nets tied to a collection glass bottle. Their biomass is estimated in terms of number per unit volume of water, after sedimentation. It is expressed as average biomass/cubic meter of water.

Consumers

They are heterotrophs which depend for their nutrition on the organic food manufactured by producers, the green plants. Most of the consumers are herbivores, a few as insects and some large fish are carnivores feeding on herbivores. Some fish also feed on other carnivores as well. The consumers in a pond are distinguished as follows:

1. Primary consumers (herbivores): Also known as primary macroconsumers, these are herbivores feeding directly on living plants (producers) or plant remains. These may be large as well as minute in size. The herbivores are further differentiated as:-
   (a) Benthos: These are: (i) the animals associated with living plants (producers), labelled as 'a' in diagram; and (ii) those bottom forms which feed upon the plant remains lying at the bottom of pond. These are known as detritivores, labelled as 'b' in the diagram. Benthic population include fish, insect larvae, beetles, mites, molluscs, crustaceans etc. Weight of benthic fauna is estimated in different zones of the pond and the biomass expressed as g/m^2 of water.
   Besides the above said herbivores, some mammals as cows, buffaloes etc. also visit the pond casually and feed on marginal rooted macrophytes. Some birds also regularly visit the pond feeding on some hydrophytes:
   (b) Zooplanktons: These are chiefly the rotifers as *Brachionus*, *Asplanchna*, *Lecane* etc., although some protozoans as *Euglena*, *Coleps*, *Dileptus* etc., and crustaceans like *Cyclops*, *Stenocypris* etc., are also present. They feed chiefly on phytoplanktons. In diagram these are labelled as 'c'.
2. Secondary consumers (carnivores): They are the carnivores which feed on the primary consumers (herbivores). These are chiefly insects and fish. Most insects as water beetles feed on zooplanktons. These are labelled as 'd' in the diagram.

3. Tertiary consumers (carnivores): There are some large fish as game fish that feed on the smaller fish and thus become the tertiary (top) consumers as shown in Fig. 6.20.

In a pond, fish may occupy more than one trophic levels. As shown in Fig. 6.20, the smaller fish belong to herbivores levels, feeding on phytoplanktons as well as living plant parts (labelled as 'a') and also on plant remains lying at the bottom (labelled as 'b'). Some fish may feed on some zooplanktons, thus occupying the secondary consumers level—the carnivores level (labelled as 'd'). Still, it is common to observe the large fish feeding on smaller fish and thus occupying the tertiary consumers level.

Decomposers

They are also known as micro-consumers, since they absorb only a fraction of the decomposed organic matter. They bring about the decomposition of complex dead organic matter of both—producers (plants) as well as the macro-consumers (animals) to simple forms. Thus they play an important role in the return of mineral elements again to the medium of the pond. These include a variety of heterotrophic microbes that are osmotrophs. These are sampled by the use of several isolation methods. They are chiefly bacteria, actinomycetes and fungi. Fungi are isolated by Warcup's method on Martin's medium, bacteria on Thornton's medium and actinomycetes on Jenson's medium. Among fungi, species of *Aspergillus*, *Cephalosporium*, *Cladosporium*, *Pythium*, *Rhizopus*, *Penicillium*, *Thielavia*, *Alternaria*, *Trichoderma*, *Circinella*, *Fusarium*, *Curvularia*, *Paecilomyces*, *Saprolegnia* etc., are most common decomposers in water and mud of the pond.

Ocean (Marine) Ecosystem

Major oceans of the world, Atlantic, Pacific, Indian, Arctic and Antarctic cover approximately 70 per cent of the Earth's surface. Each ocean indeed represents a very large and stable ecosystem. Marine environments, as compared with fresh water, appear to be more stable in their chemical composition due to being saline and moreover other such physico-chemical as dissolved oxygen content, light and temperature are also different.

The biotic components of an ocean ecosystem are of the following orders:

1. Producers: These are autotrophs and also designated as primary producers, since they are responsible for trapping the radiant energy of Sun with the help of their pigments. Producers are mainly the phytoplanktons, such as diatoms, dinoflagellates and some microscopic algae. Besides them, a number of macro-scopic seaweeds, as brown and red algae (members of Phaeophyceae and Rhodophyceae), also contribute significantly to primary production. These organisms show a distinct zonation at different depths of water in the sea.
2. Consumers: These all are heterotrophic macroconsumers, being dependent for their nutrition on the primary producers. These are:-
   (a) Primary consumers: The herbivores, that feed directly on producers are chiefly crustaceans, molluscs, fish etc.
   (b) Secondary consumers: These are carnivorous fish, as *Herring*, *Shad Mackerel* etc., feeding on the herbivores.
   (c) Tertiary consumers: Still in the food chain, there are other carnivorous fishes like Cod, Haddock, Halibut etc. that feed on other carnivores of the secondary consumers level. Thus, these are the top carnivores in the food chain.
3. Decomposers: The microbes active in the decay of dead organic matter of producers and macro-consumers are chiefly bacteria and some fungi.

Grassland Ecosystem

This is a type of terrestrial ecosystem. Grasslands occupy a comparatively fewer area, roughly 19 per cent of the Earth's surface. The various components of a grassland ecosystem are as follows:

Abiotic component

These are the nutrients present in soil and the aerial environment. Thus the elements like C, H, O, N, P, S etc., are supplied by carbon dioxide, water, nitrates, phosphates and sulphates etc., present in air and soil of the area. Moreover, in addition to the above, some trace elements are also present in soil.

Biotic component

These may be categorised as:

1. Producers: They are mainly grasses, as species of *Dichanthium*, *Cynodon*, *Desmodium*, *Digitaria*, *Dactyloctenium*, *Brachiaria*, *Setaria*, *Sporobolus* etc. Besides them a few forbs and shrubs also contribute to primary production.
2. Consumers: These occur in the following sequence:-
   (a) Primary consumers: The herbivores feeding on grasses are mainly such grazing animals as cows, buffaloes, deers, sheep, rabbit, mouse etc. Besides them, there are also present some insects as *Leptocorisa*, *Dysdercus*, *Oxyrhachis*, *Cicincella*, *Coccinella*, some termites and millipeds etc., that feed on the leaves of grasses.
   (b) Secondary consumers: These are the carnivores feeding on herbivores. These include the animals like fox, jackals, snakes, frogs, lizards, birds etc. Sometimes the hawks feed on the secondary consumers, thus occupying tertiary consumers level in the food chain.
3. Decomposers: The microbes active in the decay of dead organic matter of different forms of higher life are fungi, as species of *Mucor*, *Aspergillus*, *Penicillium*, *Cladosporium*, *Rhizopus*, *Fusarium* etc., and some bacteria and actinomycetes. They bring about the minerals back to the soil, thus making them available to the producers.

Forest Ecosystem

Forests occupy roughly 40 per cent of the land. In India, the forests occupy roughly one-tenth of the total land area. The different components of a forest ecosystem, like others, are as follows:

Abiotic component

These are the inorganic as well as organic substances present in the soil and atmosphere. In addition to the minerals present in forests we find the dead organic debris—the litter accumulation, chiefly in temperate climate. Moreover, the light conditions are different due to complex stratification in the plant communities.

Biotic component

The living organisms present in the food chain occur in the following order:

1. Producers: These are mainly trees that show much species diversity and greater degree of stratification especially in tropical moist deciduous forests. The trees are of different kinds depending upon the kind of the forest formation developing in that climate. Besides trees, there are also present shrubs and a ground vegetation. In these forests, dominant members of the flora, the producers, are such trees as *Tectona grandis*, *Butea frondosa*, *Shorea rubusta* and *Lagerstroemia parvifiora*. In temperate coniferous forests, shrubs and ground flora are

insignificant. In temperate deciduous forests the dominant trees are species of *Quercus*, *Acer*, *Betula*, *Thuja*, *Picea* etc., whereas in a temperate coniferous forests, the producer trees are species of *Abies*, *Picea*, *Pinus*, *Cedrus*, *Juniperus Rhododendron* etc.

2. Consumers: These are as follows:-
   (a) Primary consumers: These are the herbivores that include the animals feeding on tree leaves as ants, flies, beetles, leafhoppers, bugs and spiders etc., and larger animals grazing on shoots and/or fruits of the producers, the elephants, nilgai, deer, moles, squirrels, shrews, flying foxes, fruit bats, mongooses etc.
   (b) Secondary consumers: These are the carnivores like snakes, birds, lizards, fox etc., feeding on the herbivores:
   (c) Tertiary consumers: These are the top carnivores like lion, tiger etc., that eat carnivores of secondary consumers level.
3. Decomposers: These are wide variety of micro-organisms including fungi (species of *Aspergillus*, *Coprinus*, *Polyporus*, *Ganoderma*, *Fusarium*, *Alternaria*, *Trichoderma* etc.), bacteria (species of *Bacillus*, *Clostridium*, *Pseudomonas*, *Angiococcus* etc.) and actinomycetes, like species of *Streptomyces* etc. Rate of decomposition in tropical and subtropical forests is more rapid than that in the temperate ones.

Desert Ecosystem

Deserts occupy about 17 per cent of land, occurring in the regions with an annual rainfall of less than 25 centimeters. The species composition of such ecosystem is much more varied and typical due to extremes of both, temperature and water factors. The various biotic components are:

1. Producers: These are shrubs, especially bushes, some grasses and a few trees. The shrubs have widespread, branched root system with their stems and branches vigrously modified. Sometimes a few succulents like cacti are also present. Some lower plants like lichens and xerophytic mosses may also be present.
2. Consumers: The most common animals are reptiles and insects, able to live under xeric conditions. In additions to them, there are also found some nocturnal rodents and birds. The 'ship of desert', camels feed on tender shoots of the plants.
3. Decomposers: These are very few, as due to poor vegetation the amount of dead organic matter is correspondingly less. They are some fungi and bacteria, most of which are thermophilic.

Cropland Ecosystem

The ecosystems, described so far, are natural in the sense that they all operate as self-regulating systems without much direct interference and manipulations by man. However, in nature we also find another kind of ecosystems, where man is very much involved in their operation. These are the cropland ecosystems, that are artificial or man-engineered, where, in order to obtain more food, cloth, timber, medicines and other useful products, man becomes responsible for the replacement of natural systems. To secure maximum production, man makes much planned manipulations in the physico-chemical environment. These include addition of fertilisers to soil, use of chemicals for disease control, proper irrigation practices etc. Thus, a cropland ecosystem is an artificial system aimed primarily to grow a single species of one's choice. We have ecosystems of dominant crop species like wheat, maize, jowar, paddy, sugarcane, vegetable etc., under most suitable conditions of their growth and yield. It is generally argued that in a natural ecosystem, the nature maximises for gross production, whereas in an artificial ecosystem, man maximises for net production. In a cropland ecosystem, there is not necessarily an increase in the total dry matter production of the whole plants, but generally most of the production goes into grain and less into leaves, stems and roots. Thus, in agriculture (where all the systems are artificial), there is an objective to achieve high rates or production (P) of readily harvestable products with little standing crop (biomass—B) left in the field for accumulation or we may say that there is a high P/B efficiency. Nature (natural ecosystem), on the other hand, follows just a reverse efficiency—i.e., a high B/P ratio, where the standing crop is generally accumulated to its maximum. The following are the chief components of a maize cropland ecosystem.

Abiotic component

These include the climatic conditions of the region, where the crop may grow most successfully and the various minerals particularly C, H, O, N, P, K in soil and atmosphere. In such field man generally makes additions of a number of chemical fertilisers to soil. Maize generally grows best in slightly alkaline soil with good aeration.

Biotic component

The various living organisms in the food chain occur as follows:

1. Producers: The dominant plant species would naturally be *Z. ι mays*. Besides maize, a number of weeds like *Cynodon dactylon*, *Launaea nudicaulis*, *Euphorbia hirta*, *Cyperus rotundus*, *Digitaria* spp. and *Alysicarpus* sp. also contribute to primary production of the field.
2. Consumers: These are (i) Primary consumers. These are the herbivores represented by a variety of animals, big as well as small. The smaller animals include chiefly the insects as aphids, thrips, beetles etc., which feed and lay their eggs on maize leaves. Larger animals are rabbits, rats, birds and man feeding on leaves, flower and fruits of maize; (ii) secondary consumers. These are carnivores like frogs and some birds that eat insect; and (iii) tertiary consumers. These are carnivores like snakes and hawks which feed on the secondary consumers, frogs and smaller birds.
3. Decomposers: These are microbes present in soil as well as air, that decompose the dead organic matter of plants and animals. These are chiefly bacteria, actinomycetes and fungi, responsible for decay, decomposition and humification, making the minerals available again to the producers.

Productivity of Different Ecosystems

In different kinds of productivity and primary production differs in different kinds of ecosystems. As primary production depends upon the producers their nature in a particular ecosystem would naturally determine the rates of primary production. Large-sized trees with dense population does not necessarily mean that the forests would be more efficient in terms of primary production than a pond, where the producers are smaller forms of plants, mainly the phytoplanktons. In this case the size of the producer is in no way related with the efficiency of primary production. This is indeed the rate of fixation of radiant energy in an ecosystem and not the size of the producers that finally determines the efficiency of production. We shall now consider briefly the efficiency of different kinds of ecosystem, in terms of their primary production.

PRIMARY PRODUCTION IN TERRESTRIAL ECOSYSTEMS

Average net primary production (in terms of dry weight) for terrestrial ecosystems is presented in Table 6.2. It is clear that more than 60 per cent of terrestrial primary production occurs in the tropics. Thus, tropical ecosystems generally have greater production than comparable ones at higher altitudes. However, there is much variation within the substantial overlap between categories. For example, primary production in tropical forests ranges from approximately 1000–3000 g/m^2/year, temperate forests 600–2500 g/m^2/year and boreal forests 400–2000 g/m^2/year. Some tropical forests are less productive than some of their temperate or local counterparts. Similar variations in plant production of savannas and grasslands are also found.

The highest rate of primary production occurs in tropical rain forests and yet the average for these ecosystems, 2300 g/m^2/year, represents conversion of less than 0.001 per cent of the total solar radiation reaching 1 m^2 of the planet. Agricultural experiments have yielded a maximum efficiency of only 0.2 per cent. What are the possible reasons of so little of the Sun's energy converted to plant production? The possible reasons are analysed by Monteith, which are summarised below:

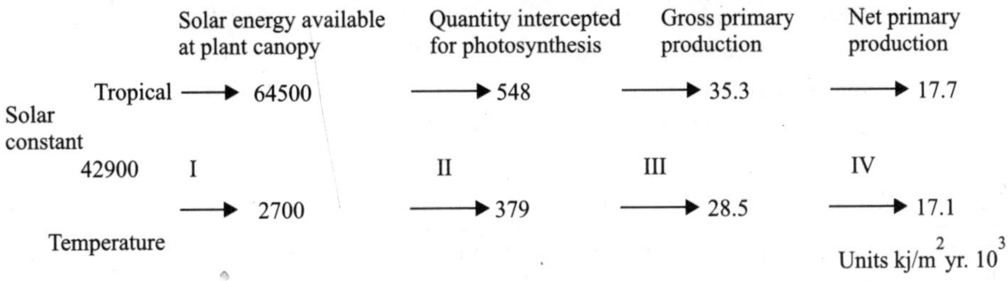

Table 6.2. Standing crop and annual net primary production (dry wt.) for terrestrial ecosystems. Values are approximate (2001).

Ecosystem type			*Standing crop (g m^{-2})*	*Net primary production (g m^{-2})*	*Per cent of total land area**	*Per cent of total primary production*
			Natural			
Forests	Tropical	humid	42000	2300	6.7	17.3
		seasonal	25000	1600	3.0	5.4
		mangrove	30000	1000	0.2	0.2
	Temperate	deciduous/mixed	28000	1300	2.0	2.9
		evergreen/coniferous	30000	1500	2.0	3.4
	Boreal	closed canopy	25000	850	4.4	4.2
		open canopy	17000	650	1.7	1.2
Woodlands and Shurblands	Tropical	wooded savanna	9000	1500	11.0	19.2
	Temperate	woodland	18000	1500	1.3	2.3
		scrubland	7000	800	1.7	1.5
Grasslands	Tropical	grass savanna	2200	2300	4.0	10.4
	Temperate	moist grassland	2100	1200	3.3	4.5
		dry grassland	1300	500	5.0	2.8

(Contd...)

Ecosystem type		*Standing crop ($g\ m^{-2}$)*	*Net primary production ($g\ m^{-2}$)*	*Per cent of total land area**	*Per cent of total primary production*
Deserts (mainly tropical)	Desert and semidesert scrub	1100	200	6.0	1.4
	Extreme desert: hot and dry	60	10	5.4	negligible
	cold and dry	300	50	0.7	negligible
Tundra	Low arctic/aline	2300	350	2.9	1.2
	High arctic/alpine	750	150	2.4	0.4
	Polar desert	150	25	1.0	negligible
Swamps	Trophical	15000	4000	1.0	4.5
	Temperature	7500	2500	0.3	0.9
Peatlands and bogs (mainly temperate)		5000	1000	1.0	1.1
		Managed			
Forest plantation		20000	1750	1.0	2.0
Agricultural	Tropical: annuals +	60	700	6.0	4.7
	perennials	6000	1600	0.3	0.3
	Temperature: annuals +	100	1200	4.0	5.4
	perennials	5000	1500	0.3	0.6
Desertified areas		500	100	8.0	0.9
Human area		4000	500	1.3	0.3

* 10 per cent of the land surface is perpetual ice with zero production; + averaged over the whole year, including times when no crop is being grown.

Above sketch shows energy losses in primary-production between solar radiation and net primary production. These are comparisons for tropical and temperate intensive agriculture. Losses are due to:

1. Geometry and rotation of the Earth and transmission through the atmosphere.
2. Energy not absorbed by chlorophyll.
3. Rate of carbon dioxide diffusions and photo-chemical efficiency.
4. Plant respiration.

Thus tropical forests, particularly tropical rain forests have maximum primary production among terrestrial ecosystems of the world. But these too have a very poor efficiency of conversion of less than 0.001 per cent of the total solar radiation reaching 1 m^2 of the Earth. Following are the possible reasons for this poor efficiency of net primary production.

Energy Losses Between the Outer Atmosphere and Plant Canopy

Only 15 per cent and 8 per cent of the solar radiation reaching the Earth remains in tropical and temperate regions respectively.

Interception or Light by Plants

Annual average leaf area indices (leaf area index is the area of leaf (m^2) arranged vertically above each square meter of ground; a useful measure of canopy density) are at their highest in evergreen forests (10 or more) and their lowest in arid deserts (0.5 or less). In seasonal environments as an Indian grassland this index value was 0.5 in dry season when the plants had died back and 5.0 in the wet season when they were actively growing.

No overall estimate can be given for interception efficiency. It can be defined as the fraction of incident light intercepted. On an estimated basis, values of annual average of interception efficiency are 20 per cent for tropical intensive agriculture, 5 per cent for tropical subsistence agriculture and 33 per cent for temperate intensive agriculture. This greater value for temperate one leads to tropical production being only slightly greater than the temperate estimate in summary values given above.

Plant Biochemistry

On an average, only about 30 per cent of gross primary production is available as net primary production in tropical rain forests compared with 40 per cent in temperate forests. This difference is partly due to faster respiration at higher temperatures than is photosynthesis. Respiration is also greater in plant communities with a high ratio of standing crop to production. In forests the photosynthetic leaves support a large non-photosynthetic biomass of wood while in herbaceous vegetation most of the above ground tissues are photosynthetic.

Another physiological factor responsible for such differences is difference in biochemical pathways. Table 6.3 summarises the differences in gross physiology of the three systems of carbon fixation. Following conclusions may be drawn:

1. C_4 plants are the most efficient photo-synthetically in optimum conditions and are adapted to hot, bright and fairly dry climates.
2. C_3 plants are adapted to cool, moist and shady environments.
3. CAM plants are adapted to very dry environments.

Table 6.3. Gross physiological attributes of the three carbon fixation pathways.

Attribute	*Photosynthetic pathway* C_3	C_4	*CAM (crasssulacean acid metabolism)*
Optimum temperature for CO_2 fixation (°C)	15–25	30–45	Approximately 35
Photosynthetic rate under optimal conditions (mg CO_2 dm^{-2} $hour^{-1}$)	15–35	40–80	0.5–0.7
Light saturation as per cent of full sunlight	10–25	100	–
Water use efficiency (no. of gram of H_2O to produce 1 gram plant matter)	380–900	250–350	50

Such distributional effects do occur as illustrated by the change from 10 per cent C_4 grasses in hot, dry lowland areas to 100 per cent C_3 grasses in cool, moist montane environments of Kenya. However, most tropical trees have C_3 pathway irrespective of macroclimate, possibly as a reflection of seedling and sapling growth in cooler, shadier microclimate beneath the canopy. The precise conditions under which C_3 and C_4 pathways are advantageous result from quite subtle interactions between light, temperature and moisture that are yet to be fully understood. CAM path is characteristic of desert succulents as cacti and euphorbias as do some orchids and bromeliads, the canopy epiphytes of torpical rain forests. Although photosynthetic rate is greater in C_4 plants, there is no evidence that C_4 dominated natural communities are more productive than C_3 counterparts. Most plant communities are of such low efficiency; compared with their chemical potential under optimum conditions, that the ecological significance of the C_3 and C_4

pathways is more apparent in the distribution of species than in net primary production. However, the most productive agricultural crops like sugarcane and maize are C_4 species.

Even if light and CO_2 are used with maximum efficiency, other constraints might be limiting in primary production in many environments. For example in humid forests, water is rarely limiting, but in seasonal areas rainfall becomes important. In West African savannas, inadequate supply of nitrogen and phosphorous are limited factors.

Below-Ground Production

Tropical vegetation translocates a smaller proportion of net production below-ground (to support growth of roots and other underground organs) than do comparable temperate plants.

On an average tropical forests retain 90 per cent and savannas 50 per cent above-ground compared with 75 per cent in temperate forests and 40 per cent in temperate grasslands.

To summarise, given sufficient nutrients and precipitation, tropical ecosystems have greater primary production than temperate ones but to a less extent than would be expected from the differences in solar radiation. This is because of lower interception efficiency and higher respiratory demand at low latitudes. Some tropical plants have compensated to some extent by C_4 pathway, which is more efficient at high temperatures and high light intensities. Many ecosystems do not reach the primary production suggested by the light climate due to other factors limiting their growth.

SECONDARY PRODUCTION

Like primary production, secondary production can be described by transfer efficiencies between the various steps in energy utilisation. From the energy budget of a population of animals, two key efficiencies have been isolated. These are the: (i) assimilation efficiency (A/C%), which is a measure of the amount of ingested energy left after waste production; and (ii) production efficiency (P/A%), the proportion of assimilated energy used to increase the energy content of the population. The energy budgets for herbivorous grasshopper population in Indian fields and carnivorous lizard population in Lamto savanna of the Ivory Coast are worked out as follows:

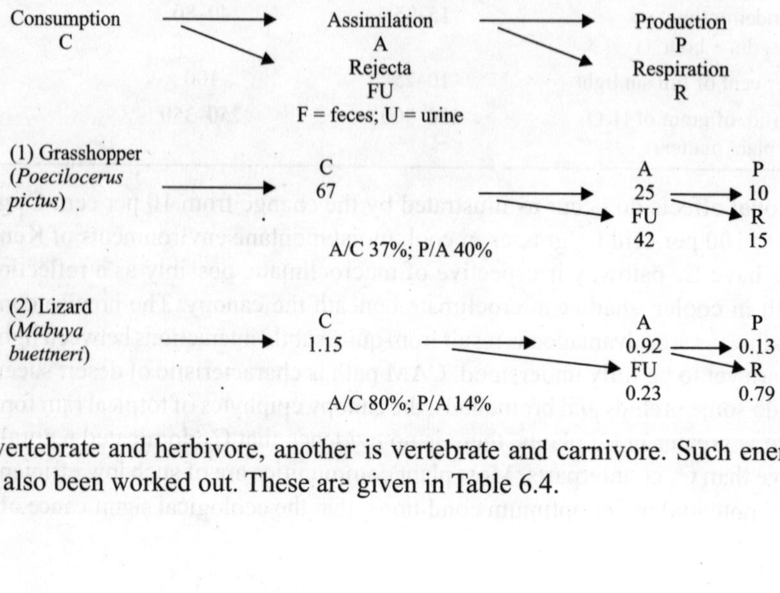

One is invertebrate and herbivore, another is vertebrate and carnivore. Such energy data for other animals have also been worked out. These are given in Table 6.4.

Table 6.4. Assimilation efficiency (A/C) and production efficiency (P/C) as percentages for heterotrophic population. Blank indicates types of population do not exist or are of minor significance in most ecosystems.

	Herbivore		*Carnivore*		*Microbivore*		*Decomposer*	
	A/C	*P/A*	*A/C*	*P/A*	*A/C*	*P/A*	*A/C*	*P/A*
Micro-organisms	–	–	–	–	–	–	–	40
Invertebrates	40	40	80	20	30	40	20	30
Vertebrates								
Homeotherms	50	2	80	2	–	–	–	–
Heterotherms	50	10	80	10	–	–	–	–

1. Assimilation efficiency (A/C): The proportion of energy assimilated depends on the quality of food ingested. Biophages are more efficient than saprophages. Animal food is usually of higher quality than plant food, due to its high protein and low fiber content. A/C in savanna-grazing mammals can be as high as 80 per cent in wet season but goes down to 20 per cent during dry season. Actually such herbivores are decomposers during dry season when they intake mainly dead grass. Most animal decomposers are simultaneously microbivores. Termites have exceptionally high A/C compared with decomposers, ranging from 50 per cent to more than 90 per cent. Thus, they have much impact on tropical ecosystems where they are abundant.
2. Production efficiency (P/A): Homeotherms vertebrates have a lower P/A than other heterotrophs. Consequently, food chains including birds and mammals tend to be shorter than those involving heterotherms because more assimilated energy is dissipated as respiration at each stage. More energy is consumed in maintaining body temperature, thus leaving less energy for production. Thus birds and mammals have lower rates of population growth. Detailed analysis of P/A by Humphreys shows more variation than given in Table 6.4. Rate of production is also correlated with body size in animals such that the rate of production per unit biomass (P/B) in population decreases as the body size of individuals increases. Rodents have a P/B ratio of around 3:1, whereas an African buffalo only 0.1:1.
3. Consumption efficiency (C.E.): This is a third transfer efficiency of importance to heterotrophs. It measures the amount of energy transferred from one trophic level to the next. For instance C.E. of herbivores is the total annual consumption of all herbivore population as a percentage of annual net primary production. The food intake is closely related to body size. The C.E. of herbivore trophic level determines what proportion of energy enters the herbivore food web and what proportion passes to decomposers. This is expressed with the help of energy flow models in ecosystem.

We have pointed out earlier that there are two kinds of basic food chains, the grazing (herbivore) and the detritus (decomposer) food chains. These are infact interlinked to form food webs.

Herbivore Food Web

The amount of net primary production consumed by herbivores usually averages less than 10 per cent in terrestrial ecosystems and there are no much differences between tropical and higher latitude environments (Table 6.5). Herbivore consumption efficiency as high as 60 per cent in savanna appear as exceptional. Terrestrial green plants are depleted seriously only in exceptional cases as during locust plagues. Why do herbivores not increase to consume more of the energy available to them? Answers are very complex. Reasons may be:

1. Herbivores may be limited by predators.
2. Plants may defend themselves from herbivores by physical and chemical means.
3. Primary production may be seasonal.
4. Quality rather than quantity of food may be limiting.

Table 6.5. Percentage of above-ground net primary production consumed by herbivores in different ecosystems (i.e., consumption efficiency). Note that this rarely exceeds 10 per cent.

Ecosystem	*Consumption efficiency (%)*
Forests	
Tropical rain forests (various)	7–13
Margrove forest (New Guinea)	5–15*
Temperate deciduous forests (various)	1–11*
Savannas and grasslands	
Lomato savanna, Ivory cost (large herbivore absent)	1.1
Savannas of east and southern Africa (large herbivores only)	2–15
North American grasslands (large herbivores absent)	1.5–5
North American grasslands (large herbivores present)	5–15

* Consumption of leaves only.

Herbivores may destroy much more in their foraging activities. Little is known about the wastage.

Decomposer Food Web

Since herbivores consume only small proportion of primary production, little less of it goes as production of dead plant material. Decomposition of below ground parts is only little studied much information is available on litter. Table 6.6 shows annual litter fall in forests.

Table 6.6. Litter fall in forests of the world. Only 'small litter' (leaf, twig, fruit, flower etc.) is included. Note greater litter production in tropical forests.

	Litter fall ($gm^{-2}\ a^{-1}$)	
Forest type (no. of sites)	*Mean*	*Range*
Lowland tropical rain forests	890	560–1330
Montane tropical rain forests	810	680–1010
Temperate deciduous forests	540	340–1600
Temperate, evergreen conifer forests	440	280–610
Boreal evergreen conifer forests	32	29–53

Maximum litter fall in wet forests occur in winter period of year, whereas deciduous tropical trees shed their leaves in dry season.

Rate of decomposition is most rapid in moist tropical environments. Alfalfa lost 75–95 per cent of its weight in six months in Costa Rica, compared with 43–73 per cent in 12 months in California USA. In moist tropics litter decomposition continues throughout the year, may be very low during winter due to low temperature. In seasonal tropical areas, decomposition may cease entirely during dry season. Rates of litter decomposition in different environments are shown in Table 6.7.

Table 6.7. Annual decay rates or leaf litter (rates more than 100 per cent are extrapolations of rapid rates, which were measured for periods of less than one year).

	Loss of dry weight (%)	
Ecosystem type	*Mean*	*Range*
Lowland tropical rain forest	200	35–550
Montane tropical rain forest	55	25–100
Temperate deciduous forest	80	20–200
Tropical savanna grasses	65	30–95
Temperate grasses in natural grasslands	30	10–60

Three major factors influence the rate of decomposition in tropical and temperate ecosystems: (i) the quality of the material to be decomposed (wood decomposes more slowly than leaves); (ii) the abiotic environment which operates through its effects on third factor; and (iii) the decomposer organisms whose total activity is generally indicated by respiration rates.

Much less is known about the decomposition of standing dead to ecologists, though microbial ecologists have worked out this aspect, but very little from energetics point of view.

7

Biotic Diversity and Community Stability

INTRODUCTION

In any closed ecosystem at equilibrium, all the energy of net production must be used up by consumers and decomposers in order for the system to have a balanced energy budget. Such an idealised system can be thought of as being saturated with individual organisms because all available energy is used and no more organisms could be supported. However, predators, by reducing densities of organisms at lower trophic levels, can prevent their prey populations from reaching otherwise maximal stable densities and thereby effectively preclude true saturation within that lower trophic level. If this is the case, only top predator populations reach a sort of 'complete' saturation. Furthermore, antiherbivoure defenses of plants render much of the net primary productivity unusable by animal consumers and thus require that many plant tissues be routed directly through a community's decomposers.

Communities, or portions thereof, can be kept from reaching saturation with individuals in other ways. Real ecological systems are almost never truly closed; instead, they usually both receive materials and energy from other systems and lose them to others. A community or a portion of a community that is not closed may be rarefied by continual or sporadic removal of organisms. As a hypothetical example, consider a lake along a river. Both lake and river contain communities of phytoplankton and zooplankton, but the lake receives an inflow of river water containing no members of the lake community, while losing water containing members of its community. Such a system can never become truly saturated because of the continual removal of organisms from it. Moreover, because the physical environment is changing continually and because it takes time to respond to these changes, populations and communities probably seldom reach equilibrium, although some very *K-selected* organisms may occasionally approach it.

How much does the degree of saturation with individuals vary within and between communities? And how does efficiency of transfer of energy change with degree of saturation with individuals? It is known that the turnover rate of prey is usually highest at intermediate prey densities; moreover, such prey populations often support higher predator densities than larger prey populations. The immense practical value of such knowledge is readily apparent.

Can communities be saturated with species? That is, is there a maximum number of different species that can exist within an ecological system? If so, a new species introduced into such a community must either go extinct or cause the extinction of another species (which it then replaces). Conversely, the successful invasion of a new species into a community without the extermination of an existing species would imply that the original community was not saturated with species.

Limited evidence suggests that portions of some communities may indeed be saturated with species, at least within habitats. R.H. MacArthur and his colleagues demonstrated that bird species diversity is strongly correlated with foliage height diversity (Fig. 7.1) in a remarkably similar way on three continents: North America, South America and Australia. Habitats with equal amounts of foliage (measured by leaf surface) in three layers (0 to 0.6, 0.6 to 8 and over 8 meters above ground) are richer in bird species than are habitats with unequal proportions of foliage in the three layers. The diversity of bird species is lowest in habitats with only one of these layers of vegetation, such as a grassland. Interestingly enough, knowledge of plant species diversity does not allow an improvement in the prediction of bird species diversity which suggests that birds recognise the structure rather than the type of vegetation. Despite the fact that avian niche space is partitioned in a fundamentally different way in Australia, bird species diversity within a given habitat on that continent is very close to what it is in a habitat of similar structure in North America. In addition to illustrating that spatial heterogeneity regulates bird species diversity, the convergence of these data suggests that these avifaunas are saturated with species. However, such neat convergences in species densities of plants, insects and desert lizards do not occur, which suggests that these groups may not always be saturated with species.

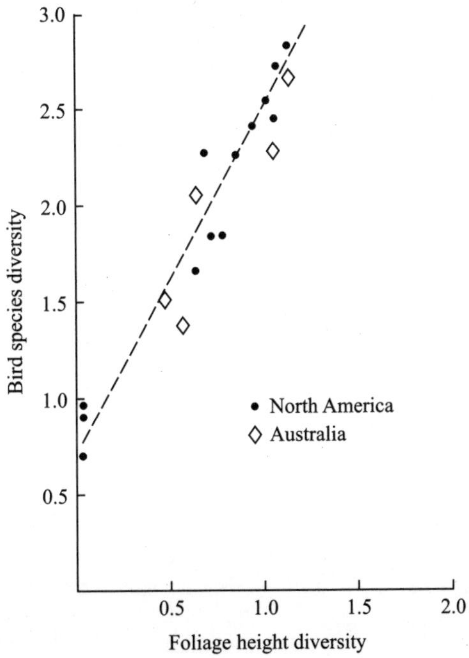

Fig. 7.1. Correlation between foliage height diversity and bird species diversity. North American habitats are shown as solid circles; Australian habitats are represented by diamond. This is perhaps the best evidence that communities may sometimes be saturated with species.

The number of species that can coexist at any point in space may have a distinct upper limit, as previously suggested; however, there is no obvious limitation on the number of species that can occur in a given area, because horizontal replacement of species can allow coexistence of many more species than actually share the use of a common point in space within that area. Indeed, MacArthur has suggested that the horizontal component of diversity ('between-habitat' diversity) may be increasing continually during evolutionary time, whereas point diversities remain nearly constant. Some upper limit on the horizontal turnover of species also seems likely.

SPECIES DIVERSITY

Why does one community contain more species than another? Some complex communities, such as tropical rainforests, consist of many thousands of different plant and animal species, whereas other communities, such as tundra communities, support perhaps only a few hundred species. The number of species often varies greatly even at a local level; thus, a grassland habitat typically contains many fewer species of birds than does an adjacent forest. Indeed, different forest communities in the same general region usually vary in numbers as well as in types of plant and animal species. The number of species is referred to as species richness or more frequently, as species density.

Communities with similar species densities often differ in yet another way. Some contain a few very common species and many rare ones, whereas others support no very common species but many of intermediate abundance. Abundance is only one way of estimating the *relative importance* of various component species within a community; other measures frequently employed include both the biomass of and the energy flow through various species' populations. The relative importance of species varies within and between communities and considerable effort has been expended in attempts to document such differences and to understand why they occur. Importances of different species within a community (or a portion of one) can be depicted conveniently by species importance curves (Fig. 7.2). Several hypothetical distributions of the relative importance of species within communities have been suggested that generate different shaped species importance curves (see Fig. 7.2).

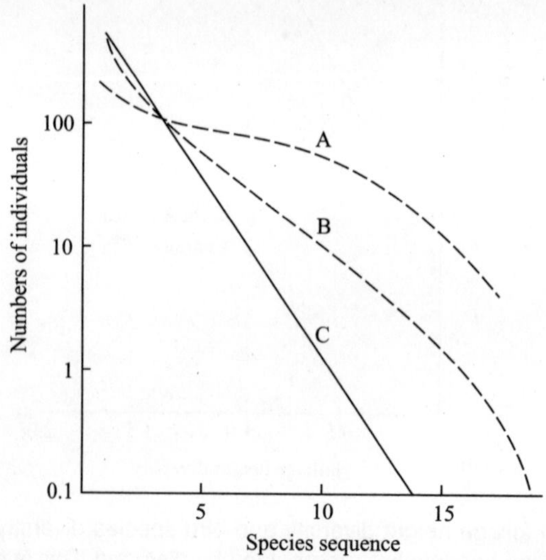

Fig. 7.2. Species importance curves, with species ranked from most important to least important. The line and curves illustrate three different hypothetical distributions based on (A) random niche boundaries (B) multiple niche dimensions, which generate a lognormal distribution of species importances and (C) niche pre-emption, which leads to a geometric series. Data from various real communities fit each hypothetical distribution reasonably well. For a given number of species, diversity would be highest in A and lowest in C.

Species density and relative importance have been combined in the concept of species diversity, which increases with both increasing species density and increasing equality of importance among the members of a community. Species diversity is high when it is difficult to predict the species of a randomly chosen individual organism and low when an accurate prediction can be made. For example, an organism chosen at random from a cornfield would probably be a stalk of corn, whereas one would not venture to guess the probable identity of a randomly chosen organism from a tropical rainforest. We are currently trying to determine not only why different communities contain different numbers of species with differing relative importances, but how such differences in species richness and importance affect other community properties such as trophic structure and community stability.

Communities can differ in species diversity in several ways. First, more diverse communities may contain a greater range of available resources (i.e., a larger total niche hypervolume space) and second, their component species may, on the average, have smaller niche breadths (i.e., each species might exploit a smaller fraction of the total niche hypervolume). The former corresponds roughly to 'more niches' and the latter to 'smaller niches.' Third, two communities with identical niche space and mean niche breadth can still differ in species diversity if they differ in the average degree of niche overlap, because greater niche overlap means that more species can exploit any particular resource (this situation is described as greater resource sharing or 'smaller exclusive niches'). Fourth, alternatively, in communities that do not contain all the species they could conceivably support (i.e., those that are 'unsaturated' with species), species diversity can vary with the extent to which all available resources are exploited by as many different species as possible (i.e., with the degree of saturation with species or with the number of 'empty niches'). Resources are seldom if ever wasted, even in communities that do not contain their full quota of species, because those species that do occur in such communities generally expand their activities and exploit nearly all the available resources, although their efficiency of exploitation may be less than that of some better-adapted species. (Thus, most communities are probably effectively saturated with individuals even if they are not saturated with species).

Because thorough understanding of community species diversity inevitably requires analysis of niche structure and diversification in component populations, investigations of species diversity have usually gone hand in hand with the study of niches. In practice, one is seldom able to study the species diversity of an entire community and usually attention is focused on a portion of a community (an 'assemblage') such as trees, ants, lizards, seed eaters or birds. Using the three major niche dimensions we can partition total species diversity of an area into its spatial, temporal and trophic components. Species replace one another along each of these three niche dimensions and diversity is generated by separation along each.

Study has been conducted on species diversity and niche relations of desert lizards in certain deserts of western North America, Southern Africa and Western Australia. Lizard niches in these deserts differ in the three fundamental dimensions (place, time and food). Moreover, the variety of resources actually used by lizards along each niche dimension, as well as the amount of niche overlap along them, differs markedly among desert systems, Food is a major dimension separating niches of North American lizards; in the Kalahari, food niche separation is slight and differences in the place and time niches are considerable. In the most diverse Australian lizard 'communities,' all three niche dimensions are important in resource partitioning and niche overlap is distinctly reduced. Differences between deserts in lizard species diversity are not accompanied by conspicuous differences in niche breadths; rather, they stem primarily from differences in the variety of resources used by lizards. Moreover, niche overlap does not increase with diversity but varies inversely with species density and is lowest in the most diverse lizard communities of Australia.

The spatial component of diversity is due to differential use of space by different populations; for convenience, it can be broken down into horizontal and vertical components. At a gross geographic level, species replace one another horizontally as one moves from one habitat to another (this is the between habitat component of overall diversity). Similar replacement of species occurs both horizontally and vertically within habitats. For instance, birds often tend to partition a given habitat by occupying different vertical strata, such as low bushes, tree trunks, lower foliage and high canopy. Ground-dwelling mammals and lizards generally partition microhabitats horizontally, with some using open spaces between shrubs and others exploiting the ground beneath or near specific types of vegetation such as grasses, shrubs and trees. Different populations, by occupying different microhabitats, are thus able to coexist within a given habitat and contribute to within-habitat diversity. The within-habitat component of diversity is most easily distinguished from the between-habitat component in relatively homogeneous communities (heterogeneous communities, such as edge communities and ecoclines, include both components). However, even a homogeneous community has an internal structure in that it consists of a mosaic of repeatable horizontal and vertical patches. Because communities and habitats frequently blend into one another, it is often difficult to distinguish between-habitat from within-habitat diversity. Where does one habitat 'stop' and another 'begin'? A sand ridge gradually gives way to a sand plain and the intertidal grades into the deeper benthic zone. The problem of defining a habitat can be overcome by the use of 'point diversities,' which consist of the species diversity occurring at a point in space. Point diversities are difficult to estimate (one might have to wait a very long time to see all the species that use a particular point!), but point diversity should invariably be lower than any areal estimate of diversity, because the different species in a community have each specialised somewhat as to the microhabitats they use.

Provided that different resources are utilised, temporal separation, both daily and seasonally, of species' populations can allow coexistence of more species and hence may add to community diversity. Many instances of subtle differences in times of activity between populations are known, in addition to such conspicuous distinctions as that between nocturnal and diurnal animals.

Yet another means by which community diversity may be enhanced is by trophic differences. Here again, in addition to the conspicuous differences between trophic levels (such as herbivorous, omnivorous and carnivorous), there are more subtle but nevertheless important differences between species even within a given trophic level in the prey they eat. Thus, different species of predators living in the same area tend to eat prey of different sizes and types, with the larger species taking larger prey items (this generalisation applies to most fish, lizards, carnivorous mammals and hawks). Moreover, the composition of the diet often varies markedly among potential competitors. Finally, diversity of plant defensive chemicals doubtless creates numerous and different potential food niches for herbivorous, especially insects thereby greatly facilitating trophic diversity at higher trophic levels.

LATITUDINAL GRADIENTS IN DIVERSITY

A prevalent global pattern of species diversity is of some interest. The diversity of living organisms is usually high near the equator and decreases rather gradually with increasing latitude, both to the north and to the south (Fig. 7.3). Such latitudinal gradients in diversity are widespread among different plant and animal groups and it is likely that a general explanation underlies these ubiquitous patterns. One reason species diversity is higher at lower latitudes than it is in the temperate zones is that often there are more habitats in the tropics. At high altitudes in the tropics, habitats similar to but richer in species than those in temperate zones often occur, whereas true tropical habitats are seldom found in temperate areas. However, the fact that there is a great variety of species where there are more habitats is neither surprising nor theoretically very interesting. Diversity differences within a given habitat type are of much greater interest, for they reflect the partitioning of available niche space within habitats.

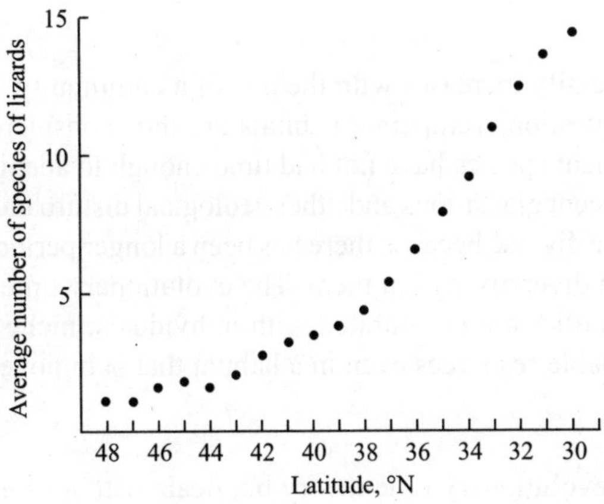

Fig. 7.3. The latitudinal gradient in the average number of species of lizards per degree square in the continental United States.

Considerable speculation about the causes of both local and latitudinal patterns in species diversity has generated many theories and hypotheses for their explanation (Table 7.1), all of which probably operate in some situations. Various mechanisms for determination of diversity are clearly not independent and several may often act in concert or in series in any given case. Each hypothesis or theory is briefly outlined subsequently and some ways in which they could interact are considered.

Table 7.1. Various hypothetical mechanisms for the determination of species diversity and their proposed modes of action.

Level	*Hypothesis or theory*	*Mode of action*
Primary	Evolutionary time	Degree of unsaturation with species
Primary	Ecological time	Degree of unsaturation with species
Primary	Climatic stability	Mean niche breadth
Primary	Climatic predictability	Mean niche breadth
Primary or secondary	Spatial heterogeneity	Range of available resources
Secondary	Productivity	Especially mean niche breadth, but also range of available resources
Secondary	Stability of primary production	Mean niche breadth and range of available resources
Tertiary	Competition	Mean niche breadth
Primary, secondary, or tertiary	Disturbance	Degree of allowable niche overlap and level of competition
Tertiary	Predation	Degree of allowable niche overlap and level of competition

These mechanisms can be classified as primary, secondary or tertiary, depending on whether they act mainly through the physical environment alone, through a mixture of both the physical and the biotic environments or through the biological environment alone, respectively. Ultimately, a thorough understanding of patterns in diversity requires knowledge of primary-level mechanisms.

Evolutionary Time

This theory assumes that diversity increases with the age of a community, although the validity of this assumption is still open to question. Temperate habitats are thus considered to be impoverished with species because their component species have not had time enough to adapt to or to occupy completely, their environment since the recent glaciations and other geological disturbances. However, more 'mature' tropical communities are more diverse because there has been a longer period without major disturbances for organisms to speciate and diversify within them. The evolutionary time theory does not necessarily imply that temperate communities are unsaturated with individuals; niche expansion may often allow nearly full utilisation of available resources even in a habitat that is impoverished in species.

Ecological Time

This theory is similar to the evolutionary time theory but deals with a shorter, more recent, time span. Here we are concerned primarily with time available for dispersal rather than with time for speciation and evolutionary adaptation. Newly opened or remote areas of suitable habitat, such as a patch of forest burned by lightning, an isolated lake or a patch of sand dunes, may not have their full complement of species because there has been inadequate time for dispersal into these areas. Dispersal powers of most organisms are good enough that this mechanism may be of relatively minor importance in most communities.

Climatic Stability

A stable climate is one that does not change much with the seasons. Successful exploitation of environments with unstable climates often requires that organisms have broad tolerance limits to cope with the wide range of environmental conditions they encounter. Thus, by demanding generalisation, such variable environments favour organisms with broad niches. Conversely, environments with more constant climates allow finer specialisation and narrower niches. For example, plants and animals in the relatively constant tropics are often highly specialised in both the places they forage and the foods they eat. Obviously, in two habitats with the same range of available resources, the one whose component species each use a smaller fraction of these resources will support more species. By these means, the number of species should increase with climatic stability.

Climatic Predictability

Many aspects of climate, although temporally variable, are nevertheless highly predictable in that they repeat themselves fairly exactly from day to day and year after year. Such cyclical predictability can allow organisms to evolve some degree of dependence on, as well as to specialise on, particular environmental conditions and temporal patterns of resource availability, thereby enhancing daily or seasonal replacement of species and the temporal component of total diversity. For example, deep freshwater lakes in the temperate zones typically have a consistent annual succession of primary producers; a major causal factor is nutrient availability, which changes markedly during the year, being greatest during the spring and fall turnovers. Thus, different species of phytoplankton have adapted to exploit lakes under particular environmental conditions that recur regularly every year. Annual plants in Arizona's Sonoran Desert have adapted to the marked bimodal annual precipitation pattern with distinct rainfall peaks in winter and summer. There are two distinct sets of species, one whose seeds germinate under wet and cooler conditions (winter annuals) and another set that germinate when conditions are wet but warmer (summer annuals that bloom in late summer after the flash floods).

Spatial Heterogeneity

A forest contains more different species of birds than a grassland, an arboreal desert generally supports more species of lizards than one without trees and a tidal flat with a great variety of particle sizes and substrate types has more species of mud-dwelling invertebrates than a homogeneous mud flat. Structurally complex habitats obviously offer a greater variety of different microhabitats than simple habitats do. Because there are more different ways of exploiting them, such spatially heterogeneous habitats usually support more species than homogeneous ones do; thus, species replace one another in space with greater frequency and the spatial component of diversity is higher. Correlations between the structural complexity of a habitat and the species diversity of its biota are widespread (one discussed earlier in this chapter is illustrated in Fig. 7.1).

Productivity

In habitats with little food, foraging animals cannot afford to bypass many potential prey items; where food is abundant, however, individuals can be more selective and confine their diets to better prey items. Thus, more productive habitats or those where food is dense, offer more prey choice and hence allow greater dietary specialisation than do less productive habitats. Because each species uses less of the total range of available foods, the same spectrum of food types will support more species in a more productive environment (Fig. 7.4). In addition, productive habitats may support more species than similar, less productive ones do, by virtue of the fact that certain resources, too sparse to support a species in unproductive habitats, are dense enough to be successfully exploited in productive habitats. An open desert with only one ant nest per hectare might not support a population of specialised ant-eating lizards, whereas another, richer area with several ant nests per hectare would.

Stability of Primary Production

Just as more stable and more predictable climates support more species, areas with temporally stable and predictable patterns of production should often allow coexistence of more species than would be possible in areas with more variable or erratic productivity. This secondary- or tertiary-level mechanism differs from those of climatic stability and climatic predictability in that plants themselves react to climatic conditions and hence can alter temporal variability with their own homeostatic adaptations and storage capacities. Plants both buffer and enhance physical fluctuations, either by releasing the products of primary production (flowers or seeds) gradually and continuously or by expending them in temporally erratic blooms. Thus, climatic stability, climatic predictability and stability of primary production could be combined under a more inclusive heading of 'temporal heterogeneity' to parallel 'spatial heterogeneity'.

Competition

In many diverse communities, such as tropical rainforests, populations are thought to be often near their maximal sizes, with the result that intraspecific and interspecific competition are frequently keen. Selection for competitive ability (K selection) is therefore strong and most successful organisms in these communities have their own zone of competitive superiority. Thus, organisms that have specialised as to foods or habitats are at a competitive advantage and the resulting narrow niches make high diversity possible. By way of contrast, populations in many less diverse communities, such as temperate and polar ones, are thought to be less stable and often well below their maximal sizes. As a result, portions of the community are often unsaturated with individuals and intraspecific and interspecific competition are frequently relatively lax. In such communities, it is often the physical world rather than the biotic one that demands adaptation. Selection for competitive ability is weak, whereas selection for rapid reproduction (r selection) is strong. Such r-selected populations typically have broad tolerance limits and relatively broad niches.

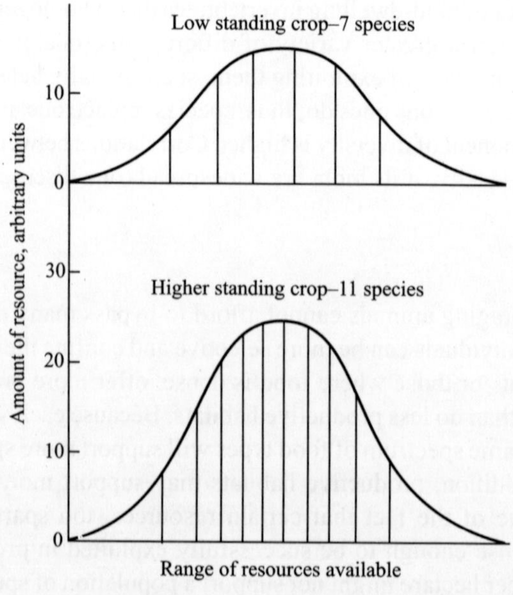

Fig. 7.4. Graphic portrayal of the way in which more abundant resources can support a greater number of species. The horizontal axis represents the different kinds of available resources (ranked in any convenient order). The vertical axis is the amount of each resource type. Both curves cover the same range of resource types, but the lower curve is exactly twice the height of the upper one. All segments under both curves, except the tails, contain about the same area and therefore approximately the same amount of resource. With a low standing crop, component species must have broad niches and only seven species can coexist; when standing crop is doubled, component species can have narrower niches and 11 species may exist in the community.

Disturbance

Disturbances can result in the continual density-independent removal of organisms from a community. The disturbance hypothesis is essentially an alternative to the competition hypothesis; these two mechanisms would seem to be mutually exclusive. In communities that are not fully saturated with individuals, competition is reduced and coexistence is possible without competitive exclusion. Thus, this hypothesis suggests that communities (or portions of communities) can in some sense be oversaturated with species in that more species coexist than would be possible if the system were allowed to become truly saturated with individuals. Disturbance may operate through primary-, secondary-or tertiary-level mechanisms (see also mechanism 10, Predation). Catastrophic winter cold snaps and subsequent density-independent kills illustrate primary-level disturbance.

An important variant of this hypothesis, termed the intermediate disturbance hypothesis, recognises that, even though infrequent to moderately frequent disturbances enhance diversity, extremely frequent disturbances can be so decimating that they operate in reverse to reduce diversity. Thus, diversity may actually peak at intermediate levels of disturbance.

Predation

By either selective or random removal of individual prey organisms, predators can act as rarefying agents and effectively reduce the level of competition among their prey. Indeed, predators can allow the local coexistence of species that are eliminated by competitive exclusion in the absence of a predator. Because many predators prey preferentially upon more abundant prey types, predation is often frequency dependent, which promotes prey diversity. Clearly, several of these mechanisms may often act together to determine the diversity of a given community and the relative importance of each mechanism doubtless varies widely from community to community. A multitude of various possible ways in which these mechanisms could interact have been suggested; one is shown in Fig. 7.5.

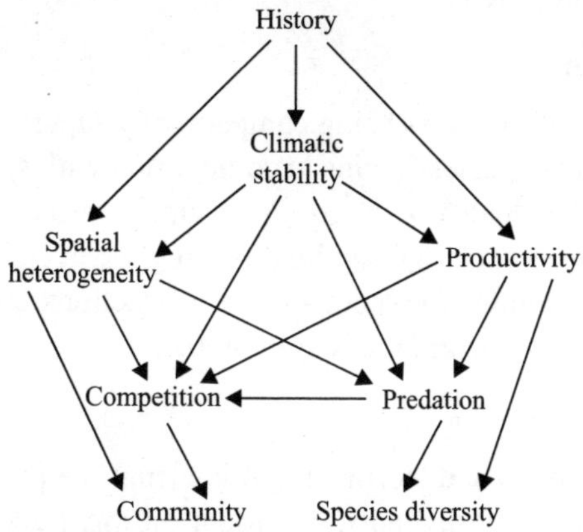

Fig. 7.5. One way in which various mechanisms might interact to determine community diversity.

Tree Species Diversity in Tropical Rainforests

In the lowland tropics, between 50 and 100 different species of trees usually occur together on a single hectare. Although tree species diversity is thus exceedingly high, many of these trees are phenotypically almost identical, with broad evergreen leaves and smooth bark (indeed, only an expert can distinguish among most species). Many species are quite rare with densities below one tree per hectare. How can so many similar species, apparently all light limited, coexist at such low densities? Explaining why tropical tree diversity is so high is among the most challenging questions facing ecologists. Numerous hypotheses have been proposed but relevant data on this fascinating phenomenon remain distressingly scant.

Seed predation hypothesis

Because seed predation is intense in the tropics, Janzen argued that seedlings cannot establish themselves in the vicinity of parental trees since the high densities of seeds attract many seed predators. This argument predicts that successful recruitment of seeds to seedlings will occur in a ring around (but at some distance from) the parental tree. Inside and outside this ring, other tree species can establish themselves. The variety of seed protection tactics (such as toxic matrices) has forced many seed predators to specialise on the seeds of particular species. Heavy seed predation coupled with species-specific seed predators holds down densities of various tree species and creates a mosaic of conditions for the establishment of seedlings. Hubbell examines data relevant to this hypothesis and concludes that factors other than seed predation limit the abundance of tropical tree species and prevent single-species dominance.

Nutrient mosaic hypothesis

The number of ways in which plants can differ is decidedly limited, especially in the wet tropics where variation in soil moisture is relatively slight. One mechanism that could help to maintain high plant species diversity involves differentiation in the use of various materials, such as nitrogen, phosphorus, potassium, calcium, various rare earth metals and so on. According to this argument, each tree species has its own peculiar set of requirements; the soil underneath each species becomes depleted of those particular resources, making it unsuitable for seedlings of the same species. (Eventually, after the tree falls and is decomposed, these materials reenter the nutrient pool and that species grows again.) Thus, like the seed predation hypothesis, this hypothesis predicts a 'shadow' around a parent tree where seedlings of that species will be rare or nonexistent.

Circular networks hypothesis

In this mechanism, species A is envisioned as being competitively superior to species B, while species B in turn excludes species C, whereas species C wins in competition with species A. Under such a circular hierarchy of competitive ability, the identity of the species occurring at a particular spot will be continually changing from C to B to A and then back to C again, repeating the cycle. Circular networks with many more species could exist. Such nontransitive competitive interactions could help to maintain the high diversity of tropical trees and corals in coral reef communities.

Disturbance hypotheses

Frequent disturbances by fires, floods and storms might interrupt the process of competitive exclusion locally and allow maintenance of high diversity. A variant on this hypothesis involving epiphytes as agents of disturbance (many more epiphytes occur in the tropics than in the temperate zones) has been developed by Strong who suggests that tree falls (due to epiphyte loads) are frequent in the tropics, continually opening up patches in the forest and fostering local secondary succession.

The preceding hypotheses barely begin to address the question of why hardly any temperate forests support more than a dozen species of trees. If there are more species-specific seed predators in the tropics, why? Why should nutrient differentiation be more pronounced in the tropics? Why doesn't the circular network mechanism foster higher diversity in temperate forests? Are disturbances more frequent in the tropics and, if so, why? Ultimately, latitudinal variation in any of the previously proposed mechanisms will have to be related to underlying variation in primary physical variables such as climate.

A tropical dry forest in Costa Rica was subjected to fairly detailed scrutiny by Hubbell's who mapped 13.44 hectares of forest and analysed dispersion patterns of the 61 species of trees occurring on this study plot. Prior to Hubbell's work, the traditional generalisation had been that tropical trees tend to occur at low densities, more or less uniformly distributed in space (evenly spread out). However, in this Costa Rican dry forest, most tree species, especially rarer ones, were either clumped or randomly dispersed. Moreover, Hubbell found that among the 30 commonest tree species, densities of juvenile trees decreased approximately exponentially with distance away from adults (only 5 of the 30 species displayed sapling 'rings' as predicted by Janzen's seed dispersal/predation model and these rings were very close to adult trees, essentially at the edge of the crown canopy, where a high concentration of seeds might be expected to fall). Hubbell stresses that his results strongly imply that relatively simple physical and biotic factors must govern seed dispersal. Hubbell does not reject the notion of intense seed predation thinning (particularly by bruchid weevil seed predators) among the large-seeded tree species, but he claims that his evidence of clumped or random spacing patterns refutes seed predation by host-specific herbivorous as a general mechanism for coexistence of tropical trees. Thus, there seems to be more to high tropical tree species diversity than tree spacing constraints. Hubbell suggests that periodic disturbances are crucial and proposes a random walk to extinction model, which generates patterns of relative abundances similar to those observed (along a 'disturbance' gradient). In support of this argument, rare species of trees on the study plot had very low reproductive success-they were not replacing themselves locally, although they may have been invading from nearby.

COMMUNITY STABILITY

Numerous concepts of stability can be and have been applied to communities, including constancy, variability, predictability, persistence, resistance, resilience and others. Too frequently the term 'stability' is left vague and undefined. The stability metric of choice is resilience, the rate at which a system returns to an original state following a perturbation. A fundamental distinction is made between local or neighborhood stability and global stability.

Traditional ecological 'wisdom' holds that more diverse communities are in some sense more stable than simpler ones. MacArthur suggested that the stability of populations in a community should increase both with the number of different trophic links between species and with the equitability of energy flow up the various food chains. He postulated that a community with many trophic links provides greater possibilities for checks and balances to operate between and among various species populations. Should anyone population begin to increase markedly, its predators, by changing their diets and feeding selectively on this abundant prey type ('predator switching'), would exert disproportionate negative density-dependent effects on its population growth. Prey refuges result in what Pimm terms 'donor-controlled' population growth, which also enhances stability. Another way in which diversity could confer stability is by means of compensatory indirect interactions that nullify unstable direct interactions.

The 'commonsense' generalisation that diversity begets stability was challenged by May, who showed that the probability of finding a stable system actually decreases with complexity in randomly constructed model communities. However, real communities are far from random in construction and must obey various constraints; there can be no more than five to seven trophic levels, no three-species food-chain 'loops' can occur, there must be at least one producer in the system and so on. Lawlor argues that it is extremely unlikely that any of May's random systems even remotely resemble real ecological systems and asserts that the relationship between complexity and stability in large-scale model ecosystems remains unresolved. An effort must be made to perform a comparable analysis on biologically realistic networks to determine whether diversity in such systems enhances or reduces stability.

Empirical studies have produced conflicting results. Watt found that herbivorous insect species that feed on a wide variety of tree species in Canada actually have less stable populations than do similar insects with more restricted diets. Watt does not indicate whether insects with stable populations have a greater variety of potential predators, but population stability did increase with increasing numbers of competing species as would be expected clearly the relationship between diversity and stability remains an important but unresolved problem in community ecology.

8

Interface Between Climate and Vegetation

INTRODUCTION

Climate is the major determinant of vegetation. Plants in turn exert some degree of influence on climate. Both climate and vegetation profoundly affect soil development and the animals that live in an area. Here we examine some ways in which climate and vegetation interact. More emphasis is given to terrestrial ecosystems than to aquatic ones, although some aquatic analogues are briefly noted.

PLANT LIFE FORMS AND BIOMES

Terrestrial plants adapted to a particular climatic regime often have similar morphologies or plant growth forms. Thus, climbing vines, epiphytes and broad-leafed species characterise tropical rainforests. Evergreen conifers dominate very cold areas at high latitudes and/or altitudes, whereas small frost-resistant tundra species occupy still higher latitudes and altitudes. Seasonal temperate zone areas with moderate precipitation usually support broad-leafed, deciduous trees, whereas tough-leafed (sclerophyllous) evergreen shrubs or so-called chaparral-type vegetation, occur in regions with winter rains and a pronounced long water deficit during spring, summer and fall. Chaparral vegetation is found wherever this type of climate prevails, including southern California, Chile, Spain, Italy, southwestern Australia and the northern and southern tips of Africa, although the actual plant species comprising the flora usually differ. Areas with very predictable and stable climates tend to support fewer different plant life forms than regions with more erratic climates. In general, there is a close correspondence between climate and vegetation; indeed, climatologists have sometimes used vegetation as the best indicator of climate. Thus, rainforests occur in rainy tropical and rainy warm-temperature climates, forests exist under more moderate mesic climates, savannas and grasslands prevail in semiarid climates and deserts characterise still drier climates. Of course, topography and soils also play a part in the determination of vegetation types, which are sometimes termed plant formations. Such major communities of characteristic plants and animals are known as biomes. Classification of natural communities is discussed later in this chapter.

MICROCLIMATE

Even in the complete absence of vegetation, major climatic forces or macroclimates, are expressed differently at a very local spatial level, which has resulted in the recognition of so-called microclimates. Thus, the surface of the ground undergoes the greatest daily variation in temperature and daily thermal flux is progressively reduced with both increasing distance above and below ground level (Fig. 8.1). During daylight hours the surface intercepts most of the incident solar energy and rapidly heats up, whereas at night this same surface cools more than its surroundings. Such plots of temperature versus height above and below ground are called thermal profiles. An analogous type of graph, called a *bathythermograph*, is often made for aquatic ecosystems by plotting temperature against depth.

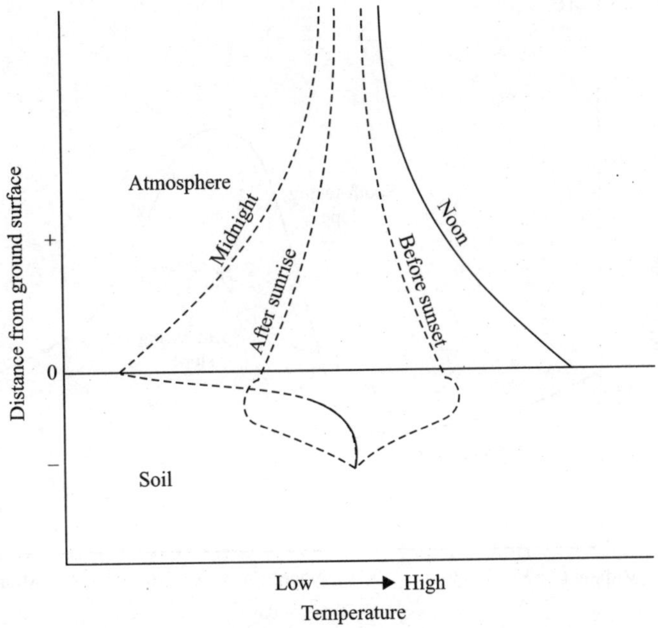

Fig. 8.1. Idealised thermal profile showing temperatures at various distances above and below ground at four different times of day.

Daily temperature patterns are also modified by topography even in the absence of vegetation. A slope facing the sun intercepts light beams more perpendicularly than does a slope facing away from the sun; as a result, a south-facing slope in the Northern Hemisphere receives more solar energy than a north-facing slope and the former heats up faster and gets hotter during the day (Fig. 8.2). Moreover, such a south-facing slope is typically drier than a north-facing one because it receives more solar energy and therefore more water is evaporated.

By orienting themselves either parallel to or at right angles to the sun's rays, organisms (and parts of organisms such as leaves) may decrease or increase the total amount of solar energy they actually intercept. Leaves in the brightly illuminated canopy often droop during midday, whereas those in the shaded understory typically present their full surface to incoming beams of solar radiation. Similarly, many desert lizards position themselves on the ground perpendicular to the sun's rays in the early morning when environmental temperatures are low, but during the high temperatures of midday, these same animals reduce their heat load by climbing up off the ground into cooler air temperatures and orienting themselves parallel to the sun's rays by facing into the Sun.

The major effect of a blanket of vegetation is to moderate most daily climatic changes, such as changes in temperature, humidity and wind. (However, plants generate daily variations in concentrations of oxygen and carbon dioxide through their photosynthetic and respiratory activities.) Thermal profiles at midday in cornfields at various stages of growth are shown in Fig. 8.3, demonstrating the marked reduction in ground temperature due especially to shading. In the mature field, air is warmest at about a meter above ground. Similar vegetational effects on microclimates occur in natural communities, for example, a patch of open sand in a desert might have a daily thermal profile somewhat like that shown in Fig. 8.1, whereas temperatures in the litter underneath a nearby dense shrub would vary much less with the daily march of temperature.

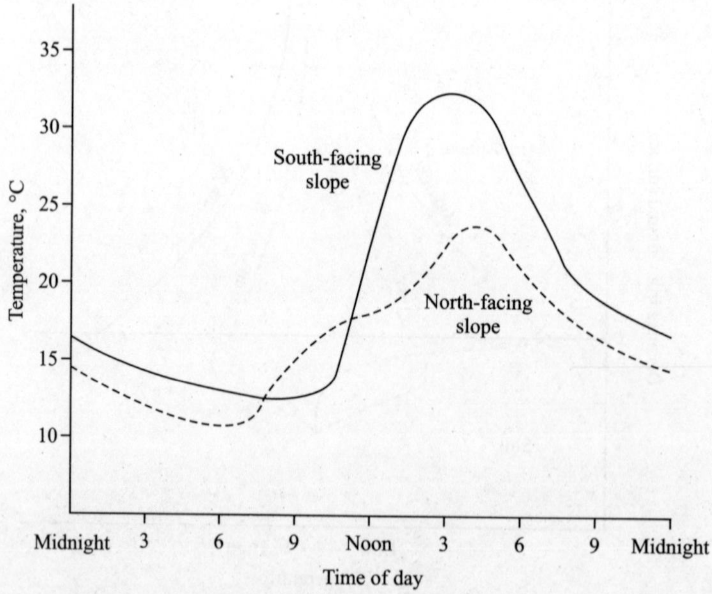

Fig. 8.2. Daily marches of temperature on an exposed south-facing slope (solid line) and on a north-facing slope (dashed line) during late summer in the Northern Hemisphere.

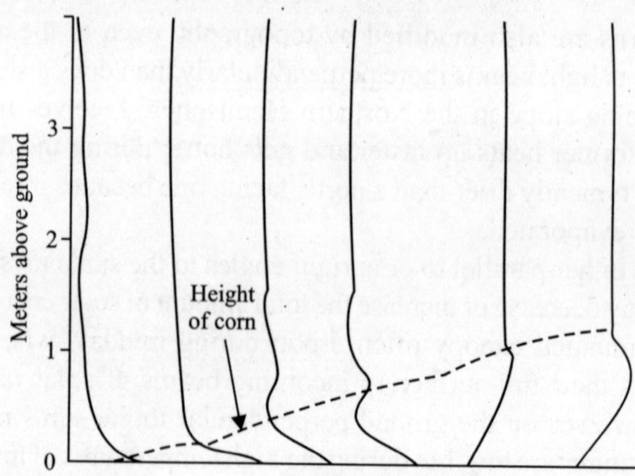

Fig. 8.3. Temperature profiles in a growing cornfield at midday; showing the effect of vegetation on thermal microclimate.

Humidities are similarly modified by vegetation, with relative humidities within a dense plant being somewhat greater than those of the open air adjacent to the plant. An aphid may spend its entire lifetime in the very thin zone (only about a millimeter thick) of high humidity that surrounds the surface of a leaf.

Moisture content is more stable and therefore more dependable, deeper in the soil than it is at the surface, where high temperatures periodically evaporate water to produce a desiccating effect. Wind velocities are also reduced sharply by the vegetation and are usually lowest near the ground (Figs. 8.4 and 8.5). Moving currents of air promote rapid exchange of heat and water; hence an organism cools or warms more rapidly in a wind than it does in a stationary air mass at the same temperature. Likewise, winds often carry away moist air and replace it with drier air, thereby promoting evaporation and water loss. The desiccating effects of such dry winds can be extremely important to an organism's water balance.

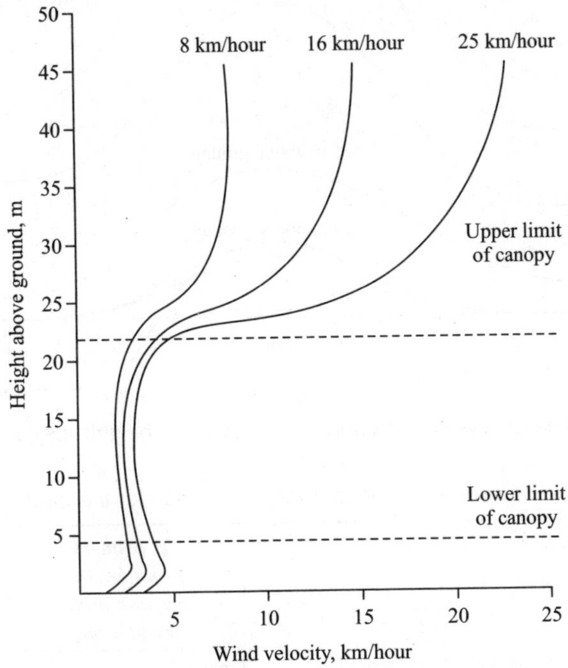

Fig. 8.4. Wind velocities within a forest vary relatively little with changes in the wind velocity above the canopy.

In aquatic systems, water turbulence parallels wind in many ways and rooted vegetation around the edges of a pond or stream reduces water turbulence. At a more microscopic level, algae and other organisms that attach themselves to underwater surfaces (so-called *periphyton*) create a thin film of distinctly modified microenvironment in which water turbulence, among other things, is reduced. Localised spatial patches with particular concentrations of hydrogen ions (pH), salinities, dissolved nitrogen and phosphorus and the like form similar aquatic microhabitats.

By actively or passively selecting such microhabitats, organisms can effectively reduce the overall environmental variation they encounter and enjoy more optimal conditions than they could without microhabitat selection. Innumerable other microclimatic effects could be cited, but these should serve to illustrate their existence and their significance to plants and animals.

PRIMARY PRODUCTION AND EVAPOTRANSPIRATION

In terrestrial ecosystems, climate is by far the most important determinant of the amount of solar energy plants are able to capture as chemical energy or the gross primary productivity (Table 8.1). In warm arid regions, water is a master limiting factor and in the absence of runoff, primary production is strongly positively correlated with rainfall in a linear fashion.

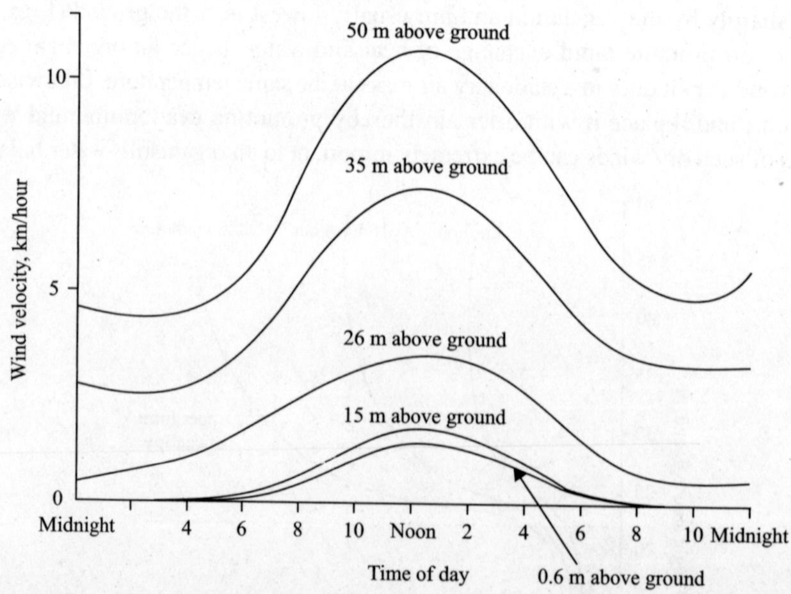

Fig. 8.5. Daily march of average wind velocities during June at various heights inside a coniferous forest in Idaho.

Table 8.1. Net primary productivity and world net primary production for the major ecosystems.

Description	*Area*[a] *(10^6 km^2)*	*Net primary productivity per unit area*[b] *(dry g/m^2/yr.)* Normal range	*Mean*	*World net primary production*[c] *(10^9 dry tons/yr.)*
Lake and stream	2	100–1500	500	1.0
Swamp and marsh	2	800–4000	2000	4.0
Tropical forest	20	1000–5000	2000	40.0
Temperate forest	18	600–2500	1300	23.4
Boreal forest	12	400–2000	800	9.6
Woodland and shrubland	7	200–1200	600	4.2
Savanna	15	200–2000	700	10.5
Temperate grassland	9	150–1500	500	4.5
Tundra and alpine	8	10–400	140	1.1
Desert scrub	18	10-250	70	1.3
Extreme desert, rock and ice	24	0–10	3	0.07
Agricultural land	14	100–4000	650	9.1
Total land	149		730	109.0

(Contd...)

		Net primary productivity per unit area[b] (dry g/m²/yr.)		
Description	*Area[a] (10^6 km²)*	*Normal range*	*Mean*	*World net primary production[c] (10^9 dry tons/yr.)*
Open ocean	332	2–400	125	41.5
Continental shelf	27	200–600	350	9.5
Attached algae and estuaries	2	500–4000	2000	4.0
Total ocean	361		155	55.0
Total for Earth	510		320	164.0

[a] Square kilometers × 0.3861 = square miles.

[b] Grams per square meter × 0.01 = t/ha, × 0.1 = dz/ha or m centn/ha (metric centers, 100 kg, per hectare, 10^4 square meters), × 10 = kg/ha, × 8.92 = lb/acre.

[c] Metric tonnes (10^6 g) × 1.1023 = English short tonnes.

Above about 80 centimeters of precipitation per year, primary production slowly decreases with increasing precipitation and then levels off (asymptotes; Fig. 8.6). Notice that some points fall below the curve in this figure, presumably because there is some water loss by runoff and seepage into groundwater supplies.

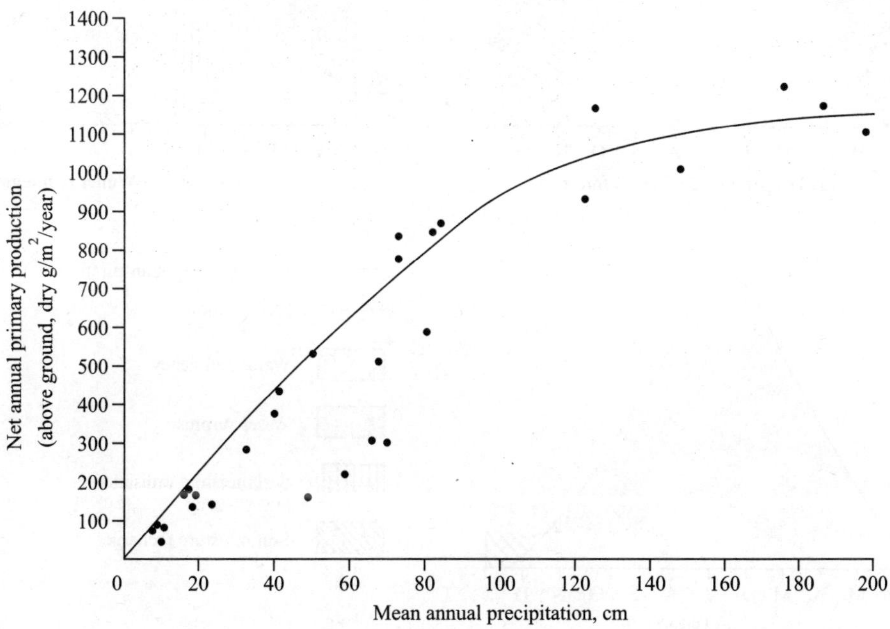

Fig. 8.6. Net primary productivity (above ground) plotted against average annual precipitation.

Evapotranspiration refers to the release of water into the atmosphere as water vapour, both by the physical process of evaporation and by the biological processes of transpiration and respiration. The amount of water vapour thus returned to the atmosphere depends strongly on temperature, with greater evapotranspiration at higher temperatures. The theoretical temperature-dependent amount of water that could be 'cooked out' of an ecological system, given its input of solar energy and provided that much water fell on the area, is called its potential evapotranspiration (PET). In many ecosystems, water is frequently in short supply, so actual evapotranspiration (AET) is somewhat less than potential (clearly, AET can never exceed PET and is equal to PET only in water-saturated habitats). Actual evapotranspiration can be thought of as the reverse of rain, for it is the amount of water that actually goes back into the atmosphere at a given spot.

The potential evapotranspiration for any spot on Earth is determined by the same factors that regulate temperature, most notably latitude, altitude, cloud cover and slope (topography). There is a nearly one-to-one correspondence between PET and temperature and an annual march of PET can be plotted in centimeters of water. By superimposing the annual march of precipitation on these plots (Fig. 8.7), seasonal changes in water availability can be depicted graphically. A water deficit occurs when PET exceeds precipitation; a water surplus exists when the situation is reversed.

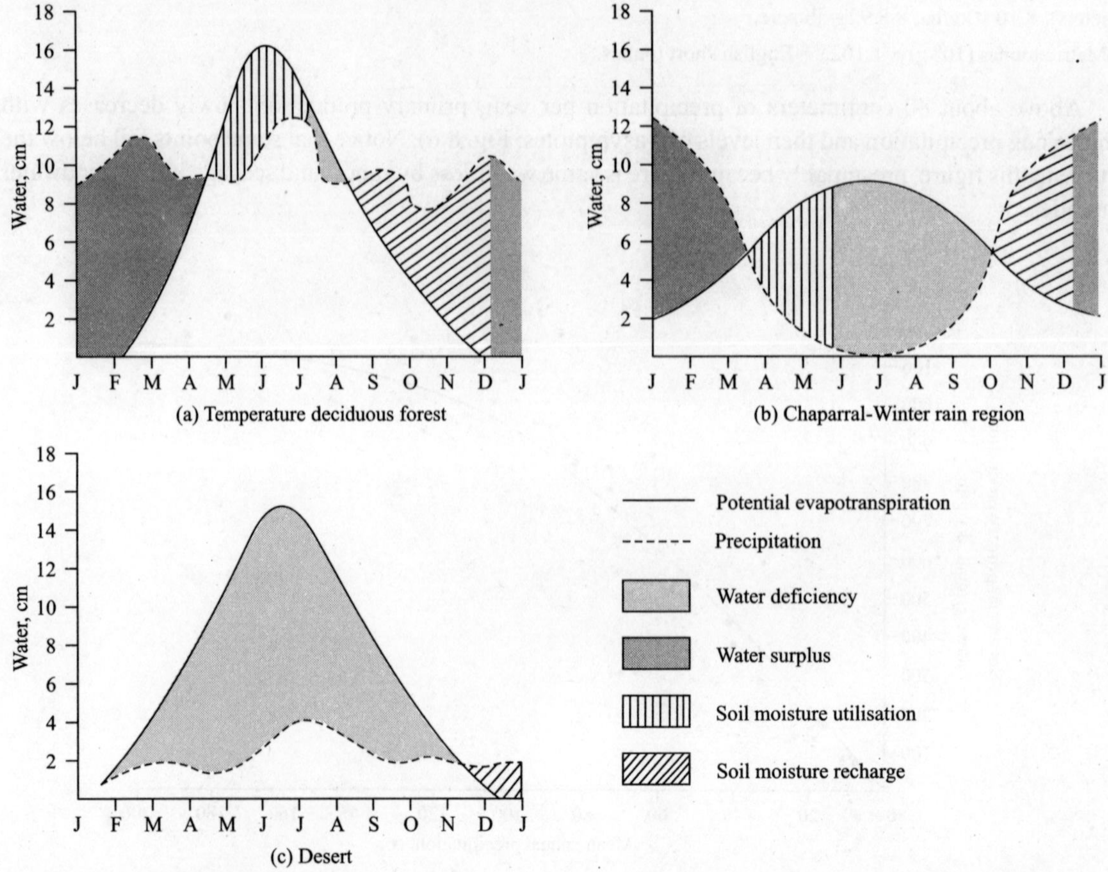

Fig. 8.7. Plots of annual march of potential evapotranspiration superimposed on the annual march of precipitation for three ecologically distinct regions showing the water relations of each. (a) Temperate deciduous forest; (b) chaparral with winter rain; and (c) desert.

During a period of water surplus, some water may be stored by plants and some may accumulate in the soil as soil moisture, depending on runoff and the capacity of soils to hold water; during a later water deficit, such stored water can be used by plants and released back into the atmosphere. Winter rain is generally much less effective than summer rain because of the reduced activity (or complete inactivity) of plants in winter; indeed, two areas with the same annual march of temperature and total annual precipitation may differ greatly in the types of plants they support and in their productivity as a result of their seasonal patterns of precipitation. An area receiving about 50 centimeters of precipitation annually supports either a grassland vegetation or chaparral, depending on whether the precipitation falls in summer or winter, respectively.

Net annual primary production above ground is strongly correlated with actual evapotranspiration or AET (Fig. 8.8). This correlation is remarkable in that AET was crudely estimated using only monthly macroclimatic statistics with no allowance for either runoff and water-holding capacities of soils or for groundwater usage.

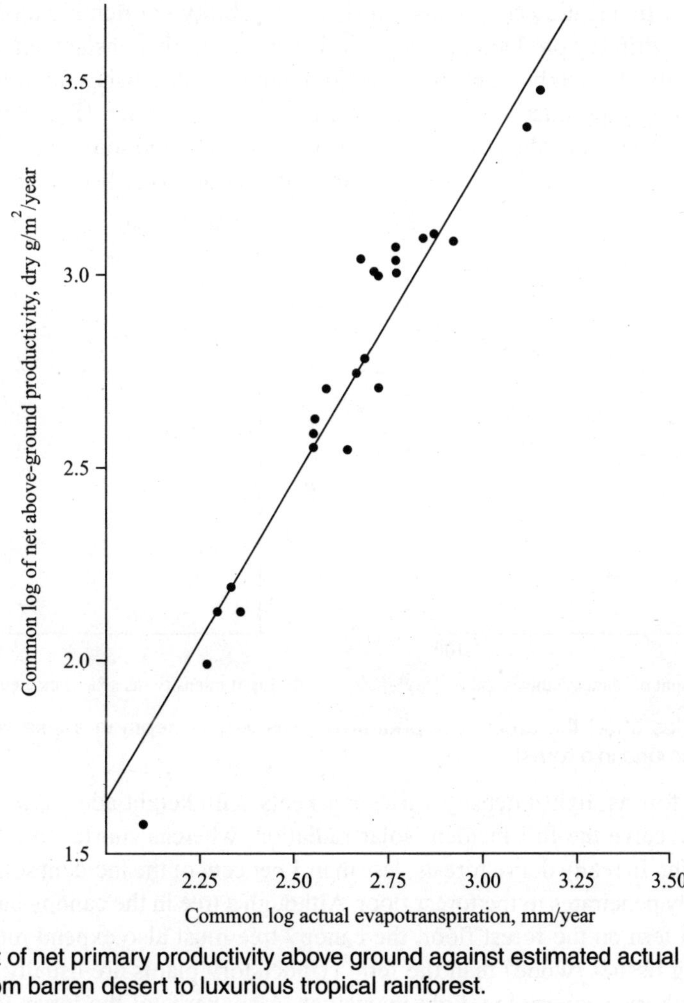

Fig. 8.8. Log-log plot of net primary productivity above ground against estimated actual evapotranspiration for 24 areas, ranging from barren desert to luxurious tropical rainforest.

Rosenzweig suggested that the reason for the observed correlation is that AET measured simultaneously two of the most important factors limiting primary production on land: water and solar energy. Photosynthesis, that fundamental process on which nearly all life depends for energy, is represented by the chemical equation

$$6CO_2 + 12H_2O \rightarrow C_6H_{12}O_6 + 6O_2 + 6H_2O$$

where $C_6H_{12}O_6$ is the energy-rich glucose molecule. Carbon dioxide concentration in the atmosphere is fairly constant at about 0.03 to 0.04 per cent and does not strongly influence the rate of photosynthesis, except under unusual conditions of high water availability and full or nearly full sunlight, when CO_2 is limiting. Rosenzweig notes that, on a geographical scale, each of the other two requisites for photosynthesis, water and solar energy, is much more variable in its availability and more often limiting; moreover, AET measures the availability of both. Temperature is often a rate-limiting factor and markedly affects photosynthesis; it, too, is presumably incorporated into an estimated AET value. Primary production may be influenced by nutrient availability as well, but in many terrestrial ecosystems, these effects are often relatively minor. In aquatic ecosystems, nutrient availability is often a major determinant of the rate of photosynthesis. Primary production in aquatic systems of fairly constant temperature (such as the oceans) is often strongly affected by light. Because light intensity diminishes rapidly with depth in lakes and oceans, most primary productivity is concentrated near the surface (Fig. 8.9a). However, short wavelengths (blue light) penetrate deeper into water than longer ones and some benthic (bottom-dwelling) marine 'red' algae have evolved unique photosynthetic adaptations to utilise these wavelengths.

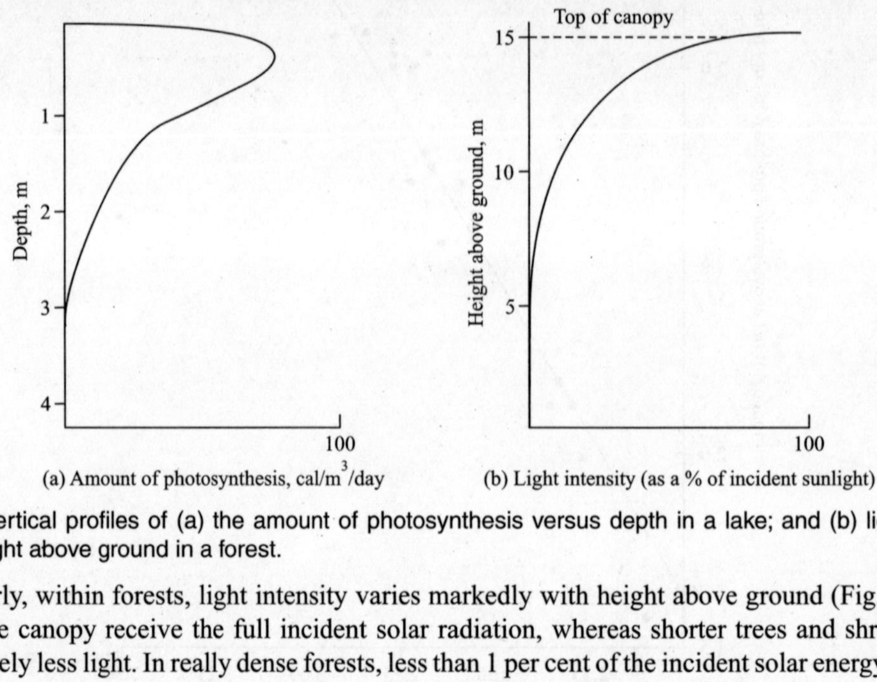

Fig. 8.9. Vertical profiles of (a) the amount of photosynthesis versus depth in a lake; and (b) light intensity versus height above ground in a forest.

Similarly, within forests, light intensity varies markedly with height above ground (Fig. 8.9b). Tall trees in the canopy receive the full incident solar radiation, whereas shorter trees and shrubs receive progressively less light. In really dense forests, less than 1 per cent of the incident solar energy impinging on the canopy actually penetrates to the forest floor. Although a tree in the canopy has more solar energy available to it than a fern on the forest floor, the canopy tree must also expend much more energy on vegetative supporting tissues (wood) than the fern. (Understory plants are usually very shade tolerant and able to photosynthesise at very low light intensities. Also, herbs of the forest floor often grow and flower early in the spring before deciduous trees leaf out.) Hence, each plant life form and each growth strategy has its own associated costs and profits.

SOIL FORMATION AND PRIMARY SUCCESSION

Soils are a key part of the terrestrial ecosystem because many processes critical to the functioning of ecosystems occur in the soil. This is where dead organisms are decomposed and where their nutrients are retained until used by plants and, indirectly, returned to the remainder of the community. Soils are essentially a meeting ground of the inorganic and organic worlds. Many organisms live in the soil; perhaps the most important are the decomposers, which include a rich biota of bacteria and fungi. Also, the vast majority of insects spend at least a part of their life cycle in the soil. Certain soil-dwelling organisms, such as earthworms, often play a major role in breaking down organic particles into smaller pieces, which present a larger surface area for microbial action, thereby facilitating decomposition. Indeed, activities of such soil organisms sometimes constitute a 'bottleneck' for the rate of nutrient cycling and as such they can regulate nutrient availability and turnover rates in the entire community. In aquatic ecosystems, bottom sediments like mud and ooze are closely analogous to the soils of terrestrial systems.

Much of modern soil science or *pedology*, was anticipated by the prominent Russian pedologist V. V. Dokuchaev. He devised a theory of soil formation or pedogenesis, based largely on climate, although he also recognised the importance of time, topography, organisms (especially vegetation) and parent materials (the underlying rocks from which the soil is derived). The relative importance of each of these five major soil-forming factors varies from situation to situation. Figure 8.10 shows how markedly the soil changes along the transition from prairie to forest in the midwestern United States, where the only other conspicuous major variable is vegetation type.

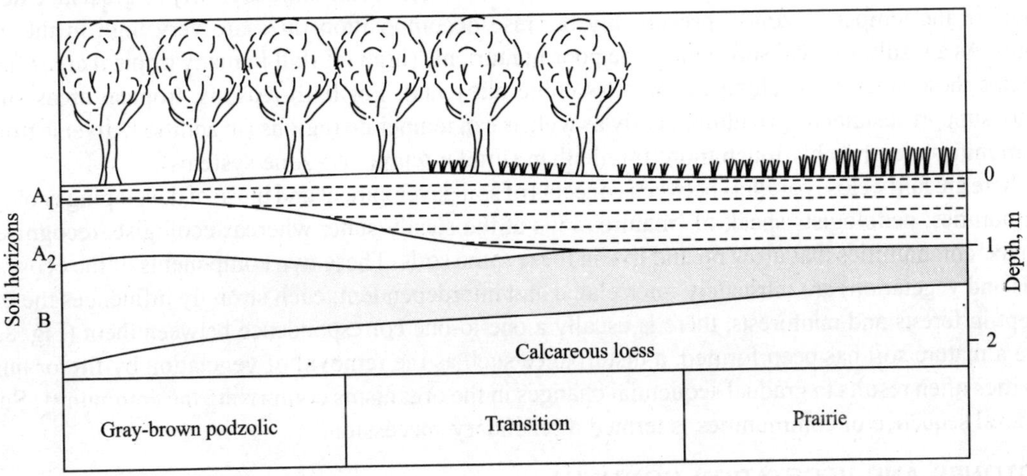

Fig. 8.10. Diagram of typical soil changes along the transition from prairie (deep black topsoil) to forest (shallow topsoil) at the edge of the North American great plains.

Grasses and trees differ substantially both in mineral requirements and in the extent to which the products of their own primary production contribute organic materials to the soil. Typically, natural soils underneath grasslands are considerably deeper and richer in both organic and inorganic nutrients than are the natural soils of forested regions.

The marked effects some soil types can have on plants are well illustrated by so-called *serpentine soils*, which are formed over a parent material of serpentine rock. These soils often occur in localised patches surrounded by other soil types; typically, the vegetation changes abruptly from non-serpentine to serpentine soils. Serpentine soils are rich in magnesium, chromium and nickel, but they contain very little calcium, molybdenum, nitrogen and phosphorus. They usually support a stunted vegetation and are relatively less productive than adjacent areas with different, richer soils. Indeed, entire floras of specialised plant species have evolved that are tolerant of the conditions of serpentine soils (particularly their low calcium levels). Introduced Mediterranean 'weeds' have replaced native Californian coastal grasses and forbs almost everywhere except on serpentine soils, where the native flora still persists.

Soil development from bare rock or primary succession, is a very slow process that often requires centuries. Rock is fragmented by temperature changes, by the action of windblown particles and in colder regions by the alternate expansion and contraction of water as it freezes and thaws. Chemical reactions, such as the formation of carbonic acid (H_2CO_3) from water and carbon dioxide, may also help to dissolve and break down certain rock types like limestones. Such weathering of rocks releases inorganic nutrients that can be used by plants. Eventually, lichens establish themselves and as other plants root and grow, root expansion further breaks up the rock into still smaller fragments. As these plants photosynthesise, they convert inorganic materials into organic matter. Such organic material, mixed with inorganic rock fragments, accumulates and soil is slowly formed. Early in primary succession, production of new organic material exceeds its consumption and organic matter accumulates; as soil 'maturity' is approached, soil eventually ceases to accumulate (see also Fig. 8.14). Whereas most organic material is contributed to soils from above as leaf and litter fall, mineral inorganic components tend to be added from the underlying rocks below. These polarised processes thus generate fairly distinct layers, termed soil horizons.

Even though litter fall is high in tropical forests, it does not accumulate to nearly as great an extent as it does in the temperate zones, presumably because decomposition rates are very high in the warm tropics. As a result, tropical soils tend to be poor in nutrients (high rainfall in many tropical areas further depletes these soils by leaching out water-soluble nutrients). For both reasons, tropical areas simply cannot support sustained agriculture nearly as well as can temperate regions (in addition, diverse tropical communities are probably much more fragile than simpler temperate-zone systems).

There are fairly close parallels between concepts of soil development and those of developing ecological communities; pedologists speak of 'mature' soils at the steady state, whereas ecologists recognise the 'climax' communities that grow on and live in these same soils. These two components of the ecosystem (soils and vegetation) are intricately interrelated and interdependent; each strongly influences the other. Except in forests and rainforests, there is usually a one-to-one correspondence between them (Fig. 8.11). Once a mature soil has been formed, a disturbance such as the removal of vegetation by fire or human activities often results in gradual sequential changes in the organisms comprising the community. Such a temporal sequence of communities is termed a secondary succession.

ECOTONES AND VEGETATION CONTINUA

Communities are seldom discrete entities; in fact, they usually grade into one another in both space and time. A localised 'edge community' between two other reasonably distinct communities is termed an ecotone. Typically, such ecotonal communities are rich in species because they contain representatives from both parent communities and may also contain species distinctive of the ecotone itself. Often a series of communities grade into one another in an almost continuous fashion (Fig. 8.12); such a gradient community is called an ecocline. Ecoclines may occur either in space or in time.

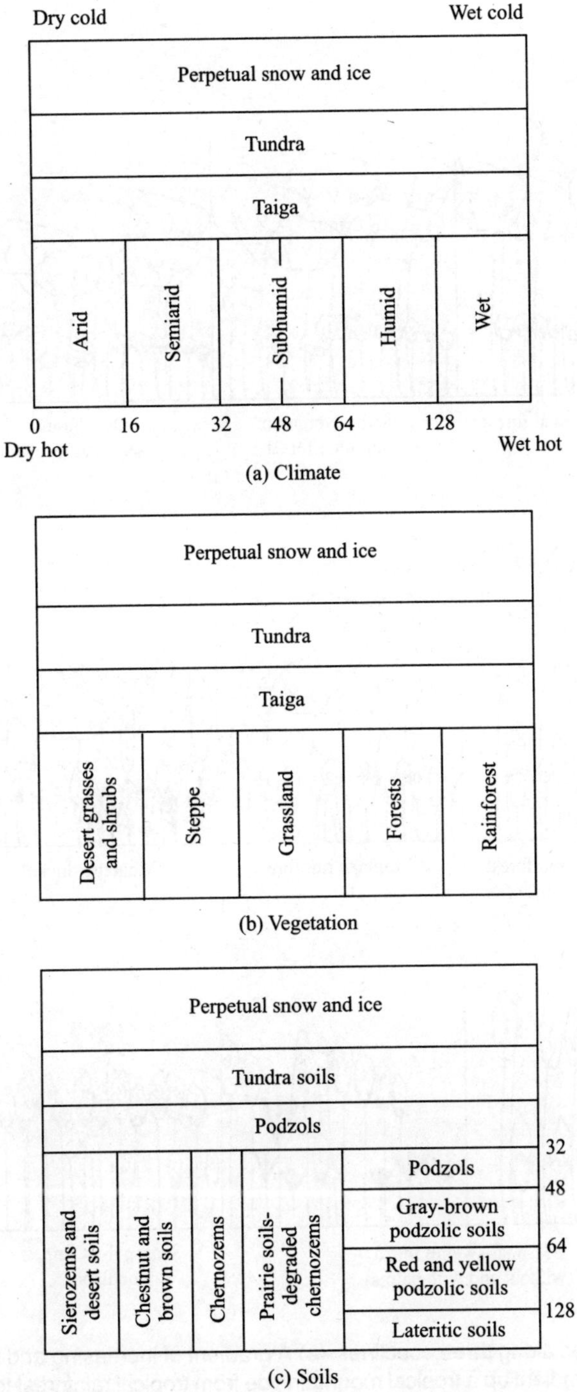

Fig. 8.11. Relationships between temperature and precipitation and (a) climatic types; (b) vegetational formations; and (c) major zonal soil groups.

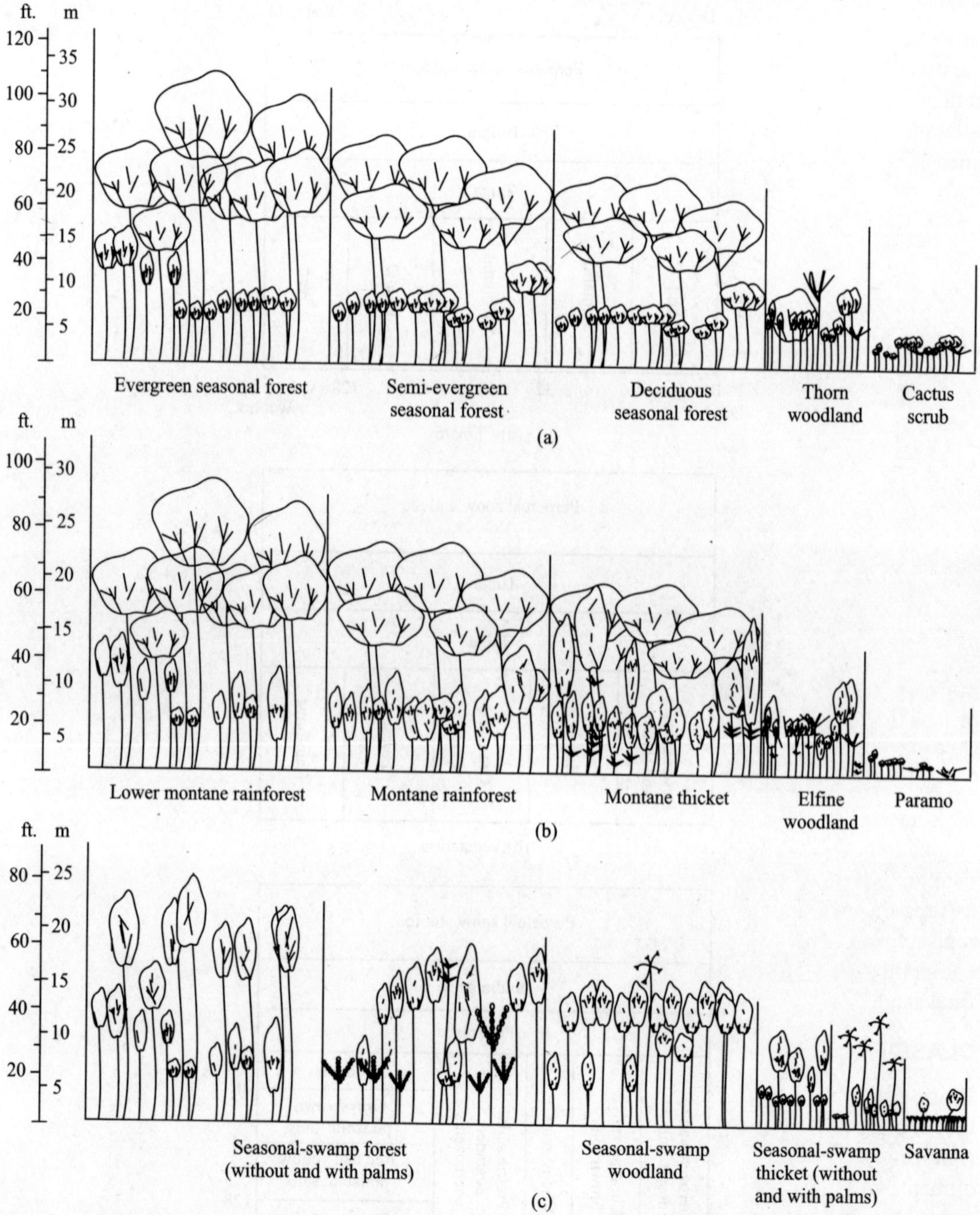

Fig. 8.12. Vegetation profiles along three ecoclines. (a) A gradient of increasing aridity from seasonal rainforest to desert; (b) an elevation gradient up a tropical mountainside from tropical rainforest to alpine meadow (paramo); and (c) A humidity gradient from swamp forest to savanna.

Spatial ecoclines on a more local scale have led to so-called gradient analysis. The abundance and actual distribution of organisms along many environmental gradients have shown that the importances of various species along any given gradient typically form bell-shaped curves reminiscent of tolerance curves. These curves tend to vary independently of one another and often overlap broadly (Fig. 8.13), indicating that each species population has its own particular habitat requirements and width of habitat tolerance—and hence its own zone of maximal importance. Such a continuous replacement of plant species by one another along a habitat gradient is termed a vegetational continuum (Fig. 8.13).

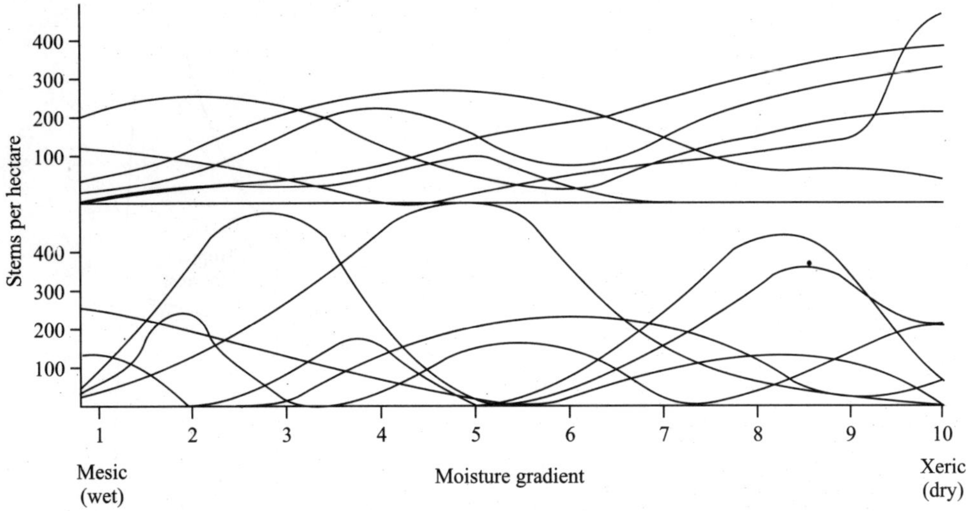

Fig. 8.13. Actual distributions of some populations of plant species along moisture gradients from relatively wet ravines to dry southwest-facing slopes in the Siskiyou mountains of northern California (above) and the Santa Catalina mountains of Arizona (below).

A temporal ecocline or a change in community composition in time, both by changes in the relative importance of component populations and by extinction of old species and invasion of new ones, is termed a succession. Primary succession, as we have just seen, is the development of communities from bare rock; secondary successions are changes that take place after destruction of the natural vegetation of an area that already has soil present. Secondary succession is often a more or less orderly sequential replacement of early succession species, typically rapidly growing, colonising species, by other more competitive species that succeed in later stages, usually slow-growing and shade-tolerant species. The final stage in succession is termed the climax.

CLASSIFICATION OF NATURAL COMMUNITIES

Biotic communities have been classified in various ways. An early attempt to classify communities was that of Merriam, who recognised a number of different 'life zones' defined solely in terms of temperature (ignoring precipitation). His somewhat simplistic scheme is no longer used, but his approach did link climate with vegetation in a more or less predictive manner. However some scientists have taken a somewhat different approach to the classification of natural communities that does not attempt to correlate climate with the plants and animals occurring in an area. Rather, they classify different natural communities into a large number of so-called biomes and associations, relying largely upon the characteristic plant and animal species that compose a particular community. As such, this scheme is descriptive rather than predictive. Such massive descriptions of different communities can often be quite useful in that they allow one to become familiar with a particular community with relative ease.

In a plot space of average annual precipitation versus average annual temperature (Fig. 8.14), various vegetational formations typically occurring under various climatic regimes can be plotted. Macroclimate determines the vegetation of an area-these correlations are not hard and fast and assume that local vegetation type depends on other factors such as soil types, seasonality of rainfall regime and frequency of disturbance by fires and floods.

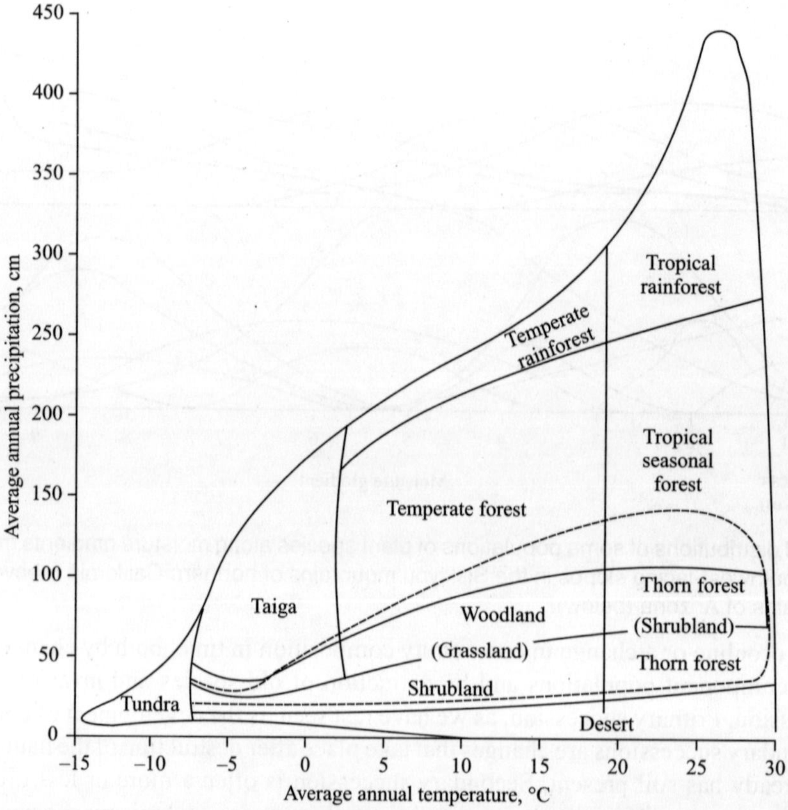

Fig. 8.14. Diagrammatic representation of the correlation between climate, as reflected by average annual temperature and precipitation and vegetational formation types. Boundaries between types are approximate and are influenced locally by soil type, seasonality of rainfall and disturbances such as fires. The dashed line encloses a range of climates in which either grasslands or woody plants may constitute the prevailing vegetation of an area, depending seasonality of precipitation.

SOME CONSIDERATIONS OF AQUATIC ECOSYSTEMS

Although the same ecological principles presumably operate in both aquatic and terrestrial ecosystems, there are striking and interesting fundamental differences between these two ecological systems. For example, primary producers on land are sessile and many tend to be large and relatively long-lived (air does not provide much support and woody tissues are needed), whereas, except for kelp, producers in aquatic communities are typically free-floating, microscopic and very short-lived (the buoyancy of water may make supportive plant tissues unnecessary; a large planktonic plant might be easily broken by water turbulence). Most ecologists study either aquatic or terrestrial systems, but seldom both. Various aquatic subdisciplines of ecology are recognised, such as aquatic ecology and marine ecology. Limnology is the study of freshwater ecosystems (ponds, lakes and streams); oceanography is concerned with bodies of salt water. Because the preceding part of this chapter and most of the remainder of the book emphasise terrestrial ecosystems, certain salient properties of aquatic ecosystems, especially lakes, are briefly considered in this section. Lakes are particularly appealing subjects for ecological study in that they are self-contained ecosystems, discrete and largely isolated from other ecosystems. Nutrient flow into and out of a lake can often be estimated with relative ease.

Water has peculiar physical and chemical properties that strongly influence the organisms that live in it. As indicated earlier, water has a high specific heat; moreover, in the solid (frozen) state, its density is less than it is in the liquid state (i.e., ice floats). Water is most dense at 4°C and water at this temperature 'sinks'. Furthermore, water is nearly a 'universal solvent' in that many important substances dissolve into aqueous solution.

A typical, relatively deep lake in the temperate zones undergoes marked and very predictable seasonal changes in temperature. During the warm summer months, its surface waters are heated up and because warm water is less dense than colder water, a distinct upper layer of warm water, termed the *epilimnion*, is formed (Fig. 8.15a). (Movement of heat within a lake is due to water currents produced primarily by wind). Deeper water, termed the *hypolimnion*, remain relatively cold during summer, often at about 4°C; an intermediate layer of rapid temperature change, termed the *thermocline*, separates the epilimnion from the hypolimnion (Fig. 8.15a). (A swimmer sometimes experiences these layers of different temperatures when diving into deep water or when in treading water his or her feet drop down into the cold hypolimnion.) A lake with a thermal profile or bathythermograph, like that shown in Fig. 8.15a is said to be 'stratified' because of its layer of warm water over cold water. Typically, there is little mixing of the warm upper layer with the heavier deeper water.

With the decrease in incident solar energy in autumn, surface waters cool and give up their heat to adjacent landmasses and the atmosphere (Fig. 8.15b). Eventually, the epilimnion cools to the same temperature as the hypolimnion and the lake becomes isothermal (Fig. 8.15c). This is the time of the 'fall turnover'. With winter's freezing temperatures, the lake's surface turns to ice and its temperature-versus-depth profile looks something like that in Fig. 8.15d. Finally, in spring the ice melts and the lake is briefly isothermal once again (it may have a spring turnover) until its surface waters are rapidly warmed, when it again becomes stratified and the annual cycle repeats itself.)

Because prevailing winds produce surface water currents, a lake's water circulate. In stratified lakes, the epilimnion constitutes a more or less closed cell of circulating warm water, whereas the deep cold water scarcely moves or mixes with the warmer water above it. During this period, as dead organisms and particulate organic matter sink into the noncirculating hypolimnion, the lake undergoes what is known as summer stagnation. When a lake becomes isothermal, its entire water mass can be circulated and nutrient-rich bottom water brought to the surface; limnologists thus speak of the spring and the fall 'turnover'. Meteorological conditions, particularly wind velocity and duration, strongly influence such turnovers; indeed, if there is little wind during the period a lake is isothermal, its water might not be thoroughly mixed and many nutrients may remain locked in its depths. After a thorough turnover, the entire water mass of a lake is equalised and concentrations of various substances, such as oxygen and carbon dioxide, are similar throughout the lake.

Lakes differ in their nutrient content and degree of productivity and they can be arranged along a continuum ranging from those with low nutrient levels and low productivity (*oligotrophic* lakes) to those with high nutrient content and high productivity (*eutrophic* lakes). Clear, cold and deep lakes high in the mountains are usually relatively oligotrophic, whereas shallower, warmer and more turbid lakes such as those in low-lying areas are generally more eutrophic.

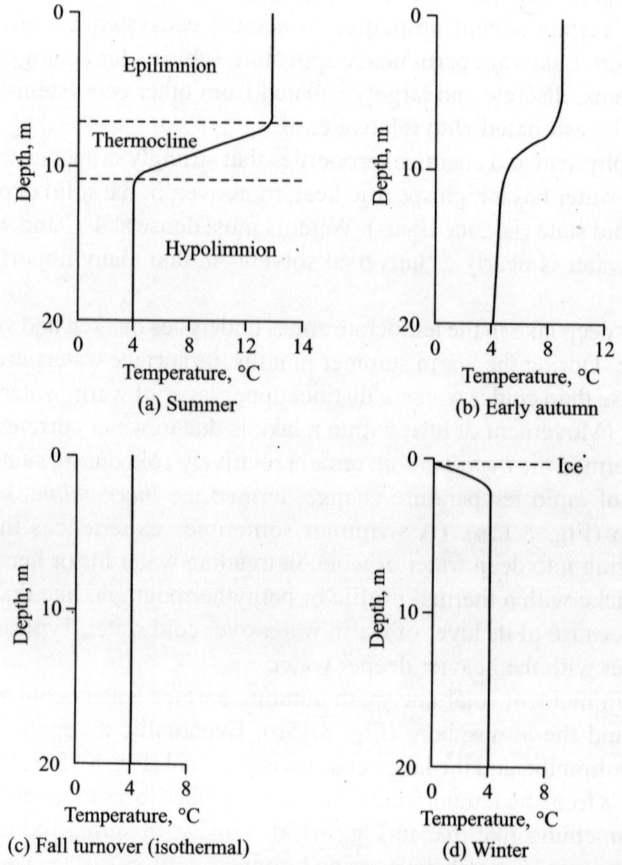

Fig. 8.15. Hypothetical bathythermographs showing seasonal changes typical of a deep temperate zone lake. (a) A stratified lake during summer; (b) in early autumn, upper water cool; (c) in late autumn or early winter, the lake's water are all at exactly the same temperature, here 4°C. (The lake is 'isothermal'); and (d) during the freezing winter months, a layer of surface ice chills the uppermost water.

Oligotrophic lakes typically support game fish such as trout, whereas eutrophic lakes contain 'trash' fish such as carp. As they age and fill with sediments, many lakes gradually undergo a natural process of eutrophication, steadily becoming more and more productive. People accelerate this process by enriching lakes with wastes and many oligotrophic lakes have rapidly become eutrophic under our influence. A good indicator of the degree of eutrophication is the oxygen content of deep water during summer. In a relatively unproductive lake, oxygen content varies little with depth and there is ample oxygen at the bottom of the lake. In contrast, oxygen content diminishes rapidly with depth in productive lakes and anaerobic processes sometimes characterise their depths during the summer months (an example of a succession). With the autumn turnover, oxygen-rich water again reach the bottom sediments and aerobic processes become possible.

However, once such a lake becomes stratified, the oxygen in its deep water is quickly used up by benthic organisms (in the dark depths there is little or no photosynthesis to replenish the oxygen).

These seasonal physical changes profoundly influence the community of organisms living in a lake. During the early spring and after the fall turnover, surface water are rich in dissolved nutrients such as nitrates and phosphates and temperate lakes are very productive, whereas during midsummer, many nutrients are unavailable to phytoplankton in the upper water and primary production is much reduced.

Organisms within a lake community are usually distributed quite predictably in time and space. Thus, there is typically a regular seasonal progression of planktonic algae, with diatoms most abundant in the winter, changing to desmids and green algae in the spring and gradually giving way to blue-green algae during summer months. The composition of the zooplankton also varies seasonally. Such temporal heterogeneity may well promote a diverse plankton community by periodically altering competitive abilities of component species, hence facilitating coexistence of many species of plants and animals in the relatively homogeneous planktonic environment. Although plankton are moved about by water currents, many are strong enough swimmers to select a particular depth. Such species often actually 'migrate' vertically during the day and with the seasons, being found at characteristic depths at any given time (Fig. 8.16). Some zooplankters sink into the dark depths during the daylight hours (probably an adaptation to avoid visually hunting predators such as fish), but ascend to the surface water at night to feed on phytoplankton.

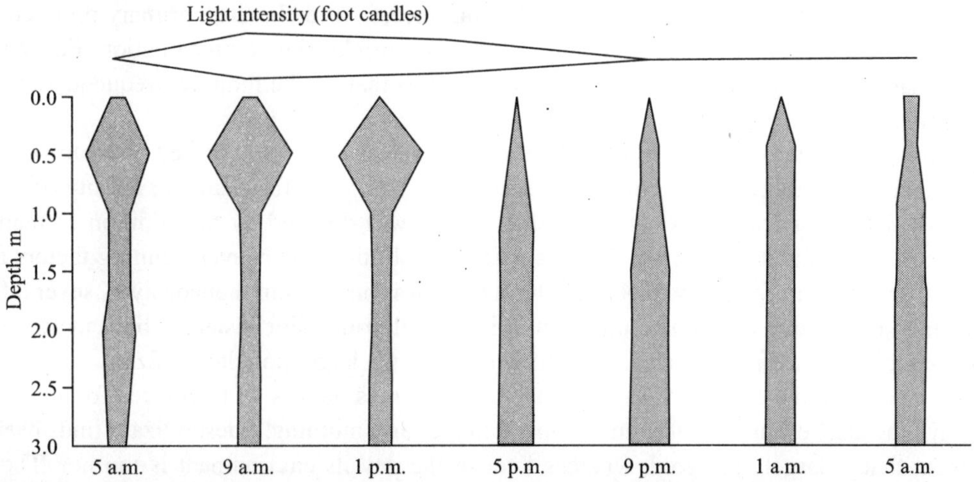

Fig. 8.16. Many small freshwater planktonic animals move vertically during the daily cycle of illumination somewhat as shown here (widths of bands represent the density of animals at a given depth at a particular time).

9

Resource Acquisition and Allocation

INTRODUCTION

Ecological events and their outcomes, such as growth, reproduction, photosynthesis, primary production and population size, are often regulated by the availability of one or a few factors or requisites in short supply, whereas other resources and raw materials present in excess may go partially unused. This principle has become known as the law of the minimum. For instance, in arid climates, primary production (the amount of solar energy trapped by green plants) is strongly correlated with precipitation (Fig. 9.1); here water is a 'master limiting factor'. Of many different factors that can be limiting, frequently among the most important are various nutrients, water and temperature.

When considering population, we often speak of those that are food limited, predator-limited or climate-limited. Population may be limited by other factors as well; for example, density of breeding pairs of blue tits (*Parus caeruleus*) in an English woods was doubled by the addition of many new nesting boxes, providing an indication that nest sites were limiting. However, limiting factors are not always so clear-cut but may usually interact so that a process is limited simultaneously by several factors, with a change in any one of them resulting in a new equilibrium. For instance, both increased food availability and decreased predation pressures might result in a larger population size.

A related concept, developed by Shelford, is now known as the law of tolerance. Too much or too little of anything can be detrimental to an organism. In the early morning, a desert lizard finds itself in an environment that is largely too cold, whereas later in the day its environment is too hot. The lizard compensates somewhat for this by spending most of its time during the early morning in sunny places, whereas later on most of its activities take place in the shade. Each lizard has a definite optimal range of temperature, with both upper and lower limits of tolerance. More precisely, when measures of performance (such as fitness, survivorship or foraging efficiency) are plotted against important environmental variables, bell-shaped curves usually result (e.g., Figs. 9.2 and 9.8).

Organisms can be viewed as simple input-output systems, with foraging or photosynthesis providing an input of materials and energy that are in turn 'mapped' into an output consisting of progeny. Fairly extensive bodies of theory now exist both on optimal foraging and reproductive tactics. In optimal foraging theory, the 'goal' usually assumed to be maximised is energy uptake per unit time (successful offspring produced during an organism's lifetime would be a more realistic measure of its foraging ability, but fitness is exceedingly difficult to measure). Similarly, among organisms without parental care, reproductive effort has sometimes been estimated by the ratio of calories devoted to eggs or offspring over total female calories at any instant (rates of uptake versus expenditure of calories have unfortunately not yet infiltrated empirical studies of reproductive tactics). To date, empirical studies of resource partitioning have been concerned largely with 'input' phenomena such as overlap in and efficiency of resource utilisation and have neglected to relate these to 'output' aspects. In contrast, empirical studies of reproductive tactics have done the reverse and almost entirely omitted any consideration of foraging. Interactions and constraints between foraging and reproduction have barely begun to be considered. A promising area for future work will be to merge aspects of optimal foraging with optimal reproductive tactics to specify rules by which input is translated into output; optimal reproductive tactics ('output' phenomena) surely must often impose substantial constraints upon 'input' possibilities and vice versa.

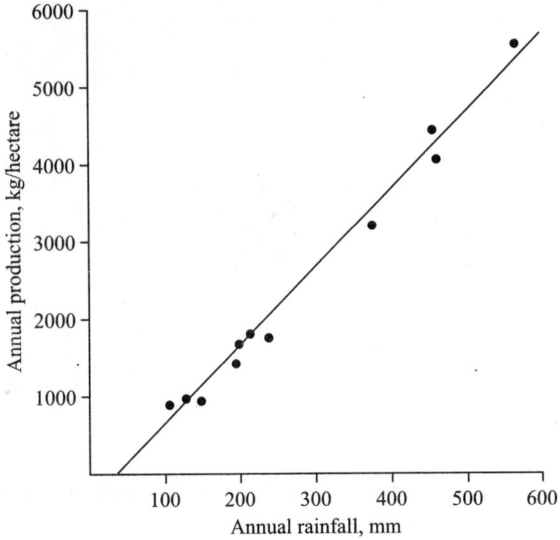

Fig. 9.1. An example of the strong correlation between annual rainfall and primary production along a precipitation gradient in a desert region of Namibia.

RESOURCE BUDGETS: THE PRINCIPLE OF ALLOCATION

Any organism has a limited amount of resources available to devote to foraging, growth, maintenance and reproduction. The way in which an organism allocates its time and energy and other resources among various conflicting demands is of fundamental interest because such apportionments provide insight into how the organism copes with and conforms to its environment. Moreover, because any individual has finite resource and energy budgets, its capacity for regulation is necessarily limited. Organisms stressed along anyone environmental variable are thus able to tolerate a lesser range of conditions along other environmental variables. Various tolerance and performance curves (Fig. 9.2) are presumably subject to certain constraints. For example, their breadth (variance) usually cannot be increased without a simultaneous reduction in their height and vice versa. This useful notion of trade-offs, known as the principle of allocation, has proven to be quite helpful in interpreting and understanding numerous ecological phenomena.

As an example of allocation, imagine an animal of a given size and mouth part anatomy. A certain size of prey item is optimal, whereas other prey are suboptimal because they are either too large or too small for efficient capture and swallowing. Any given animal has its own utilisation curve that indicates the actual numbers of prey of different sizes taken per unit time under particular environmental conditions. In an idealised, perfectly stable and infinitely productive environment, a utilisation curve might become a spike with no variance, with the organism using only its most optimal prey resource type. In actuality, limited and changing availabilities of resources, in both time and space, result in utilisation curves with **breadth** as well as height. In terms of the principle of allocation, an individual with a generalised diet adapted to eat prey of a wide range of available sizes presumably is not so effective at exploiting prey of intermediate size as another, more specialised, feeder. In other words, a jack-of-many-trades is a master of none. We will consider this subject in more detail later.

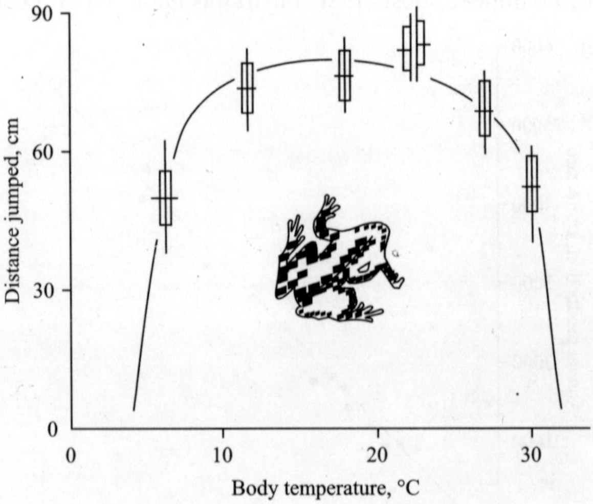

Fig. 9.2. Distance jumped by a frog as a function of its body temperature. Notice that performance diminishes at both low and high temperatures.

TIME, MATTER AND ENERGY BUDGETS

Time, matter and energy budgets vary widely among organisms. For example, some creatures allot more time and energy to reproduction at any instant than do others. Varying time and energy budgeting is a potent means of coping with a changing environment while retaining some degree of adaptation to it. Thus, many male songbirds expend a great deal of energy on territorial defense during the breeding season but little or none at other times of the year. Similarly, in animals with parental care, an increasing amount of energy is spent on growing offspring until some point when progeny begin to become independent of their parents, whereupon the amount of time and energy devoted to them decreases. Indeed, adult female red squirrels, *Tamiascurus*, at the height of lactation consume an average of 323 kilocalories of food per day compared with an average daily energy consumption of a similar-sized adult male of only about 117 kilocalories. The time budgets of these squirrels also vary markedly with the seasons. In a bad dry year many annual plants 'go to seed' while still very small, whereas in a good wet year, these plants grow to a much larger size before becoming reproductive; presumably more seeds are produced in good years, but perhaps none (or very few) would be produced in a bad year if individuals attempted to grow to the sizes they reach in good years.

An animal's time and energy budget provides a convenient starting point for clarifying some ways in which foraging influences reproduction and vice versa. Any animal has only a certain finite period of time available in which to perform all its activities, including foraging and reproduction. This total time budget, which can be considered either on a daily basis or over the animal's lifetime, is determined both by the diurnal rhythm of activity and by the animal's ability to 'make time' by performing more than one activity at the same time (such as a male lizard sitting on a perch, simultaneously watching for potential prey and predators while monitoring mates and competing males). Provided that a time period is profitable for foraging (expected gains in matter and energy exceed inevitable losses from energetic costs of foraging), any increase in time devoted to foraging clearly will increase an animal's supply of matter and energy. However, necessarily accompanying this increase in matter and energy is a concomitant decrease in time available for nonforaging activities such as mating and reproduction. Thus, profits of time spent foraging are measured in matter and energy, while costs take on units of time lost. Conversely, increased time spent on nonforaging activities confers profits in time, while costs take the form of decreased energy availability. Hence, gains in energy correspond to losses in time, while dividends in time require reductions in energy availability. (Of course, risks of foraging and reproduction also need to be considered).

The preceding arguments suggest that optimal allocation of time and energy ultimately depend on how costs in each currency vary with profits. However, because units of costs and profits in time and energy differ, one would like to be able to convert them into a common currency. Costs and profits in time might be measured empirically in energetic units by estimating the net gain in energy per unit of foraging time. If all potential foraging time is equivalent, profits would vary linearly with costs; under such circumstances, the loss in energy associated with nonforaging activities would be directly proportional to the amount of time devoted to such activity. Optimal budgeting of time and energy into foraging versus nonforaging activities is usually profoundly influenced by various circadian and seasonal rhythms of physical conditions, as well as those of predators and potential prey. Clearly, certain time periods favourable for foraging return greater gains in energy gathered per unit time than other periods. Risks of exposure to both harsh physical conditions and predators must often figure into the optimal amount of time to devote to various activities. Ideally, one would ultimately like to measure both an animal's foraging efficiency and its success at budgeting time and energy by its lifetime reproductive success, which would reflect all such environmental 'risks'. Foraging and reproductive activities interact in another important way. Many organisms gather and store materials and energy during time periods that are unfavourable for successful reproduction but then expend these same resources on reproduction at a later, more suitable, time. Lipid storage and utilisation systems obviously facilitate such temporal integration of uptake and expenditure of matter and energy. This temporal component greatly complicates the empirical measurement of reproductive effort.

Prey density can strongly affect an animal's time and energy budget. Gibb watched rock pipits, *Anthus spinoletta*, feeding in the intertidal along the English seacoast during two consecutive winters. The first winter was relatively mild; the birds spent 6½ hours feeding, 1¾ hours resting and ¾ hour fighting in defense of their territories (total daylight slightly exceeded 9 hours). The next winter was much harsher and food was considerably scarcer; the birds spent 8¼ hours feeding, 39 minutes resting and only 7 minutes on territorial defense! Apparently, the combination of low food density and extreme cold (endotherms require more energy in colder weather) demanded that over 90 per cent of the bird's waking hours be spent feeding and no time remained for frivolities. This example also illustrates that food is less defendable at lower densities, as indicated by reduced time spent on territorial defense. Obviously, food density in the second year was near the lower limit that would allow survival of rock pipits. When prey items are too sparse, encounters may be so infrequent that an individual cannot survive. Gibb calculated that, to balance their energy budget during the winter in some places, English tits must find an insect on the average once every 2½ seconds during daylight hour.

Time and energy budgets are influenced by a multitude of other ecological factors, including body size, mode of foraging, mode of locomotion, vagility, trophic level, prey size, resource density, environmental heterogeneity, rarefaction, competition, risks of predation and reproductive tactics.

LEAF TACTICS

Leaves take on an almost bewildering array of sizes and shapes (Fig. 9.3). Some leaves are deciduous, others evergreen; some are simple, others compound; and their actual spatial arrangement on a given plant differs considerably both within and between species. Some leaves are much more costly to produce and maintain than others (elementary economic considerations dictate that any given leaf must pay for itself plus generate a net energetic profit). Presumably, this great diversity of leaf tactics is a result of natural selection maximising the lifetime reproductive success of individual plants under diverse environmental conditions. Leaf tactics are influenced by many factors that include light, water availability, prevailing winds and herbivores. When grown in the shade, individuals of many species grow larger, less dissected leaves than when grown in the sun. Similarly, shade-tolerant plants of the understory usually have larger and less lobed leaves than canopy species. Similar types of leaves often evolve independently in different plant lineages subjected to comparable climatic conditions at different geographic localities, especially among trees. Compound leaves, thought to conserve woody tissue, with small leaflets are found in hot dry regions, whereas those with larger leaflets occur under warm moist conditions. Lowland wet tropical rainforest trees have large evergreen leaves with nonlobed or continuous margins, chaparral plants tend to have small sclerophyllous evergreen leaves, arid regions tend to support leafless stem succulents such as cacti or plants with entire leaf margins (especially among evergreens), plants from cold wet climates often have notched or lobed leaf margins and so on. Such repeated patterns of leaf size and shape suggest that a general theory of leaf tactics is possible.

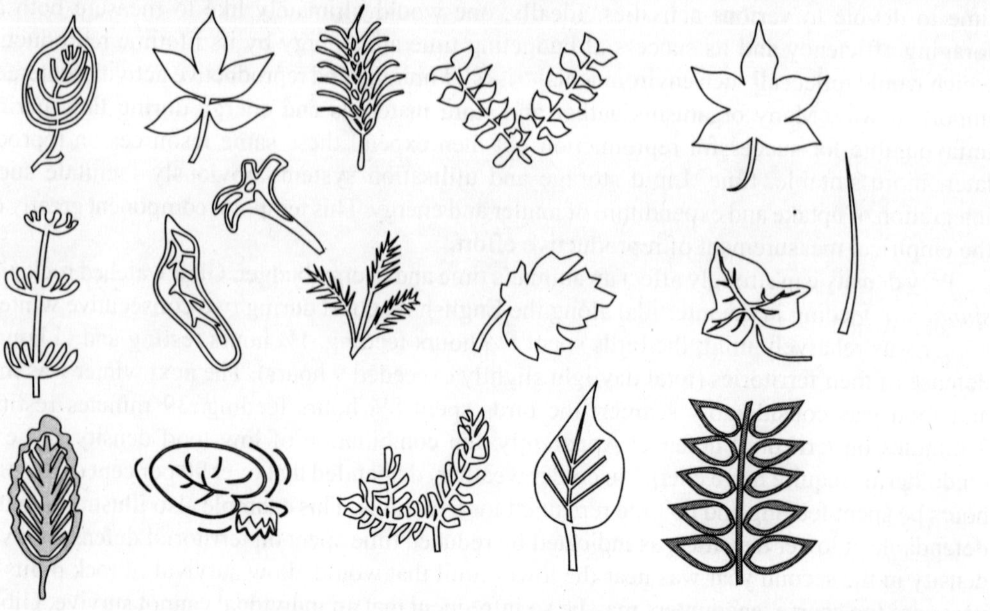

Fig. 9.3. Leaves have evolved a spectacular variety of shapes and sizes and yet repeatable patterns do occur.

Several models for optimal leaf size under differing environmental conditions have been developed. Efficiency of water use (grams of carbon dioxide assimilated per gram of water lost) was the measure of plant performance maximised by Parkhurst and Loucks. A similar model for size and shape of vine leaves was developed by Givnish and Vermeij. Even these relatively simple models predict several observed patterns in leaf size, such as large leaves in warm shady wet places and small leaves in colder areas or warmer and sunnier locales (Fig. 9.4).

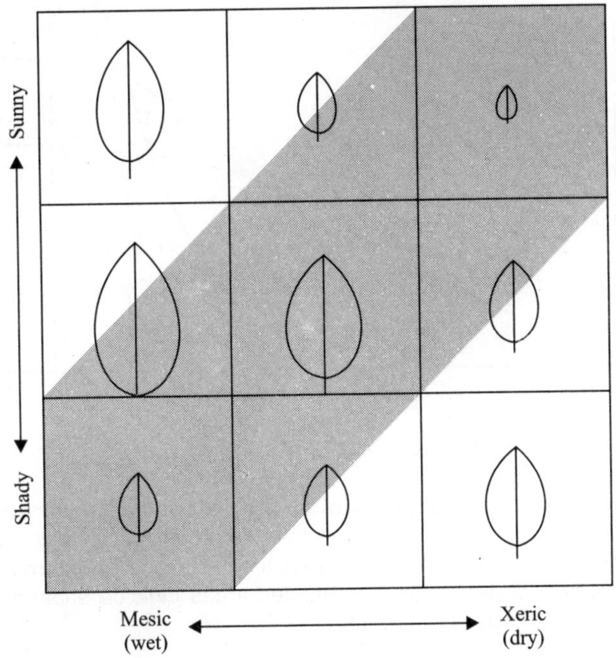

Fig. 9.4. Patterns of leaf sizes predicted under differing environmental conditions from wet to dry (mesic to xeric) and shady to sunny. The shaded region represents conditions likely to prevail in nature.

The evergreen versus deciduous dichotomy can be approached similarly using cost-benefit arguments. In considering leaf tactics of desert plants, Orians and Solbrig contrast leaf types along a continuum ranging from the relatively inexpensive deciduous 'mesophytic' leaf to the more costly evergreen 'xerophytic' leaf. Mesophytic leaves photosynthesise and transpire at a rapid rate and hence require high water availability (low 'soil water potential'). In deserts, such plants grow primarily along washes. In contrast, xerophytic leaves cannot photosynthesise as rapidly when abundant water is available, but they can extract water from relatively dry soil. Each plant leaf tactic has an advantage either at different times or in different places, thereby promoting plant life form diversity. During wet periods, plants with mesophytic leaves photosynthesise rapidly, but under drought conditions, they must drop their leaves and become dormant. During such dry periods, however, the slower photosynthesisers with xerophytic leaves are still able to function by virtue of their ability to extract water from dry soils. Of course, all degrees of intermediate leaf tactics exist, each of which may enjoy a competitive advantage under particular conditions of water availability (Fig. 9.5). In a predictable environment, net annual profit per unit of leaf surface area determines the winning phenotype. Even a relatively brief wet season could suffice to give mesophytic leaves a higher annual profit (which accounts for the occurrence of these plant life forms in deserts).

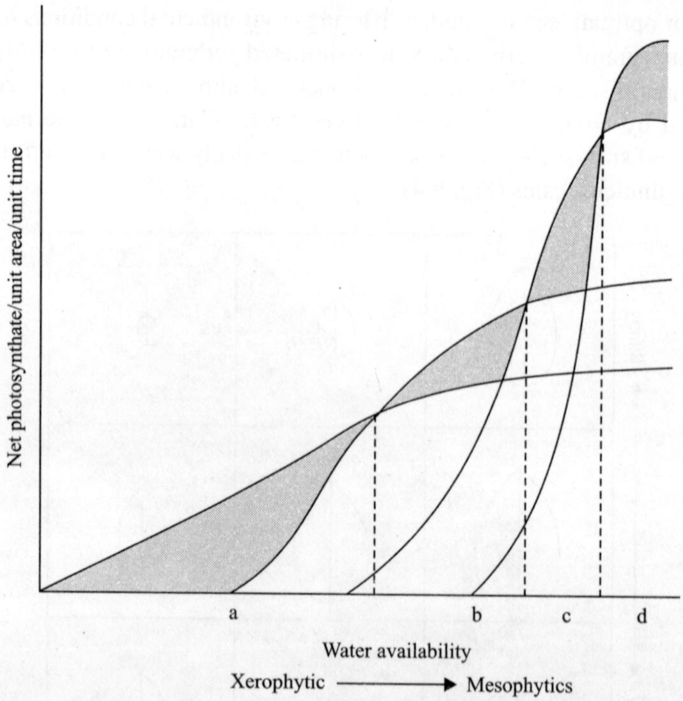

Fig. 9.5. Probable relationship between efficiency of photosynthesis and water availability among different types of leaves. The most xerophytic leaf performs best under conditions of low water availability (zone a), whereas the most mesophytic leaf does best when soils are wet (zone d). Intermediate types are superior under intermediate conditions of water availability. Shaded areas indicate superiority of various leaf types under different conditions of soil moisture availability.

In an interesting discussion of leaf arrangement and forest structure, Horn distinguished 'monolayers' from 'multilayers'. Each plant in the multilayer of a forest (usually sunnier places such as the canopy) has leaves scattered throughout its volume at several different levels, whereas monolayer plants have essentially a single blanket or shell of leaves. Plants in the multilayer gain from a geometry that allows some light to pass through to their own leaves at lower levels. Horn points out that lobing facilitates passage of light and that such plants do well in the sun (in the shade, inner leaves may respire more than they photosynthesise). In contrast, the optimal tree design in the shade is a monolayer in which each leaf typically intercepts as much light as possible (leaves are large and seldom lobed). Moreover, slowgrowing monolayered plants eventually outcompete fast-growing multilayered plants that persist by regular colonisation of newly vacated areas created by continual disturbance.

FORAGING TACTICS AND FEEDING EFFICIENCY

Foraging tactics involve the ways in which animals gather matter and energy. As explained above, matter and energy constitute the profits gained from foraging, in that they are used in growth, maintenance and reproduction. But foraging has its costs as well; a foraging animal may often expose itself to potential predators and much of the time spent in foraging is rendered unavailable for other activities, including reproduction. An optimal foraging tactic maximises the difference between foraging profits and their costs. Presumably, natural selection, acting as an efficiency expert, has often favoured such optimal foraging behaviour. Consider, for example, prey of different sizes and what might be termed 'catchability'. How great an effort should a foraging animal make to obtain a prey item of a given catchability and of a particular size (and therefore matter and energy content)? Clearly, an optimal consumer should be willing to expend more energy to find and capture food items that return the most energy per unit of expenditure upon them. Moreover, an optimal forager should take advantage of natural feeding routes and should not waste time and energy looking for prey either in inappropriate places or at inappropriate times. What is optimal in one environment is seldom optimal in another and an animal's particular anatomy strongly constrains its optimal foraging tactic. There is considerable evidence that animals actually do attempt to maximise their foraging efficiencies and a substantial body of theory on optimal foraging tactics exists.

Numerous aspects of optimal foraging theory are concisely summarised by MacArthur. He makes several preliminary assumptions: (a) environmental structure is repeatable, with some statistical expectation of finding a particular resource (such as a habitat, microhabitat and/or prey item); (b) food items can be arranged in a continuous and unimodal spectrum, such as size distributions of insects. (This assumption is clearly violated by foods of some animals, such as monophagous insects or herbivores generally, because plant chemical defenses are typically discrete; (c) similar animal phenotypes are usually closely equivalent in their harvesting abilities; an intermediate phenotype is thus best able to exploit foods intermediate between those that are optimal for two neighbouring phenotypes. Conversely, similar foods are gathered with similar efficiencies; a lizard with a jaw length that best adapts it to exploit 5-mm-long insects is only slightly less efficient at eating 4 and 6-mm insects; (d) the principle of allocation applies and no one phenotype can be maximally efficient on all prey types; improving harvesting efficiency on one food type necessitates reducing the efficiency of exploiting other kinds of items; and (e) finally, an individual's economic 'goal' is to maximise its total intake of food resources. (Assumptions b, c and d are not vital to the argument).

MacArthur then breaks foraging down into four phases: (i) deciding where to search; (ii) searching for palatable food items; (iii) upon locating a potential food item, deciding whether or not to pursue it; and (iv) pursuit itself, with possible capture and eating. Search and pursuit efficiencies for each food type in each habitat are entirely determined by the preceding assumptions about morphology (assumption c) and environmental repeatability (assumption a); moreover, these efficiencies dictate the probabilities associated with the searching and pursuing phases of foraging: (ii) and (iv) respectively. Thus, MacArthur considers only the two decisions: where to forage and what prey items to pursue phases: (i) and (ii) of foraging. Clearly, an optimal consumer should forage where its expectation of yield is greatest—an easy decision to make, given knowledge of the previous efficiencies and the structure of its environment (in reality, of course, animals are far from omniscient and must make decisions based on incomplete information). The decision as to which prey items to pursue is also simple. Upon finding a potential prey item, a consumer has only two options: either pursue it or go on searching for a better item and pursue that one instead. Both decisions end in the forager beginning a new search, so the best choice is clearly the one that returns the greatest yield per unit time. Thus, an optimal consumer should opt to pursue an item only when it cannot expect to locate, catch and eat a better item (i.e., one that returns more energy per unit of time) during the time required to capture and ingest the first prey item.

Many animals, such as foliage-gleaning insectivorous birds, spend much of their foraging time searching for prey but expend relatively little time and energy pursuing, capturing and eating small sedentary insects that are usually easy to catch and quickly swallowed. In such 'searchers', mean search time per item eaten is large compared to average pursuit time per item; hence, the optimal strategy is to eat essentially all palatable insects encountered. Other animals ('pursuers') that expend little energy in finding their prey but a great deal of effort in capturing it (such as, perhaps, a falcon or a lion) should select prey with small average pursuit times (and energetic costs). Thus, pursuers should generally be more selective and more specialised than searchers. Moreover, because a food-dense environment offers a lower average search time per item than does a food-sparse area, an optimal consumer should restrict its diet to only the better types of food items in the former habitat. To date, optimal foraging theory has been developed primarily in terms of the rate at which energy is gathered per unit of time. Limiting materials such as nutrients in short supply and the risks of predation have so far been largely neglected.

Carnivorous animals forage in extremely different ways. In the 'sit-and-wait' mode, a predator waits in one place until a moving prey item comes by and then 'ambushes' the prey; in the 'widely foraging' mode, the predator actively searches out its prey. The second strategy normally requires a greater energy expenditure than the first. The success of the sit-and-wait tactic usually depends on one or more of three conditions: a fairly high prey density, high prey mobility and low predator energy requirements. The widely foraging tactic also depends on prey density and mobility and on the predator's energy needs, but here the distribution of prey in space and the predator's searching abilities assume paramount importance. Although these two tactics are endpoints of a continuum of possible foraging strategies (and hence somewhat artificial), foraging techniques actually employed by many organisms are rather strongly polarised. The dichotomy of sit-and-wait versus widely foraging therefore has substantial practical value. Among snakes, for example, racers and cobras forage widely when compared with boas, pythons and vipers, which are relatively sit-and-wait foragers.

Among hawks, accipiters such as Cooper's hawks and goshawks often hunt by ambush using a sit-and-wait strategy, whereas most buteos and many falcons are relatively more widely foraging. Web-building spiders and sessile filter feeders such as barnacles typically forage by sitting and waiting. Many spiders expend considerable amounts of energy and time building their webs rather than moving about in search of prey; those that do not build webs, such as wolf spiders, forage much more widely. Some general correlates of these two modes of foraging are listed in Table 9.1.

Table 9.1. Some general correlates of forageing mode.

Description	*Sit-and-wait*	*Widely foraging*
Prey type	Eat active prey	Eat sedentary and unpredictable (but clumped or large) prey
Volume prey captured/day	Low	Generally high, but low in certain species
Daily metabolic expense	Low	High
Types of predators	Vulnerable primarily to widely foraging predators	Vulnerable to both sit-and-wait and to widely forageing predators
Rate of encounters with predators	Probably low	Probably high
Morphology	Stocky (short tails)	Streamlined (generally long tails)
Probably physiological correlates	Limited endurance (anaerobic)	High endurance capacity (aerobic)
Relative clutch mass	High	Low
Sensory mode	Visual primarily	Visual or olfactory
Learning ability	Limited	Enhanced learning and memory, larger brains
Niche breadth	Wide	Narrow

Similar considerations can be applied in comparing herbivores with carnivores. Because the density of plant food almost always greatly exceeds the density of animal food, herbivores often expend little energy, relative to carnivores, in finding their prey (to the extent that secondary chemical compounds of plants, such as tannins and other antiherbivore defenses reduce palatability of plants or parts of plants, the effective supply of plant foods may be greatly reduced). Because cellulose in plants is difficult to digest, however, herbivores must expend considerable energy in extracting nutrient from their plant food. (Most herbivores have a large ratio of gut volume to body volume, harbour intestinal micro-organisms that can digest cellulose and spend much of their time eating or ruminating-envision a cow chewing its cud.) Animal food, composed of readily available proteins, lipids and carbohydrates, is more readily digested; carnivores can afford to expend considerable effort in searching for their prey because of the large dividends obtained once they find it. As would be expected, the efficiency of conversion of food into an animal's own tissues (assimilation) is considerably lower in herbivores than it is in carnivores).

Many carnivores have extremely efficient prey-capturing devices; often the size of a prey object markedly influences this efficiency. Using simple geometry (Fig. 9.6), Holling estimated the diameter of a prey item that should be optimal for a praying mantid of a particular size. He then offered five hungry mantids prey objects of various sizes and recorded percentages attacked (Fig. 9.7). Mantids were noticeably reluctant to attack prey that were either much larger or much smaller than the estimated optimum! Thus, natural selection has resulted in efficient predators both by producing efficient prey-capturing devices and by programming animals so that they are unlikely to attempt to capture decidedly suboptimal items. Larger predators tend to take larger prey than smaller ones. It may in fact be better strategy for a large predator to overlook prey below some minimal size threshold and to spend the time that would have been spent in capturing and eating such small items in searching out larger prey (see also above). Similarly, the effort a predator will expend on any given prey item is proportional to the expected return from that item (which usually increases with prey size). Thus, a lizard waiting on a perch will not usually go far for a very small prey item but will often move much greater distances in attempts to obtain larger prey.

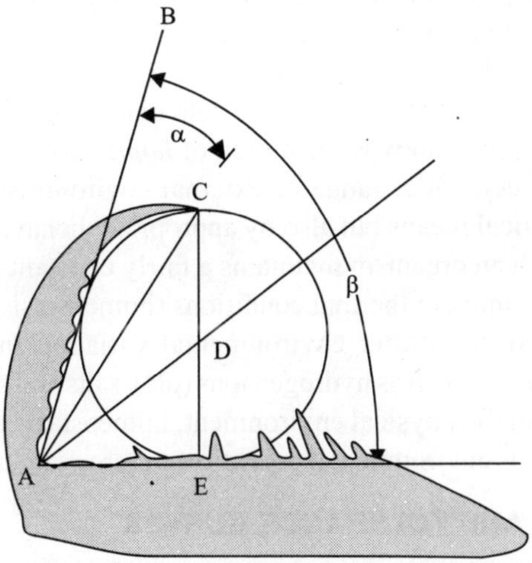

Fig. 9.6. Diagram showing how Holling used geometry to calculate an estimated optimal size of a prey item from the anatomy of a mantid's foreleg. Optimal prey diameter, D, is simply T $\sin(\beta - \alpha)$, where T is the distance A to C.

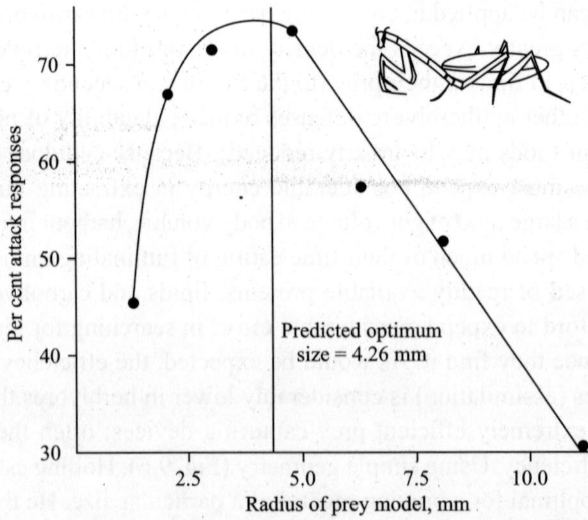

Fig. 9.7. The percentage of prey items actually attacked by hungry mantids versus prey size.

Because small prey are generally much more abundant than large prey, most animals encounter and eat many more small prey items than large ones. Small animals that eat small prey items encounter prey of suitable size much more frequently than do larger animals that rely on larger prey items; as a result, larger animals tend to eat a wider range of prey sizes. Because of such increased food niche breadths of larger animals, size differences between predators increase markedly with increasing predator size.

PHYSIOLOGICAL ECOLOGY

The concern of environmental physiology or ecophysiology, is how organisms function within, adapt and respond to and exploit their physical environments. Physiological ecologists are interested primarily in the immediate functional and behavioural mechanisms by which organisms cope with their abiotic environments. Physiological mechanisms clearly must reflect ecological conditions; moreover, mutual constraints between physiology and ecology dictate that both must evolve together in a synergistic fashion.

A fundamental principle of physiology is the notion of *homeostasis*, the maintenance of a relatively stable internal state under a much wider range of external environmental conditions. Homeostasis is achieved not only by physiological means but also by appropriate behavioural responses. An example is temperature regulation in which an organism maintains a fairly constant internal body temperature over a considerably greater range of ambient thermal conditions (homeostasis is never perfect). Homeostatic mechanisms have also evolved that buffer environmental variation in humidity, light intensity and concentrations of various substances such as hydrogen ions (pH), salts and so on. By effectively moderating spatial and temporal variation in the physical environment, homeostasis allows organisms to persist and be active within a broad range of environmental conditions, thereby enhancing their fitness.

PHYSIOLOGICAL OPTIMA AND TOLERANCE CURVES

Physiological processes proceed at different rates under different conditions. Most, such as rate of movement and photosynthesis, are temperature dependent (Fig. 9.8). Other processes vary with availability of various materials such as water, carbon dioxide, nitrogen and hydrogen ions (pH). Curves of performance, known as tolerance curves, are typically bell shaped and unimodal, with their peaks representing optimal conditions for a particular physiological process and their tails reflecting the limits of tolerance. Some individuals and species have very narrow peaked tolerance curves; in others these curves are considerably broader. Broad tolerance curves are described with the prefix *eury* (e.g., eurythermic, euryhaline), whereas stenois used for narrow ones (e.g., stenophagous). An organism's use of environmental resources such as foods and microhabitats can profitably be viewed similarly and performance can be measured in a wide variety of units such as survivourship, reproductive success, foraging efficiency and fitness.

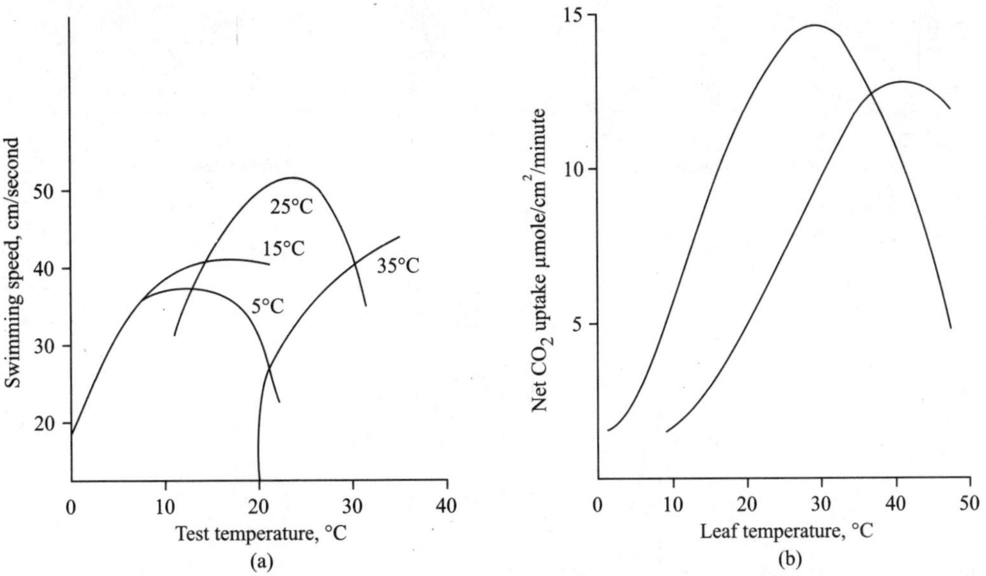

Fig. 9.8. Two plots of performance against temperature; (a) Goldfish swimming speed versus temperature for fish acclimated to different thermal conditions; in most cases, performance peaks near the temperature to which fish are accustomed; and (b) Photosynthetic rate versus leaf temperature in the plant *Atriplex lentiformis* at two different localities.

Performance curves can sometimes be altered during the lifetime of an individual as it becomes exposed to different ambient external conditions. Such short-term alteration of physiological optima is known as acclimation or acclimatisation (Fig. 9.8). Within certain design constraints, tolerance curves clearly must change over evolutionary time as natural selection moulds them to reflect changing environmental conditions. However, very little is known about the evolution of tolerance; most researchers have merely described the range(s) of conditions tolerated or exploited by particular organisms. Tolerance curves are often taken almost as given and immutable, with little or no consideration of the ecological and evolutionary forces that shape them.

Performance or tolerance is often sensitive to two or more environmental variables. For example, the fitness of a hypothetical organism in various microhabitats might be a function of relative humidity (or vapour pressure deficit), somewhat as shown in Fig. 9.9a. Assume that fitness varies similarly along a temperature gradient (Fig. 9.9b). Figure 9.9c combines humidity and temperature conditions to show variation in fitness with respect to both variables simultaneously (a third axis, fitness, is implicit in this figure). The range of thermal conditions tolerated is narrower at very low and very high humidities than it is at intermediate and more optimal humidities. Similarly, an organism's tolerance range for relative humidity is narrower at extreme temperatures than it is at more optimal ones. The organism's thermal optimum depends on humidity conditions (and vice versa). Fitness reaches its maximum at intermediate temperatures and humidities. Hence, temperature tolerance and tolerance of relative humidities interact in this example. The concept of a single fixed optimum is in some ways an artifact of considering only one environmental dimension at a time.

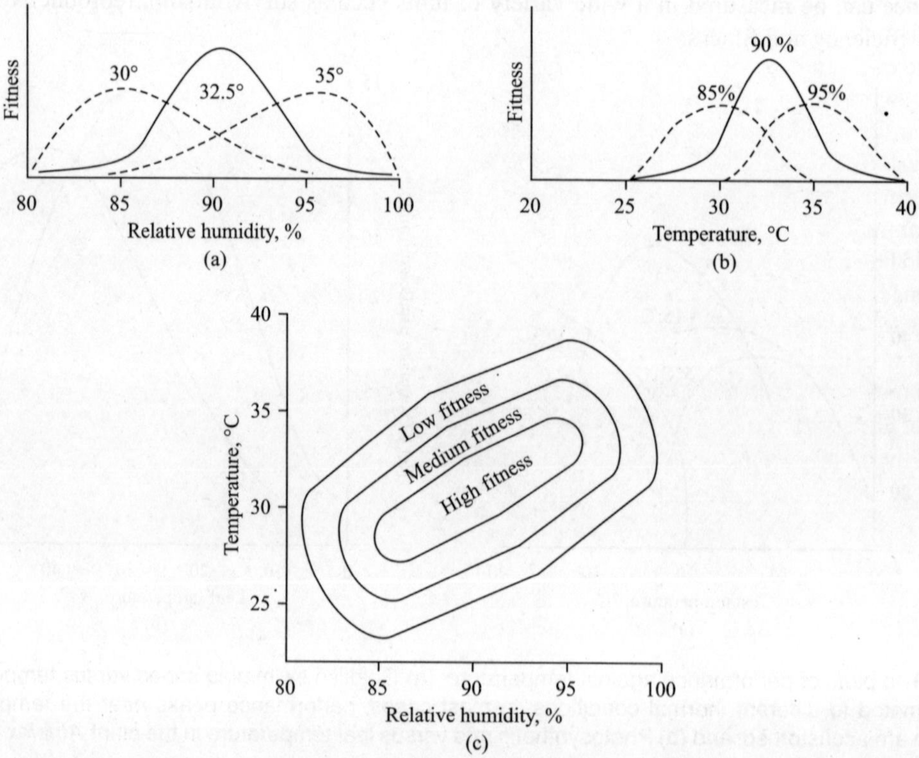

Fig. 9.9. Hypothetical response curves showing how two variables can interact to determine an organism's fitness. Fitness is reduced at extremes of either temperature or humidity and the range of humidities tolerated is less at extreme temperatures than it is at intermediate ones.

ENERGETICS OF METABOLISM AND MOVEMENT

Some of the food ingested by any animal passes through its gut unused. Such egested material can be as high as 80 to 90 per cent of the total intake in some caterpillars. The food that is actually digested is termed *assimilation*: a fraction of this must be used in respiration to support maintenance metabolism and activity. The remainder is incorporated into the animal concerned as secondary productivity and ultimately can be used either in growth or in reproduction. These relationships are summarised as:

Ingestion = Assimilation + Egestion

Assimilation = Productivity + Respiration

Productivity = Growth + Reproduction

The total amount of energy needed per unit time for maintenance increases with increasing body weight (Fig. 9.10). However, because small animals have relatively high ratios of body surface to body volume, they generally have much higher metabolic rates and hence have greater energy requirements per unit of body weight than larger animals (Fig. 9.11). Animals that maintain relatively constant internal body temperatures are known as *homeotherms*; those whose temperatures vary widely from time to time, usually approximating the temperature of their immediate environment, are called *poikilotherms*. These two terms are sometimes confused with a related pair of useful terms. An organism that obtains its heat from its external environment is an ectotherm; one that produces most of its own heat internally by means of oxidative metabolism is known as an *endotherm*. All plants and the vast majority of animals are ectothermic—the only continuously endothermic animals are found among birds and mammals (but even some of these become ectothermic at times). Some poikilotherms (some large reptiles and certain large fast-swimming fish such as tuna) are at times at least partially endothermic. Certain ectotherms (many lizards and temperate-zone flying insects) actually regulate their body temperatures fairly precisely during periods of activity by appropriate behavioural means. Thus, ectotherms at times can be homeotherms! An active bumblebee or desert lizard may have a body temperature as high as that of a bird or mammal (the layperson's terms 'warm-blooded' and 'cold-blooded' can thus be quite misleading and should be abandoned). Because energy is required to maintain a constant internal body temperature, endotherms have considerably higher metabolic rates, as well as higher energy needs (and budgets), than ectotherms of the same body weight. There is a distinct lower limit on body size for endotherms about the size of a small hummingbird or shrew (2 or 3 grams). Indeed, both hummingbirds and shrews have very high metabolic rates and hence rather precarious energetic relationships; they depend on continual supplies of energy-rich foods. Small hummingbirds would starve to death during cold nights if they did not allow their body temperatures to drop and go into a state of torpor.

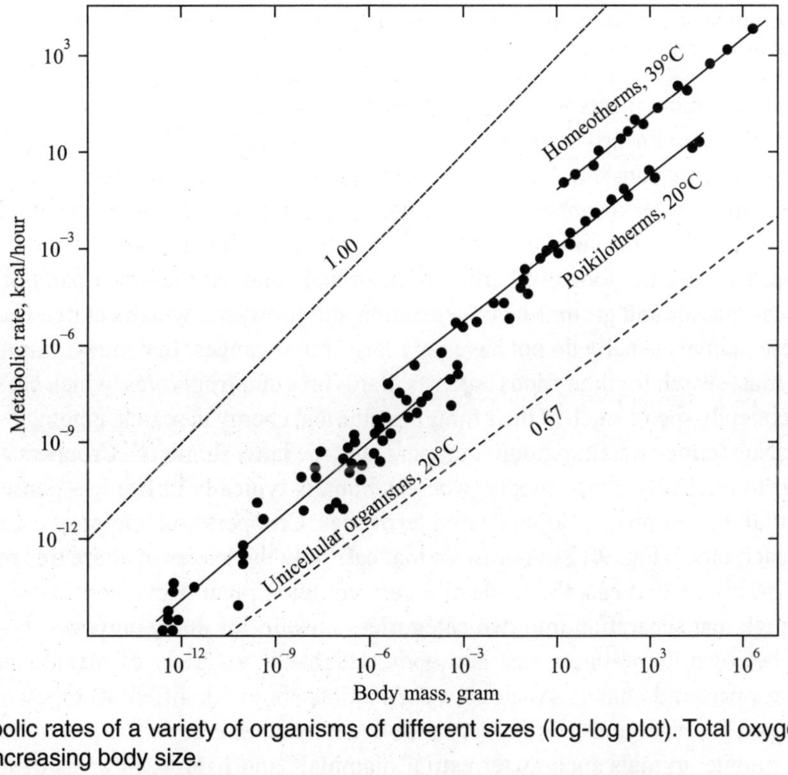

Fig. 9.10. Metabolic rates of a variety of organisms of different sizes (log-log plot). Total oxygen consumption increases with increasing body size.

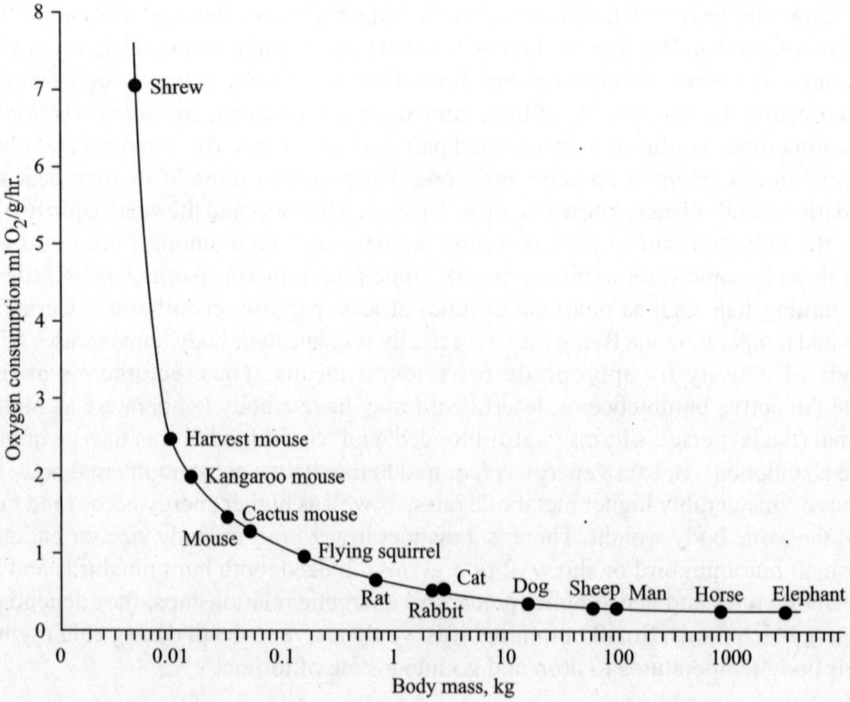

Fig. 9.11. Semilogarithmic plot of rates of oxygen consumption per unit of body mass for a wide variety of mammals plotted against body mass.

Body size, diet and movements are complexly intertwined with the energetics of metabolism. Energy requirements do not scale linearly with body mass, m, but instead as a fractional exponent: $E = cm^{0.67}$, where c is a taxon-specific constant. Large animals require more matter and energy for their maintenance than small ones and in order to obtain it they usually must range over larger geographical areas than smaller animals with otherwise similar food requirements. Food habits also influence movements and home range size. Because the foods of herbivorous animals that eat the green parts of plants (such as grazers, which eat grasses and ground-level vegetation and browsers, which eat tree leaves) are usually quite dense, these animals usually do not have very large home ranges. In contrast, carnivores and those herbivores that must search for their foods (such as granivores and frugivores, which eat seeds and fruits, respectively) frequently spend much of their foraging time and energy in search, ranging over considerably larger areas. McNab termed the first group 'croppers' and the latter 'hunters'. Croppers generally exploit foods that occur in relatively dense supply, whereas hunters typically utilise less dense foods. Hunters are often territorial, but croppers seldom defend territories. Croppers and hunters are not discrete but in fact grade into each other (Fig. 9.12). A browser that eats only the leaves of a rare tree might be more of a hunter than a granivore that eats the seeds of a very common plant. However, such intermediates are uncommon enough that separation into two categories is useful for many purposes. Figure 9.13 shows the correlation between home range size and body weight for a variety of mammalian species, here separated into croppers and hunters. Analogous correlations, but with different slopes and/or intercepts, have been obtained for birds and lizards. Very mobile animals, like birds, frequently range over larger areas than less mobile animals such as terrestrial mammals and lizards. In areas of low productivity (e.g., deserts), most animals may be forced to range over a greater area to find adequate food than they would in more productive regions. Large home ranges or territories usually result in low densities, which in turn markedly limit possibilities for evolution of sociality. Thus, McNab points out that complex social behaviour has usually evolved only in croppers and among exceptionally mobile hunters.

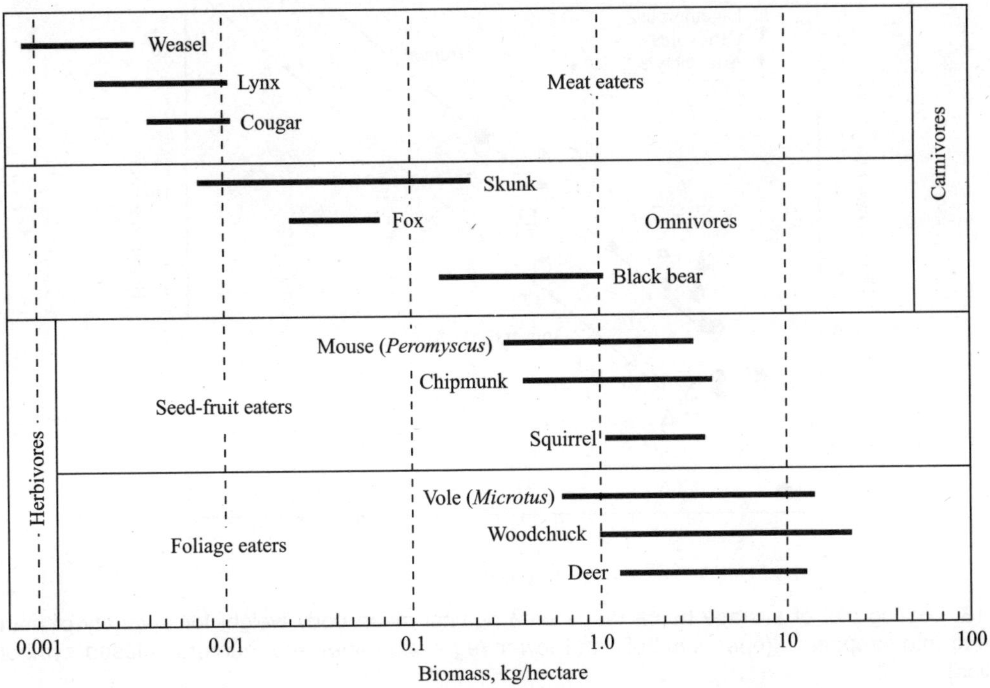

Fig. 9.12. Biomass, in kilograms/hectare, of various mammals, arranged according to their food habits. Although mammals as a class vary over five orders of magnitude, the range among those that eat any given type of food is considerably smaller. Meat eaters and omnivores are much less dense than herbivores.

The metabolic cost of movement varies with both an animal's body size and its mode of locomotion. The cost of moving a unit of body weight some standard distance is actually less in larger animals than in smaller ones (Fig. 9.14). Terrestrial locomotion is the most expensive mode of transportation, flight is intermediate in cost and swimming is the most economical means of moving about—provided body shape is fusiform and buoyancy is neutral (Fig. 9.14).

Physiologists have documented numerous consistent size-related trends in organs and metabolic properties. For example, among mammals, heart weight is always about 0.6 per cent of total body weight, whereas blood volume is almost universally about 5.5 per cent of body mass over a great range of body sizes (these organ systems are thus directly proportional to size). Other physiological attributes, such as lung surface in mammals, vary directly with metabolic rate rather than with size. However, some organ systems, such as the kidney and liver, are not scaled directly either to size or metabolic rate. Such 'physiological rules' or design constraints apparently dictate available avenues for physiological change, thereby constraining possible ecological adaptations.

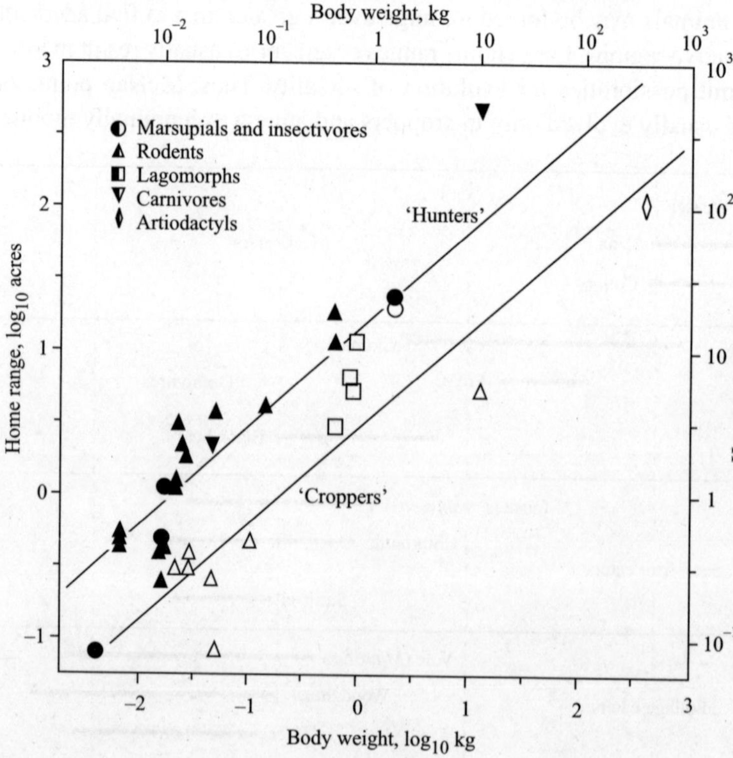

Fig. 9.13. Log-log plot of average home range area against mean body weight for a variety of mammals, separated into 'croppers' (open symbols and lower regression line) and 'hunters' (closed symbols and upper line).

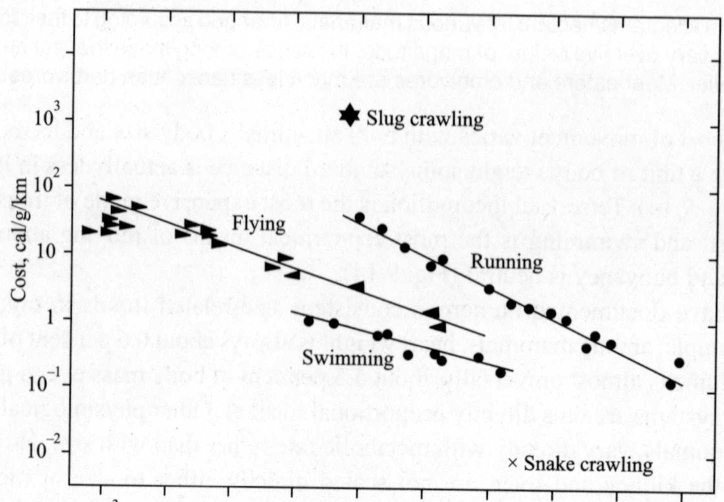

Fig. 9.14. Comparison of the energetic cost of moving a unit of body weight one kilometer for several different modes of locomotion.

ADAPTATION AND DETERIORATION OF ENVIRONMENT

Organisms are adapted to their environments in that, to survive and reproduce, they must meet their environment's conditions for existence. Evolutionary adaptation can be defined as conformity between the organism and its environment. Plants and animals have adapted to their environments both genetically and by means of physiological, behavioural and/or developmental flexibility. The former includes instinctive behaviour and the latter learning. Adaptation has many dimensions in that most organisms must conform simultaneously to numerous different aspects of their environments. Thus, for an organism to adapt, it must cope not only with various aspects of its physical environment, such as temperature and humidity conditions, but also with competitors, predators and escape tactics of its prey. Conflicting demands of these various environmental components often require that an organism compromise in its adaptations to each. Conformity to any given component takes a certain amount of energy that is then no longer available for other adaptations. The presence of predators, for example, may require that an animal be wary, which in turn is likely to reduce its foraging efficiency and hence its competitive ability.

Organisms can conform to and cope with highly predictable environments relatively easily, even when they change in a regular way, as long as the changes are not too extreme. Adaptation to an unpredictable environment may usually be much more difficult; adapting to extremely erratic environments may even prove impossible. Many organisms have evolved dormant stages that allow them to survive unfavourable periods, both predictable and unpredictable. Annual plants everywhere and brine shrimp in deserts are good examples. Brine shrimp eggs survive for years in the salty crust of dry desert lakes; when a rare desert rain fills one of these lakes, the eggs hatch, the shrimp grow rapidly to adults and they produce many eggs. Some seeds known to be many centuries old are still viable and have been germinated. Changes in the environment that reduce overall adaptation are collectively termed the deterioration of environment; such changes cause directional selection, resulting in accommodation to the new environment.

A simple but elegant model of adaptation and undirected environmental deterioration was developed by Fisher. He reasoned that no organism is 'perfectly adapted'—all must fail to conform to their environments in some ways and to differing degrees. However, a hypothetical, perfectly adapted organism can always be imagined (actually this reflects the environment) against which existing organisms may be compared. Fisher's mathematical argument is phrased in terms of an infinite number of 'dimensions' for adaptation (only three are used here for ease of illustration). Imagine an adaptational space of three coordinates representing, respectively, the competitive, predatory and physical environments (Fig. 9.15). An ideal 'perfectly adapted' organism lies at a particular point (say A) in this space, but any given real organism is at another point (say B), some distance, d, away from the point of perfect adaptation. Changes in the position of A correspond to environmental changes, making the optimally adapted organism different; changes in B represent changes in the organisms concerned, such as mutations. The distance between the two points, d, represents the degree of conformity between the organism and environment or the level of adaptation. Fisher noted that very small undirected changes in either organism or environment have a 50:50 chance of being to the organism's advantage (i.e., reducing the distance between A and B). The probability of such improvement is inversely related to the magnitude of the change (Fig. 9.16). Very great changes in either organism or environment are always maladaptive because even if they are in the correct direction, they 'overshoot' points of closer adaptation. (Of course, it is remotely possible that such major environmental changes or 'macromutations' could put an organism into a completely new adaptive realm and thereby improve its overall level of adaptation.) Fisher makes an analogy with focusing a microscope. Very fine changes are as likely as not to improve the focus, but gross changes will almost invariably throw the machine further out of focus. Organisms may be thought of as 'tracking' their environments in both ecological and evolutionary time, changing as their environments change; thus, as point A shifts because of daily, seasonal and long-term environmental fluctuations, point B follows it. Such environmental tracking may be physiological (as in acclimation), behavioural (including learning) and/or genetic (evolutionary), depending on the time scale of environmental change. Reciprocal counterevolutionary responses to other species (prey, competitors, parasites and predators) constitute examples of such evolutionary tracking,' and have been termed coevolution.

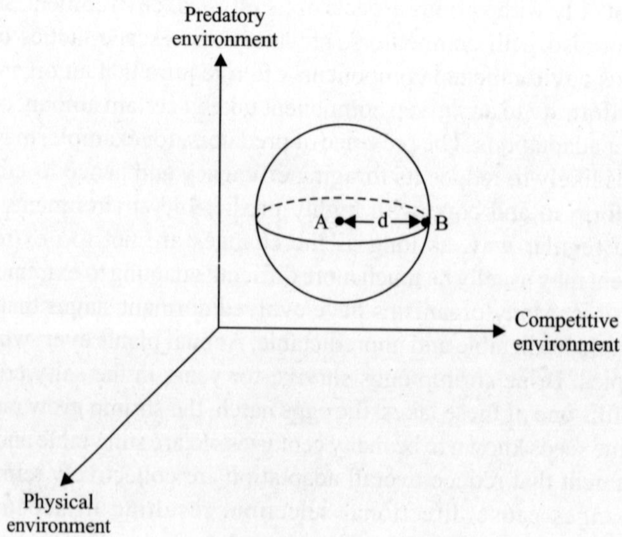

Fig. 9.15. Fisher's model of adaptation and deterioration of environment. Point A represents a hypothetical 'perfectly adapted' organism; an actual organism (point B) is never perfectly adapted and thus lies at some distance, d, from point A. The surface of the sphere represents all possible points with a level of adaptation equal to the organism under consideration. Very small undirected changes in either organism (point B) or environment (point A) are equally likely to increase or to decrease the level of adaptation, d. Notice the duality of the model (make $A \rightarrow B$ and $B \rightarrow A$).

Individual organisms with narrow tolerance limits, such as highly adapted specialists, generally suffer greater losses in fitness due to a unit of environmental deterioration than generalised organisms with more versatile requirements, all else being equal. Thus, more specialised organisms or those with restricted homeostatic abilities cannot tolerate as much environmental change as generalists or organisms with better developed homeostasis (Fig. 9.16). Fisher's model applies only to nondirected changes in either party of the adaptational complex—such as mutations and perhaps certain climatic fluctuations or other random events. However, many environmental changes are probably nonrandom. Changes in other associated organisms, especially predators and prey, are invariably directed so as to reduce an organism's degree of conformity to its environment; thus, they inevitably constitute a deterioration of that organism's environment. Directed changes in competitors can either increase or decrease an organism's level of adaptation, depending on whether they involve avoidance of competition or improvements in competitive ability per second. Directed changes in mutualistic systems would usually tend to improve the overall level of adaptation of both parties.

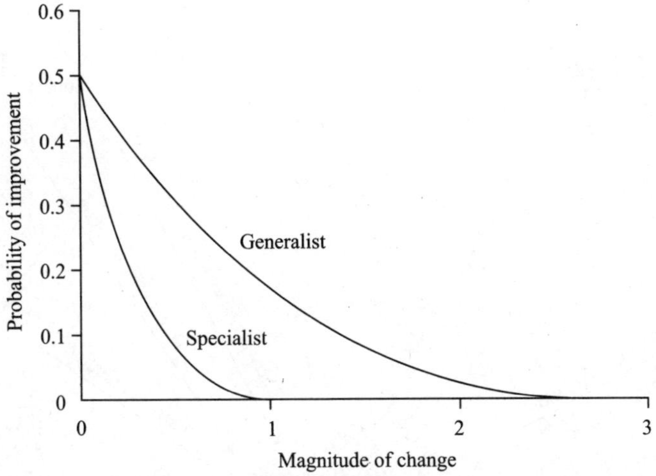

Fig. 9.16. The probability of improvement of the level of adaptation (i.e., of reducing d) is plotted against the magnitude of an undirected change when the number of dimensions is large. Two hypothetical organisms are shown, one highly adapted such as a specialist with narrow tolerance limits and one less highly adapted such as a generalist with broader tolerance limits or a greater number of niche dimensions or both. A random change of a given magnitude is more likely to improve the level of adaptation of the generalist than the specialist.

HEAT BUDGETS AND THERMAL ECOLOGY

When averaged over a long enough period of time, heat gained by an organism must be exactly balanced by heat lost to its environment; otherwise, the plant or animal would either warm up or cool off. Many different pathways of heat gains and heat losses exist (Fig. 9.17). The notion of a heat budget is closely related to the concept of an energy budget; balancing a heat budget requires very different adaptations under varying environmental conditions. At different times of day, ambient thermal conditions may change from being too cold to being too warm for a particular organism's optimal performance. Organisms living in hot deserts must avoid overheating by being able to minimise heat loads and to dissipate heat efficiently; in contrast, those that live in colder places such as at high altitudes or in polar regions must avoid overcooling—and they have hence evolved efficient means of heat retention, such as insulation by blubber, feathers or fur, that reduce the rate of heat exchange with the external environment.

Since, environmental temperatures fluctuate in characteristic ways at different places over the Earth's surface, both daily and seasonally. In the absence of a long-term warming or cooling trend, environmental temperatures at any given spot remain roughly constant when averaged over an entire annual cycle. Recal that the range in temperature within a year is much greater at high latitudes than it is nearer, the equator. An organism could balance its annual heat budget by being entirely passive and simply allowing its temperature to mirror that of its environment. Such a passive thermoregulator is known as a *thermoconformer* (Figs. 9.18 and 9.19). Of course, it is also an ectotherm. Another extreme would be to maintain an absolutely constant body temperature by physiological or behavioural means, dissipating (or avoiding) excess bodily heat during warm periods but retaining (or gaining) heat during cooler periods (in endotherms, energy intake is often increased during cold periods and more metabolic heat is produced to offset the increased heat losses).

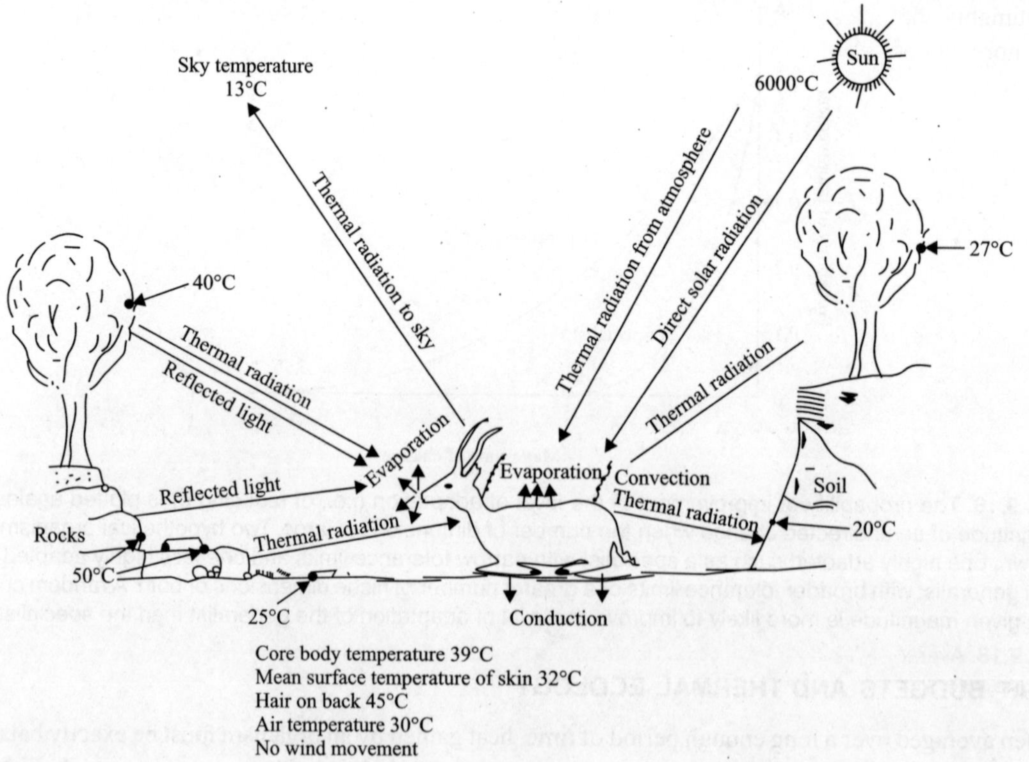

Fig. 9.17. Diagrammatic representation of various pathways of heat gains and losses between an animal and its environment under moderately warm conditions. Heat exchange between an animal and its environment is roughly proportional to its body surface: Because small animals have a larger surface area relative to their weight than larger ones, the former gain or lose relatively more heat than the latter.

Organisms that carefully regulate their internal temperatures are called *thermoregulators* or *homeotherms*. (Recall that both endotherms and ectotherms may regulate their body temperatures.) There is, of course, a continuum between the two extremes of perfect conformity and perfect regulation. Homeostasis, remember, is never perfect. Because regulation clearly has costs and risks as well as profits, an emerging conceptual framework envisions an optimal level of regulation that depends on the precise form of the constraints and interactions among the costs and benefits arising from a particular ecological situation. Thermoregulation often involves both physiological and behavioural adjustments; as an example of the latter, consider a typical terrestrial diurnal desert lizard. During the early morning, when ambient temperatures are low, such a lizard locates itself in warmer microclimates of the environmental thermal mosaic (e.g., small depressions in the open or on tree trunks), basking in the sun with its body as perpendicular as possible to the Sun's rays and thereby maximising heat gained. With the daily march of temperature, ambient thermal conditions quickly rise and the lizard seeks cooler shady microhabitats. Individuals of some species retreat into burrows as temperatures rise; others climb up off the ground into cooler air and orient themselves by facing into the Sun's rays, thereby reducing heat load. Many lizards change colours and their heat reflectance properties, being dark and heat absorbent at colder times of day but light and heat reflectant at hotter times. Such adjustments allow individual lizards to be active over a longer period of time than they could be if they conformed passively to ambient thermal conditions;

presumably, they are also more effective competitors and better able to elude predators as a result of such thermoregulatory behaviours.

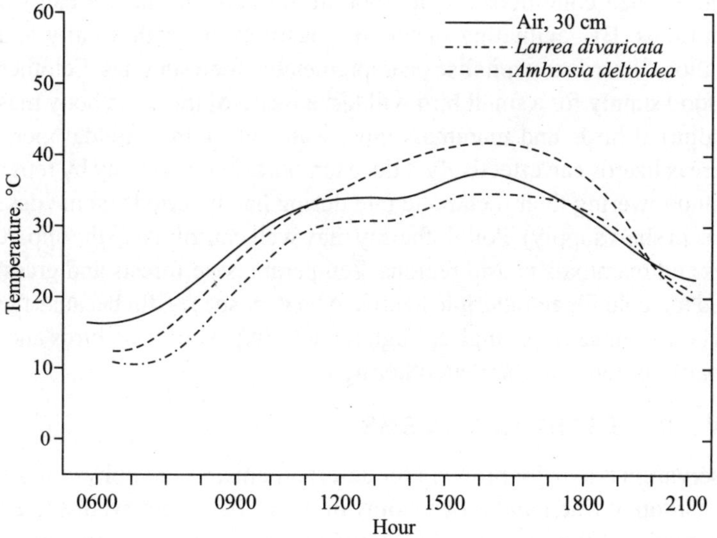

Fig. 9.18. Average temperatures of leaves of *Larrea divaricata* and *Ambrosia deltoidea* during three sunny days in June with slight breezes prevailing.

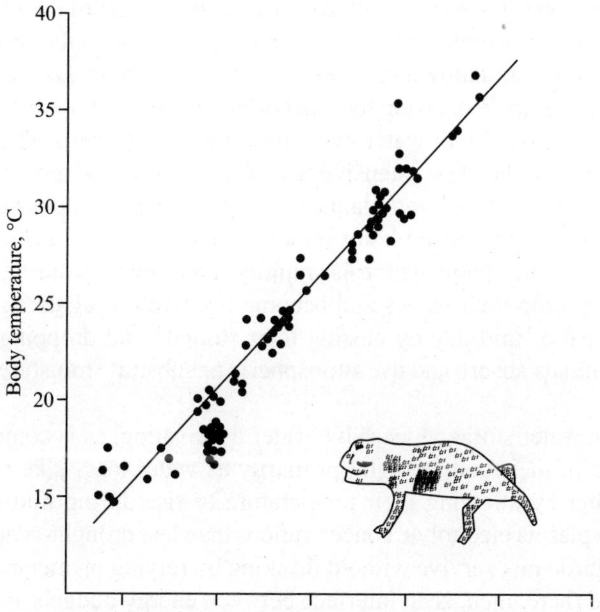

Fig. 9.19. Body temperatures of 86 active Australian nocturnal gekkonid lizards (*Nephrurus laevissimus*) plotted against ambient air temperature. The line represents body temperatures equal to air temperatures. Like most nocturnal lizards, these animals are nearly perfect thermoconformers.

Hot, arid regions typically support rich lizard faunas, whereas cooler forested areas have considerably fewer lizard species and individuals. Lizards can enjoy the benefits of a high metabolic rate during relatively brief periods when conditions are appropriate for activity and yet can still become inactive during adverse conditions. By facilitating metabolic inactivity on both a daily and a seasonal basis, poikilothermy thus allows lizards to capitalise on unpredictable food supplies. Ectotherms are low-energy animals; one day's food supply for a small bird will last a lizard of the same body mass for a full month! Most endothermic diurnal birds and mammals must wait out the hot midday period at considerable metabolic cost, whereas lizards can effectively reduce temporal heterogeneity by retreating underground, becoming inactive and lowering their metabolic rate during harsh periods (some desert rodents estivate when food or water is in short supply). Poikilothermy may well contribute to the apparent relative success of lizards over birds and mammals in arid regions. Temperate zone forests and grasslands are probably simply too shady and too cold for ectothermic lizards to be very successful because these animals depend on basking to reach body temperatures high enough for activity; in contrast, birds and mammals do quite well in such areas partly because of their endothermy.

WATER ECONOMY IN DESERT ORGANISMS

Because water conservation is a major problem for desert organisms, their physiological and behavioural adaptations for acquisition of water and for economy of its use have been well studied. These interesting adaptations are quite varied. Like energy and heat budgets, water budgets must balance; losses must be replaced by gains. For examples of water acquisition mechanisms, consider rooting strategies. Desert plants usually invest considerably more in root systems than plants from wetter areas; one study showed that perennial shrubs in the Great Basin Desert allocate nearly 90 per cent of their biomass to underground tissues, whereas roots apparently represent a much smaller fraction (only about 10 per cent) of the standing-crop biomass of a mesic hardwood forest. The creosote bush *Larrea divaricata* has both a surface root system and an extremely deep tap root that often reaches all the way down to the water table. This long tap root provides *Larrea* with water even during long dry spells when surface soils contain little moisture. Cacti, in contrast, have an extensive but relatively shallow root system and rely on water storage to survive drought. Many such desert plants have tough sclerophyllous xerophytic leaves that do not allow much water to escape (they also photosynthesise at a low rate as a consequence). Mesophytic plants occur in deserts too, but they photosynthesise rapidly and grow only during periods when moisture is relatively available; they drop their leaves and become inactive during droughts. Plants also reduce water losses during the heat of mid-day by closing their stomata and dropping their leaves (wilting). Many desert plants and animals absorb and use atmospheric or substrate moisture; most can also tolerate extreme desiccation.

Camels do not, rely on water storage to survive water deprivation, as is commonly thought, but can lose as much as a quarter of their body weight, primarily as water loss. Like many desert organisms, camels also conserve water by allowing their temperature to rise during mid-day. Moreover, camels tolerate greater changes in plasma electrolyte concentrations than less drought-adapted animals. In deserts, small mammals like kangaroo rats survive without drinking by relying on metabolic water derived from the oxidation of their food (here, then, is an interface between energy budgets and water budgets). Most desert rodents are nocturnal and avoid using valuable water for heat regulation by spending hot daytime hours underground in cool burrows with high relative humidity, thereby minimising losses to evaporation (most desert organisms resort to evaporative cooling mechanisms such as panting only in emergencies). The urine of kangaroo rats is extremely concentrated and their feces contain little water. Most other desert animals minimise water losses in excretion, too. Birds and lizards produce solid uric acid wastes rather than urea, thereby requiring little water for excretion. Desert lizards also conserve water by retreating to burrows and lowering their metabolic rate during the heat of the day.

OTHER LIMITING MATERIALS

Numerous other materials, including calcium, chloride, magnesium, nitrogen, potassium and sodium, may be in short supply for particular organisms and must therefore be budgeted. Neural mechanisms of animals depend on sodium, potassium and chloride ions, which are sometimes available in limited quantities. Because many herbivorous mammals obtain little sodium from their plant foods (plants lack nerves and sodium is not essential to their physiology), these animals must conserve sodium or find supplemental sources at salt licks—indeed, Feeny suggests that plants may actually withhold sodium as an antiherbivore tactic! Similarly, amino acids are in short supply for many insects, such as in Heliconius butterflies, which supplement their diets with protein-rich pollen.

An organism's nutrient and vitamin requirements are strongly influenced by the evolution of its metabolic pathways; likewise, these same pathways may themselves determine certain of the organism's nutritional needs. To illustrate: Almost all species of vertebrates synthesise their own ascorbic acid, but humans and several other primate species that have been tested cannot; they require a dietary supplement of ascorbic acid, known as vitamin C. Of thousands of other species of mammals, only two—the guinea pig and an Indian fruit-eating bat—are known to have lost the ability to synthesise their own ascorbic acid. A few species of birds must supplement their diets with ascorbic acid, too. Thus, a frog, a lizard, a sparrow or a rat can make its own ascorbic acid, but we cannot. Why should natural selection favour the loss of the ability to produce a vital material? Pauling suggests that species of animals that have lost this capacity evolved in environments with ample supplies of ascorbic acid in available foods. It might actually be advantageous to dismantle a biochemical pathway in favour of another once it becomes redundant. Conversely, natural selection should favour evolution of the ability to synthesise any necessary materials that cannot be predictably obtained from available foods where this is possible (clearly organisms cannot synthesise elements; e.g., herbivores cannot make sodium).

SENSORY CAPACITIES AND ENVIRONMENTAL CUES

Animals vary tremendously in their perceptive abilities. Most (except some cave dwellers and deep-sea forms) use light to perceive their environments. But visual spectra and acuity vary greatly. Some, such as insects, lizards and birds, have colour vision, whereas others (most mammals, except squirrels and primates) do not. Ants, bees and some birds can detect polarised light and exploit this ability to navigate by the Sun's position; pigeons have poorly understood backup systems that enable them to return home remarkably well, even when their vision is severely impaired with opaque contact lenses. Many temperate-zone species of plants and animals rely on changes in day length to anticipate seasonal changes in climatic conditions. This, in turn, requires an accurate 'biological clock'. (Some organisms may also use barometric pressure to anticipate climatic changes.) Certain snakes, such as pit vipers and boas, have infrared receptors that allow them to locate and capture endothermic prey in total darkness. Most animals can hear of course, although response to different frequencies varies considerably (some actually perceive ultrasonic sounds). Bats and porpoises emit and exploit sonar signals to navigate by echolocation. Similarly, nocturnal electric fish perceive their immediate environments by means of self-generated electrical fields. Certain bioluminescent organisms, such as 'fireflies' (actually beetles), produce their own light for a variety of purposes, including the attraction of mates and prey as well as (possibly) the evasion of predators (in some cases, a mutualistic relationship is formed with a bioluminescent bacteria). Certain deep-sea fish probably use their 'headlights' to find prey in the dark depths of the ocean. Pigeons can detect magnetic fields. Although a few animals have only a relatively feeble sense of smell (e.g., birds and human), most have keen chemoreceptors and olfactory abilities. Certain male moths can detect exceedingly dilute pheromones released by a female a full kilometer upwind, allowing these males to find females at considerable distances. Similarly, dung beetles use a zigzag flight to 'home in' on upwind fecal material with remarkable precision.

Various environmental cues clearly provide particular animals with qualitatively and quantitatively different kinds and amounts of information. Certain environmental cues are useful in the context of capturing prey and escaping predators; others may facilitate timing of reproduction to coincide with good conditions for raising young. Some environmental signals are noisier and less reliable than others. Moreover, the ability to process information received from the environment is limited by a finite neural capacity. The principle of allocation and the notion of trade-offs dictate that an individual cannot perceive all environmental cues with high efficiency. If ability to perceive a broad range of environmental stimuli actually requires lowered levels of performance along each perceptual dimension, natural selection should improve perceptual abilities along certain critical dimensions at the expense of other less useful ones. Clearly, echolocation has tremendous utility for a nocturnal bat, whereas vision is relatively much less useful. In contrast, the values of these two senses are reversed for a diurnal arboreal squirrel. Within phylogenetic constraints imposed by its evolutionary history, an animal's sensory capacities can be viewed as a bioassay of the importance of particular perceptual dimensions and environmental cues in that animal's ecology.

ADAPTIVE SUITES

A basic point of this chapter is that any given organism possesses a unique coadapted complex of physiological, behavioural and ecological traits whose functions complement one another and enhance that organism's reproductive success. Such a constellation of adaptations has been called an optimal design or an adaptive suite.

Consider the desert horned lizard *Phrynosoma platyrhinos* (Fig. 9.20). Various features of its anatomy, behaviour, diet, temporal pattern of activity, thermoregulation and reproductive tactics can be profitably interrelated and interpreted to provide an integrated view of the ecology of this interesting animal. Horned lizards are ant specialists and usually eat essentially nothing else. Ants are small and contain much undigestible chitin, so that large numbers of them must be consumed. Hence, an ant specialist must possess a large stomach for its body size. When expressed as a proportion of total body weight, the stomach of this horned lizard occupies a considerably larger fraction of the animal's overall body mass (about 13 per cent) than do the stomachs of all other sympatric desert lizard species, including the herbivorous desert iguana *Dipsosaurus dorsalis* (herbivores typically have lower assimilation rates and larger stomachs than carnivores). Possession of such a large gut necessitates a tanklike body form, reducing speed and decreasing the lizard's ability to escape from predators by flight. As a result, natural selection has favoured a spiny body form and cryptic behaviour rather than a sleek body and rapid movement to cover (as in most other species of lizards). Risks of predation are likely to be increased during long periods of exposure while foraging in the open. A reluctance to move, even when actually threatened by a potential predator, could well be advantageous; movement might attract attention of predators and negate the advantage of concealing colouration and contour. Such decreased movement doubtless contributes to the observed high variance in body temperature of *Phrynosoma platyrhinos*, which is significantly greater than that of all other species of sympatric lizards.

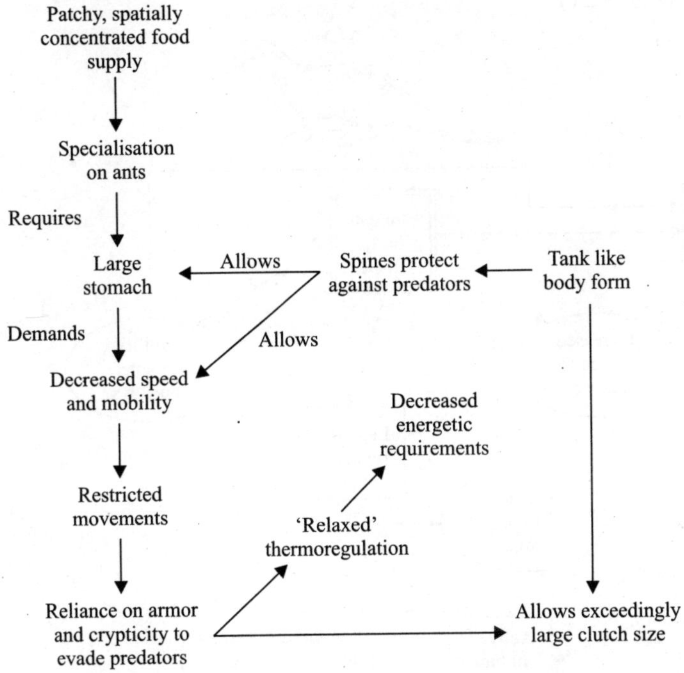

Fig. 9.20. Diagrammatic portrayal of factors influencing the ecology and body form of the North American desert horned lizard, *Phrynosoma platyrhinos*.

Phrynosoma platyrhinos are also active over a longer time interval than any sympatric lizard species. Wide fluctuations in horned lizard body temperatures under natural conditions presumably reflect both the long activity period and perhaps their reduced movements into or out of the Sun and shade (most of these lizards are in the open Sun when first sighted). More time is thus made available for activities such as feeding. A foraging anteater must spend considerable time feeding. Food specialisation on ants is economically feasible only because these insects usually occur in a clumped spatial distribution and hence constitute a concentrated food supply. To make use of this patchy and spatially concentrated, but at the same time not overly nutritious, food supply, *P. platyrhinos* has evolved a unique constellation of adaptations that include a large stomach, spiny body form, an expanded period of activity and 'relaxed' thermoregulation (eurythermy). The high reproductive investment of adult horned lizards is probably also a simple and direct consequence of their robust body form. Lizards that must be able to move rapidly to escape predators, such as racerunners (*Cnemidophorus*), would hardly be expected to weight themselves down with eggs to the same extent as animals like horned lizards that rely almost entirely upon spines and camouflage to avoid their enemies.

Energetics of metabolism of weasels provide another, somewhat more physiological, example of a suite of adaptations. Due to their long thin body shape, weasels have a higher surface-to-volume ratio than mammals with a more standard shape and as a consequence, they have an increased energy requirement. Presumably, benefits of the elongate body form more than outweigh associated costs; otherwise, natural selection would not have favoured evolution of the weasel body shape. Brown and Lasiewski speculate that a major advantage of the elongate form is the ability to enter burrows of small mammals (weasel prey), which results in increased hunting success and thus allows weasels to balance their energy budgets (Fig. 9.21).

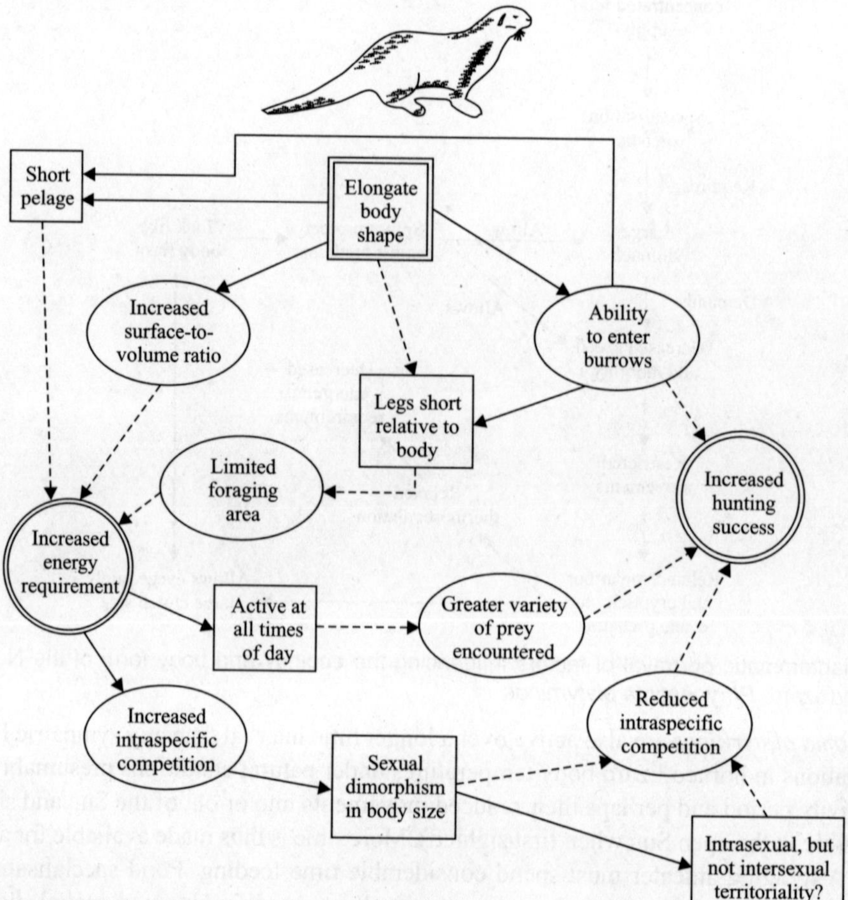

Fig. 9.21. A schematic representation of factors involved in the evolution of elongate body shape in weasels. Circles indicate primary consequences of evolving a long, thin body configuration; ellipses show secondary consequences; rectangles indicate phenotypic characteristics of weasels affected by evolution of this body shape. Unbroken arrows indicate selective pressures and dashed arrows show causal sequences. Changes proceed in the direction of the arrows as long as selection favours a more elongate shape.

A further spinoff of the elongate shape is the evolution of a pronounced sexual dimorphism in body size, which allows male and female weasels to exploit prey of different sizes and hence reduces competition between the sexes (related mustelids such as skunks and badgers do not have the marked sexual size dimorphism characteristic of weasels).

DESIGN CONSTRAINTS

Most biologists are acutely aware that possible evolutionary pathways are somehow constrained by basic body plans. Although natural selection has 'invented', developed and refined a truly amazing variety of adaptations, selection is clearly far from omnipotent. Wheels might be a desirable solution to certain environmental contingencies and yet they have not been evolved. Such 'design constraints' are usually elusive and not easily demonstrable.

Students of thermoregulation have often noted an apparent upper thermal limit of about 40°C for most of the earth's eukaryotic creatures (most plants, invertebrates and vertebrates). This thermal 'lid' has frequently been used as evidence for some extremely archaic and inflexible fundamental physiological process (perhaps an enzyme basic to life processes, such as a dehydrogenase, denatures above this limit). The major exceptions are certain heat-tolerant bacteria and blue-green algae, inhabitants of hot springs and oceanic volcanic vents. These prokaryotic organisms may well have arisen before the origin of the heat-sensitive metabolic pathway that seems to limit the eukaryotes.

An example of such a physiological design constraint involves the thermal relationships of vertebrates, spanning classes from reptiles to mammals. Detailed consideration of behavioural thermoregulation in lizards enables a fairly accurate prediction of the active body temperatures of mammalian homeotherms. A provocative biological 'constant' can thus be identified that suggests a substantial degree of physiological inertia.

An intriguing hypothesis for the evolution of homeothermy was offered by Hamilton, who suggested that homeothermy is a by-product of advantages gained from maintaining maximum body temperatures in the face of such an innate physiological ceiling. Ecologically optimal temperatures need not coincide with physiological optima.

Remember that not all homeotherms are endotherms; many ectotherms have attained a substantial degree of homeothermy by means of behavioural thermoregulation. Typically, such organisms actively select thermally suitable microhabitats, orient their bodies (or parts thereof) to control heat exchange and/or shuttle between Sun and shade as necessary to maintain a more or less constant internal body temperature.

Thermoregulation in lizards is not nearly as simple as it might appear to be at first glance, but rather encompasses a wide diversity of very different thermoregulatory tactics among species ranging from ectothermic poikilothermy to and including ectothermic homeothermy. Even a casual observer quickly notices that various species of desert lizards differ markedly in their times and places of activity. Some are active early in the morning, but other species do not emerge until late morning or midday. Most geckos and pygopodids and some Australian skinks are nocturnal. Certain species are climbers, others subterranean, while still others are strictly surface dwellers. Among the latter, some tend to be found in open areas, whereas others frequent the edges of vegetation. Thermal relations of active lizards vary widely among species and are profoundly influenced by their spatial and temporal patterns of activity. Body temperatures of some diurnal heliothermic species average 38°C or higher, whereas those of nocturnal thigmothermic species are typically in the mid-twenties, closely paralleling ambient air temperatures.

Interesting interspecific differences also occur in the variance in body temperature as well as in the relationship between body temperatures and air temperatures. For example, among North American lizards, two arboreal species (*Urosaurus graciosus* and *Sceloporus magister*) display narrower variances in body temperature than do terrestrial species. Presumably, arboreal habits often facilitate efficient, economic and rather precise thermoregulation. Climbing lizards have only to shift position slightly to be in the Sun or shade or on a warmer or cooler substrate and normally do not move through a diverse thermal environment. Moreover, arboreal lizards need not expend energy making long runs as do most ground dwellers and thus climbing species do not raise their body temperatures metabolically to as great an extent as do terrestrial lizards.

Such differences in temporal patterns of activity, the use of space and body temperature relationships are hardly independent. Rather, they complexly constrain one another, sometimes in intricate and obscure ways. For example, thermal conditions associated with particular microhabitats change in characteristic ways in time; a choice basking site at one time of day becomes an inhospitable hot spot at another time. Perches of arboreal lizards receive full Sun early and late in the day when ambient air temperatures tend to be low and basking is therefore desirable, but these same tree trunks are shady and cool during the heat of midday when heat-avoidance behaviour becomes necessary. In contrast, the fraction of the ground's surface in the Sun is low when shadows are long early and late, but reaches a maximum at midday. Terrestrial heliothermic lizards may thus experience a shortage of suitable basking sites early and late in the day; moreover, during the heat of the day, their movements through relatively extensive patches of open Sun can be severely curtailed. Hence, ground-dwelling lizards encounter fundamentally different and more difficult thermal challenges than do climbing species.

Radiation and conduction are the most important means of heat exchange for the majority of diurnal lizards, although the thermal background in which these processes occur is strongly influenced by prevailing air temperatures. Ambient air temperatures are critical to nocturnal lizards as well as to certain, very cryptic diurnal species.

In an analysis of the costs and benefits of lizard thermoregulatory strategies, Huey and Slatkin identified the slope of the regression of body temperature against ambient environment temperature as a useful indicator (in this case, an inverse measure) of the degree of passiveness in regulation of body temperature. On such a plot of active body temperature versus ambient temperature, a slope of one indicates true poikilothermy or totally passive thermoconformity (a perfect correlation between air temperature and body temperature results), whereas a slope of zero reflects the other extreme of perfect thermoregulation. Lizards span this entire thermoregulation spectrum. Among active diurnal heliothermic species, regressions of body temperature on air temperature are fairly flat (for several species, including some quite small ones, slopes do not differ significantly from zero); among nocturnal species, slopes of similar plots are typically closer to unity. Various other species (nocturnal, diurnal and crepuscular), particularly Australian ones, are intermediate, filling in this continuum of thermoregulatory tactics.

A straight line can be represented as a single point in the coordinates of slope versus intercept; these two parameters are plotted for linear regressions of body temperatures on air temperatures among some 82 species of lizards in Fig. 9.22. Each data point represents the least-squares linear regression of body temperature against air temperature for a given species of desert lizard. Interestingly enough, these data points fall on yet another, transcendent, straight line. The position of any particular species along this spectrum reflects a great deal about its complex activities in space and time. The line plotted in Fig. 9.22 thus offers a potent linear dimension on which various species can be placed in attempts to formulate general schemes of lizard ecology. Various other ecological parameters, including reproductive tactics, can be mapped on to this emergent spatial-temporal axis.

The intriguing 'intercept' of the intercepts (38.8°C) approximates the point of intersection of all 82 regression lines and presumably represents an innate design constraint imposed by lizard physiology and metabolism. It is presumably not an accident that this value also corresponds more or less to the body temperature of homeotherms, particularly mammals.

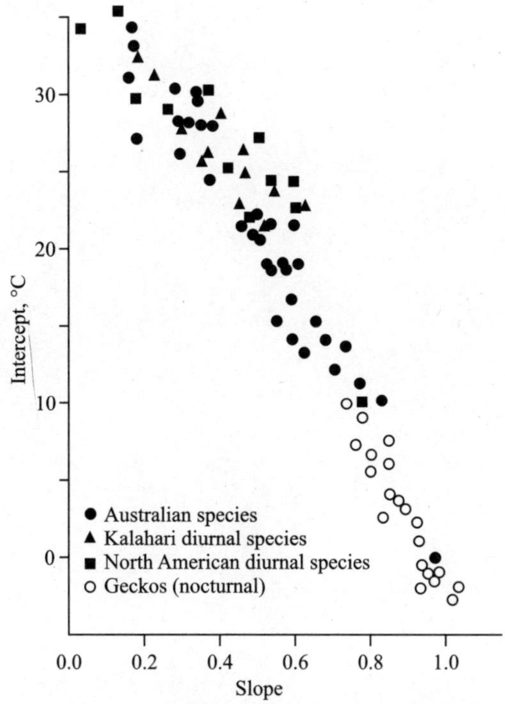

Fig. 9.22. Each data point represents the least-squares linear regression of body temperature against air temperature for a given species of desert lizard. Sample sizes are usually substantial (average is 145). The horizontal axis represents the spectrum of thermoregulatory tactics ranging from active thermoregulators (slope of zero) to entirely passive thermoconformity (slope of one). The intriguing 'intercept' of the intercepts (38.8°C) approximates the point of intersection of all 82 regression lines and presumably represents an innate design constraint imposed by lizard physiology and metabolism.

Birds, which maintain slightly higher body temperatures than mammals, descended from another reptilian stock, the archosaurs, represented today by the crocodilians. Would a comparable study of crocodilian thermoregulation yield a higher intercept of the intercepts? (This prediction could be doomed to failure by the mere fact that crocodilians are aquatic and very large—yet they obviously thermoregulate when out of the water.) Although most insects are so small that convective heat exchange prevents them from attaining body temperatures much higher than that of ambient air, some, such as bumblebees and butterflies, do exhibit behavioural thermoregulation. Would a plot for insects show more scatter and a different intercept?

SECTION III

Population

10

Individual

INTRODUCTION

Although ecologists are often interested in the complex interactions between species, it is worth remembering that it is individual organisms that are the products of natural selection. This chapter deals with the biology of individuals. Individuals are the fundamental units of population, communities, ecosystems and biomes which are discussed in later chapters. In this chapter we will look at individuals from an ecological perspective. We will start with the essentials of how they obtain their energy and nutrients and then consider how these are allocated to maintenance, growth and reproduction.

ORGANISMS

All organisms need energy to live and different organisms obtain this energy in different ways. There are many approaches to classifying the ways in which individuals obtain their food. A useful one is to divide organisms into autotrophs and heterotrophs. Autotrophs obtain only the simplest inorganic substances from their environment. Green plants are the most obvious autotrophs. These need only visible light, water, carbon dioxide and inorganic ions such as nitrate (NO_3^-) to survive, grow and reproduce. The process of photosynthesis enables most plants, the photosynthetic bacteria and some protoctists to synthesise all the complex organic molecules that they require from these simple building blocks. Because these organisms use light as their energy source, they are called photoautotrophs. Chemoautotrophs are autotrophs that obtain their energy not from sunlight but from certain specific chemical reactions involving only inorganic substrates. Animals, fungi, most bacteria, many protoctists and a few plants cannot synthesise their organic molecules from inorganic ones. They need to take in an external source of organic carbon. These organisms are heterotrophs.

Terms Associated With Heterotrophic Nutrition

While autotrophs such as green plants and chemosynthetic bacteria obtain the few nutrients that they require from the environment around them, heterotrophs have first to ingest their food, if necessary egesting some of it too and then to digest what they have ingested. Digestion is followed by absorption and absorption by assimilation. Finally, some matter is excreted.

Ingestion

Ingestion is the process by which heterotrophs take in their food. For most animal species it is useful to distinguish two stages to ingestion. In the first, the food is captured; in the second, it is brought within the alimentary canal.

Egestion

In some animals not all the food that is ingested is eventually digested. For example, some time after an egg-eating snake has swallowed an egg it regurgitates the crushed shell. The eggshell is said to have been egested. Owls also egest pellets containing the fur, feathers and bones of the small animals they have eaten. In general, the parts of the food ingested by a heterotroph which have not been digested sufficiently to allow them to be absorbed into the tissues are disposed of by egestion. The faeces produced by animals also contain egested material. For example, the plant fiber in our diet, though important for the proper functioning of our large intestine, is not digested, but egested via the anus.

Digestion

Digestion is the process by which heterotrophs break down the food they have ingested into particles small enough to be absorbed. In most cases, this means that mastication and hydrolytic enzymes are used to break down food particles to their constituent monomers. Proteins are broken down to amino acids; carbohydrates to monosaccharides; fats to tiny fat globules or even to fatty acids, glycerol and other simple molecules. Salts and vitamins can be absorbed without needing to be digested.

Absorption

Absorption is the uptake from the gut of vitamins, salts and the products of digestion. In mammals these substances are absorbed by certain of the villi of the small intestine. The products of absorption are either respired or assimilated.

Assimilation

Assimilation occurs when the products of absorption are taken up by cells and synthesised into macromolecules. These macromolecules may be stored or used for repair, growth and reproduction.

Excretion

Excretion is a characteristic of all organisms, not just heterotrophs and occurs when the waste products of metabolism are expelled from an organism. Nitrogen is an important excretory product of heterotrophic organisms. When proteins are broken down, ammonia (NH_3) is produced. Ammonia is very toxic and needs to be diluted in a large volume of water. Freshwater fish, which obviously have water in abundance, are able to dilute and then excrete their ammonia directly. Other vertebrates convert the ammonia to compounds which can be concentrated so as to be less wasteful of water, such as trimethylamine oxide, urea ($CO(NH_2)_2$ or uric acid.

The excretory products of one organism are almost invariably utilised by other organisms. For instance, the nitrogenous waste products of heterotrophs are typically converted to nitrates by soil bacteria and taken up by plant roots. Oxygen is an excretory product of plants yet is vital for all aerobic organisms.

Ingestion by Heterotrophs

The way an animal feeds profoundly influences many aspects of its ecology. Heterotrophs differ greatly in the sorts of food they ingest and in the ways they obtain their food. Three main types of heterotrophic nutrition are commonly recognised.

Holozoic nutrition

Holozoons feed on relatively large pieces of dead organic material. Most of the animals with which we are familiar can therefore be described as holozoic. Carnivorous, such as the fox (*Vulpes vulpes*), feed on prey which they have caught (Fig. 10.1). Herbivorous, such as sheep, cattle and goats, feed on vegetation. Animals which feed on a mixture of plant and animal food, such as the pig and most humans, are called omnivorous.

Fig. 10.1. A five month old fox cub (*Vulpes vulpes*) feeding on a hen pheasant.

Each of these terms may be subdivided. For example, herbivorous include granivorous (animals such as the world's most abundant bird, the redbilled quelea (*Quelea quelea*) which feed on grain or seeds), frugivores (which feed on fruit), folivorous (which feed on the leaves of shrubs or trees), grazers (which feed on herbs and grasses) and browsers (which feed on the leaves, young shoots and fruit of shrubs and trees). Animals have anatomical, physiological and behavioural adaptations which are associated with their diets. Consider an animal such as a deer or antelope which grazes grass. Grasses contain very fine particles of silica (SiO_2), the same hard substance of which sand is composed. This silica wears away teeth and many grazers, if predation is infrequent, die of starvation when their teeth literally wear away. Over the course of evolution grazers which can maintain the grinding surfaces of their teeth have been favoured by natural selection. The teeth of many grazers grow from their roots throughout their life. Elephants have responded to the problem of tooth abrasion in a remarkable manner. At anyone point in their life they only have four grinding teeth in use, one in the upper and one in the lower jaw on each side. Each tooth, as it is worn down, is slowly replaced by another from behind. Should an elephant get to be more than about 60 years old, it will have used up all of its 24 teeth in this way.

Parasitic nutrition

Parasites, unlike holozoons, feed off matter that is still alive. Because of this, parasites are usually much smaller than their hosts. This means that they generally have shorter lifespans than their hosts and are more numerous than them. As well as being smaller than their hosts, parasites harm their hosts. The host gains nothing from the relationship and is therefore constantly selected to avoid being parasitised.

Parasitism is an evolutionary trait which has evolved independently in a huge number of different taxa. Indeed, parasites may be found in all five kingdoms—bacteria, protoctists, fungi, plants and animals. Parasites can conveniently be divided into endoparasites, such as tapeworms, which live inside their hosts and ectoparasites, such as fleas, which live outside them. Whether endoparasitic or ectoparasitic, most parasites obtain their food in liquid form. Parasites of plants may obtain their nourishment from the phloem, as aphids do. Animal parasites may live in the gut or tap into the lymph or blood system.

Saprotrophic nutrition

Saprotrophs feed on dead organic matter which they either absorb in solution or ingest as very small pieces. Many fungi, for instance, are saprotrophs, living off dead organisms. Most saprotrophs obtain their food by extracellular digestion—enzymes are secreted on to the food source and the soluble products are then absorbed. This is the technique used by bacteria, fungi and the house fly (*Musca domestica*) for instance. On the other hand, earthworms obtain their food in solid form including small bits of dead leaves and soil invertebrates. Animals which feed off dead organic matter, but ingest large pieces, may be classified as scavengers. Spotted hyaenas, for example, often scavenge, though they obtain a large amount of their energy by hunting and killing prey.

METABOLIC RATE

One of the most useful single pieces of information an ecologist can have about an individual is its metabolic rate. The metabolic rate of an organism is the amount of energy it needs per unit time. This indicates what the demands of that organism are on its environment. For instance, animals with large metabolic rates are generally found near the top of food webs and at quite low population densities. A convenient unit for measuring metabolic rate is the number of joules of energy an organism uses each day. For organisms that rely on aerobic respiration to supply their energy needs, oxygen consumption is directly proportional to energy requirements and metabolic rate may be measured in units of hourly oxygen consumption as this is fairly easy to measure.

Organisms require energy in order to replace their tissues and make new ones for growth and reproduction. Organisms which burn up a lot of energy for their size, such as birds, are said to have a high metabolic rate. Organisms such as snakes, which can survive on much less food, are said to have a low metabolic rate. Many measures of metabolic rate have been made, of which perhaps the best known is basal metabolic rate. The basal metabolic rate of an organism is the minimum amount of energy it needs to respire to maintain life when the body is at rest and just 'ticking over'. Three conditions need to be fulfilled for the metabolic rate of an organism to be its the basal metabolic rate. First, the organism needs to be in a thermoneutral environment (so that no energy has to be used for thermoregulation). Second, the organism must be at rest (so that no energy is needed for locomotion or any other activities). Third, the organism has to be in a post-absorptive state (that is some time after food has been eaten) because the metabolic rate of an organism rises shortly after it eats. To an ecologist, measurements of basal metabolic rates are not especially useful. For one thing, the concept of a thermoneutral environment only applies to homoiotherms (endotherms). Anyway, ecologists are more interested in what an organism's energy expenditure is in real life.

FACTORS AFFECTING METABOLIC RATE

Size

Perhaps the most important factor affecting the metabolic rate of an individual is its size. Overall, larger birds, need more energy each day to live.

However, metabolic rates are not directly proportional to body mass. A bird ten times heavier than another one does not need ten times as much food each day.

The small animals, such as shrews, have such a high metabolic rate relative to their body size that they may need to eat more than their body weight in food each day! An elephant, on the other hand, takes about three months to eat its own body weight of food.

Life Style

Even among organisms of the same size, there is a great deal of variation in their metabolic rates. Hemmingsen compared the basal metabolic rates of unicellular organisms, poikilothermic animals and homoiothermic animals. As you might expect, homoiotherms, which maintain a constantly high body temperature, need a lot more energy than poikilotherms. It is perhaps surprising just how much energy they need. It is a lot cheaper to feed a pet snake weighing 5 kg, than a pet dog weighing 5 kg.

A poikilotherm has a metabolic rate some 8–10 times higher than a unicellular organism of the same size. Evidently there is quite an energetic cost to being multicellular.

Also much of the variation is found within the three groups—unicellular organisms, poikilotherms and homoiotherms—can be accounted for by body size. However, there is considerable scatter about the three lines. Attempts to identify the reasons for this scatter have mainly focused on birds and mammals. Essentially, two reasons for the variation can be suggested: life style and phylogeny. For example, mammals that feed on vertebrates have high basal metabolic rates relative to other mammals. It might therefore be that a mammal that eats vertebrates needs a high basal metabolic rate because of its life style, perhaps because it must constantly be ready to rush after its prey. As Elgar and Harvey point out, however, the association might have a quite different explanation. It turns out that the correlation among mammals of high basal metabolic rate with the eating of vertebrates is mainly due to the possession of both of these traits by whales and dolphins (Cetacea) and seals (Pinnipedia). Perhaps the high basal metabolic rates of these, animals reflect their marine life styles rather than their feeding habits. Of course, this too would be an ecological explanation for their relatively high basal metabolic rates. Alternatively, perhaps all cetaceans and pinnipeds have high basal metabolic rates irrespective of their ecology. This would mean that their high basal metabolic rates might be due to phylogeny rather than to life style. Another possible explanation arises from the fact that it is extremely difficult to measure the basal metabolic rates of seals and dolphins under laboratory conditions. They stress easily and often show abnormally high 'basal' metabolic rates. It might therefore be that the high values are an artefact.

SIZE DETERMINES MORE THAN METABOLIC RATE

As we have seen, size is a very important determinant of an organism's metabolic rate. However, there are many other features of an organism's life that are strongly affected by size.

An organism's size greatly influences its ecology. Knowing an individual's size immediately tells us something about how it interacts with other organisms and with the rest of its environment. In particular, the amount of energy an organism takes in each day is closely related to its body size. But what do organisms do with this energy? How much of it can they apportion to growth or to reproduction and how much do they need just to remain alive? These questions can be answered by looking at the energy budgets of organisms.

ENERGY BUDGETS

Organisms take in food so that energy and nutrients can be channelled towards the production of offspring.

Individuals cannot devote all their resources directly to reproduction. Some of their energy intake is needed to keep themselves alive and in good condition. In fact, around 10 to 30 per cent of the energy absorbed from food ends up being used to digest food. When individuals are juvenile, they devote some of their energy to growth. Presumably, in most cases, a juvenile individual that tried to reproduce would not be very successful at it. For one thing, it would be too small to leave many offspring. It is more likely to maximise its lifetime reproductive success by waiting until it has grown more.

Assimilation Efficiency

When an organism ingests food, only some of it is assimilated. The term assimilation efficiency refers to the percentage of the energy that an organism ingests that is assimilated rather than egested. As one might expect, organisms differ greatly in their assimilation efficiencies depending on the type of food they eat. Carnivorous feeding on vertebrates may have assimilation efficiencies in excess of 90 per cent; insectivores typically have assimilation efficiencies of 70–80 per cent, while most herbivorous have assimilation efficiencies of 30–60 per cent, though zooplankton feeding on phytoplankton have assimilation efficiencies of 50–90 per cent and the giant panda (*Ailuropoda melanoleuca*) has the lowest assimilation efficiency of any mammal yet measured at 20 per cent. Assimilation efficiencies can be much lower than 20 per cent.

Production and Respiration

Once a heterotroph has assimilated its food, the products of assimilation can either be respired to provide the energy needed to maintain existence and to repair old and damaged body tissue or they can be diverted to growth and reproduction. Together, growth and reproduction are called production.

The percentage of the energy assimilated that an organism diverts to growth is called its growth efficiency. Growth efficiencies are economically very important to farmers. Juvenile pigs have growth efficiencies of up to 20 per cent, which is very high for a farm animal. This means that pigs are very efficient at converting their feed into pigs meat. The efficiency of this conversion is affected by many factors. As one might expect, their growth efficiency is highest when they do not have to expend any energy on thermoregulation and movement. Under intensive production some farmers therefore keep pigs warm and in almost total darkness—pigs move less when it is dark.

For any species, growth efficiencies are higher the smaller the individuals are, relative to their size at maturity. Figures 10.2a and 10.2b show growth efficiencies as functions of size for Leach's storm-petrel (*Oceanodroma leucorhoa*) and the fish *Ophiocephalus striatus*. In each case the growth efficiency decreases as the individuals mature. Because of this, it is rather difficult to make comparisons between species about their growth efficiencies. Nevertheless, as a general rule, juvenile poikilotherms have higher growth efficiencies than juvenile homoiotherms. However, if one looks at the growth efficiencies of very young organisms, one finds that homoiotherms may have growth efficiencies as high as 50–70 per cent, almost the same as those of poikilotherms which may be as high as 50–80 per cent. By comparison, bacteria, protoctists and metazoan cells in culture have growth efficiencies of about 60 per cent.

Allocation to reproduction

Perhaps rather surprisingly, comparatively few studies appear to have been done which look at the allocation of assimilated food to reproduction. It would be very useful to have data on what might be termed reproductive efficiencies for different species. In animals it may be the case that homoiotherms allocate a smaller percentage of their assimilated food to reproduction than do poikilotherms.

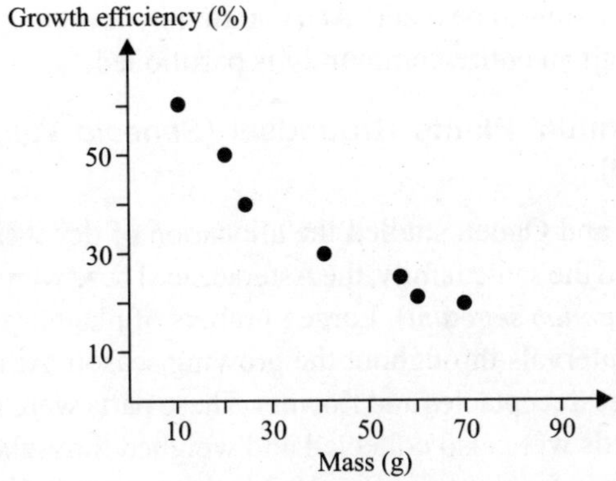

Fig. 10.2a. Growth efficiency as a function of size in juvenile Leach's storm-petrel (*Oceanodroma leucorhoa*).

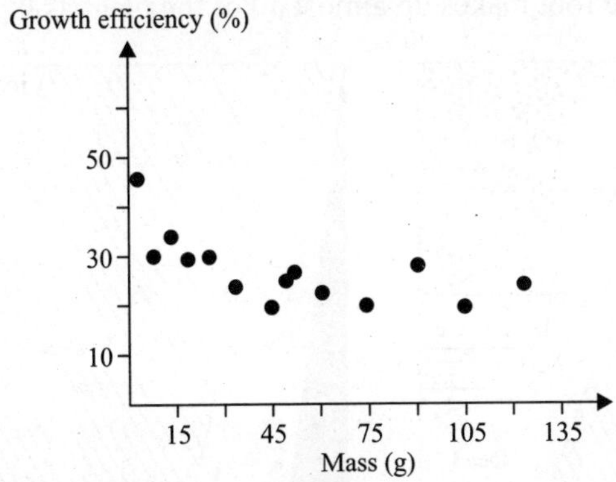

Fig. 10.2b. Growth efficiency as a function of size in juveniles of the fish *Ophiocephalus striatus.*

Some plant ecologists have looked at the percentage of the annual net assimilation, that is photosynthesis minus respiration, allocated by a plant to reproduction. Measuring the prcportion of an organism's energy budget that it devotes to reproduction may not be a true index of reproductive effort. Energy may not be the resource that is limiting reproduction. For many plants, available nitrogen or phosphorus are more likely to be in short supply than sunlight. So far, however, most workers have looked at energy budgets rather than nutrient budgets. It may turn out to be the case that for most species the two are closely related anyway, so that measurements of energy allocations adequately reflect the evolutionary pressures facing organisms.

Drawing up a complete energy budget

A number of studies have been carried out to determine the energy budgets of individual species. Although they often take a long time to perform and are difficult to carry out with great accuracy, energy budgets are valuable because they provide an insight into the ecological pressures operating on a species. We will look at one animal and two plant examples.

Energy budgets allow comparisons to be made between males and females of the way they allocate what they ingest to processes such as growth and reproduction. They can also be done for individuals of different ages and sizes allowing developmental changes to be seen. If carried out on several species, they allow interspecific comparisons to be made. At a community level they enable an investigator to see how the energy flowing through an entire community is partitioned.

Energy Budgets of the Annual Plants Groundsel (*Senecio Vulgaris*) and Corn Marigold (*Chrysanthemum Segetum*)

In two classic studies, Harper and Ogden studied the allocation of dry weight to different stages of the life cycle in two annual plants in the same family, the Asteraceae. These were groundsel (*Senecio vulgaris*) and corn marigold (*Chrysanthemum segetum*). Large numbers of plants were grown in glasshouses and samples harvested at regular intervals throughout the growing season. At most harvests each plant was divided into roots, stems, leaves, receptacles and flowers. These parts were then weighed separately after drying to constant weight. Seeds were also collected and weighed throughout the seeding period.

The results for the two species are shown in Fig. 10.3 in a form which allows the life cycles of the two species to be compared visually. These diagrams show how the compositions of the plants change as they grow. In both species, the root makes up almost all of the plant as it grows from the small seed.

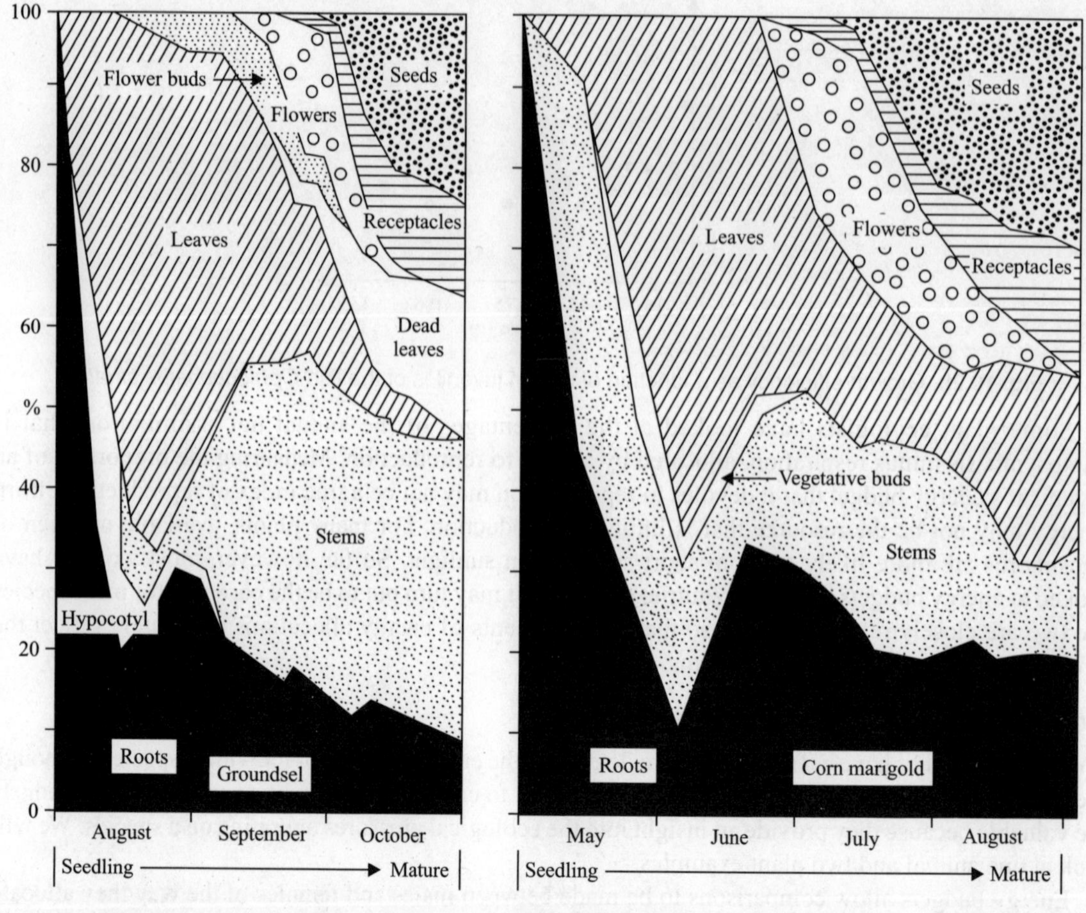

Fig. 10.3. Allocation of dry weight to different structures throughout the life cycles of the plants groundsel (*Senecio vulgaris*) and corn marigold (*Chrysanthemum segetum*).

Then the first leaves develop and grow and a stem is produced to support these leaves. In both cases flowers start to be formed about half-way through the life cycle as the plants begin to shift their resources from leaf production to flower and then seed production.

It would be valuable to have this sort of information for a large number of plant species with differing life histories as such information tells us a great deal about how the life cycle of the plant has been shaped by natural selection to adapt it to the environment in which it lives.

DISTINGUISHING BETWEEN GROWTH AND REPRODUCTION

So far we have assumed that growth and reproduction are two entirely separate processes. In many cases, however, this is not the case. Consider, for instance, a grass that produces tillers. In one sense this is growth: in another it is reproduction as the tillers soon grow roots and may even become detached from the parent plant. Perhaps the solution to the problem of what is growth and what is reproduction is that we should not try to divide all production into these two categories. Little is to be gained by forcing terms devised to describe higher animals to fit the life styles of all other organisms.

Many species show colonial growth—corals and bryozoans, for instance. Here, countless genetically identical multicellular subunits remain attached to one another. Essentially this is little different from the aggregation of genetically identical cells into a multi cellular organism.

Fungi can also defy classification as individuals. Alan Rayner argues that fungi and many other microbes challenge the whole notion of the individual as the unit of reproduction. Some fungi exist as 'diffuse individuals'—there is one such 'individual' that extends over 15 hectares of forest in Montana, weighs more than 100 tonnes and is thought to be over a thousand years old. Other fungi exist not as discrete, bounded entities but as dynamic networks. We, as mammals, tend to think of discrete organisms as the norm, but in fact fungi have twice the total biomass of all the animals in the world. Perhaps discrete individuals are the odd life form in a world of diffuse and colonial organisms.

Recently, many botanists have begun to think of plants as modular organisms where each unit such as a branch acts rather like a whole plant with leaves, flowers, seeds and so on. In most vascular plants growth is indeterminate in the sense that meristems (regions of actively dividing cells) continue to produce new shoots and roots until the opportunities for further local growth are exhausted. The individual and repeated sub-units of which a plant may be composed are sometimes referred to as modules. Growth is often decentralised in modular organisms, with each sub-unit capable of considerable regeneration.

11

Population Ecology and Population Dynamics

INTRODUCTION

The foregoing chapters provide a background of the various ecological factors as they might govern the distribution of organisms and detatils of the requirements and reactions of an individual species (autecology). However, organisms in nature rarely grow as separated from each other. Invariably, organisms grow organised as populations, communities and ecosystems. We would consider the basic concepts applicable at each of these three levels of organisation. And we begin with population level. In this chapter, we would consider properties of populations as though largely isolated from populations of other species.

Population ecology is the study of individuals of the same species where the processes as aggregation, interdependencies between individuals etc. and the various factors governing such processes are emphasised. This has been an important development in modern ecology. The foundations of population ecology were laid by animal ecologists in the first half of this century. It is only relatively recently that detailed studies of plant populations have been made. The increasing population of the world could attract the attention of not only of ecologists but also of behaviourial scientists and it led to the development of such new facets in ecology. Population ecology thus has to play a vital role in a number of socio-economic problems both at national as well as international levels.

A population is generally a group of individuals of a particular species occupying a particular area at a specific time. Some of the ecologists, however, recognise two types of population:

1. Monospecific population .It is the population of individuals of only one species.
2. Mixed or polyspecific population. It is the population of individuals of more than one species. However, in ecology polyspecific population is generally referred to as a community and the term population is used for group of individuals of any kind of organism.

BASIC CONCEPTS OF POPULATION ECOLOGY

Populations are characterised with such characteristics as dispersion, fluctuation in numbers (density), sex ratio, birth rate and death rate etc. The population of a species generally arises as a result of reproduction, active transport of individuals or their passive transport by such agencies as wind, water etc. All the three means of population growth are influenced by several factors of the environment as well as by the characteristics of the individuals of the species itself. Under favourable conditions, the group of individuals increases in numbers. However, environment is never static and keeps on changing from time to time i.e., is dynamic. Thus, environment acts as a natural check on population. An increase in the number of individuals of a species results in the consequences of concern to the species itself and also to other interdependent species growing in the area. Such an increase brings about harmful as well as beneficial effects (interactions) on the species. The former type of effects is mainly due to increased competition, particularly intraspecific for nutrition, space etc. The beneficial effects include the protection, influences on reproduction and origin of division of labour.

The subject of population ecology, may be discussed as:

1. Describing a population (population characteristics). The various characteristics which are unique to the group (population) as a whole and not to an individual in the group are described including some of their numerical patterns.
2. Population dynamics. The various theories to explain the population growth.
3. Regulation of population. The factors which govern population density.

POPULATION CHARACTERISTICS

The population—a collective group of organisms of the same species occupying a particular space, has the following characteristics:

1. Population size and density.
2. Dispersion.
3. Age structure.
4. Natality (Birth rate).
5. Mortality (Death rate).
6. Life tables.

Size and Density

Total size is generally expressed as the number of individuals in a population. More informative are estimates of density, the number per unit area (or volume) of environment. Larger organisms as trees may be expressed as 500 trees per hectare, whereas smaller ones like phytoplanktons (as algae) as 2 million cells per cubic meter of water. In terms of weight, it may be 100 pounds of fish per hectare of water surface.

Since, the patterns of dispersion of organisms in nature are different it becomes important to distinguish between crude density and specific (ecological) density:

1. Crude density: It is the density (number or biomass) per unit total space.
2. Specific or ecological or economic density. It is the density (number or biomass) per unit of habitat space i.e., available area or volume that can actually be colonised by the population.

This distinction becomes important due to the fact that organisms in nature grow generally clumped into groups and rarely as uniformly distributed. For example, in plant species like *Cassia tora, Oplismenus burmanni,* etc., individuals are found more crowded in shady patches and few in other parts of same area. Thus, density calculated in total area (shady as well as exposed) would be crude density, whereas the density value for only shady area (where the plants actually grow) would be ecological density.

Another contrast between crude density and ecological density is shown in Fig. 11.1 which is based upon Kahl's study on fish density. As shown in the figure the crude density of small fish in the area as a whole goes down as the water level drops during the winter dry season, but the ecological density in the contracting pools of water increases as fish are crowded into smaller and smaller water area. Thus with changing time, water level and total number of fish, the two densities are different. Not only with changing but at a given time also the two would differ.

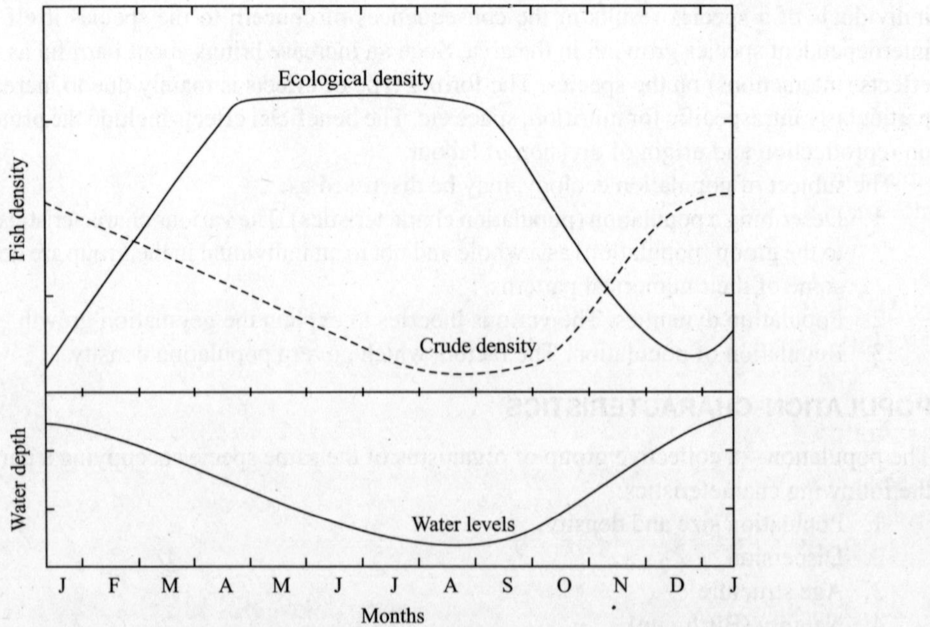

Fig. 11.1. Ecological and crude density of fish prey in relation to the breeding of stork predator.

Often it is more important to know if the population is changing with time (increasing or decreasing) than to know its size at any moment. The two densities will differ in each case. In case of changing population indices of relative abundance are useful. These indices may be 'time relative' as for example the number of animals (birds) seen per hour or they may be percentages of various kinds used widely in vegetation studies.

Concepts of growth rates

Before dealing with other characteristics of a population, it is necessary to have idea of growth rates. Since populatiuon is a changing entity, we are interested not only in its size and composition but also in nature of its change. As varying from place to place, population density also varies in time. Population may remain constant, they may fluctuate or they may steadily increase or decrease. Such charges are the main focus of population ecology. They are brought about by the interplay between four factors. Natality (production of offsprings) and immigration (individuals entering a population from elsewhere) lead to an increase in density, while mortality (death of individuals) and emigration (moving out of population) lead to a decrease in density. A rate is generally obtained by dividing the change by the period of time elapsed during the change. Thus, growth rate of a population is the number of organisms added to the population per time. It is customary to abbreviate 'the change in something' by writing the symbol Δ (delta) in front of the letter representing the thing changing. Thus, if N represents the number of organisms and t the time, then

ΔN = the change in number of organisms.

$\frac{\Delta N}{\Delta t}$ = the average rate of change in the number of organisms per time. This is growth rate.

$\frac{\Delta N}{N\,\Delta t}$ = the average rate of change in the number of organisms per time, per organisms (the growth rate divided by the number of organisms initially present or alternately by the average number of organisms during the period of time). This is often called specific growth rate.

If multiplied by 100 i.e. $\Delta N/(N\Delta t) \times 100$, it becomes the percent growth rate. Suppose a population of 50 amoebae in a pool increased to 150 after an hour. Then,

N	(the initial number)	= 50
ΔN	(change in number)	= 100
$\Delta N/\Delta t$	(average rate of change per time i.e. growth rate)	= 100 per hour
$\Delta N/(N\Delta t)$	(average rate of change per time per individual i.e. specific growth rate)	= 2 amoebae per hour per individual (a 200 per cent increase per hour)

Dispersion

Dispersion is the spatial pattern of individuals in a population relative to one another. Figure 11.3 shows the three basic patterns that occur.

1. Regular dispersion: Here the individuals are more or less spaced at equal distance from one another (Fig. 11.2A). This is rare in nature but is common in managed systems (cropland). Animals with territorial behaviour tend toward this dispersion.
2. Randam dispersion: Here the position of one individual is unrelated to the positions of its neighbours (Fig. 11.2B) This is also relatively rare in nature.
3. Clumped dispersion: Most populations exhibit this dispersion to some extent, with individuals aggregated into patches interspersed with no or few individuals (Fig. 11.3c). Such aggregations may result from social aggregations, such as family groups or may be due to certain patches of the environment being more favourable for the population concerned. The relative frequency of these three pattern is well shown by trees in a semideciduous forest. Of sixteen species, 12 were clumped, 3 were random and 1 was regular in dispersion.

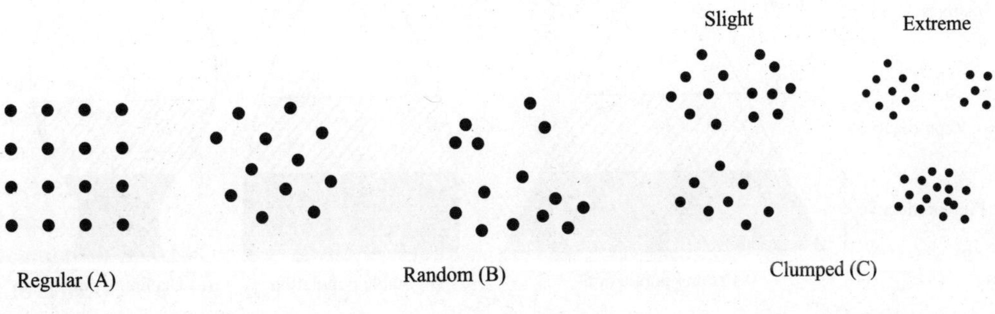

Fig. 11.2. Dispersion patterns in popualtions. In clumped, the clumps may themselves be regularly or randomly dispersed.

Age structure

In most populations, individuals are of different ages. The proportions of individuals in each age group is called age structure of that population. For instance an understory palm tree population (*Astrocayum mexicanum*) in an evergreen forest of Mexico, had 50 per cent individuals as seedlings (less than

2-year-old), 19 per cent as saplings (8-year-old), 5 per cent as 30-year-old-adults and so until 70-year old trees made up less than 2 per cent of the population.

Age distribution is important, as it influences both, natality and mortality of the population. The ratio of the various age groups in a population determines the current reproductive status of the population, thus anticipating its future. From an ecological view point there are three major ecological ages (age groups) in any population. These are, prereproductive, reproductive and postreproductive. The relative duration of these age groups in proportion to the life span varies greatly with different organisms. In man, the three 'ages' are relatively equal in length, about a third of his life falling in each class. Many plants and animals have a very long prereproductive period. Some animals, particularly insects, have extremely long prereproductive periods, a very short reproductive period and no post-reproductive period.

Age pyramids

The model representing geometrically the proportions of different age groups in the population of any organism is called age pyramid. The three hypothetical pyramid types are shown in Fig. 11.3 which are:

Pyramid with broad base

It indicates a high percentage of young individuals. In rapidly growing young populations birth rate is high and population growth may be exponential as in yeast, house-fly, paramecium etc. Under such conditions, each successive generation will be more numerous than the preceding one and thus a pyramid-with broad base would result (Fig. 11.3a).

Bell-shaped polygon

It indicates a moderate proportion of young to old. As the rate of growth becomes slow and stable, i.e., the prereproductive and reproductive age groups become more or less equal in size, post-reproductive group remaining as the smallest, there results a bell-shaped structure (Fig. 11.3b).

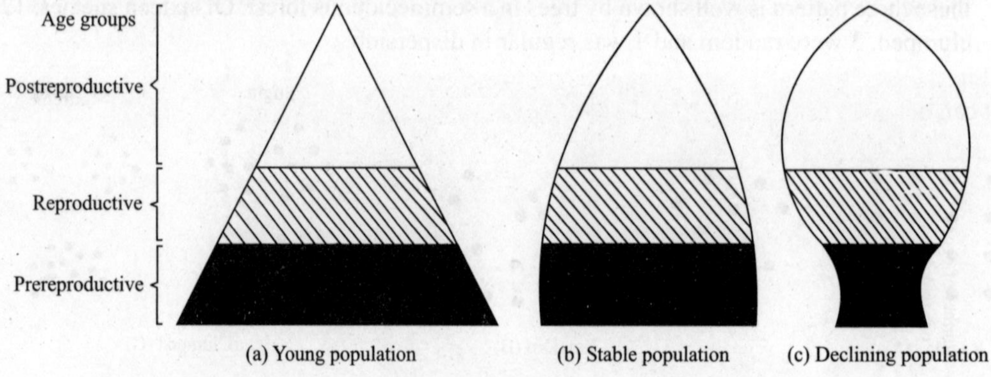

Fig. 11.3. Hypothetical diagram showing different types of age pyramids.

An urn-shaped figure

It indicates a low percentage of young individuals. If the birth rate is drastically reduced the prereproductive group dwindles in proportion to the other two groups and it results in an urn-shaped (see Fig. 11.4c) which indicates that population is dying off.

While density reveals little about differences between populations in developed and developing countries, age structure highlights some of the causes of rapid population growth in the latter. Developing countries have a much greater preponderance of young people and the side of pyramids become concave. Such a shape means high population growth, not only at present but also likely in future.

Natality (Birth Rate)

It is simply a broader term covering the production of new individuals of any organism. These new individuals are born, hatched, germinated, arise by division etc. In human population, however, the natality rate is equivalent to the 'birth rate'. Natality rate is the number of offsprings produced per female per unit time. There are distinguished two types of natality:

1. Maximum (absolute or potential or physiological) natality. It is the theoretical maximum production of new individuals under ideal conditions (i.e., no ecological limiting factors, reproduction being limited only by physiological factors). It is a constant for a given population. This is also called fecundity rate.
2. Ecological or relised natality. It is also known simply as natality, which refers to population increase under an actual, existing specific condition. Thus it takes into account all possible existing environmental conditions. This is also designated as fertility rate.

Natality is expressed as:

$\Delta N_n/\Delta t$ = the absolute natality rate (B)

$\Delta N_n/N\Delta t$ = the specific natality rate (b) (natality rate per unit of population)

where,

N = initial number of organisms

n = new individuals in the population

t = time

Three population characteristics determine the rate at which females produce offspring:

1. Clutch size or the number of young produced on each occasion.
2. The time between one reproductive event and the next.
3. The age of first reproduction.

Natality rate usually increases during the period of maturity and then falls again as the organism gets older (Fig. 11.4). Some trees seem to continue to increase fruit production as they get older, but this at present can not be explained due to lack of information we have at present.

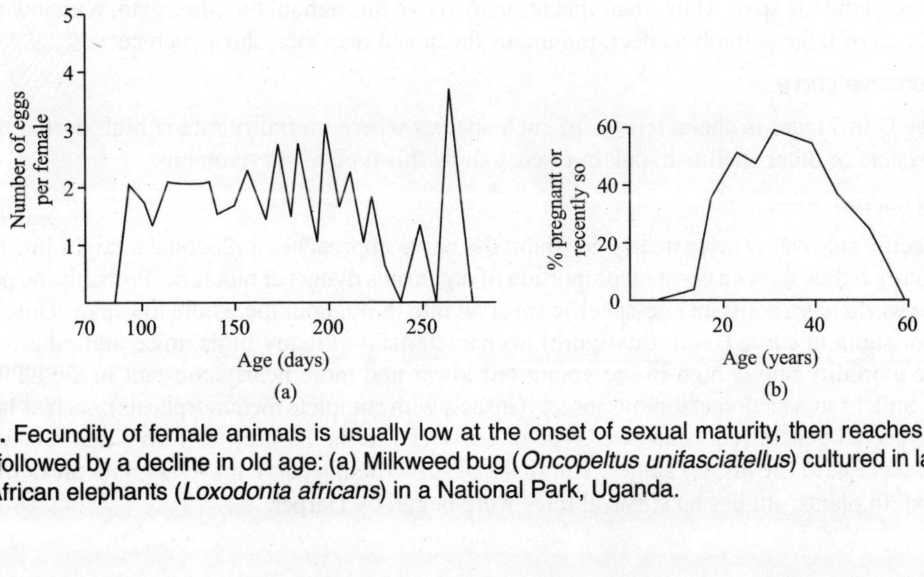

Fig. 11.4. Fecundity of female animals is usually low at the onset of sexual maturity, then reaches a lengthy plateau, followed by a decline in old age: (a) Milkweed bug (*Oncopeltus unifasciatellus*) cultured in laboratory; and (b) African elephants (*Loxodonta africans*) in a National Park, Uganda.

Natality patterns differ in tropical and temperate populations. Breeding time and clutch size are two important criteria. In tropical areas with dry periods, breeding is at least as seasonal as in temperate areas. In continually tropical humid areas breeding may occur throughout the year, with some population showing a seasonal peak. Clutch size in tropical environment is smaller than in temperate zone. Many birds, some plants, some insects and some small mammals exhibit this trend.

Mortality (Death Rate)

It refers to death of individuals in the population. Like natality, mortality may be:

1. Minimum mortality: Also called specific or potential mortality, it represents the theoretical minimum loss under ideal or non-limiting conditions. It is a constant for a population. Thus, even under the best conditions, individuals would die of 'old age' determined by their physiological longevity.
2. Ecological or realised mortality: It is the actual loss of individuals under a given environmental condition. It is, like ecological natality, not a constant and varies with population and environmental conditions.

Like natality, mortality may be expressed as the number of individuals dying in a given period (deaths per time) or as specific rate in terms of units of the total population or any part thereof.

A birth-death ratio $\left(100 \times \frac{\text{births}}{\text{deaths}}\right)$ is called vital index.

For a population, the important thing is not which members die but which members survive. Thus survival rates are of much interest than the death rates. Survival rates are generally expressed by survivorship curves.

Survivorship curves

The pattern of mortality with age is best illustrated by survivorship curves which plot the numbers surviving to a particular age. There are three general types of survivorship curves (Fig. 11.5) which represent the different nature of survivors in different types of populations. These are as follows:

Highly convex curves

Curve A in the figure, is characteristic of the species in which the population mortality rate is low until near the end of the life span. Thus, such species tend to live throughout their life span, with low mortality. Many species of large animals as deer, mountain sheep and man etc., show such curves.

Highly concave curve

This curve C in Figure is characteristic of such species where mortality rate is high during the young stages. Oysters or other shell fish, oak trees etc., show this type of survivorship.

Diagonal curves

If age-specific survival is more nearly constant, the curve approaches a diagonal straight line (curve B_2 in the Figure). It thus shows a constant proportion of organisms dying per unit time. Probably, no population in the real world has a constant age-specific survival rate throughout the whole life span. Thus a slightly concave or sigmoid curve (B_3 in the figure) is characteristic of many birds, mice and rabbits. In these cases, the mortality rate is high in the young but lower and more nearly constant in the adult (1 year or older). Still, in some halometabolous insects (insects with complete metamorphosis), such as butterflies, there is expected generally a 'stair step' type of curve (B_1 in the Figure). In such cases survival differs greatly in successive life history stages. The shape of survivorship curve may vary with the density of the population. In plants, studies on survival rates were begun by Harper.

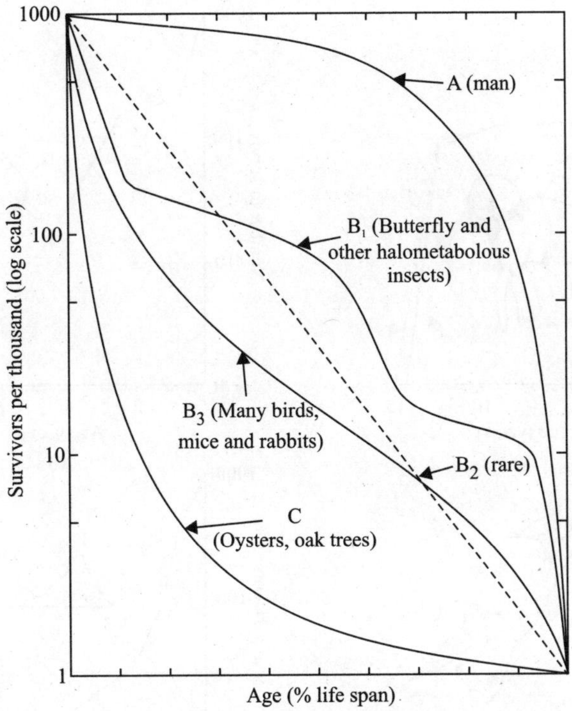

Fig. 11.5. Different types of survivorship curves plotted on the basis of survivors and age.

Survivorship of a variety of tropical populations shows that few match exactly the idealised curves. A convex pattern is shown by some large mammals, particularly primates and humans (Fig. 11.6a). Approximately diagonal curves are found in some rodents and birds (Fig. 11.6b, c) while many plants and insects tend to concave patterns (Fig. 11.6d, e). The termite curve is rather different (Fig. 11.6d) than the others, as it shows survivorship of whole colonies rather than of individual in a population. Survivorship may vary in the same species in different environments. Do patterns of survivorship differ in tropical and temperate populations? Convex curves are more probable in fairly constant climates as found in moist tropics. Sometimes convex curves come from populations in highly variable environments. Tropical populations seem to have enhanced survivorship compared with their temperate relatives. Tropical birds live longer than temperate ones. Cottontail rabbits in tropics show lower rates of survival than those in the temperate zone.

Biotic potential

Each population has the inherent power to grow. When the environment is unlimited (space, food, other organisms not exerting a limiting effect), the specific growth rate (i.e., the population growth rate per individual) becomes constant and maximum for the existing conditions.

The value of the growth rate under these favourable conditions is maximal, is characteristic of a particular population age structure and is a single index of the inherent power of a population to grow. It may be designated by 'the symbol r, which is the exponent in the differential equation for population growth in an unlimited environment under specific physical conditions.

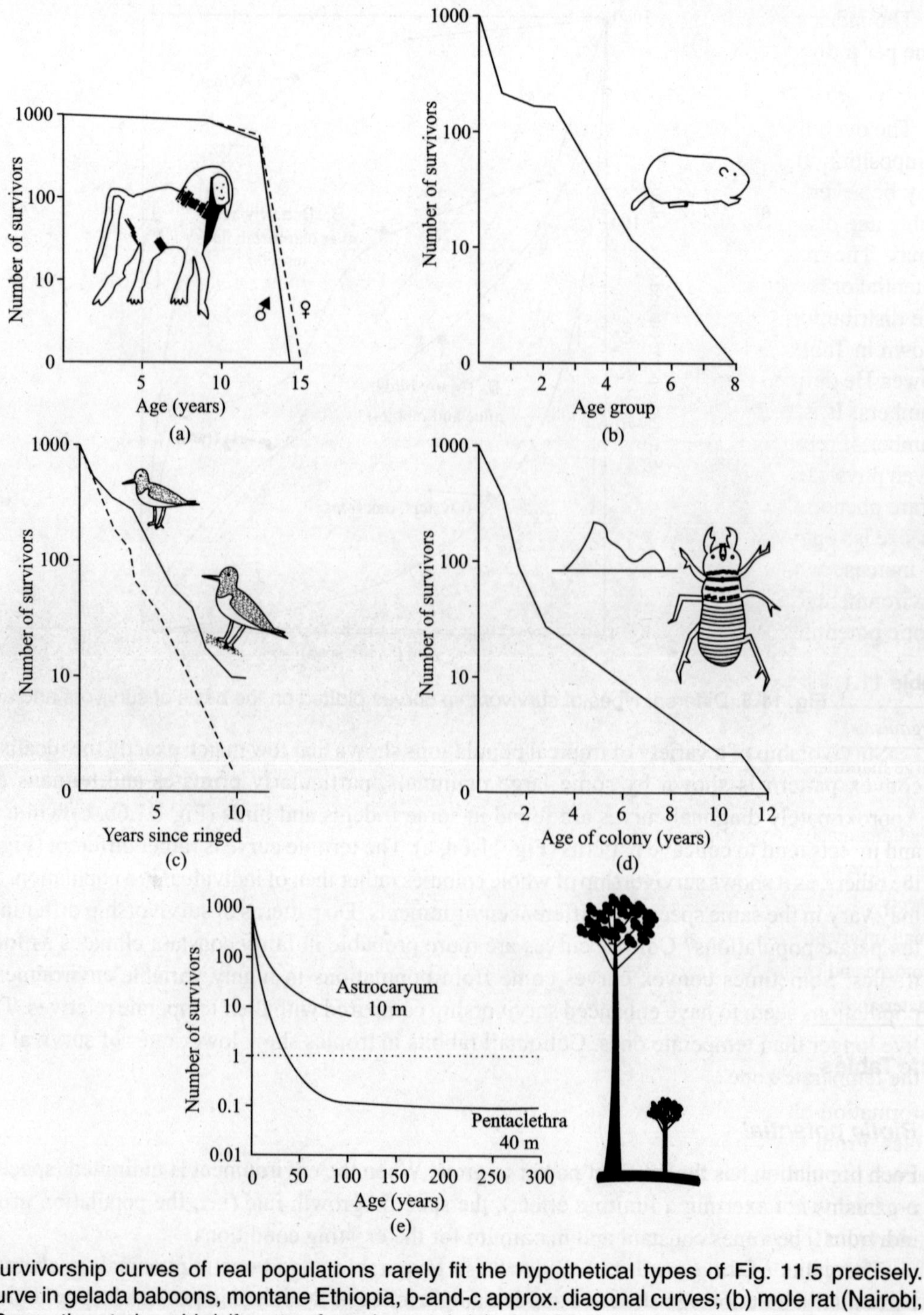

Fig. 11.6. Survivorship curves of real populations rarely fit the hypothetical types of Fig. 11.5 precisely. (a) convex curve in gelada baboons, montane Ethiopia, b-and-c approx. diagonal curves; (b) mole rat (Nairobi, Kenya); (c) Brown-throated sunbird (forests of south east Asia). d and e concave curves; (d) whole termite colonies (Nigerian savanna); and (e) two rainforest trees (from neotropics). Solid line is for understory palm in Mexico; broken line for a canopy tree in Costa Rica.

The index r is actually the difference between the instantaneous specific natality rate (i.e., rate per time per individual) and the instantaneous specific death rate and may thus be expressed:

$$r = b - d$$

The overall population growth rate under unlimited environmental conditions (r) depends on the age composition and the specific growth rates due to reproduction of component age groups. Thus, there may be several values of r for a species depending upon population structure. When a stationary and stable age distribution exists, the specific growth rate is called the intrinsic rate of natural increase or r max. The maximum value of r is often called by the less specific but widely used expression biotic potential or reproductive potential. Thus with this term, one is able to put together natality, mortality and age distribution. The ranges of the intrinsic rate of natural increase for various kinds of organism are shown in Table 11.1. Chapman proposed the term biotic potential to designate maximum reproductive power. He defined it as 'the inherent property of an organism to reproduce, to survive, i.e., to increase in numbers. It is sort of the algebraic sum of the number of young produced at each reproduction, the number of reproduction in a given period of time, the sex ratio and their general ability to survive under given physical conditions.' For details, reader may refer to Graham. But under natural conditions, this is a rare phenomenon, since environmental conditions do not permit unlimited growth of any population. Its size is kept under natural check. The difference between the maximum r (biotic potential) and the rate of increase which occurs in an actual laboratory or field condition is often taken as a measure of the environmental resistance, which is the sum total of environmental limiting factors which prevent the biotic potential from being realised.

Table 11.1. Intrinsic rate of natural increase.

Organisms	*Approximate biotic potential, r (per year)*
Large mammals	0.02–0.5
Birds	0.05–1.5
Small mammals	0.3–8
Larger invertebrates	10–30
Insects	4–50
Small invertebrates (including large protozoans)	30–800
Protozoa and Unicellular algae	600–2000
Bacteria	3000–20000

Life Tables

Information on natality and mortality in different ages and sexes can be combined in the form of life tables. From these it is possible to estimate the growth or decline of a population. As with survivorship curves, life tables are standardised to follow the progress of a cohort. In each table there are columns for age of individuals; number surviving to each age; the number dying in each age group; the proportion dying from the previous age category; fertility rate; and the number of young born by each age group. The information obtained from these figures provides the net reproductive rate of the population i.e., offspring left by each individual.

Similarly from life table, mortality in a logarithmic form is also obtained. These are then used to calculate the rate of population growth.

POPULATION DYNAMICS

After having studied the various characteristics of a population, we attempt to follow the change in it. Population have characteristic patterns of increase which are called population growth forms. Such growth forms represent the interaction of biotic potential and environmental resistance. What are the theoretical bases that attempt to account for such changes?

Approaches to study of population dynamics are: (i) mathematical models; (ii) laboratory studies; and (iii) field studies. Mathematical models are of two types: theoretical models and simulation models. The former derive equations that may describe population changes; and the behaviour of such models can be tested by comparing these with real populations. The latter (simulation models) use detailed data from real populations and try to predict future behaviour of these populations under specific environmental conditions.

Models are of use only if they explain facets of population dynamics of real populations. Their predictions are often tested experimentally using laboratory populations. Mathematical and laboratory studies are of value only if they help to explain the behaviour of natural populations. Field studies are made, though these are complex.

Theory of Population Growth

There may be assumed a model of geometric increase that assumes that there is no environmental constraint on population growth. A geometrical series is one with a constant multiplier. The simplest example is binary fission in Protozoa, where each individual gives two each time it reproduces. Starting with one individual (generation O), there will be 2,4,8,16,32,64..in succeeding generations if fission is synchronous. To determine number (N_t) present in any generation (t), the following equation is used:

$Nt = N_o (2^t)$, Number is no present in generation O. Thus the number in generation 3 will be $N_3 = 1 (2^3) = 8$

This equation is for a population that doubles in each generation. A more generation equation would be, $N_t = N_o (e^{rt})$.

This is the exponential growth equation, where e^r is substituted for 2; e is a universal constant, the base of natural logarithms and has a value of 2.718; r is a constant for a particular population under specific environmental conditions and is called the intrinsic rate of natural increase. The significance of logarithms in exponential growth is shown in Fig. 11.7 which compares the predictions of the binary fission equation on arithmetic and logarithmic scales. The logarithmic plot is a straight line, meaning that the number added in any time period is proportional to the number present at the beginning of the period (compare with surviorship curve (diagonal) in Fig. 11.5 which is an exponential decline).

The intrinsic rate of natural increase (r) is easy to calculate if natality and mortality rates are known. For example, for each of the 14 million people present in a country in 1980, 0.028 were born and 0.008 died during the year; r = natality rate minus mortality rate or 0.028–0.008 per year. Substituting in the exponential growth equation, the number present in is predicted as:

$$N_t = N_o (e^{rt})$$
$$N_{10} = 14 \text{ million } x2.718^{0.02 \times 10} = 17 \text{ million}$$

Note this is only strictly correct if the population is growing exponentially with constant birth and death rates and age structure and there is no immigration or emigration.

Does it occur in nature? Most populations do not show exponential increase because their environment prevents this. In an ecosystem, other species present, a variety of abiotic conditions etc., act in a way that a given population will have an upper size limit that cannot be exceeded. Such a limit is called carrying capacity of the environment for that population. Carrying capacity is determined by the availability of resource as food and space.

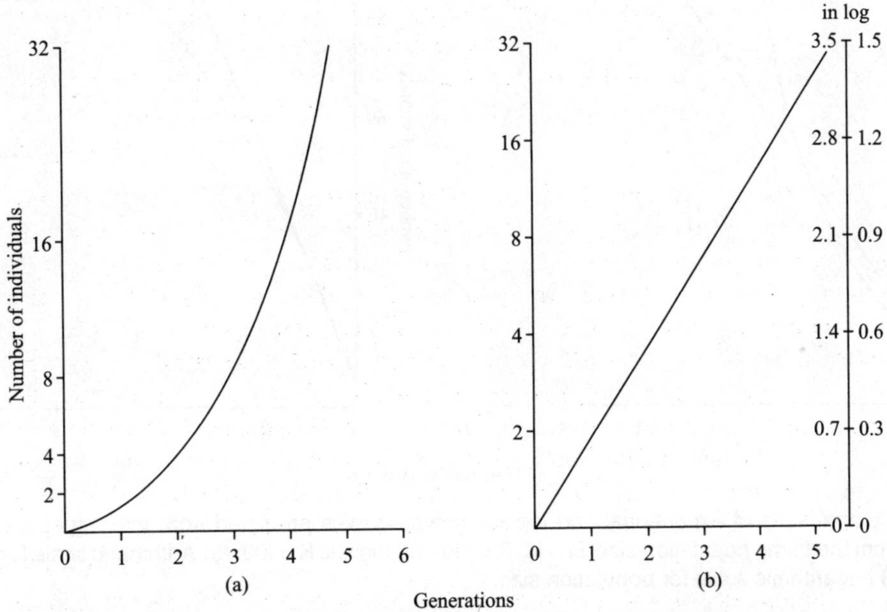

Fig. 11.7. The exponental growth curve corresponding to synchronous binary fission: (a) Arithmetic scale for population size; and (b) Logarithmic scale for population size; both common logarithms (log) and natural logarithms are shown.

The logistic equation was devised by Verhulst to describe population growth with an upper limit. Suppose, environment has a carrying capacity (K) for a particular population. Then the logistic equation assumes that the intrinsic rate of natural increase (r) is progressively reduced as population size increases toward that carrying capacity. If there are 10 individuals already present (N = 10) and K = 100 individuals, the remaining resources can support a further K – N = 90 individuals and the proportion of remaining resources is $\frac{K-N}{K} = 0.9$

The logistic equation simply assumes that r is reduced in relation to proportion of remaining resources. The equation is derived from the exponential such that:

$N_t = N_0 e^r \left(\frac{K-N}{K}\right)^t$ –the logistic growth equation. Suppose r = 1; then when N = 10, the growth rate actually achieved, $r\left(\frac{K-N}{K}\right)$, is 1 x 0.9. When N = 90, $\left(\frac{K-N}{K}\right) = 0.1$ and the growth rate is 1 x 0.1 and so on, giving the progressive decline in realised growth rate. The logistic curve is S-shaped on arithmetic coordinates and is always less steep than its exponential equivalent (Fig. 11.8a). A comparison of the same curve on semilogarithmic coordinates is given in Fig. 11.8b, using differential calculus, the logistic equation becomes

$$\frac{dN}{dt} = rN\left(\frac{K-N}{K}\right)$$

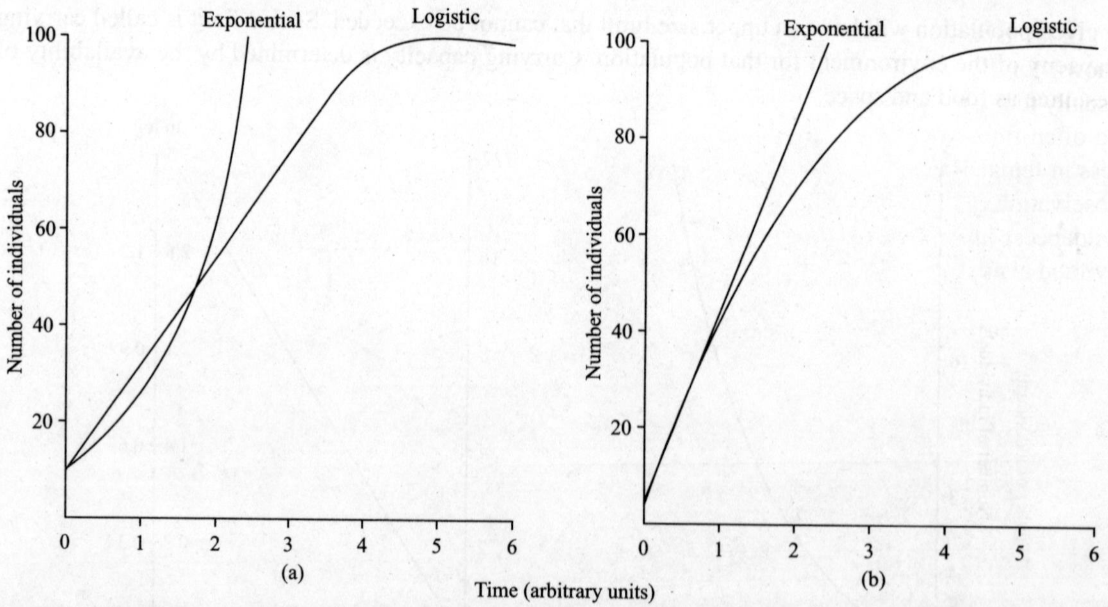

Fig. 11.8. Comparisons of exponential and logistic growth curves predicted from the respective equations, beginning from the same population size. N = l0, R = l for the logistic K = l00: (a) Arithmetic scale for population size; and (b) Logarithmic scale for population size.

Growth of laboratory and field populations

From the above account of carrying capacity, one would expect exponential growth only under circumstances of no environmental resistance. Such a growth is possible in laboratory culture if excess organisms are removed to avoid over crowding. However, in nature, exponential increase in population experiences catastrophic decline as shown in Fig. 11.9 for insect population. Under adequate food and climate, population increased, but then declined when these turned unfavourable.

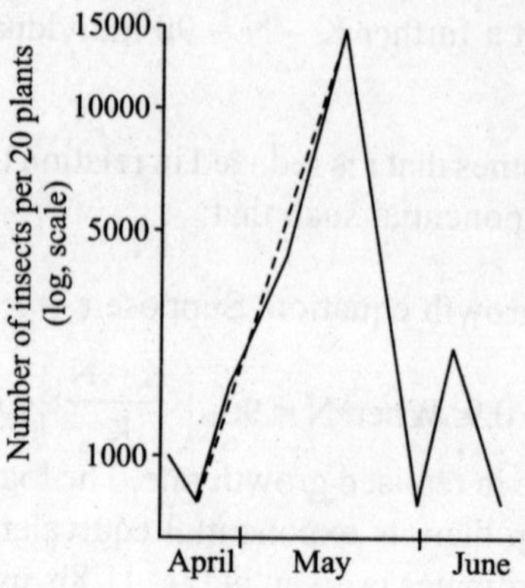

Fig. 11.9. The population of thrips living on Cassava grows approximately exponentially in Brazilian fields (note log scale). Solid line is observed density; dashed line is exponential prediction. Exponential growth is usually followed by catastrophic decline.

In the laboratory logistic growth has been observed in bacteria, yeasts, algae and protozoa. This is shown for *Paramecium* in Fig. 11.10a. In laboratory cultures, somewhat much less fit to logistic is usually found in metazoans as beetle, like azuki bean weevil (*Callosobruchus chinensis*), but populations do often fluctuate around a carrying capacity (Fig. 11.10b). There are several reasons as competition, loss in female fecundity, increased egg mortality etc., at high density of this beetle. In the field, direct observation of variatation and limitation of a population are rare. However, there is a logistic model of wildebeest (*Connochaetes taurinus*) in Tanzania and Kenya. The population levels off at K determined by food availability following its low density.

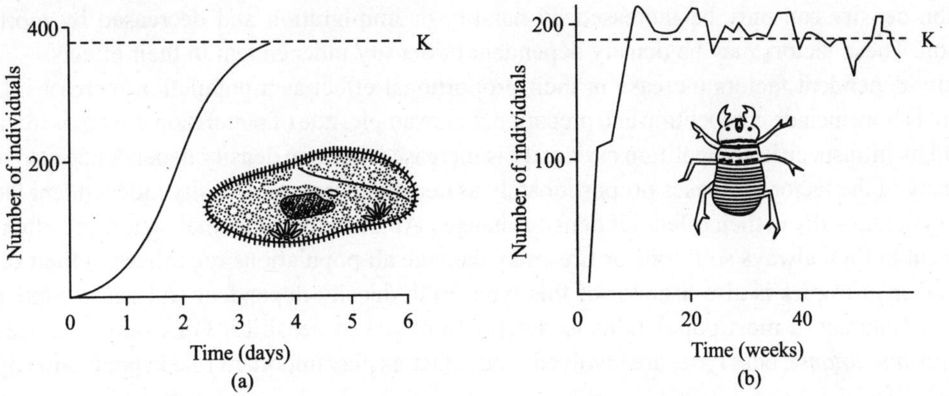

Fig. 11.10. Approximations to logistic growth observed in real populations: (a) Classical experiment of Gause on Paramecium in laboratory culture; and (b) azuki bean weevil in laboratory culture; as with most metazoa, the population fluctuates about K.

Modifications of logistic theory

Following assumptions are implicit in the logistic equation: (i) environment is constant such that r and K do not charge; (ii) there are no time lags in response of the population to change in its density; (iii) individuals have identical ecological characteristics in terms of their responses to density, regardless of age or sex; (iv) growth is continuous; and (v) age structure does not change in successive generations. These assumptions are unlikely to be met in laboratory and will never obtain in the field. Therefore, logistic theory has been suitably modified to make it biologically reasonable. These are: (i) inclusion of a time lag there; (ii) there are fluctuations in the carrying capacity; (iii) there may be chance events for surrival and reproduction of an individual as well as for their death, due to landslide etc.; and (iv) discontinous growth of some populations which grow in discrete steps with little or no overlap between successive generations.

Keeping in view the above, it is natural to ask 'Are, then mathematical models of any use?'. The answer is 'Yes' because:

1. As theoretical abstractions, they set the crude limits to possible patterns of population growth. Observations on real populations can be assessed within these limits.
2. As models become more refined, they more closely fit the behaviour of real populations. One can make increasingly accurate predictions of population behaviour.
3. From above 1 and 2, it follows that population models can be applied to real practical problems as human population growth and the management of plant and animal populations.

Regulation of Population Density

The logistic model and its derivatives assume that a population will level off at its carrying capacity, that there is an upper limit to population density set by the environment and that the population is regulated at or around that level. The significance of regulation in population has been a subject of controversy among ecologists for last more than 50 years. What regulates population? Krebs presented an account of this subject.

Nature of factors that influence population density

Population density can only be increased by natality or immigration and decreased by mortality or emigration. These factors may be density dependent or density independent in their effects.

Density-dependent factors increase in their proportional effect as a population increases. Density-dependent factors include competition and predation. For example, rate of population growth is increasingly depressed by intraspecific competition as density is increased. Inverse density dependence also occurs if the intensity of the factor decreases proportionately as density increases. Density independent factors do not vary systematically in their effects as density changes. Abiotic environemntal factors are often density independent but not always so floods or fires may damage all populations regardless of their densities. Cold weather in tropics is also a factor of this type. Both density-dependent and density-independent factors may interact in most populations to determine observed densities. For example in tsetse flies (*Glossimia morsiotans*), both types are involved. These factors play important role in population dynamics also.

Key factor analysis

This is the method to analyse mortality factors, to find out which may be regulatory. The k-values for each mortality factor together with total k (the sum of all factors), are plotted for several successive generations. The k-factor that most closely follows the pattern of k is called the key factor. In insect population, loss of adults through migration or death is the key factor.

Self-regulation of populations

Intraspecific density-dependent interactions regulate many populations in the laboratory and in that sense these are selfregulating. Accumulation of waste products may depress population growth. This is also self-regulation. Social behaviour may also be self-regulatory. Many species have individuals or groups of individuals that defend food or breeding partners in a territory. Death or failures to mate are density-dependent and can regulate population density.

Immigration, emigration and population dynamics

Besides natality and mortality, immigration and emigration also affect population density. These both are features of disposal. Whereby individuals (singly or in groups) move from one population and die if no suitable environment is found; establish a new population or join an existing one at a new locale.

Plant population dynamics

In many respects they behave like animals. However, some features are unique to plant populations.

What are these?

1. Most higher plants are modular organisms, developing from a single zygote but producing an indeterminate number of repeatitive structures (modules) vegetatively. A clump of herbs, grasses or trees may be the product of one zygote.
2. In plants, there are two levels of population structure: (i) a genet (equivalent to N in the population equation), which is the individual produced from a single zygote; and (ii) ramet or tiller, the vegetative offshoots.
3. Plants can not move to mate or disperse. Thus they have evolved means as gravity, wind, water flow or animals for dispersal of pollen, seed, vegetative parts etc.
4. Most aspects of growth of population are density related. One important generalisation applied is the $\frac{3}{2}$ thinning law.

If we plot the relationship between dry weight and density of shoots (known number of individuals of) in plant populations, the line relating weight of each individual to density has a slope of –1.5 (or $-\frac{3}{2}$).

The slope would be –1 if increasing density has exactly compensated by reduction in weight of individuals. Thinning is normally inversely density dependent, but does not always occur if the growth of plants is extremely plastic. This law has been verified for a wide variety of plants from mosses to trees. Perhaps the $\frac{3}{2}$ law is universal, though exact reason for its occurrence is yet to be known.

Population ecology

(Interactions between different populations of different species)

We have considered in the preceding sections the populations as though they existed independently of other species. Most ecosystems contain populations of many species that interact in vital ways so that changes in one population will have effect on many others. A number of such interactions have been recognised, of which the three distinct classes are, competition, predation and mutualism. Competition is usually between members of the same trophic level and predation is always and mutualism is often between populations on adjacent trophic levels.

The effects of one population on the growth of another (indicated by + and – signs) can be categorised as follows:

1. Competition	Population 1	$\overset{-}{\rightarrow}$ / $\underset{-}{\leftarrow}$	Population 2
2. Predation	Population 1 (prey)	$\overset{+}{\rightarrow}$ / $\underset{-}{\leftarrow}$	Population 2 (predator)
3. Mutualism	Population 1	$\overset{+}{\rightarrow}$ / $\underset{+}{\leftarrow}$	Population 2

The two more interaction types are:

4. Commensalism	Wildebeest	$\overset{+}{\rightarrow}$ / $\underset{0}{\leftarrow}$	Gazelles
5. Amensalism	Population 1	$\overset{0}{\rightarrow}$ / $\underset{-}{\leftarrow}$	Population 2

Population ecology and evolution

Most population ecologists use the population parameters to draw conclusions about evolution of species. The important synthesis was done by MacArthur and Wilson. According, to them, populations are outcome of r-or k-selection; r-selected populations have a high intrinsic rate of growth (r) and tend to 'boom' when environmental conditions are favourable and 'best' when those conditions deteriorate. As a result they exhibit large fluctuations in density and incidentally have the potential for large genetic change through the founder effect. K-selected populations have relatively constant density at or near the carrying capacity (K) of the environment. Many other features associated with each of these types of populations are listed in the Table 11.2.

Table 11.2. Some predicted trends in r-or k-selected populations and the environments in which they are likely to occur.

	r-selected	*k-selected*
Environment	Variable and unpredictable	Constant or predictably variable
Population characteristics	Survivorship curve concave	Survivorship curve convex.
	High fecundity	Low fecundity
	Density variable usually well below carrying capacity	Density fairly constant at or near carrying capacity
	Scramble-type intraspecific competition	Contest-type inter-specific competition
Individual characteristics	Small body size	Large body size
	Good dispersal powers	Poor dispersal powers
	Outcrossing	Parthenogenetic or vegetative reproduction
	Low level of social organisation	High level of social organisation

12

Interactions Between Populations

INTRODUCTION

The traditional approach to population interactions has been to consider just the direct pairwise interactions. In this simplistic view of things, two populations may or may not affect each other; if they do, the influence may be beneficial or adverse. By designating a detrimental effect with a minus, no effect with a zero and a beneficial effect with a plus, all possible population interactions can be conveniently classified. When neither of two populations affects the other, the interaction is designated as (0, 0). Similarly, a mutually beneficial relationship is (+, +) and a mutually detrimental one is (–, –). Other possible interactions are (+, –), (–,0) and (+, 0), making a total of six fundamentally different ways in which populations can interact (Table 12.1).

Table 12.1. Summary of the various sorts of direct pairwise interactions that may occur between two populations.

Type of interaction	*Species* *A*	*B*	*Nature of the interaction*
Competition	–	–	Each population inhibits the other
Predation, parasitism, Batesian mimicry	+	–	Population A—the predator, parasite or mimic—kills or exploits members of population B—the prey, host or model
Neutralism	0	0	Neither population affects the other
Mutualism, Mullerian mimicry	+	+	Interaction is favourable to both (can be obligatory or faculative)
Commensalism	+	0	Population A, the commensal, benefits, whereas B, the host, is not affected
Amensalism	–	0	Population A is inhibited, but B is unaffected

Competition (–, –) takes place when each of two population affects the other adversely. Typically, both require the same resource(s) that is (are) in short supply; the presence of each population inhibits the other. If the resource is another population (a prey species), competition is indirect and mediated by means of resource depression—this type of competition is termed exploitation competition. Other kinds of competition also occur. For example, competition can also be direct, as in agonistic encounters such as allelopathy or interspecific territoriality (known as interference competition). Predation (+, –) occurs when one population affects another adversely but benefits itself from the interaction. Usually, a predator kills its prey and consumes part or all of the prey organism. (Exceptions include lizards losing their tails to predators and plants losing their leaves to herbivores.) Parasitism (+, –) is essentially identical to predation, except that the host (a member of the population being adversely affected) is usually not killed outright but is exploited over some period of time. Thus, parasitism can in some ways be considered a 'weak' form of predation; Batesian mimicry and herbivory could be placed here as well. Interactions that benefit both population (+, +) are classified as mutualisms. In some mutualisms, the association is obligatory (neither population can exist without the other), but in others the interaction is facultative because it is not an essential condition for survival of either population. When one population benefits while the other is unaffected, the relationship is termed commensalism (+, 0). Amensalism (–, 0) is said to occur when one population is affected adversely by another but the second is unaffected. Neutralism (0, 0) occurs when the two populations do not interact and neither affects the other in any way whatsoever; it is thus of little ecological interest. True neutralism is likely to be very rare or even nonexistent in nature, because there are probably indirect interactions between all the population in any given ecosystem, although their significance may be minimal. Three of the six population interactions—competition, predation and mutualisms—are of overwhelming importance.

COMPLEX POPULATION INTERACTIONS

Interactions among population often become quite intricate, particularly in diverse communities. Colwell studied interactions among four species of nectar-feeding birds, four species of flowering plants and two species of mites (Fig. 12.1) in Costa Rica. Three hummingbirds, *Colibri thalassinus*, *Eugenes fulgens* and *Panterpe insignis*, compete for nectar and are pollinating vectors for the plants. Flowers of various species differ in corolla lengths and are visited differentially by hummingbirds that differ in beak lengths. Only *Eugenes*, which has the longest beak, can reach the nectar of *Centropogon talamancensis* through its very long corolla (Fig. 12.1). However, *Panterpe* hummingbirds steal nectar from this *Centropogon* species by piercing the base of flower corollas. The fourth species of bird, *Diglossa plumbea*, is a nectar thief that obtains nectar from all four plant species by breaking their corolla bases. Two species of mites live within flowers of different plant species, moving among flowers by climbing up beaks and riding in hummingbird nostrils.

Both mite species are found on *Panterpe* and *Eugenes* hummingbirds, which make legitimate pollinating visits through the corollas of flowers of three and four species of plants, respectively. One mite, *Rhinoseius colwelli*, is restricted to flowers of the two species of *Centropogon* and never occurs in either *Macleania* or *Cavendishia* flowers. The second species of mite, *R. richardsoni*, is found only in *Macleania* and *Cavendishia* flowers and never occurs in *Centropogon* flowers. Significantly, avian pollinator visits are structured so that transfers between *Macleania* and *Cavendishia* flowers are frequent, whereas transfers between flowers of these two species and those of *Centropogon* are much more infrequent. Experimental introductions of mites into flowers without mites showed that both species can live and reproduce successfully in the flowers normally occupied only by the other species.

Adult male mites are extremely aggressive, particularly in interspecific encounters and Colwell observed male *R. colwelli* killing *R. richardsoni*. Over evolutionary time, adult male mites may have reinforced the observed species-specific separation on flowers of different species by killing mites of the other species when these made the mistake of leaving their hummingbird carriers to invade a flower of the wrong species, which contained adult males of the other species of mite. Among these ten species, then, interactions include intense interference competition (between the two species of mites), exploitation competition (among nectarivorous birds), facultative mutualism (between hummingbird pollinators and plants), parasitism (between plants and nectar thieves) and commensalism (between mites and their hummingbird carriers).

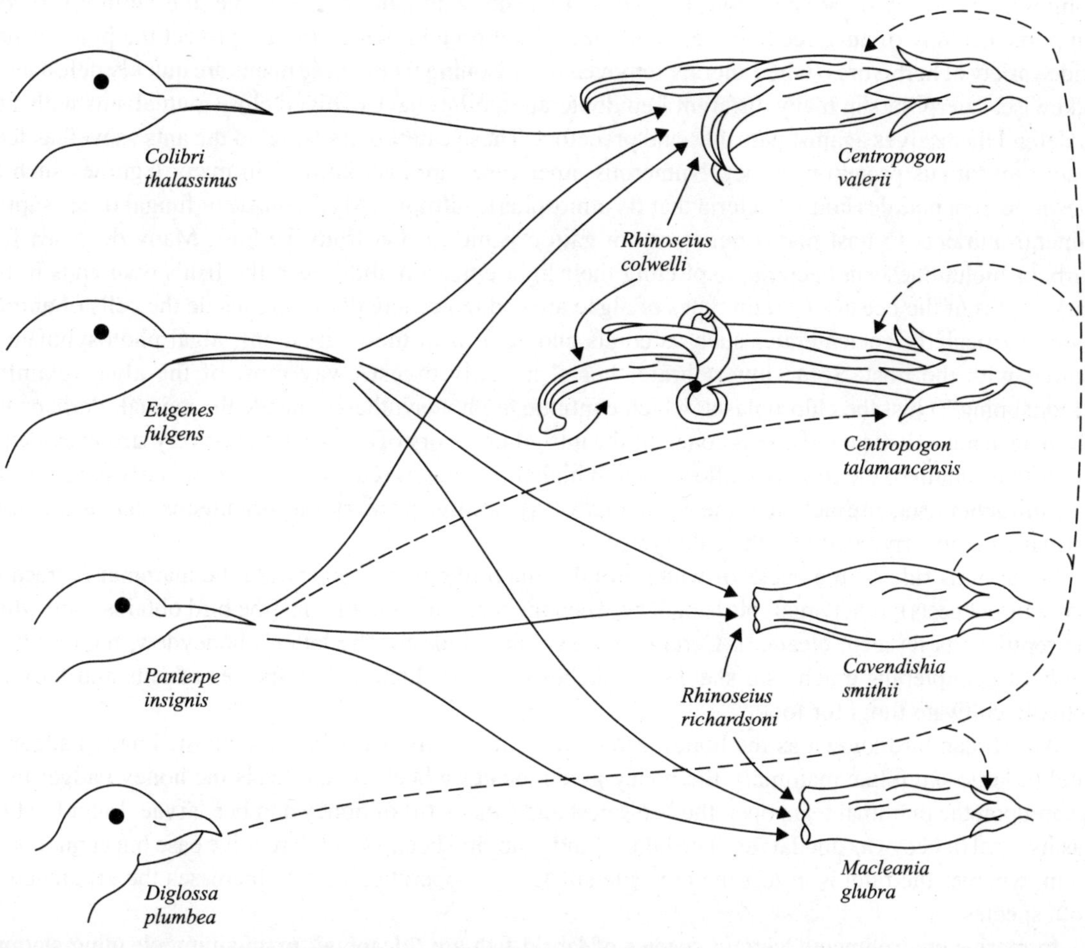

Fig. 12.1. Patterns of exploitation of flowers of four species by various birds and mites. Broken lines indicate illegitimate visits by nectar thieves, which pierce the base of the corolla.

MUTUALISTIC INTERACTIONS AND SYMBIOTIC RELATIONSHIPS

Symbiosis means living together. Usually the term is used only to describe pairs of organisms that live together without harming one another, thereby excluding parasitism (+, –) and amensalism (–, 0), in which one party is affected adversely (see Table 12.1 for explanation of symbols). Hence, symbiotic relationships include mutualism (+, +), commensalism (+, 0) and neutralism (0, 0). Obligate mutualisms can be distinguished from facultative ones. These various types of interactions can change in evolutionary time and grade into one another. Although mutualism is a symmetric relationship, there may nevertheless usually be an asymmetry in costs versus benefits to each of the parties concerned (a conflict of interests arises even in mutualistic relationships).

As pointed out earlier, true neutralism is uninteresting as well as uncommon and therefore need not be considered. However, mutualism and commensalism are fairly widespread, particularly in diverse communities. Many organisms have formed mutualisms with ants. For example, the bullhorn *Acacia* supports a colony of ants, feeding them both nectar and protein, which in turn protect the plant from a wide variety of herbivores (when ants are removed by poisoning them, these plants are quickly defoliated). Likewise, caterpillars in many different Lepidopteran families have evolved close mutualisms with ants to defend themselves against parasites and predators. These caterpillars 'sing' to the ants as well as feed them a nutritious protein-rich diet. Numerous other cases are also known. In many legumes such as mesquite, root nodules house bacteria that fix atmospheric nitrogen. Mycorrhizae or fungal roots, supply mineral nutrients to host plants but in return gain organic carbon from the host. Many deep-sea fish harbour bioluminescent bacteria, exploiting their light-emission abilities to the fish's own ends in the dark depths of the oceans. Certain types of algae are endozoic, actually living inside the cells of animal hosts, particularly coelenterates such as corals and *Hydra*. In these situations, algal photosynthate is shared with the host. Some invertebrate 'hosts' actually digest away most of the alga, retaining ('kidnapping'?) just the chloroplasts, which continue to photosynthesise inside the animal. Perhaps the ultimate in mutualistic interactions concerns the intriguing theory of endosymbiosis; very strong evidence exists that certain of the cell organelles found in higher organisms (eukaryotes), particularly chloroplasts and mitochondria, are actually the remnants of symbiotic prokaryotic organisms that have been permanently incorporated into the eukaryotes.

Some birds ride on the backs of water buffalo (the bird obtains food while the mammal is freed of many insect pests); other small birds pick between the teeth of crocodilians (the bird obtains food while the reptile gets its teeth cleaned). Certain ants exploit aphids for the latter's honeydew, tending their herds of hemipterans much as a shepherd watches over his flock. Other species of ants and termites actually cultivate fungi for food.

An African bird known as the honey guide has formed a unique alliance with the honey badger or ratel (a large skunklike mammal); the honey guide locates a beehive and leads the honey badger to it, whereupon the mammal tears open the bee's nest and eats its fill of honey and bee larvae. Later the bird has its meal of beeswax and larvae. The honey guide can find beehives with relative ease but cannot open them, whereas the ratel is in just the opposite situation; cooperation clearly increases the efficiency of both species.

In marine environments, certain species of labrid fish are 'cleaners', maintaining cleaning stations where other species of larger fish come to be cleaned of ectoparasites and bacteria, sometimes lining up rather like cars at a gas station. Interspecific displays are used in recognition. Cleaner fish are conspicuous and brightly coloured. Interestingly, an unrelated fish species has evolved in another family, the saber-toothed blenny, which mimics the cleaner fish, but brings woe to the unsuspecting large fish (these blennies eat the vascularised gill tissue of the large fish!).

Because most land plants cannot move, they often exploit animals both for pollination and for seed dispersal (some rely on wind, too). Seeds of many fruits pass unharmed through the intestines of herbivores and germinate to grow a new plant from the droppings of the animal dispersing agent. Colourful flowers with nectar and brightly coloured fruits can only be interpreted as having been evolved to attract appropriate animals. Here, as in plant-herbivore interactions, a high degree of plant-animal specificity has arisen. Animals that pollinate a particular plant are referred to as pollinating vectors. For example, in Central America different species of male euglossine bees are highly specific to particular species of tiny epiphytic orchids; male bees travel long distances between orchids. Different bee species are attracted by different orchid fragrances, as can be shown by putting out 'baits' of artificially synthesised orchid 'fragrances'. (These male bees do not obtain nectar from the orchids they visit, but only obtain orchid products that the insects use for production of their own pheromones to attract females.) These bees are probably necessary for and may have allowed the evolution of, the great diversity of tropical orchids, many of which are evidently quite rare and far apart. Such specificity of pollinating vectors assures that the plant's pollen is transmitted to the ovules of its own species. Whereas female euglossine bees are not as specific to the plant species they pollinate as males, individual females travel distances up to 23 kilometers and regularly move long distances between sparsely distributed plants in gathering nectar and pollen; thus, they probably promote outcrossing among tropical plants at low densities. Indeed, Janzen suggests that such 'traplining' by female bees may actually permit the very existence of plant species forced to very low densities by factors such as competition and predation on their seeds and seedlings.

Some pollinators, such as *Heliconius* butterflies, obtain amino acids from the pollen of plants they pollinate. Because production of nectar and pollen (and fruit) requires matter and energy, attracting animal pollinators (and seed dispersers) has its costs to the plant. Nectar and fruits are usually rich in sugars and other carbohydrates but contain relatively little protein; in contrast, pollen and seeds contain considerably greater amounts of nitrogen and other limiting materials. Due to the frequent scarcity of such vital nutrients, carbohydrates are presumably cheaper for a plant to produce than amino acids and proteins. Thus, pollen-eating pollinators presumably cost a plant considerably more than strict nectar feeders. Returns from visiting a flower (or eating a fruit) must be great enough to an animal pollinator or seed disperser to make it worthwhile, yet small enough that the animal will travel the distance necessary to disperse the pollen or seeds. This intricate energetic interplay between plants and their pollinators is reviewed by Heinrich and Raven.

Obligate mutualisms are less common than facultative ones, probably because both populations depend completely on the relationship and neither can survive without the other. A very high degree of interdependency occurs between figs and the fig wasps that pollinate them (wasp eggs are laid inside fig fruits where larvae develop). There are hundreds of species of figs, each with its own, species of wasp (this is a good example of tight or species-specific, coevolution). Similarly, termites cannot themselves produce enzymes to digest the cellulose in wood, but by harbouring in their intestines a population of protozoans that can make such enzymes, the insects are able to exploit wood successfully as a food source. Neither termite nor protozoan could survive without the other. These intestinal endosymbionts are passed on from one generation of termites to the next through exchange of intestinal contents. Large-grazing mammals have a rumen in their gut system, an anaerobic chamber that houses endosymbiotic protozoans and bacteria, which similarly assist in digestion. Another putative example of mutualism is lichens, which are composed of a fungus and an alga; the fungus provides the supportive tissue, whereas the alga performs photosynthesis. (Algae of some lichens can be grown without the fungi.)

Commensalism occurs when one population is benefited but the other is unaffected (+, 0). Small epiphytes such as bromeliads and orchids, which grow on the surfaces of large trees without obvious detriment to the tree, might be an example. A well-documented case of commensalism is the association between cattle egrets and cattle. These egrets follow cattle that are grazing in the sun and capture prey (crickets, grasshoppers, flies, beetles, lizards, frogs) that move as cattle approach. The number of cattle egrets associated with cattle is strongly dependent on the activities of the cattle; thus, Heatwole observed fewer egrets than expected on a random basis near resting cattle, but nearly twice as many egrets as expected (if the association were entirely random) accompanied cattle that were actively grazing in the sun. Since the birds seldom take prey (such as ticks and other ectoparasites) directly from the bodies of the cattle, the mammals probably benefit little from their relationship with egrets. Moreover, egret-feeding rates and feeding efficiency are markedly higher when these birds are associated with cattle.

Because of their plus-plus and symmetric nature, mutualisms exhibit positive feedback and hence can run away—for this reason, they are destabilising unless the intraspecific negative self-dumping is stronger than the interspecific positive mutualistic effects. Mutualistic relationships are easily modelled with equations similar to the Lotka-Volterra competition equations simply by changing the signs of the alphas (K's have also been changed to X's since they no longer represent maximal densities), as follows:

$$\frac{dN_1}{dt} = r_1 N_1 \left(\frac{X_1 - N_1 + \alpha_{12} N_2}{X_1} \right) \quad \text{... (12.1)}$$

$$\frac{dN_2}{dt} = r_2 N_2 \left(\frac{X_2 - N_2 + \alpha_{21} N_1}{X_2} \right) \quad \text{... (12.2)}$$

Equilibrium conditions are described by a pair of linear equations and are shown graphically in Fig. 12.2. Populations reach equilibrium at density X_1 or X_2 in the absence of the other species and each populations equilibrium density is increased by increasing the density of the other species. If both X_1 and X_2 are positive and if α_{12} and α_{21} are chosen so that isoclines cross, the joint equilibrium is stable.

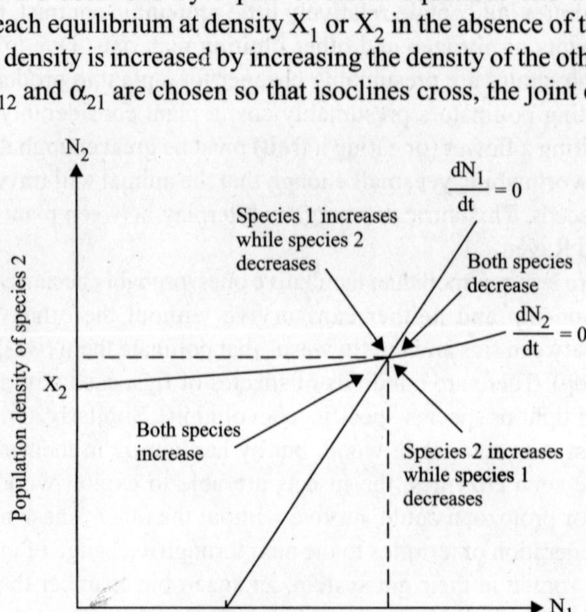

Fig. 12.2. Isoclines for two species with a mutually beneficial interaction and a stable joint equilibrium.

INDIRECT INTERACTIONS

Superimposed on direct pairwise interactions, more subtle indirect interactions are mediated through other members of the community concerned. Darwin anticipated the concept of indirect interactions and gave as an example the interactions among cats, field mice, humblebees (bumblebees) and red clover. The bees pollinate clover, but field mice raid bee nests and eat bee larvae. Lots of clover grows around villages, presumably because cats keep mice population down, allowing bumblebees to flourish, which in turn assists clover. Darwin's staunch defender Huxley carried Darwin's example further and noted that spinsters (who have lots of cats) facilitate Britain's naval prowess because strong sailors must be well fed and British beef thrives on clover. Here we have a long string with a pathlength of seven: spinsters → cats → mice → bees → clover → beef → sailors → naval prowess.

Five different sorts of indirect interactions involving three or four different species' population are depicted in Fig. 12.3. Pointed arrows indicate beneficial effects, whereas circle-headed 'arrows' depict detrimental interactions.

Indirect interactions

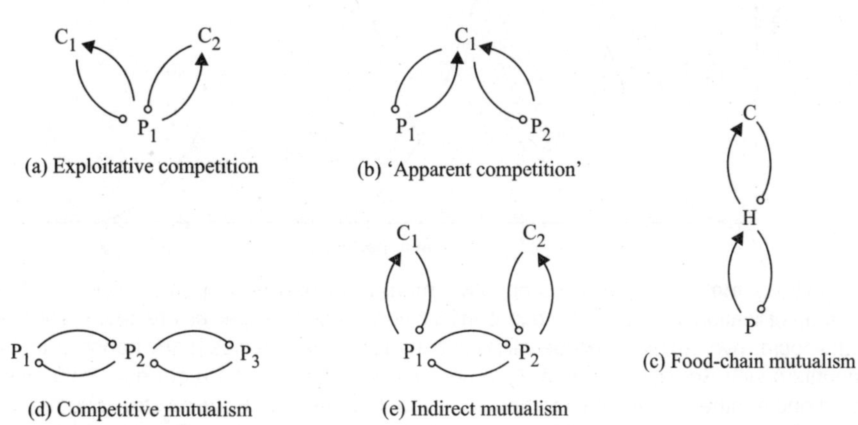

Fig. 12.3. (a) Indirect competition through classical exploitative competition resulting from resource depression; (b) apparent competition; (c) food-chain mutualism; (d) competitive mutualism; and (e) facilitation or indirect mutualism.

Two consumers sharing a common prey may compete indirectly through classical exploitative competition (resource depression). Two prey species may appear to compete because, if either increases, a shared predator also increases, which operates to the detriment of the other prey population and is called this apparent competition. Three specie's populations at three different trophic levels result in what has been termed a food-chain mutualism (such vertical interactions have also been called cascading effects or cascades). The plant and carnivore are indirect mutualists because the plant generates herbivores, which constitute food for the carnivore. The carnivore reduces herbivory, which benefits the plants. An example was provided by Power and others fish-eating bass feed upon herbivorous minnows in pools of an Oklahoma creek. When bass were removed (pools were fenced to keep these predators out) and minnow densities raised, the standing crop of algae diminished. With the readdition of bass, minnows retreated to shallow water and algal densities increased significantly over the next two weeks. Three species' population at the same trophic level, arranged so that one species (P_2) is sandwiched between two others (referred to as horizontal interactions), can also result in indirect mutualism. Population P_1 and P_3 are indirect mutualists because each inhibits the other's competitor P_2. Such a situation can also arise even when P_1 and P_3 are actually weak competitors, so long as competitive interactions with P_2 are strong (this has been called competitive mutualism—Pianka). An alternate depiction of how indirect competitive mutualism arises is shown in Fig. 12.4. A four-species system results in an indirect 'mutualism' (perhaps best termed facilitation. In this case, populations C_1 and C_2, which do not interact directly but consume different prey species, interact indirectly because their prey compete: if consumer C_1 increases, its prey P_1 decreases, which in turn reduces the competition with P_2, hence allowing an increase in this second prey population (P_2) and providing more food for consumer C_2. Numerous other sorts of indirect interactions are also possible.

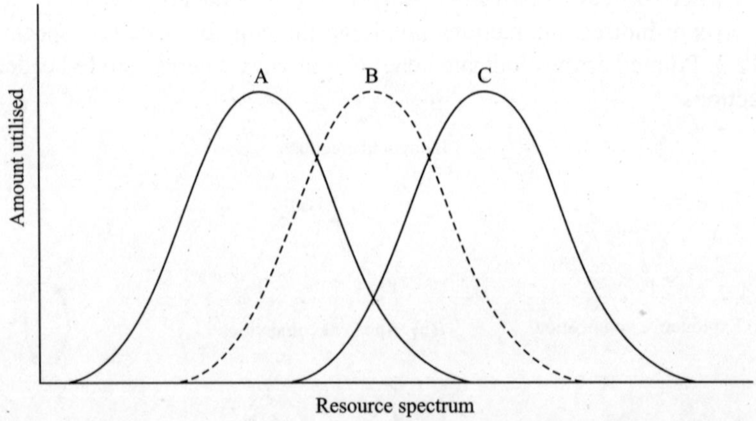

Fig. 12.4. Conditions that can lead to competitive mutualism between species. Species A and C overlap moderately in their utilisation of resources, so that, in isolation, these two species are potentially weak competitors. However, both species overlap more extensively with a third species, Species B and hence potentially experience intense competition with Species B. Since each species, A and C, exerts a stronger influence on Species B than they do on one another, when all three species occur together, each exerts strong competition on Species B, thus reducing the intensity of competition between Species B and the other. Resulting indirect effects between Species A and C, as mediated through Species B, are beneficial (each species reduces the fitness of a strong competitor of the other species). Hence the net interaction between Species A and C changes qualitatively in the presence or absence of Species B.

An indirect effect can be defined mathematically as the product of all the various direct effects along a directed series of links or a pathway, in which no species node is passed through more than once. Such a path product represents the indirect effect between two nodes, which may also be connected by a direct effect. Typically, the longer the pathway by which an indirect effect is mediated, the longer is the time lag required for the effect to be transmitted from one node to another. Thus, indirect effects typically take longer to occur than direct effects. Positive indirect effects can arise both by means of mutualistic links and by means of products of an even number of negative links. If, however, there are an odd number of negative links in a pathway, the overall indirect effect is negative.

Indirect effects are usually weaker than direct effects. However, because there are many more indirect effects than direct ones in a given system, the former can assume paramount importance even though they are weak. Indirect effects may actually oppose the direct effects and if their overall effects are intense enough, the overall net effect of one population on another, sometimes termed the community effect, can actually be reversed. Although this sort of double thinking seems circuitous and complex, it may prove to be vital to understanding community organisation; direct interactions in opposition to indirect interactions would moderate each other, leaving a target species only weakly affected.

Indeed, an interaction between any given pair of population depends vitally on the complex network of other interactions within which the pair concerned is embedded. Indirect effects render interpretation of simple experiments and observations extremely difficult.

13

Population Regulation

INTRODUCTION

As already mentioned, populations are dynamic: they are always changing. Organisms are born, die, immigrate or emigrate. But what would happen if a population grew in size without stopping? Darwin considered this situation and wrote 'The elephant is reckoned to be the slowest breeder of all known animals. . . it breeds when thirty years old and goes on breeding till 90 years old, bringing forth three pairs of young in this interval; if this be so, at the end of the fifth century there would be alive fifteen million elephants, descended from the first pair.' But this does not happen: we are not crowded out by elephants, mice or any other species. In other words populations do not usually increase to such an extent that they are out of balance with the rest of the environment. There are exceptions, where species suddenly increase in numbers and a plague follows, but numbers usually return to lower levels quite quickly.

POPULATION GROWTH WITHOUT REGULATION–EXPONENTIAL GROWTH

To understand what can happen in an unregulated population let us consider a very simple example of a hypothetical population. We can take as a start one breeding pair of animals which gives birth to two offspring every year and each offspring can reproduce when it is one year old. From this we can draw up a table of population numbers as shown in Table 13.1

Table 13.1. The increase of population size in a hypothetical population which starts from a single breeding pair that have two offspring each year. The offspring are capable of breeding after one year and also have two young each year.

Year	0	1	2	3	4	5	6	. . . n
Breeding pairs	1	1	2	4	8	16	32	2^{n-1}
Offspring	0	2	4	8	16	32	64	2^n
Population	2	4	8	16	32	64	128	2^{n+1}
Log_{10} population	0.3	0.6	0.9	1.2	1.5	1.8	2.1	0.3 (n + 1)

The data for the growth of this hypothetical population can be shown graphically as in Fig. 13.1a. As you can see the curve of population size quickly begins to rise very steeply. This population is undergoing exponential growth. If each pair of animals had given rise to four individuals each year, then the population growth would have been even faster. If we display the population increase as a semilog plot (shown in Fig. 13.1b) then the data lie on a straight line. This is because the intrinsic rate of increase (the maximum number of offspring born per individual in the population) is a constant. In the example above, the intrinsic rate of increase is one per year. If four offspring were born for each pair of animals each year then the line would still be straight on a semi-log plot, but it would be steeper: the intrinsic rate of increase would be two per year.

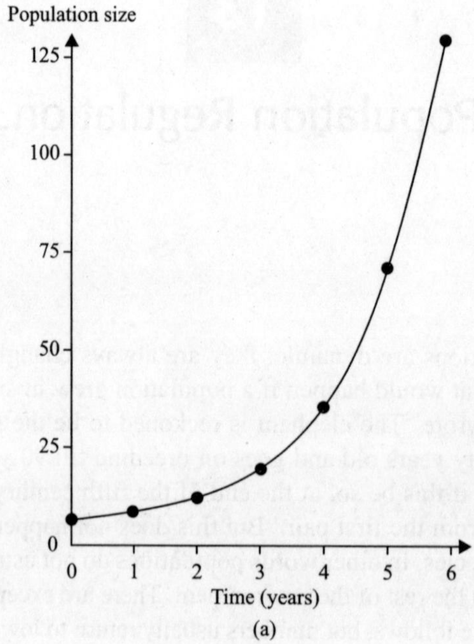

Fig. 13.1a. The increase in size of a hypothetical population through time using the data in Table 13.1 population shows exponential growth.

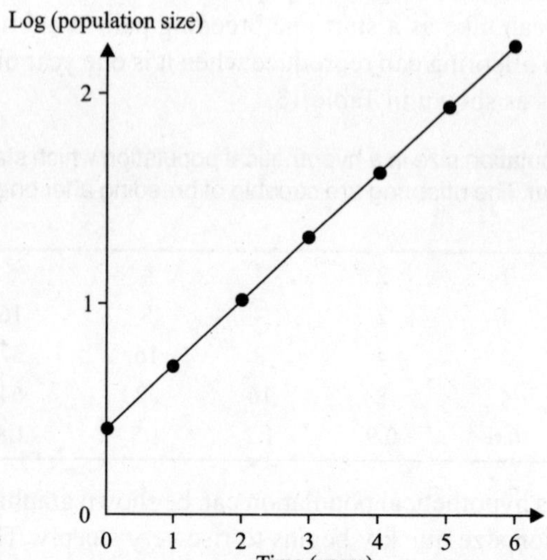

Fig. 13.1b. A semi-log plot of the data from Table 13.1 with the population size plotted as $\log_{10}$ population size.

Real populations seldom show exponential growth rates for many generations. A species introduced into a new and favourable area may well show exponential growth for a while, but some environmental factor usually begins to regulate the population when it reaches a certain size.

The population is potted as a log to the base 10, just as in Fig. 13.1b. In other words in the human population as a whole, even the *rate* of increase is increasing. This incredible rate of growth in the human population cannot continue indefinitely, we will run out of food and the space to grow it in. In other words at some stage population regulation will occur, either because we impose it ourselves or because other environmental factors take over. Even the solution most favoured by science fiction writers colonisation of other planets–would require unattainable efforts. Even now about 2,00,000 people would have to be launched into space each day just to keep the population steady!

Simple Population Regulation–the Logistic Growth Curve

Often a population invading a new area where space and food are plentiful will undergo exponential growth to begin with. Soon, however, the external constraints of the environment influence the rate of increase which begins to decline. Eventually, in most cases, the population size stabilises. The number in the population at this stage is called the carrying capacity of the environment.

Populations of protoctists kept in laboratory conditions show this type of population increase and levelling off. A classic experiment of this sort was conducted by Gause. He put 20 individuals of one species, either *Paramecium aurelia* or the larger *P. caudatum,* into a glass tube with a standard amount of bacteria as food. Every day the *Paramecium* were given food and every other day the mix was given an extra rinse to make sure toxins did not build up in the tube. The population in each test tube grew in size during the experiment and the resulting population growth curves are shown in Fig. 13.2. The *Paramecium* multiplied by binary fission, that is each organism divided into two without sexual reproduction. At first the shape of the curves in Fig. 13.2 are like that of the exponential curve of Fig. 13.1a: the *Paramecium* were dividing close to their maximum population growth rate. As the *Paramecium* numbers increased the rate of population increase slowed down. After four or five days the number of *Paramecium* in the tube no longer increased: the population had reached its carrying capacity.

If the larger species, *P. caudatum,* was in the tube then the carrying capacity was smaller (about 400 *Paramecium*/cm^3) than if *P. aurelia* was present (about 1000 *Paramecium*/cm^3). However, the volume of protoctists in each case was approximately the same.

The shape of the curves for population growth in *Paramecium* seen in Fig. 13.2 are typical of logistic growth. For *Paramecium* the rate of increase of the populations rises to a maximum after about two days and then decreases to zero when the carrying capacity is reached. It is this rise then fall in the rate of increase of population numbers which is predicted by the logistic equation:

$$\frac{dN}{dt} = rN\left(1 - \frac{N}{K}\right)$$

where,

$\frac{dN}{dt}$ = intrinsic rate of increase population

r = intrinsic rate of increase per individual of population

K = carrying capacity of population

N = number of organisms in population at time t

This equation can be represented graphically in two ways. If the number of organisms in the population N is plotted against time t (as in Fig. 13.3a) then a curve very similar in shape to the population increase in Gause's *Paramecium* is obtained (compare Fig. 13.3a with Fig. 13.2). This curve is sometimes called a sigmoid curve as it is roughly S-shaped (sigma is the Greek letter S). Most populations undergoing logistic growth tend to overshoot the carrying capacity (K), after which numbers oscillate about this number, this feature can be seen in Gause's experiments in Fig. 13.2.

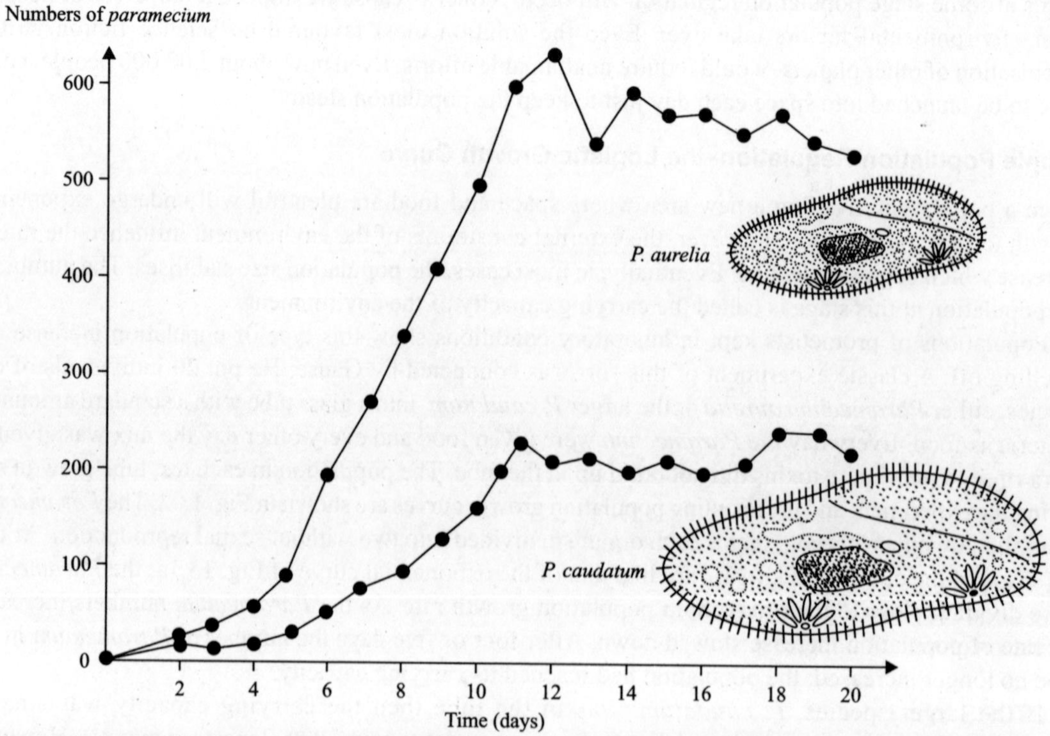

Fig. 13.2. Population growth curves for two species of *Paramecium*. The population numbers are means of three replicate experiments for each species.

If the rate of increase dN/dt, not the increase itself, is plotted against time a curve like Fig. 13.3b is obtained. When N is very small (as at the beginning of Gause's experiments or during colonisation) N/K is very small so (1–N/K) is approximately one and the logistic equation approximates to:

$$\frac{dN}{dt} = rN$$

This is the equation of exponential growth. As the carrying capacity is reached, N/K tends to one so the logistic equation becomes:

$$\frac{dN}{dt} = rN(1-1) = 0$$

In other words, the rate of population growth becomes zero.

Both these curves, Figs. 13.3a and 13.3b, can be referred to as 'the logistic curve' which may be rather confusing, but as already mentioned, both represent different features of the logistic equation.

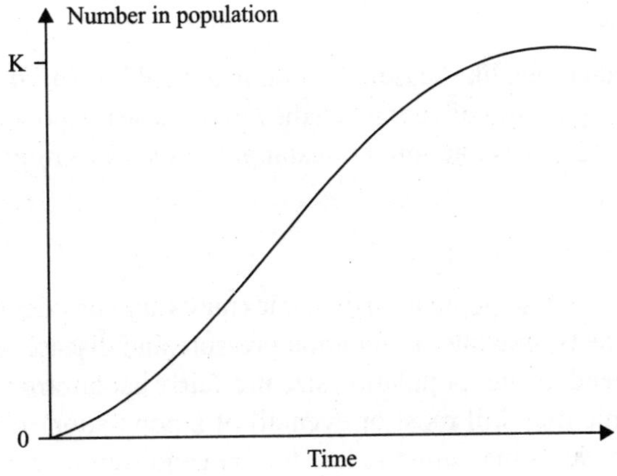

(a)

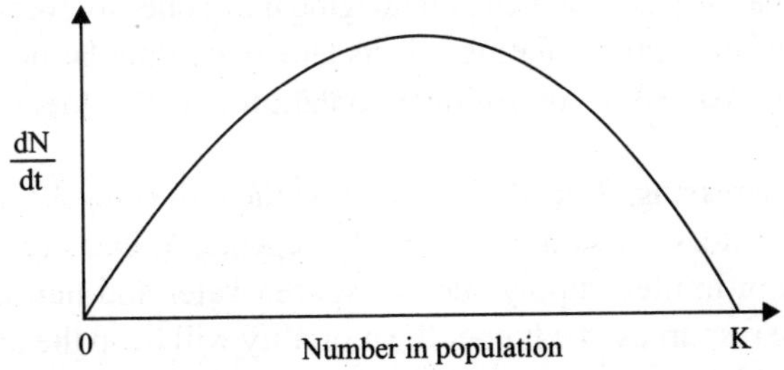

(b)

Fig. 13.3. Two ways of illustrating the changes in a population undergoing logistic growth where dN/dt =rN(1– N/K). Figure 13.3a is the change in population size with time. Figure 13.3b is the rate of change of population growth with population size.

FACTORS WHICH REGULATE POPULATION SIZE

Types of Regulation

As has been already noted, few species show exponential growth for many generations. Populations seem to fail to achieve their maximum potential for growth.

The environmental factors which are responsible for regulating populations are often the ones that affect the growth of the individual. This is because populations are made up of individuals. Population increase depends on the reproductive fitness and life span of these individuals. Factors such as energy and nutrient availability, flood, drought, predators and disease will all affect population size.

Environmental factors can affect populations in various ways depending on whether the factor is variable or not and how predictable the variation is. It is worth classifying these environmental factors here so they can be borne in mind for the rest of the section.

Constantly limiting factors

These are always in short supply, but are relatively constant so that a population is limited to a certain fairly constant size by the factor. Individuals may have to compete for the resource. For example, plants compete for space and light, birds for nesting territories and many heterotrophs for food. These factors do not usually produce large changes in population.

Variably limiting factors

These are variable but predictable, like seasonal drought and cold or variation in food availability. They are only an influence at certain times of year when they may cause the population to crash. Some species avoid much factors by taking evasive action. For example, some birds migrate and deciduous trees drop their leaves.

Unpredictable factors

These change in an irregular way so population dynamics may vary considerably over time. Unpredictable factors include weather effects, grazing or predation pressure and disease levels.

Unpredictable factors tend to alter population size in a fairly haphazard way: a volcanic eruption, fire or virulent disease epidemic may kill most or even all of a population. If the population has died out completely, then new individuals may migrate into the area and start a new population. It can be argued that such limiting effects are not really population regulators. This is because such factors are not constantly or repeatedly limiting population size, but merely damaging it in a one-off event, after which the population will recover up to a level where real regulating effects take over. Population regulation implies that one or more factors are always involved in controlling population size, whether they are constant or variable in operation.

Many factors have an increasing effect on a population the more closely spaced the individuals in the population become. Such factors are said to be density dependent. Density-dependent influences often include resources that are in limited supply such as space, water and nutrients. The more individuals there are competing for the resources, the higher the mortality will be in the population. Disease can also be density dependent as epidemics spread more rapidly if organisms are close together and in frequent contact rather than sparsely distributed. Factors which do not show this effect are said to be density independent and include fire, volcanic eruption and other catastrophes, extreme weather conditions like heavy frosts and high winds and toxic pollution.

Some of the major factors which alter population size, both unpredictable hazards and population regulators, are discussed in this section. Spatial heterogeneity may mean that more than one factor regulates a single population and that density-independent factors can be important in population regulation. Experiments or observations of natural populations are included which give examples of how these factors can regulate populations. Although these are examples where one main factor is involved in regulation, the size of most populations is influenced by a whole complex of interactive factors. Different factors may be important at different times of year, in different years or on individuals of different ages and in different parts of the geographical range of the population.

Space

The most basic requirement an organism has is space. Protoctists and most animals, except for stationary ones like scale insects, corals or barnacles, are not restricted by lack of physical space. They may have problems when finding the right place to breed, but this is a form of population restriction which falls under the heading of territories.

Plants, on the other hand, are often crowded closely together. As each individual grows it spreads out roots and branches and takes up more space. Unless it is an isolated plant, at some point in growth it starts to interfere with or be interfered with by neighbouring plants of the same or different species. Thus a mature plant will probably have displaced several others.

Tree plantations are good examples of groups of plants of the same age and species growing up and competing for space. The results of several studies of Ponderosa pine (*Pinus ponderosa*) in plantations are recorded diagrammatically in Fig. 13.4. The data are a combination from plantations of several ages from seedlings to adults, but we can go back a stage further to imagine the situation for seeds too. To consider how a stand of trees might develop, start at the right-hand side of the graph in Fig. 13.4 with the seeds which have a very low mean plant weight and are small and well spaced out. At first the germinating seeds are not close enough to one another to compete for space. The seedlings can therefore increase in volume (and weight, which is easier to measure and so is recorded here) while remaining at the same density. This is represented by the dotted vertical line on the lower right of Fig. 13.4).

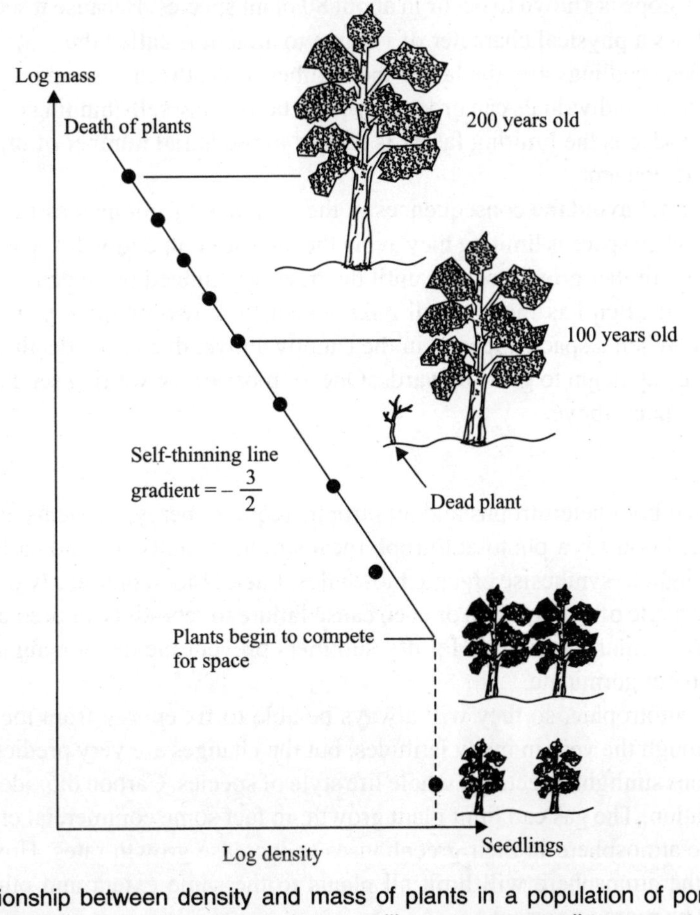

Fig. 13.4. The relationship between density and mass of plants in a population of ponderosa pine (*Pinus ponderosa*). The dotted line shows the phase when seedlings are too small to compete with each other for space at the density they are planted. The solid line shows density-dependent growth, where the plants can grow in size only at the expense of others which die, hence the density of plants decreases.

As the seedlings get larger, however, they begin to compete for space. Some plants are out-competed and die as others take over their space. Because plants are dying the density of plants (that is the number of plants per unit area) falls and the line begins to slope upwards to the left. This process of plant death and takeover is called self-thinning. As the pines grow, each one covers more ground area at the expense of other individuals which die. As each successful individual grows it increases the surface area of the ground it covers. The increase in ground surface area (measured in units squared, e.g. m^2) covered by each tree means fewer trees per unit area.

As each tree increases its area, it also increases the volume it fills (measured in cubic units). The plant matter in the tree is proportional to its volume and therefore the weight of the tree is also proportional to the volume. This means that if the data on increase in weight and decrease in density are represented on a log-log plot (as in Fig. 13.2), then the values are predicted to fall on a straight line with a gradient of $-3/2$. The $3/2$ figure is the ratio of the power of volume (units cubed) to area (units squared). The minus occurs because the density is falling as volume and mass are increasing, so the slope is a negative one. This gradient of slope is known to occur in about 80 plant species. Because it seems to be a common occurrence and follows a physical character of volume to area, it is called the $-3/2$ power law.

The more crowded seedlings are, the larger the number of deaths in a given area. If there are only a few seedlings, then those individuals can grow very large before any self-thinning occurs. The amount of self-thinning where space is the limiting factor is related to the initial number of organisms present and is therefore density dependent.

Plants can sometimes avoid the consequences of the $-3/2$ law by not increasing their mass with age. If they remain small when space is limiting they avoid the intense competition for space. Often a plantation stagnates and no more timber growth occurs until the trees are thinned by selective felling. The floor of a deciduous woodland often has many small oaks which look two or three years old but which are actually much older. When a space appears in the canopy above, due to the death of a mature tree, the seedlings close to the gap begin to grow upwards. One or more of the waiting seedlings may succeed in reaching the canopy space above.

Food and water

As already mentioned both heterotrophs and autotrophs require energy, nutrients and water to survive, grow and reproduce. Food for a photo-autotroph means mineral nutrients and carbon dioxide and the presence of enough light to synthesise organic molecules. Lack of food or scarcity of particular nutrients can reduce the growth rate of an organism or even cause failure to reproduce or premature death. Drought conditions can cause similar hardships. Hot dry summers prevent the developing seeds from ripening fully, so that they do not germinate.

Most plants are autotrophic, so they will always be able to fix energy from the sun. Sunlight does vary in intensity through the year in many latitudes, but the changes are very predictable, so rather than regulating populations sunlight affects the whole life style of species. Carbon dioxide is also unimportant in population regulation. The gas can limit plant growth, in fact some commercial crop growers create a high carbon dioxide atmosphere in their greenhouses to increase growth rates. However, the levels of carbon dioxide in the atmosphere will limit all plants to the same extent and other, less predictable factors are likely to be more important.

The vital nutrients for plant growth, especially nitrogen, potassium and phosphorus, are often in short supply soils. Species are known to be affected differently by various soils. Many species of lowgrowing plants favour poor soils where they are not overgrown by the lush foliage of the larger plants that grow well in fertile soil. Hence soil nutrient content may limit the population size of certain species, but population changes from year to year will be regulated by other factors. One of the few times when soil nutrient content changes sufficiently to affect populations dynamics is after a fire. The fire clears vegetation and converts it into high-nutrient ash and smoke.

For some plants, lack of nutrients is less important than lack of water. Some plant populations are completely regulated by rainfall or lack of it: many desert or semi-desert areas 'bloom' with vivid carpets of flowers after heavy rain which may only occur very ten years or so. The dormant seeds in the dry soils germinate and grow rapidly after the rain, then flower, set seed and die as the area again dries out. In other plant communities rainfall is greater and more frequent, but occasional dry periods can be a considerable problem. The effects of drought conditions are obvious when plants wilt or die outright, but more subtle effects of water shortage include failure to ripen seed (as in the beech) or loss of leaves which weakens the plant as it has to replace them later.

The direct effects of food and water on the health of animals and on their potential to raise young, are more obvious then for plants and it is easy to see how these influence population size. In fact the lack of sufficient food to maximise reproductive potential may be the most important regulator of population size in animals. If food is rare, animals have to spend considerable time finding it. If it is of poor nutrient quality then there may not be enough time available to eat a sufficient amount. Food availability is often seasonal and unpredictable. In temperate regions, food may abound in summer and autumn as young shoots and fruits, but foliage becomes rare and of low quality in winter or during summer drought. Hence herbivores alter their life style to migrate, hibernate, survive as a dormant phase or live as best they can, often with considerable loss of life. Because there is a lack of herbivore prey in winter, insectivores and carnivores also have difficulties at these times.

It is often difficult to measure quantitively the relative abundance of the food available for a population. Many observations show that when food is scarce animals reproduce less successfully then when food is more abundant, but the details of the way food regulates a population are difficult to disentangle. In some species, the mode of life of the animal makes it a suitable candidate for experimental study, in such cases, the data comes from laboratory or field experiments rather than natural situations. However, such experiments can give us a better idea of how populations are regulated.

Territories

A territory is an area which contains a resource of sufficient value to an animal that it defends the area. The animal holds or defends the territory by marking the boundaries in some way and challenging strangers if they approach too closely. Territories may vary considerably in size depending on what resource they contain. Also they may be held for a very short time or for longer, depending on how long the resource is useful to the animal. Many species are known to hold territories, including mammals, birds, reptiles, amphibians, fish, insects and crustaceans.

Territories are usually held because they contain a food source or because they provide some site needed in breeding, such as an area to attract mates or a nest site. Often these two reasons are connected and a breeding territory will contain the nest site and sufficient area to supply food to the growing offspring. Food territories may be held by a solitary animal. A good example is the tiger which hunts alone and has a very large territory. Often the territory of a male tiger will overlap that of one or more females. Some territories support whole social groups of animals like the meerkats (*Suricata suricatta*) where a few individuals keep watch for danger or strangers and the whole adult troop goes to defend the area if necessary.

Territories which are not held for food are usually very small. The male starling, defends potential nest sites where he sings to attract a female. Animals which have a lek, that is an area where males gather and display to attract females, have minute territories. The best males hold tiny areas at the very center of the display area, less good males have larger territories around the edges of the lek. Many birds, like the

European ruff (*Philomachus pugnax*) and the American sage grouse (*Centrocercus urophrasianus*) have leks.

The size and number of territories in an area affects how many breeding individuals can live in the area and so influences population size. The small breeding territories do not usually regulate populations, unless actual nesting sites are limiting, as may be the case in cliff-dwelling seabirds where food is abundant, but safe ledges for nests may not be. It is the larger food territories which probably have most effect on population size: so what is it that regulates the size of these territories?

In a territory defended for food the amount the animal needs and the amount and quality of the food in the defended area will all affect territory size. Observation of the rufus hummingbird (*Selasphorus rufus*) which feeds on nectar from flowers, a relatively easy resource to quantify, has given interesting insights into territory size. Rufus hummingbirds breed in northern America, even as far north as Alaska. For the winter they migrate south to Mexico and Guatemala. On migration the hummingbirds stop in mountain meadows to rest and feed. They defend feeding territories for short periods of one to seven days before moving on to another meadow further south.

In the meadows of California. Gass and others found that the hummingbirds held territories where they fed mainly from two species of nectar-producing flower, the red columbine (*Aquilegia formosa*) and the Indian paintbrush (*Castilleja miniata*). They measured the nectar production and sugar concentration in the nectar of the two flowers to obtain the value of each flower to the hummingbirds; they then converted their results into a measure called a floral unit. They also observed the birds to see what territories they defended and how many flowers these territories contained. For example, a territory of 250 floral units contained 250 columbine flowers or 780 paintbrush flowers or a mixture of the two. Fig. 13.5 shows the relationship Gass and others found between territory size and floral unit density. They showed that territories were maintained to contain a certain number of floral units. If flowers were sparse, then the hummingbirds held large territories; if the flowers were abundant, territory size was much smaller.

There are times when it is not worth the hummingbirds defending a territory. If flowers and nectar are very plentiful in the meadows, then the birds abandon their territories as there is no need to defend food. If nectar is very sparse, then the birds again abandon territories as the areas they need to defend to include enough flowers is far too large.

The most obvious way in which territories can regulate population size is in species which hold dual purpose territories for food and nesting. The is the case in many bird species. After a re-shuffle of areas by remaining birds, new birds moved in a take over the remaining vacant territories. These new birds had come from hedgerows outside the wood where they has been holding poor territories. The reproductive success of birds in the wood was higher than those in the hedgerow.

The work on great tits shows that both number and quality of territories in an area can regulate population size. The number of territories in the wood remained fairly stable, indicating that there was a limited number of territories available to the population. This limited the total number of breeding pairs in the area and hence the potential population growth. Each year the number of breeding adults remains about the same.

Herbivores and predators

Predators obviously affect the population size of their prey, but the effect of herbivory on the population dynamics of vegetation is less apparent. If an organism eats seeds or seedlings, so killing plants, this directly affects the total plant population size. But there are many seeds and few seedlings survive to maturity because of the $-3/2$ self-thinning effect. So the effect of the herbivory on adult plant populations is lessened. Many herbivores do not even kill the plant they feed off. The effect this has on population dynamics is even harder to determine. Flower spikes may be removed so reducing seed production or defoliation or root grazing may weaken the plant and prevent it reproducing. Defoliation experiments show that plants respond differently to this depending on species, stage of reproduction reached and amount of leaf area lost. Leaf removal can delay flowering by several weeks, reduce seed numbers to zero or reduce seed size.

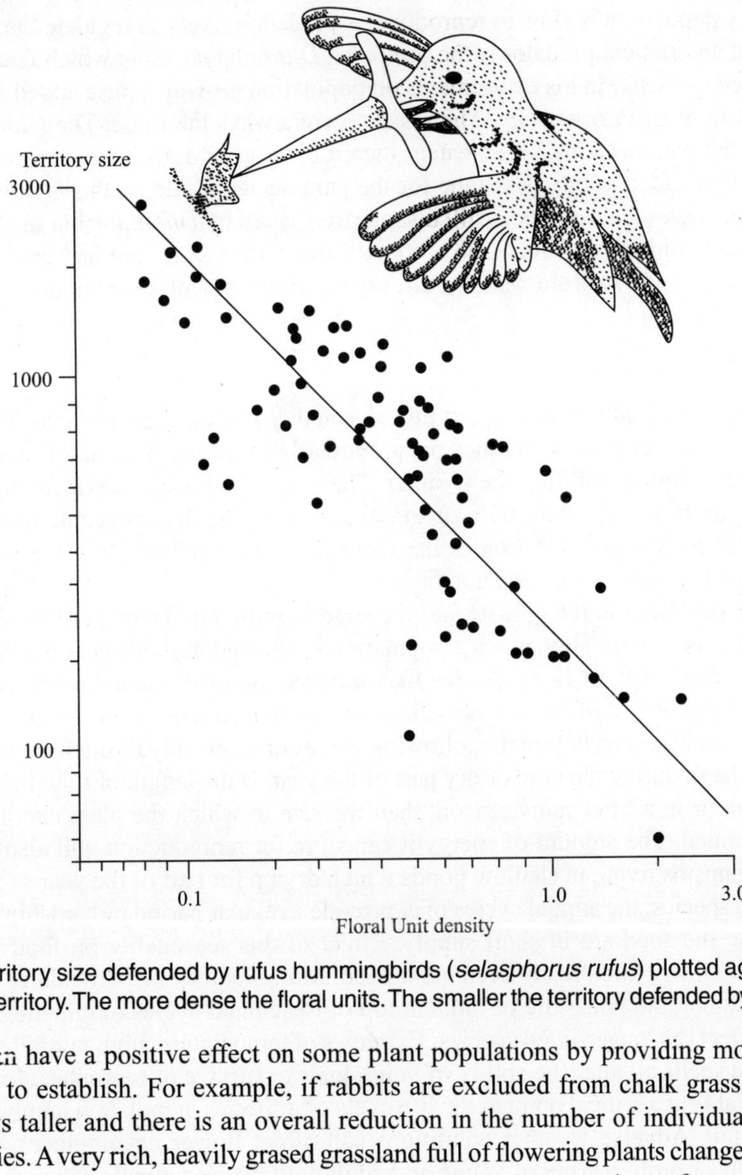

Fig. 13.5. The territory size defended by rufus hummingbirds (*selasphorus rufus*) plotted against the density of floral units in the territory. The more dense the floral units. The smaller the territory defended by the hummingbirds.

Herbivory can have a positive effect on some plant populations by providing more opportunity for new individuals to establish. For example, if rabbits are excluded from chalk grassland, the ungrased vegetation grows taller and there is an overall reduction in the number of individual plants and in the diversity of species. A very rich, heavily grased grassland full of flowering plants changed to one dominated by species of grass with very few herbs. Rabbits evidently eat grasses in preference to herbs, so that when rabbits are present, the herbs can flower, set seed and increase. Once rabbits are excluded, the grass crowds out the other species.

Predators, despite their obvious effect on prey numbers, may not always have a long-term effect on population regulation. If numbers of prey are influenced mainly by food availability, then the predator may not be regulating the population at all. When prey numbers are high, because of abundant food, any prey the predator takes is likely to have died of starvation when food was no longer plentiful. When food is short, so that the prey population is small, each death from predation leaves a little more food for survivors and the population will increase in size once food is again plentiful. Only if predators are very efficient or the prey population is slow to reproduce, is predation likely to regulate the prey population.

One example of an efficient predator is the protoctist *Didinium nasutum* which feeds on the species of *Paramecium* used by Gause in his experiments on population growth. Gause added small numbers of *Didinium* to populations of *Paramecium*. The results were always the same. The *Didinium* population increased until all the paramecia had been eaten; then it died out in turn, unless more paramecia were added. If a refuge was added to the glass tube for the paramecia to hide in, the situation became more complicated. In some cases the paramecia still disappeared, eaten by *Didinium*, but in other experiments some of the paramecia hid in the refuge and were still there after *Didinium* had died of starvation. In other words, in the case of these protoctists, features of the environment were important in the predator–prey relationship.

Weather and climate

The climate of an area is usually seasonal, but this seasonality is always predictable. There may be wet and dry seasons or hot and cold ones, but they occur regularly each year. Within this predictable climate there is a shorter term unpredictability: the weather. There is considerable variation from day to day in temperature, wind speed, precipitation (rain or snow) and so on. Such changeable features occur most noticeably in temperate zones. In the tropics the weather is more stable, but there are often extreme diurnal variations such as cold nights and hot days.

Both the annual variations in the climate and the shorter, more rapid changes in weather can have a regulating effect on populations. How much a population is affected depends very much on the life span of the species concerned. A short-lived species like an insect or small annual plant may complete the whole of its life cycle in one part of the annual cycle of variation in climate. Many annual plants germinate in spring and flower and set seed before the following winter or grow only through the wet season. They survive in the seed bank during the cold or dry part of the year. If the length of time for growth is short, as it is in the tundra or in a brief rainy season, then the size to which the plant can grow in the time available will be limited. The amount of energy it can store for reproduction will also be limited. The same is true for organisms living in shallow ponds which dry up for part of the year.

For longer-lived species, the annual cycles may provide a regular period of hardship such as drought or cold when water and food are in short supply. Effect of this seasonality on food availability and therefore on population size has already been mentioned. Mortality may be particularly high during this period, especially among the immature or old. On top of the annual cycles of climate, the extremes of weather will also affect the longer-lived species. Extremes of temperature, high rainfall or drought, mild winters or very cold spells all alter the ability of individuals to survive or reproduce. In years when the weather is unfavourable at a critical point in the life cycle of a species, population numbers may crash or reproduction may fail. Adverse weather conditions can affect flower development, pollination and germination in plants or birth, rearing of young and adult mortality in animals. Climate change and the variability of weather, as it affects population regulation.

Parasites and diseases

A parasite is an organism which lives on or in another organism. The organism on which the parasite lives is called the host. Parasites depend on their hosts for food. A heavy infection of parasites, called the parasite load, may eventually cause the death of the host, directly or indirectly, by weakening it so that it succumbs to a predator or disease.

A parasite is recognised as being much smaller than its host, so that often many parasites can live on a single host. Parasites usually reproduce much faster than their host and have a much faster rate of population increase. However, as with many other definitions in ecology, there is a boundary area between these three groups where it is difficult to decide which category to use. For example, European cuckoos (*Cuculus canorus*) are much larger than their host birds, yet the baby cuckoo in the nest of a meadow pipit is considered as a parasite. On the other hand a colony of monkeys eating the fruit or leaves of a tree are much smaller than their host yet are considered to be herbivores, not parasites.

If an organism causes disease in its host, often leading to the death of the host, it is called a pathogen. These organisms are usually very small and include viruses, bacteria, fungi and protoctists. An alternative name for such small organisms is microparasites.

Both plant and animal populations can be infected with parasites and pathogens. The population dynamics of the host and the disease organism are linked. If the pathogen or parasite population is very small, then it usually has little effect on the host population. If the parasite population increases in numbers, so that the host organism has a heavy parasitic load, then this is likely to weaken or even kill the host. Usually a heavy parasite load affects population dynamics by reducing the birth rate and by increasing deaths from secondary causes.

Most population have low levels of disease present, but sometimes a disease will multiply and spread rapidly in a population causing an epidemic. Parasite infections or disease epidemics are more likely to spread if the host species is very common or closely spaced together. This makes regulation of disease and parasites important, especially in agriculture and among human populations where such conditions frequently occur. Because of the importance of such factors in crop growing and animal husbandry, there has been much investigation of the effects of pathogens and parasites on population dynamics. Several different models have been produced to explain the way hosts and parasites or pathogens regulate each other.

Natural disasters

Disasters are uncommon and unpredictable events which can occur in any area of the world. Such events include fire, floods, winds of various destructive kinds from gales to hurricanes, tidal waves, mud flows, earthquakes, volcanic eruptions and meteor impacts.

Some of these disasters are the result of geological processes such as plate tectonics or the impact of extraterrestrial debris. Such events may be very rare, but have large-scale effects. A huge meteor crashing to earth is thought to have caused the extinction or the dinosaurs and many other species at the end of the Cretaceous period, 65 million years ago. Other geological events such as volcanos may be rather local in effect covering only a few hundred square kilometres with lava or thick dust; however, atmospheric dust from volcanos can block sunlight over a much larger area.

Some disasters are caused by extreme weather conditions. These include high winds, floods and fires (influenced by hot dry periods and caused in nature by lightning strikes). They tend to occur more often than geological disasters. Their effects may be apparent over a large area, but are not global in influence. The effects of disasters on populations, are considerable. These events do not regulate populations in the continuous and fairly stable way like other factors discussed in this chapter. A fire, flood, hurricane or volcanic eruption (Fig. 13.6) may wipe out many species in the vicinity of the disaster. New populations may then be established by immigrants from neighbouring populations. Survivors in seed banks may provide the next generation for the plant populations. If these have also been destroyed, fresh incoming seed is required to colonise the area. In many cases new species will colonise the devastated area. It may then be some time before populations of the original species re-invade the site.

Fig. 13.6. The devastation which can be caused by natural disasters: about 800 square kilometres of forest on and around the slopes of Mount Saint Helens. Washington State, USA, was completely flattened by the volcanic eruption of 18 May 1980.

Fires and high winds may be extremely important factors in the variation in age structure of plant populations. Young trees and bushes are most likely to suffer fire damage, while mature trees are most likely to be blown down. In grasslands, frequent fires are caused by the presence of dry grass leaves. These fires are low in temperature and do not kill the growing points of the grass in the soil, but damage any young trees or bushes. Thus, grassland fires tend to prevent establishment of woody species. This maintains the grassland and prevents it being invaded by scrub.

In tropical and temperate forests, large gaps are occasionally created by wind damage. This provides the opportunity for young forest trees to grow and produces patchiness in the forest. Diversity in tree age and forest structure is therefore maintained. Forest fires are much hotter than grassland fires and may kill all the trees and animals in a large area. The fires cause nutrient loss to the atmosphere and release some nutrients into the soil in ash. These fires may devastate the tree populations, but some tree species will be favoured by fire. Several pine (*Pinus*) species and in Australia, Eucalyptus species have seeds whose germination is triggered by fire. The seedlings thus grow in a nutrient-rich soil and experience little competition from other species while they are growing.

Fire and high winds are probably extremely important factors in maintaining diversity in species and the structure of vegetation in an area. Because they are infrequent they do not tend to be studied as often as the other more continuous factors which regulate populations.

Self-regulation and stress

Often there is no obvious reason why a population does not increase in numbers. However, natural populations are usually difficult to study, so it may be hard to determine the exact causes of hardship or death or reasons for emigration.

Laboratory experiments can have a much tighter control over environmental factors and because the population is small and confined it is easier to study. Sometimes such experiments give extremely interesting results. One such study was carried out on rats by Calhoun. Calhoun created his population by putting 16 young female and 16 young male rats into an enclosure. They were given all the food and water they needed and left to reproduce. After one year the number of rats in the population had risen to 80 adults. The enclosure, which was made up of four areas each with a nesting unit, was designed to hold about 50 rats, so it was overcrowded with 80.

The population was kept at 80 by the removal of young rats after they had been weaned from their mothers' milk. This represented emigration from the population. The population structure at this time consisted of a couple of dominant males who held territories in the box, each with a harem of about five female rats. These females were good mothers and successfully raised about half their young. The other females, not in harems, were not so successful and about 80 per cent of their offspring died before they were fully weaned. This was because the non-harem mothers provided inadequate care for their young. Outside the territories of the two males the females were more crowded and failed to build proper nests for their young. Often they abandoned their young around the compound. The young often left the nests before they were capable of surviving alone, presumably because the poor nests offered little comfort. The behaviour of male rats also became unusual for caged animals. There was increased fighting among some males, while others showed almost total inactivity and did not appear to notice the existence of other animals. Some rats showed homosexual or hypersexual activity and there was cannibalism of the young.

Calhoun's rat experiment shows how completely the normal social behaviour of many of the rats broke down after the population became overcrowded. Other experiments have shown that such overcrowding can cause physical as well as behavioural problems. Mice developed ulcers or a lowered resistance to disease when kept in similar conditions. In natural conditions there would be more space for the population and overcrowding would be reduced by increased emigration. However, overcrowding does occur in nature, for example on islands or during widespread favourable periods when densities may reach high levels, especially in small mammals.

Thus, overcrowding in industrialised cultures can lead many humans into the aptly named 'rat race' of competitive behaviour. Being successful means having a good job and earning money. Many people in employment show stress-related conditions such as ulcers and heart attacks. At the other end of the scale, when unemployment is persistently high, there are people without an active role in society who may become over-aggressive or lethargic. Most would argue that this sort of interpretation is too simplistic. After all, human societies are much more complex and varied than the social structures developed by rats in a box.

PATTERNS IN POPULATION DYNAMICS

Some populations have regular patterns of increases and decreases in population numbers. These population cycles are found in several mammals including small mammals such as voles and lemmings and larger carnivores such as foxes and lynx. Figure 13.7 shows how the fluctuations in lemming numbers over several years follow a regular pattern. The lemming populations increase rapidly from only one or two lemmings per hectare to about 30 or 40 lemmings per hectare. This increase occurs in only a few months. Shortly afterwards the numbers fall back to the earlier low levels and remain low for the next few years.

As you can see from Fig. 13.7, the peaks and troughs of lemming numbers are regularly spaced about four years apart. This four-year cycle occurs not only in lemmings, but also in mice (Muridae) and foxes (*Alopex and Vulpes* spp.). Other species such as the snowshoe hare (*Lepus americanus*), muskrat (*Ondatra zibethica*), some foxes and the Canadian lynx (*Lynx canadensis*) have ten-year population cycles

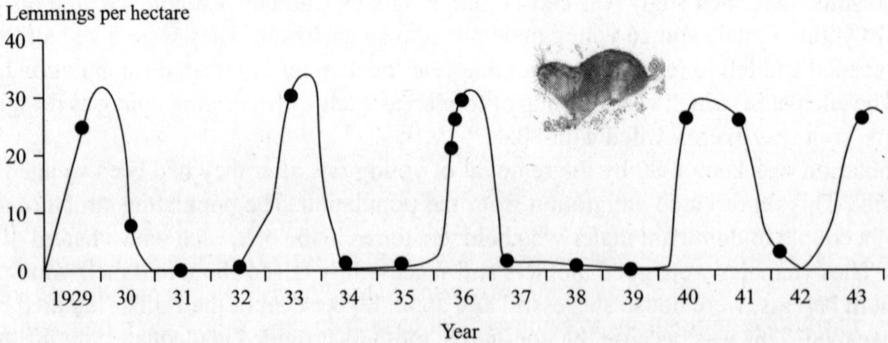

Fig. 13.7. Changes in the number of collared lemmings (*Dicrostonyx*) in northern Manitoba, Canada, from field estimates and trapping data.

So what makes these populations increase and decrease with such regularity and why should the cycles be four or ten years long?

In the case of the carnivores such as foxes and lynx the answer may be fairly simple. These species feed on lemmings, mice and hares, so when the small mammals are abundant the carnivores have plenty of food. This means they can raise and feed more young in the peak years of the cycle. When the small mammal populations crash then the carnivores starve and many die taking population levels back to the low numbers between peak years.

This still does not explain why the herbivorous mammals have regular cycles. Lemmings can raise two or three broods of young a year. They can rapidly increase their population size in favourable conditions. The sharp rise in population size in the small mammals is probably due to some favourable environmental factor which results in an increase in the birth rate.

The equally rapid fall in the population in the lemmings seems to result from a high death rate and lowering of the birth rate. The population crash does not seem to be due to disease or to starvation because of lack of food. There are many stories of emigrating lemmings going mad and committing mass suicide by plunging into rivers. The phrase 'to rush like lemmings to their fate' comes from this idea. The tales of madness appear to be unfounded. Lemmings often swim rivers during migration and inevitably some will be washed away and drowned in the process. The more lemmings there are on the move, the more accidental deaths there will be. Krebs looked at 4000 dead lemmings collected after a population peak when the death rate was high. He could find no evidence of stress in the animals and they all seemed quite well fed. Lemmings are short-lived animals and do not seem to survive for more than a year in the wild. The rapid fall in numbers seems to be due to a failure to breed coupled with the dying off of the older animals.

It was once suggested that the increase in predators such as the foxes caused the lemming population to crash. This was the case with the efficient predator *Didinium* on the laboratory populations of

Paramecium, but foxes are not such efficient predators. They do not eat sufficient lemmings to cause the crash.

The fact is that we are still not sure why lemming populations cycle with such regularity. Anything related to weather is not so regular: peaks and slumps in conditions might occur, but not on such a regular four-year pattern. The most likely suggestion is that the period between population peaks is the shortest time needed by the population to recover from the last collapse, but it does not really explain why the population collapses in the first place.

As mentioned earlier, the snowshoe hare cycles regularly too, on an 8–11-year cycle. Its predator, the Canadian lynx, as in the case of foxes feeding on lemmings, increases in numbers when the hares do and many starve when the hare population crashes. The hares eat twigs on shrubs and may exhaust the food supply, as many hares show the characteristics of starvation. The bushes respond to the heavy browsing by increasing toxins in their new shoots for two or three years after a 'hare high'. This extended period, during which food is in short supply, keeps the hare populations low and is thought to explain why the cycling is as long as ten years.

14

Habitats and Niches

INTRODUCTION

The word habitat is used extensively in ecology when describing where an organism lives i.e., habitat is the physical environment in which an organism is found. Habitats typically contain many niches and support many different species. Thus, temperate broad-leaved woodland provides a vast number of niches for a range of birds (e.g., nuthatches, great tits, woodcocks), mammals (e.g., wood mice, foxes, common shrews); insect (e.g., butterflies, moths, beetles, spiders, aphids) and plants (e.g., wood anemones, bluebells, mosses, lichens).

We may also use the word micro-habitat, which describes an organism's environment over distances of millimeters to centimeters to meters, scales smaller than those typically used for habitat. We include descriptions of micro-habitats because organisms may sometimes be affected by factors operating over small spatial scales as well as larger scales.

Within their habitats (and micro-habitats), organisms acquire the resources of food, living space, shelter and so forth that they require for growth, survival and eventual reproduction. Typically we will find some species have very broad diets and may occur in a variety of habitats—we call such organisms generalists. In contrast, we may find that other species live in only specific habitats (or even micro-habitats) and/or eat a very narrow range of foods and we call these taxa specialists. The reasons why some organisms are specialists and others generalists lie partly within events long past—in the evolutionary history of each specie. Nevertheless, we can ask: What modern-day factors restrict some organisms to living in a small range of habitats or micro-habitats? Can this limitation explain why some species are common and others rare? These are central questions in population ecology and we will now look at how ecologists have approached getting answers to them.

Organisms of a given species can maintain viable population only under a certain range of conditions and will utilise only particular resources. They also may only occur in a given environment during particular times (e.g., insectivorous bats are nocturnal, when few insectivorous birds are feeding). The intersection of these factors describes the niche, which is the position the organism fills in its environment. Organisms may change niches as they develop, for example common toads *Bufo bufo* occupy an aquatic environment (and are grazers of algae and detritus) prior to metamorphosing into adults, whence they become terrestrial (and are insectivorous).

The distinction between habitat and niche is an important one in ecology. The word niche originally meant a shallow ornamental recess formed in the wall of a building, usually for the purpose of containing a statue or other decorative object. In ecology it came to stand for the precise way in which a species fits into its environment. A habitat is a description of where an organism is found, but its niche is a complete description of how the organism relates to its physical and biological environment. The concept of the niche has been very important in ecology, but it takes a great deal of field work to determine the niche of an organism in any detail.

DETERMINING NICHES

Each condition that affects an organism or resource that it utilises can be regarded as a single axis or dimension, within which there can be defined a range that the organism will occur in. By considering a number of such dimensions at once, an increasingly refined picture of the organism's niche can be reached. For example, the temperature range which a chaffinch will tolerate will overlap with that of many other species. However, if we consider the prey size and the foraging height as further dimensions we will differentiate the chaffinch niche from that of many other species (Fig. 14.1). It is theoretically possible (though difficult to measure or represent on a page) to add every dimension of resource or condition which affects the organism, resulting in a fully defined ecological niche—the '*n*'-dimensional hypervolume' (where '*n*' is the number of axes). Simple theory suggests that this fully defined niche is expected to be unique to that one species (or even one lifestage of that species), although recent work shows this need not be so in a dynamic or patchy environment. A practical weakness in the '*n*'-dimensional hypervolume theory is that it is impossible to be sure that all dimensions have been considered, but, nevertheless, this is a useful concept.

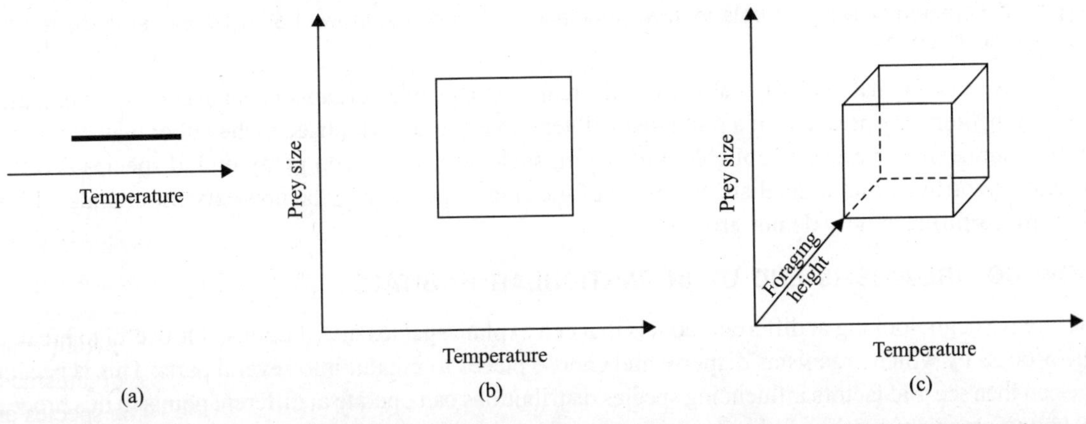

Fig. 14.1. Niche dimensions for an insectivorous bird: (a) A one-dimensional niche covering temperature tolerance; (b) a two-dimensional niche incorporating temperature and prey size; and (c) a three-dimensional niche incorporating temperature, prey size and the height a which foraging occurs.

Fundamental Niche

The niche space a species can occupy is affected by the degree of competition and predation that impact upon it. Commonly, in the absence of these stresses, a species can thrive in a wider range of resource levels and conditions. This potential niche space is the fundamental niche.

Realised Niche

Normally, species are exposed to competitors and predators and are hence limited to a more restricted niche space, known as the realised niche. The impact of competition on the fundamental niche is illustrated by a classical experiment, on two species of *Galium* spp. by the plant ecologist Tansley. Heath bedstraw (*Galium saxatile*) is found on acidic soils, whilst slender bedstraw (*G. pumilium*) is found on calcareous soil. When grown alone, both species thrive: in both soil types. However, when the species were grown together, slender bedstraw was excluded on the acidic soil, whilst heath bedstraw was excluded from the calcareous soil. Clearly, competition affects the observed niche. Experiments with guppies (*Poecilia reticulata*) and sticklebacks (*Gasterosteus aculeatus*) suggest that the proximity of a predator constrains the fish to feed in sheltered areas where predation risk is lowered. In another example, the pattern of feeding times of the three American bat species shown in Fig. 14.2 suggests that the realised temporal niche of these animals is also shaped by interspecific interactions.

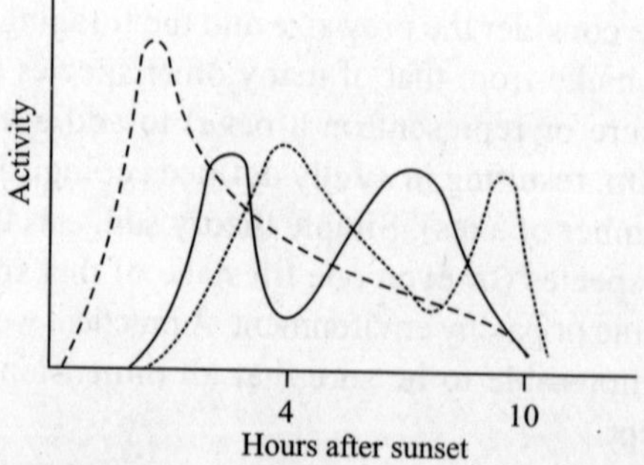

Fig. 14.2. Foraging activities of bats in Iowa woodlands—dotted line, brown bat; solid line, silver-haired bat; dashed line, hoary bat.

It may be noted that mutualists also affect the realised niche of an organism, but in contrast to predators and competitors, the presence of a mutualist will tend to expand the realised niche rather than contract it. In the more extreme cases of obligate mutualism, such as that between many orchid species and their fungal mycorrhizal root mutualists, the niche of the single species may be non-existent orchids without their mycorrhizae simply do not grow.

HOW DO ORGANISMS END UP IN PARTICULAR HABITATS?

Before we begin looking at different models that can explain species distributions, it is useful to break up the process by which organisms disperse and choose places to inhabit into several parts. This is because we can then see that factors influencing species distributions can operate at different points in this process. A young organism hatches, is born or dispersed from its parent—the first step can be dispersal away from the natal site (Fig. 14.3). As already discussed, some organisms may disperse long distances from their natal sites (perhaps travelling between population) and others very little at all. Following dispersal, organisms make contact with locations of potentially suitable habitat. They then may assess whether that habitat is suitable or not, using behaviour that is instinctive (i.e., genetically based) and choose whether to stay or to go. Plants cannot 'behave' in this sense of course, but seeds or spores may not germinate until they are in a suitable location or until a certain event has happened. Following selection of a suitable habitat spot, organisms must .then persist and survive. It is possible that they may be driven away by competing species, eaten by predators or die because the habitat subsequently becomes unsuitable, so that after some time has passed the distribution of organisms does not reflect the choices they made during colonisation. Some organisms may go through this sequence multiple times through their lives, dispersing and colonising different habitat patches each time, whereas individuals of other species may get only one opportunity. Organisms that settle permanently in one spot, like plants and some animals (e.g., sponges), disperse only during juvenile stages and habitat choices made by larvae (or germination by seeds and spores), if any, determine where adults can live.

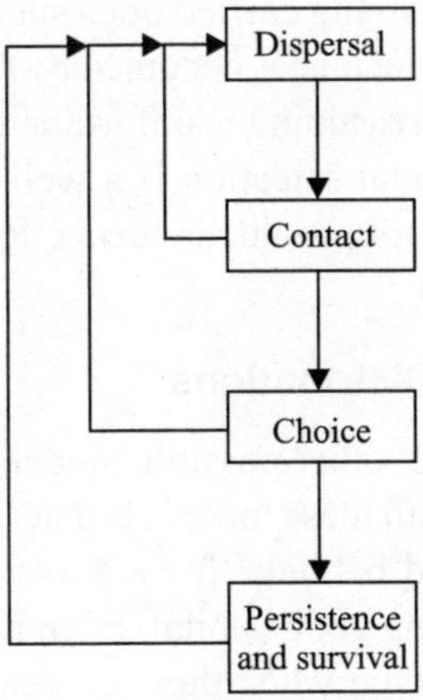

Fig. 14.3. Division of the process of habitat colonisation into the separate steps of dispersal, initial contact, choice of specific habitat location and post-choice persistence and survival. The arrows show the sequences in which events can occur. Various factors influence the success of individuals at finding and settling into habitat at each of these steps and these factors are likely to vary among taxa and types of ecosystem. Individuals of some species (e.g., sessile taxa such as plants) may disperse and colonise new habitats only once during their lifetime, whereas other, highly mobile species may disperse and colonise habitats multiple times.

Why is it useful to break up the process of how plants and animals find themselves in certain habitats into its component steps? Well, it allows us to see firstly that there is a wide array of possible factors influencing where an organism lives and secondly that these factors can be viewed as a series of explanations of increasing complexity. Hence, a species may be limited to occupying particular habitats or micro-habitats simply because individuals are unable to disperse to some suitable locations. Alternatively, it may be limited to those places not because it has any difficulty in dispersing but because a superior competitor has already occupied some suitable places. The latter model is more complex because it requires knowledge about the effects of factors in all the steps preceding choice.

A Simple Model of Habitat Selection

One of the most straightforward models we can propose to explain species distributions is that of simple habitat selection. In this model, an individual is able to disperse to and locate, all suitable locations in the area and it prefers to inhabit particular places over others. No other species prevent it from occupying its preferred choices. If this model is correct, we should be able to predict where we will find this species in greatest abundance simply by knowing what kinds of habitat preferences individuals make. A simple habitat selection model thus focuses exclusively on the step where organisms choose and assumes that factors affecting dispersal, location or post-choice persistence and survival are all unimportant.

Researchers with simple survey data commonly invoke habitat selection to explain the distribution of organisms in nature. However, that some researchers fall into a trap of making circular arguments. In these instances, researchers collect survey data to document the habitat use of species and conclude that 'preferences' determine where organisms are found. This sort of reasoning springs from an assumption that does not usually appear anywhere in their paper: that natural selection has produced behaviour that leads organisms to choose only places that provide optimal survival and reproductive success (i.e., behaviour is 'adaptive'). 'Adaptive preferences' are thus presumed to determine habitat selection and hence distribution. However, if all we have are survey data, the above 'conclusion' is nothing more than a restatement of the model. That is, having carried out a survey and shown that some species exhibit specialised habitat use, we conclude that the places occupied by these organisms must be optimal because that is where they are found. This sort of reasoning is unfortunately common. The sheer number of these studies can give the impression that habitat selection is a well-supported explanation for distribution. Unfortunately, many of these papers do not provide evidence for or against any models of distribution, including that of simple habitat selection.

Other Models that Explain Species Distributions

A perusal of Fig. 14.3 illustrates the many other possible models we can consider besides that of simple habitat selection. The common feature of all these models is that some factor or factors prevent individuals from realising all possible resources and habitats that are available. For example, organisms may be unable to disperse to some areas containing good habitat. Even if they are able to reach an area, they may not be able to find some suitable spots even when they are nearby. An organism's own behaviour can lead it to exclude some places that are suitable—possibly because that behaviour evolved under different environmental conditions. Another suite of factors preventing organisms from occupying particular locations are interactions with other species—competitors for the same food for example or predators that might be particularly effective hunters in some parts of the environment.

Judging from experiments carried out so far, the likeliest outcome is that distributions are determined by a combination of factors, not just one. The challenge now is to discover whether there are any generalities—that is, can we make predictions about the kinds of species or kinds of ecosystems where particular limiting factors will be important? This is an exciting question. We now turn to some examples that will show exactly how these different sorts of factors can work, using common stream organisms for illustration.

BLACKFLY LARVAE IN STREAMS

We shall look at a series of studies on some insects called blackflies. Blackflies are only a few millimeters long and larvae attach to hard substrata temporarily, which makes them amenable to experiments where we can manipulate and separate the effects of different factors affecting their distribution.

Blackflies (Order Diptera—the true flies) are common components of stream fauna around the world. Adult blackflies are small flies and the females of many species bite and draw blood from mammals and other vertebrates. They are sometimes serious pests of both human beings and livestock and carry parasitic organisms that cause serious illnesses (e.g., river blindness caused by the parasitic nematode *Onchocerca volvulus*) in some parts of the world. Adult females lay their eggs in flowing water and the larvae make their way to riffles (places with fast-flowing, turbulent water flow). The larvae (Fig. 14.4) have a pair of fans which they use to filter particles as small as bacteria out of the water column. Periodically, each larvae will flick each fan down and wipe particles off into its mouth. The larvae require hard substrata to attach to in order to filter feed and to attach to the substratum they spin a small pad of silk that sticks to hard surfaces. The larvae have a circlet of hooks at the end of the abdomen and they slip these hooks into the silk pad. This allows them to stand upright and face into the current. Not surprisingly, blackfly larvae require at least some water current to be able to filter out passing particles and we might hypothesise that the faster the water, the more particles are likely to pass through their fans. This then is the micro-habitat sought out by blackfly larvae: hard substrata like rocks and logs, a relatively smooth surface for attachment and fast-flowing water that delivers small particles of food.

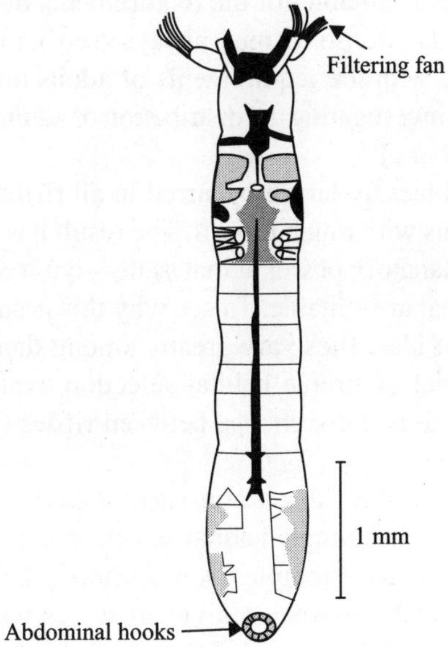

Fig. 14.4. Drawing of a blackfly (Simulidae). The fans on the head are used to filter fine particles from the water column. The abdominal hooks are used to anchor the larvae to the substrate in fast-flowing water.

Larvae can use either of two dispersal methods. First, they can use the stream current to drift downstream. They sometimes trail a thread of silk out behind them while drifting, which snags onto hard objects and allows larvae to pull themselves down to the 'rock or log' surface. Second, larvae can move around the surface of individual substrata by looping, which is where they alternatively attach the mouth and abdomen to move over the surface (a bit like some caterpillars). Looping allows larvae to move small distances (millimeters to centimeters), which might allow them to move to nearby spots with faster current.

BLACKFLIES AND SIMPLE HABITAT SELECTION

So, can we predict accurately the distribution of black fly larvae in streams? Let us start with a model of simple habitat selection. Our knowledge of their habitat requirements would lead us to predict that if blackfly distribution is determined solely by habitat preferences, we will find blackflies in all areas of the streambed where there are hard substrata and the water is relatively fast flowing. Because faster water usually means more food per unit time, we would also predict that larvae will be more abundant in fast-flowing areas than in moderate or slower flows. (Note that it is possible to get slow water in riffles because of the zones of greatly reduced velocity created by flow around large objects like boulders.)

Are these predictions successful? The answer is, like many in ecology, yes—to a point. We usually find no (or very few) blackfly larvae in slow-moving areas of rivers like pools, backwater or dead water zones and often many larvae in riffles. That is consistent with our expectation that blackfly larvae choose places with water that is, overall, flowing reasonably fast. Nevertheless, some rivers and streams may have no blackfly larvae at all, even when the habitat appears very suitable. Part of the reason may lie simply in the evolutionary history of these species—perhaps blackflies never reached that particular stream. Species with complex life-histories can have stages living in quite disparate environments. The riparian habitat around rivers may be unsuitable for the requirements of adults, which means there will be no females to lay eggs and produce larvae. So we must always keep in mind that some of the explanation for the distribution of larvae may lie with the requirements of adults and, in particular, the egg-laying needs of females. (Likewise, if we are investigating the distribution of adults, we might find their occurrence largely set by the requirements of larvae.)

However, even if we found that blackfly larvae occurred in all riffles but not pools and dead water zones, that finding does not provide us with much insight. The result has confirmed a habitat type that is unsuitable—so it provides some explanatory power at that scale—but it does not necessarily tell us what sets limits to abundances in places that are suitable. To see why this is so, suppose we sample a number of riffles and find that abundances of blackflies vary greatly among them or among fast-flowing spots within the same riffle. Can our model of simple habitat selection explain these differences? If so, it would mean that blackflies are able to and do, choose between riffles (or specific stones) differing in average water velocity.

We will first provide an example of a study that is consistent with this explanation, then a series of far more comprehensive studies showing that simple habitat selection is unlikely to be a useful model for explaining blackfly distribution. In our first example, Downes and Lake found that abundance of some Australian blackfly species varied greatly between individual stones within one riffle. They measured the average current speed over stones and then sampled them and counted the number of blackflies that were present. They found that numbers of blackflies were associated with the velocity of the water passing over those stones. Stones that were in relatively fast water had more blackflies than those that were in slow water (Fig. 14.5).

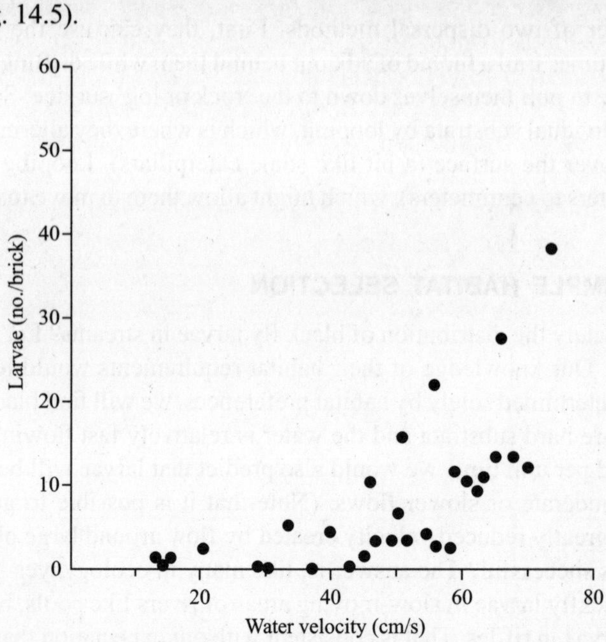

Fig. 14.5. The number of blackflies, *Austrosimulium victoriae*, occurring on substrata (bricks) having different water velocities.

So, the explanation for why blackflies are not found in pools (slow water velocities) helped explain within-riffle variability in this study. Possibly, a simple model of habitat selection (that blackflies can and do choose stones occupying relatively fast water velocities) explains the distribution of these species. We say 'possibly' because the consistency of observations with this hypothesis is not strong evidence for it. As explained above, we need to collect further data that demonstrate that blackflies can and do have such preferences before we can conclude that simple habitat selection is a useful model.

We now turn to a series of studies that demonstrates very neatly how far more complicated nature is than this.

BLACKFLY LARVAE AND OTHER MODELS OF DISTRIBUTION

Hart studied the distribution of larvae of the blackfly *Simulium vittatum* in Pennsylvania in the USA. Feeding rates of blackflies depend on the speed of water flowing through their fans. This knowledge led them to expect that, under a model of simple habitat selection, *S. vittatum* should be most abundant on stones within riffles that had fast flows. However, in contrast to the study of Lake and Downes, numbers of blackfly larvae on stones were not associated well with water velocity over those stones. Numbers of black flies were associated with flows when the latter were measured at much smaller spatial scales—at the scale of a millimeter, but this did not explain why stones in overall fast flows did not have more larvae than stones in slow flows.

Could this lack of association be a consequence of blackflies being prevented from reaching some locations during dispersal and that behaviour during dispersal changes the sorts of initial locations they are able to contact? In other words, can blackflies control their dispersal in the current and choose locations where to stop? These questions were addressed for *S. vittatum* by Fonseca, who observed the behaviour of blackfly larvae in a specially constructed flume (flow tank) where relatively laminar flows could be generated. (Laminar flows have fairly predictable velocities at different heights above the bed and hence are more tractable to work with than turbulent flows, which can be chaotic and virtually unpredictable). In the first experiment, Fonseca measured the rate at which both dead and living larvae sank in a simple glass tube, to see whether they can control their dispersal up and down through the water column. The sinking rates of dead and living larvae were the same, suggesting that blackfly larvae cannot control their positions in the water column, a behaviour that can greatly alter dispersal distances in streams. If larvae can maintain a high position in the water column, they can disperse further because velocities are generally higher at greater distances above the bed even in turbulent flows. The fact that larvae sink without being able to do much about it means that they can be modelled essentially as dispersing passive particles. We should be able to use relatively simple mathematical equations that can make predictions about how far larvae will travel when dispersing from different heights above the bed.

This is what Fonseca did in the next experiments using the flume, where she released larvae at known heights in the water column and measured how far they travelled in different rates of flow before they were able to attach themselves to the substrate.

The simple equations were largely successful at predicting these distances, except for the highest flow (Fig. 14.6). In high flows, larvae were unable to attach at all before being swept to the end of the flume, even when their silk line snagged on the flume's surface. These results suggested that larvae may be unable to attach to optimal sites in the field (remembering that fast) water flow equates with potentially greater food supply) simply because the water flow is too fast.

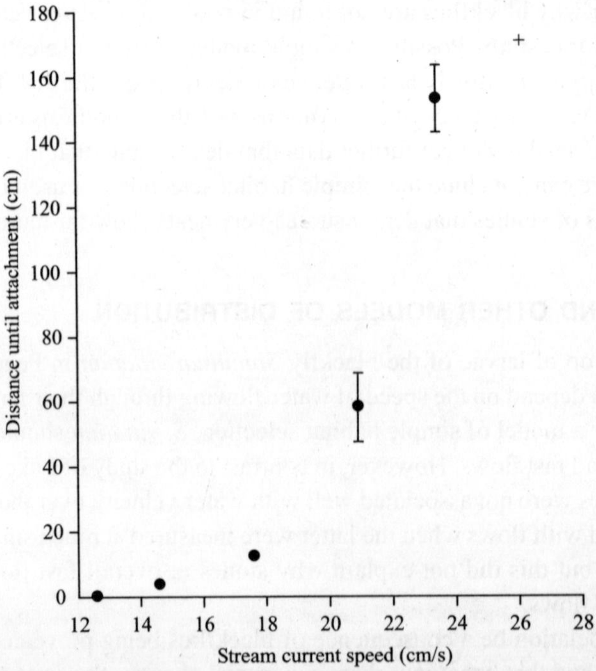

Fig. 14.6. Relationship between distance travelled in a flume by *S. vittatum* larvae until attachment (cm) and stream current speed. Distances travelled were well predicted by mathematical equations of the dispersal of passive particles in different water velocities. The distances travelled after releases at 27 cm/s (cross) were 170 cm and were not included in the analysis because no larvae were able to attach to the substrate. The error bars represent ±1 standard error and there were 20 larvae used for each current speed.

Fonseca and Hart confirmed this using a field experiment. They placed artificial substrata throughout a riffle of Taylor Run. Larvae colonised these substrata and consistent with previous observations, some substrata gained far more larvae than others, despite water flows over the substrata being similar (that is, numbers of larvae on substrata were not well explained just by water velocity). After noting which substrata obtained high numbers of larvae, Fonseca and Hart removed all blackflies from all substrata and then reattached the same number of larvae to all of them. They then observed what happened to the numbers of larvae. Their reasoning was this: if substrata that originally had low numbers of larvae were low because they offered poorer habitat quality, then larvae attached to these substrata should leave at higher rates than larvae attached to substrata that previously had higher numbers of blackflies. Thus, if the initial abundance patterns reflected habitat preferences, then the numbers of larvae should readjust to return them back to the high and low numbers originally recorded for those spots. Alternatively, if the numbers present on substrata were a reflection of differences in initial settlement rates—driven by different exposures to numbers of settling larvae for example—and unrelated to habitat quality, then emigration rates should be comparable between these two sorts of substrata. Fonseca and Hart also monitored substrata that were cleared of all larvae at the start of the experiment, which allowed them to correct for new larvae settling onto their substrata during the experiment.

Fonseca and Hart's results showed that the emigration rates of larvae from substrata that previously had many larvae were equivalent to those that had few. This suggested that the reason that numbers differed was not because of habitat quality but because the rates of dispersal and settlement originally differed, as the flume experiments had suggested. Larvae of *S. vittatum* are an example of a species where constraints during dispersal and contact with the substrate restrict the number of places larvae are able to occupy. This is because their method of dispersal over long distances—using the drift—allows them little control over where they can land. They can only exhibit choice by leaving substrates —not by choosing the best ones in the first instance.

Do post-choice factors also affect the distribution of larval blackflies? Hart and Merz examined the effects that a predatory flatworm has on the distribution of larval *S. vittatum*. The flatworms live on the bottom of substrata or within tufts of algae, but venture out onto top surfaces to feed. In a small site on Chester Creek, blackfly larvae are their major prey. Do flatworms modify the post-attachment distribution of blackfly larvae through predation? In Chester Creek, blackfly larvae are most abundant in places where the water velocity is higher. However, flatworms are less abundant in high-velocity spots. To what degree is blackfly distribution determined by larval preferences for high water velocities compared with the effects of predation? Hart and Merz carried out an experiment using tiles, which were readily colonised by both blackflies and flatworms. They removed the flatworms from half of the tiles by dislodging them. Then they reduced the water velocity over half of these tiles and half of the undisturbed tiles by placing small mesh fences upstream of them. This experiment allowed them to examine the separate effects of predation and water flow upon blackflies.

They found that blackflies had higher emigration rates from substrates where the water velocity was reduced and that abundances were also reduced where flatworms were present (Fig. 14.7). The latter only ate about half the blackflies—the rest fled into the drift to escape.

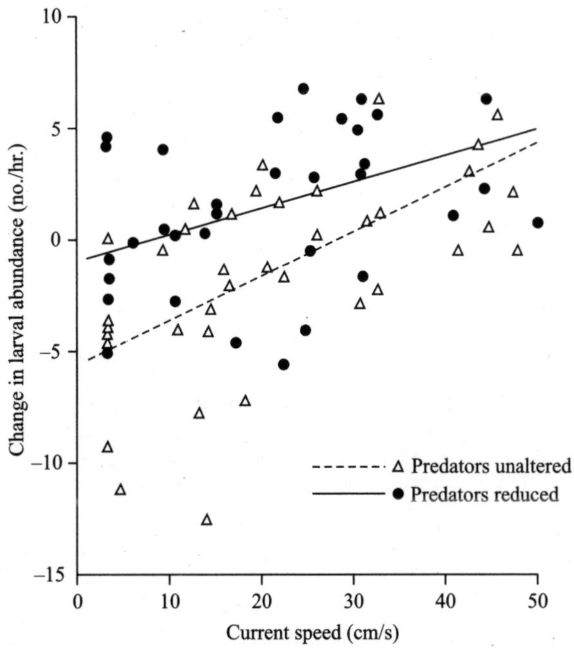

Fig. 14.7. Relationship between the rate of change in larval blackfly abundance and current speed in the reduced-predator (solid line) and unaltered predator (dashed line) treatments. The response variable was the change in larval abundance per hour. Statistical analysis showed that both predator abundance and current speed were significant predictors of changes in larval abundances.

The experiment also suggested that flatworms were less effective as predators in high-velocity treatments, because they were rarely seen on the top surfaces of tiles in fast-flowing water. Hart concluded that the negative effect of water velocity on flatworms creates a refuge from predation for blackflies.

DETERMINING NICHES

Charles Elton pointed out that the word niche should be used to indicate what an animal was actually doing in its community. A complete account of the niche of a badger would therefore include some sort of a description of what badgers eat, what, if anything, eats them, what sort of a physical environment they need in terms of temperature, soil type and so on.

Providing a quantitative description of a niche is difficult. Figure 14.8 shows an attempt to represent on a two-dimensional graph the feeding niche of the bird *Polioptila caerulea*, the blue-grey gnatcatcher. Along the horizontal axis is drawn the length of its prey. The vertical axis shows the height above ground at which blue-grey gnatcatchers feed. Also shown on the figure are contour lines showing the frequency with which the birds fed at a particular height and on a particular length of prey. We can see that the birds concentrated on prey about 4 mm in length which they caught about 10 to 15 feet off the ground.

An attempt to represent the data of Fig. 14.8 as a three-dimensional volume is given in Fig. 14.9. Here the feeding niche of the birds is represented as a hill. The species can only be found on the surface of or inside the hill; the peak of the hill shows where the birds are most likely to be found. Hutchinson proposed that the environmental variables that are relevant to a species could be conceived of as a set of independent axes. In Figs. 14.8 and 14.9 only two axes are plotted. In principle, other axes could be added. Ambient temperature might be a third, risk from predation at different times of the year a fourth and so on. It becomes almost impossible graphically to represent niches when more than two environmental variables are involved, but computers can easily hold and analyse the data. Hutchinson suggested that the niche of a species could be defined as an '*n*'-dimensional hypervolume, where the number of different environmental variables equals n – 1.

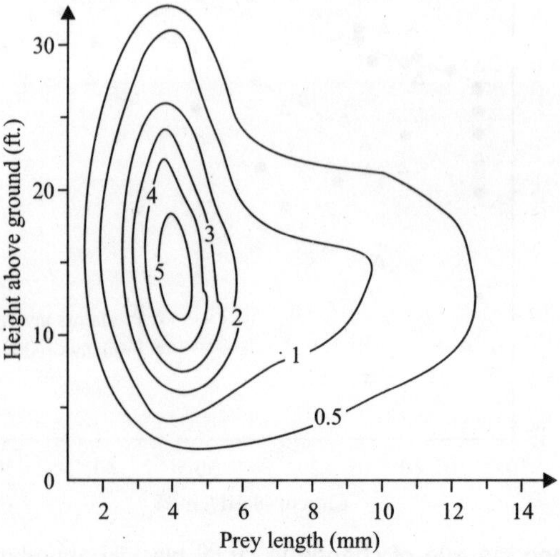

Fig. 14.8. A two-dimensional representation of the feeding niche of the blue-grey gnatcatcher (*Polioptila caerulea*), the curved contours show the feeding frequencies (equal to percentage of total diet) for adult gnatcatchers during the incubation period in July and August in oak woodlands in California.

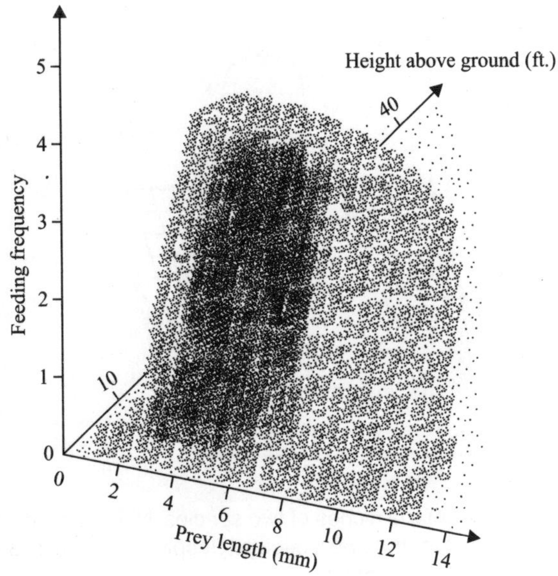

Fig. 14.9. The feeding niche of the blue-grey gnatcatcher plotted as a three-dimensional hill. (Data as for Fig. 14.8.)

Each Species has its Own Unique Niche

Ecology has few laws. One of them is that each species has its own unique niche. Suppose that two species occupied the same niche. It is almost inconceivable that they could occupy exactly the same niche and coexist. One of them, so the theory goes, would be bound to out-compete the other, driving it to extinction.

Two species may occupy separate *n*-dimensional hypervolumes, yet overlap considerably on each of the separate axes representing the environmental variables. When these axes show environmental variables such as prey length or protein concentration of food, they are often referred to as resource axes. Figure 14.10a shows how two species may overlap considerably on each of two resource axes, yet occupy almost distinct three dimensional niche volumes. Figure 14.10b shows, in a different way, how the data for several species can be represented on one graph.

GAUSE'S COMPETITIVE EXCLUSION PRINCIPLE

The notion that no two species can coexist if they occupy the same niche is now known as Gause's competitive exclusion principle, even though earlier workers such as Grinnell and even Darwin had expressed essentially this idea.

Gause first grew the protoctists *Paramecium caudatum* and *P. aurelia* in separate cultures which contained bacteria on which the *Paramecium* fed. The results displayed in Figs 14.11a and 14.11b show that each species grew in numbers according to the logistic equation. It is clear from Fig. 14.11 that *P. aurelia* grows in numbers more quickly than *P. caudatum* and ends up with more individuals in the same volume of culture medium. In other words, the carrying capacity of the culture medium is greater for *P. aurelia* than it is for *P. caudatum* and *P. aurelia* has the greater rate of natural increase of the two species. These differences make sense as *P. caudatum* is larger than *P. aurelia*.

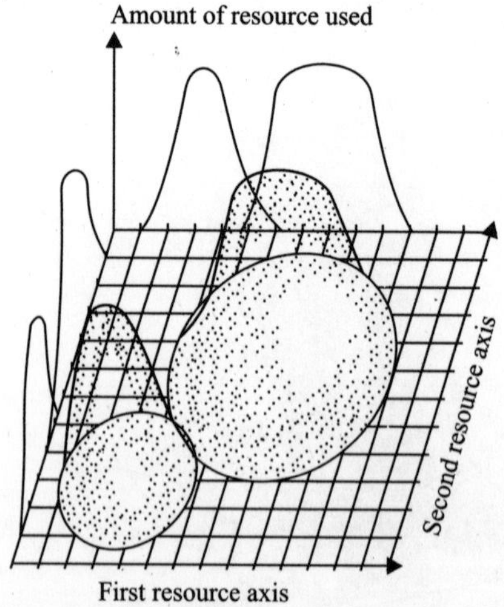

Fig. 14.10a. Diagrammatic representation of the niches of two species. Note how on each of the two resource axes there is considerable overlap between the two species. Despite this, the two species occupy three-dimensional niche volumes which overlap only slightly.

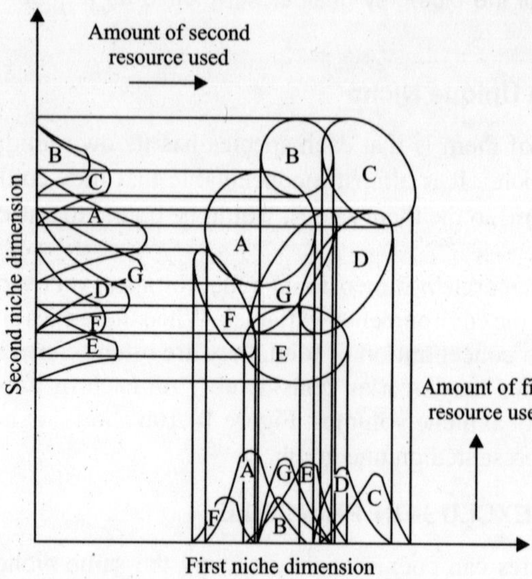

Fig. 14.10b. Diagrammatic representation of the niches of several species, showing how in two dimensions niches may overlap only slightly, despite considerable overlap along each of the one-dimensional resource axes.

Gause then grew the two species together in the same small 5 cm^3 culture volume. A new culture medium was inoculated with a mixture of both species and the numbers of both species were counted each day. The average results from three experiments are plotted in Fig. 14.12. Each of the three experiments followed the same course. Initially both species grew in numbers, but eventually *P. cauda* turn declined and became extinct. Gause attributed the inevitable extinctions to the existence of 'but a single niche in the conditions of the experiment'.

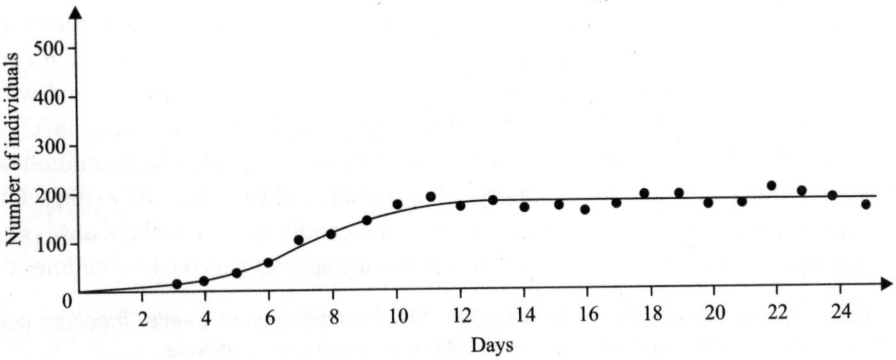

Fig. 14.11a. The growth in the population size of *Paramecium caudatum* in a laboratory culture.

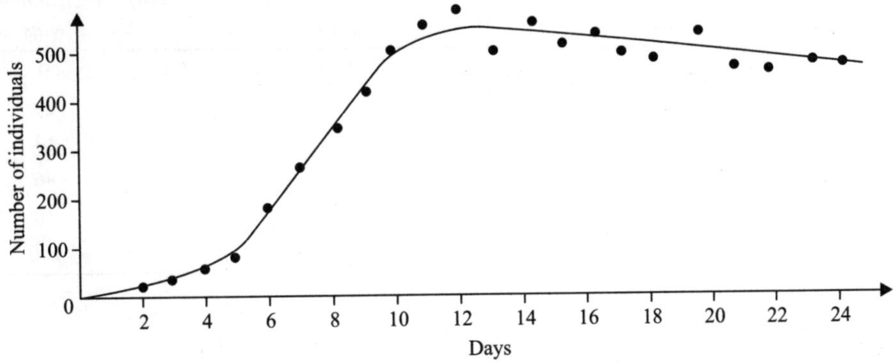

Fig. 14.11b. The growth in the population size of *Paramecium aurelia* maintained on its own in a laboratory culture.

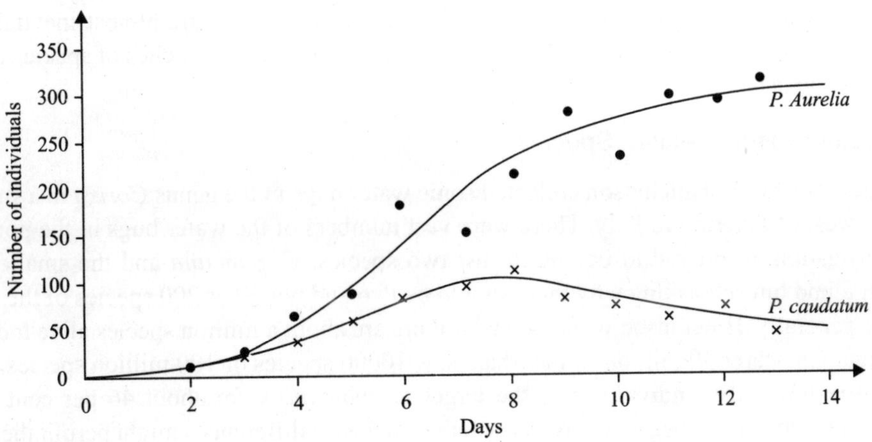

Fig. 14.12. The changes in the numbers of *Paramecium caudatum* and *P. aurelia* when grown together in a laboratory culture. At first, both species grow in numbers. Eventually, however, *P. caudatum* declines in numbers while *P. aurelia* continues to increase its population size.

In Gause's experiments, *P. aurelia* always won the competition between the two species. Park carried out a long series of laboratory experiments on two species of flour beetle, *Tribolium confusum* and *T. castaneum* and found that in mixed cultures which of the two species survived and which became extinct depended on the environmental conditions. When conditions were hot (34°C) and humid (70 per cent relative humidity), *T. confusum* always went extinct. However, when conditions were cooler (24°C) and drier (30 per cent relative humidity), *T. castaneum* always became extinct. Of additional interest was the observation that under some environmental conditions, it could not always be predicted which species would go extinct (Table 14.1). Sometimes one species survived, sometimes the other.

Table 14.1. The outcome of competition between the two species of flour beetle *Tribolium confusum* and *T. castaneum* averaged over many replicates under different laboratory conditions.

Temperature	*Relative humidity*	*% of times T. confusum wins*	*% of times T. castaneum wins*
24°C	30%	100	0
24 °C	70%	71	29
29°C	30%	87	13
29 °C	70%	14	86
34°C	30%	90	10
34 °C	70%	0	100

SPECIES COEXISTENCE

Subsequent to the work of Gause, Park and MacArthur, a great many studies of species coexistence were conducted in the field. It soon became apparent that however closely related a group of organisms was, niche differences could be found. But this does not provide a test of Gause's competitive exclusion principle. By the very fact that two species are different species, they are almost inevitably going to occupy different niches. The important question is 'How different do the niches of species have to be to allow species coexistence?'

Size Ratios in Closely Related Species

American zoologist G. E. Hutchinson collected some water bugs in the genus *Corixa* in a small pond on a hill to the west of Palermo in Italy. There were vast numbers of the water bugs in the pond which on closer investigation all proved to belong to just two species, *C. punctata* and the smaller *C. affinis*. Hutchinson found himself asking why there should be two and not 20 or 200 species of the genus in the pond. More generally, Hutchinson wondered why there are about a million species alive today (we now think the figure is nearer 30 million) rather than, say, 10000 species or 100 million species.

Hutchinson found that individuals of the larger *C. punctata* were about 46 per cent longer than individuals of *C. affinis*. He began to wonder whether such size differences might permit the two species to coexist by enabling them to feed on different prey. Hutchinson extended his studies to birds and mammals and from a few measurements found that when pairs of closely related species were examined which fed at the same level of a food web, linear measurements of skeletal characters relevant to feeding (such as skull length in mammals) differed by an average of 28 per cent. For instance, in Britain, skull lengths of the stoat (*Mustela erminea*) are on average 31 per cent longer than those of the weasel (*M. nivalis*). As closely related species are approximately the same shape, a difference of 28 per cent in linear measurements (such as skull length) between two species means that the larger species is approximately $(1.28)^3 = 2.1$ times heavier than the lighter. Subsequent to Hutchinson's paper, a great many ecologists rushed out into the field or went to museums and weighed closely related species found in the same area. Often it turned out that they differed in mass by a factor of about two. Hutchinson's tentative suggestion became elevated to the level of an immutable law. An example of the sort of data used to support Hutchinson's suggestion is given in Fig. 14.13.

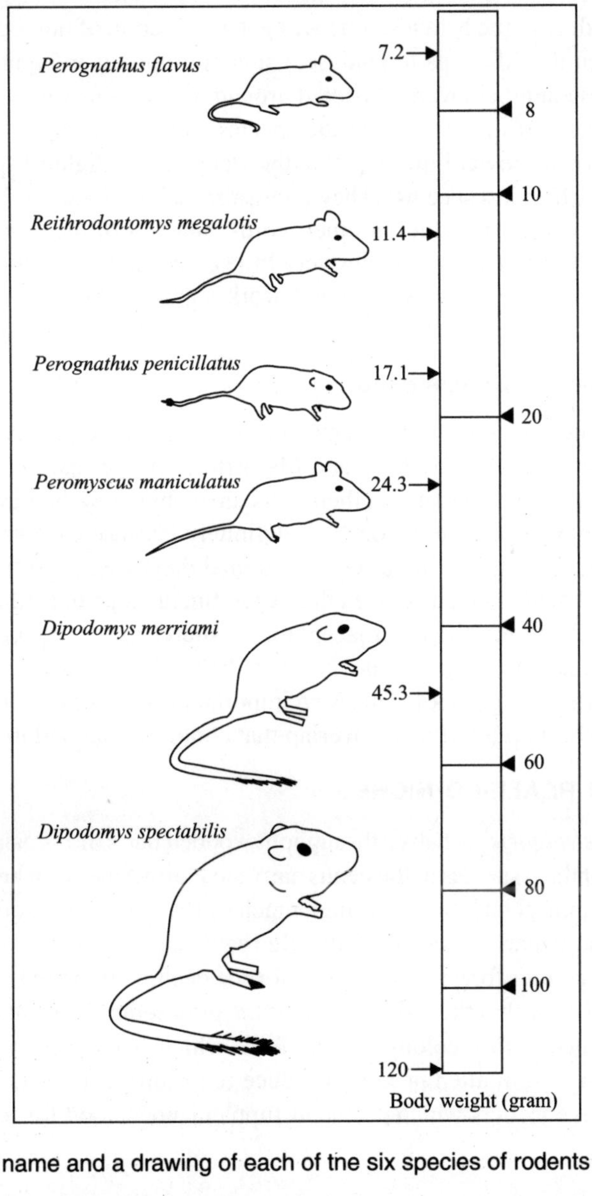

Fig. 14.13. The mass, name and a drawing of each of the six species of rodents in the Sonoran Desert.

Brown concluded that 'communities of coexisting species have distributions of body sizes with remarkably regular spacing' and at first sight the data in Fig. 14.13 appear to support this assertion. However, if one calculates the ratios of the body weights of these six rodents arranged in pairs, starting with the lightest and finishing with the heaviest, one obtains the numbers 1.58, 1.50, 1.42, 1.86 and 2.65. At the very least, these figures show considerable scatter about the 'predicted' value of 2.1. Furthermore, the value of 2.1 is itself, of course, not a prediction. It simply follows from the value of 1.28 that Hutchinson measured.

Simberloff and Boecklen pointed out that, inevitably, if one arranges a number of closely related species in order of body size and then calculates the ratios of the weights of neighbouring species starting at the lightest and ascending to the heaviest, one will get a collection of numbers greater than one. What Simberloff and Boecklen then did was to analyse, using appropriate and quite sophisticated statistical techniques, whether these numbers were clustered around a particular value or whether they varied as much as would be expected if the weights of the species were independent of each other and merely arranged at random between a lower limit (equal to the weight of the lightest species) and an upper limit (equal to the weight of the heaviest species). They found almost no evidence whatsoever that the numbers were clustered around a particular value. In other words, analysis of the size ratios of pairs of closely related species found in the same area provides very little evidence to suggest that their sizes have been affected by interspecific competition. Subsequent work supports the conclusions of Simberloff and Boecklen.

Niche Overlap and Species Coexistence

A different approach to the question of how much niche overlap can be tolerated between species before one of them goes extinct was taken by May and MacArthur. If one makes, for example, the extreme assumption that animal species never vary in their diets, then obviously, however small the difference in their feeding niches, they will be able to coexist indefinitely because each species has access to foods which the other will not take. In real life, of course, animal diets are not so fixed. May and MacArthur produced a mathematical model which showed that when fluctuations in feeding niches were taken into account, the average food size for species adjacent on a resource axis to coexist indefinitely must differ by an amount roughly equal to the standard deviation in the food size taken by either species. This result is shown diagrammatically in Fig. 14.14. This is an important result because for the first time it allowed a genuine prediction of the amount of niche overlap that could be tolerated in nature.

FUNDAMENTAL AND REALISED NICHES

The red squirrel (*Sciurus vulgaris*) is native throughout wooded parts of Eurasia. The larger grey squirrel (*S. carolinensis*) is one of three species in the genus native to North America where it is found in hardwood forests. While the grey squirrel has spread through much of Britain, the red squirrel has at the same time had its range reduced, particularly since the 1940s. Red squirrels are nowadays most abundant in Britain in large areas of mature conifer forest. The reasons for the decline of the red squirrel are not known for certain, but it does seem clear that the introduction of the grey squirrel has, at the very least, resulted in the red squirrel being unable to recolonise areas from which it has become absent through habitat destruction or disease. A careful attempt to reintroduce red squirrels to Regent's Park, London, failed despite the use of selective feeders which provided supplementary food for red squirrels, but excluded grey squirrels.

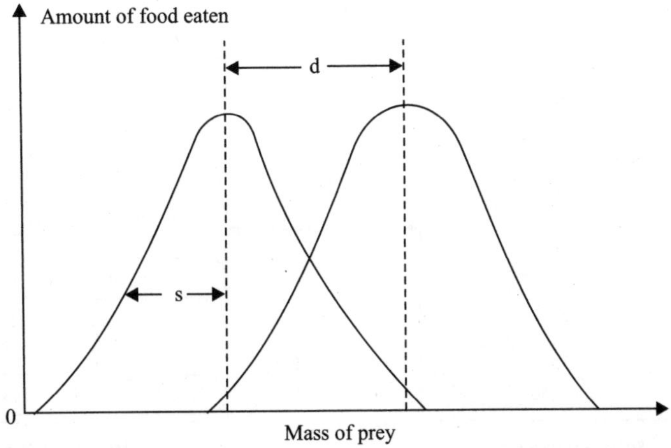

Fig. 14.14. Diagrammatic representations of the feeding niches of two species on a one-dimensional resource axis, such as prey mass. The prediction of May and MacArthur is that in order for the two species to coexist in the face of interspecific competition, the difference, d, between the means of the prey masses taken by each species must be at least as great as the standard deviation, s, of the prey mass taken by either species.

The red squirrel in Britain is still declining but it is unknown if it is in danger of becoming extinct. Grey squirrels predominate in areas of mature hardwood or mixed woodland, provided at least 25 per cent of the trees are hardwood. Red squirrels predominate in conifer woods, mixed woods dominated by conifers and places such as Brownsea Island and the Isle of Wight where the grey squirrel has never reached. Hutchinson introduced the terms 'fundamental niche' and 'realised niche'. He suggested that the niche an organism could occupy in the absence of competitors and predators be called its fundamental niche, while the role it actually played in the community should be called its realised niche. In real life, therefore, the realised niche of an organism is smaller than its fundamental niche. In the case of the two squirrels in Britain, the realised niche of the red squirrel has contracted with the establishment and spread of the grey squirrel.

MacArthur and Pianka considered what might happen when several competitors invaded a habitat. They argued that in the presence of a greater number of competing species, each species should concentrate its search efforts for food on a smaller part of the total available habitat—so-called ecological compression. On the other hand, within the reduced habitat searched by each species, a full range of food items may still be present. Consequently, MacArthur and Pianka argued, the range of food taken by a species should be independent of the number of competing species present. Similarly when interspecific competition is relaxed, species should increase the range of habitats in which they obtain food, but not the actual range of food taken, This idea is illustrated pictorially in Fig. 14.15.

Pianka found that in communities with several lizard species, the more lizard species there were, the less overlap there was between the feeding niches of the lizards. This held true for lizards in North America, Australia and Kalahari. Perhaps if evolution allows the coexistence of many species with very similar life styles, these species must have relatively narrow feeding niches. Another way of interpreting these results is to suppose that, as argued in the previous paragraph, the presence of more species of competitors leads to ecological compression. This ecological compression results in species specialising as to which habitats they occupy. Consequently, they might inevitably specialise more in terms of the food they eat. This increased specialisation might be enough to reduce niche overlap.

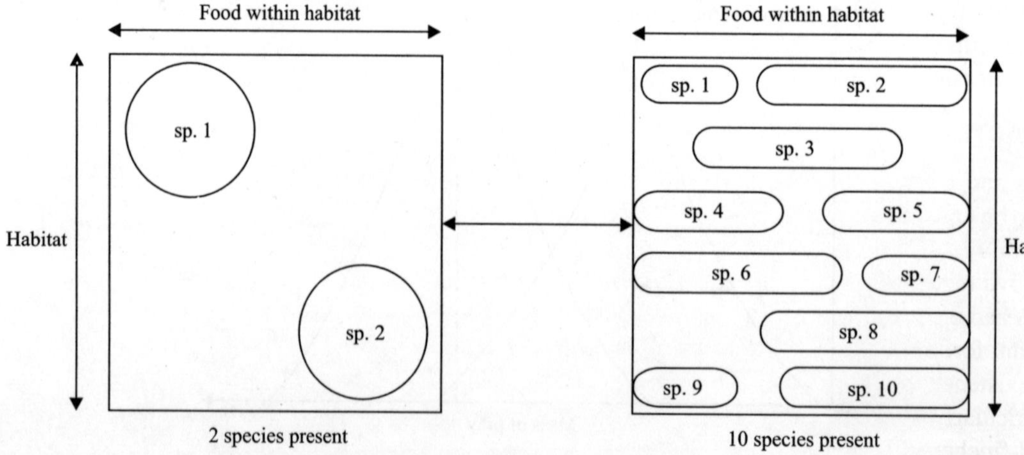

Fig. 14.15. The compression hypothesis. As more species make use of a habitat, the habitat used by each species shrinks. However, the range of food items taken by each species remains constant. The actual diet may become more concentrated, by the range of items should not change. The hypothesis only applies to short-term non-evolutionary changes.

RESOURCE PARTITIONING

Where two or more species divide out a resource such as food or nesting sites resource partitioning is said to occur. Here the partitioning is not that different plant species take different sorts of water—there is, after all, only one sort of water. Rather, the plants are dividing up access to the water. Some specialise on ephemeral sources of water, which may be trapped by surface roots; others rely on relatively permanent sources of deep water.

A classic instance of resource partitioning among animals was provided by Bell in his studies of grazing mammals on the Serengeti plain in Tanzania. Careful analyses of the stomach contents of most of the principle large herbivores on the Serengeti plain—zebra (*Equus burchelli*), wildebeest (*Connochaetes taurinus*), topi (*Damaliscus korrigum*) and Thomson's gazelle (*Gazella thomsoni*)—revealed considerable resource partitioning. Zebra were able to take the diet highest in cell wall content. This is the poorest diet and zebra were able to live on it for two reasons. First, they are the heaviest of the four species, with an average adult female having a body mass of 220 kg. The larger a species of animal, the poorer the quality of diet on which it can subsist because of its low metabolic rate, relative to its body size. The second reason why zebra can survive on low quality forage is that, unlike the other three species, they lack a rumen. The ruminant strategy is to obtain maximum nourishment by thorough digestion. The equid strategy, adopted by zebras and horses, is to pass food through the gut rapidly, getting what can easily be obtained from it, but not attempting to maximise digestibility. Zebras thus need more food for their size than do ruminants. However, this evolutionary strategy means that zebras can survive off the poorest quality herbage, unlike the other Serengeti herbivores.

The species that ate the highest quality diet was the smallest ruminant, the Thomson's gazelle. Adult female Thomson's gazelles weigh about 16 kg. Of the four herbivores. Thomson's gazelles were the only ones to take a large amount of dicotyledonous material: 40 per cent, chiefly fruit. This high-protein high-energy diet is necessary for the small ruminant with its high metabolic rate, relative to its size. Topi (110 kg) and wildebeest (160 kg) took food intermediate in quality between that taken by Thomson's gazelles and zebra.

Perhaps the most clear-cut instances of resource partitioning come from studies on host-specific parasites. Often, anyone parasite can parasitise only one or at most a few, host species. Furthermore, host species are rarely attacked by two closely related parasite species.

CHARACTER DISPLACEMENT

Charles and Darwin collected some dull-looking finches on the Galapagos Islands. They have since come to be known as 'Darwin's finches', for they were cited by him as a classic instance of the evolution of several species from a common ancestor. Just before the Second World War, David Lack was able to spend five months on the Galapagos Islands. He argued that the differences between the birds, which Darwin had observed, were adaptive. In particular, he maintained that interspecific competition was an important force structuring the Galapagos communities and that ecological isolation is as necessary for the coexistence of closely related species as is reproductive isolation.

Particularly interesting were the results of Lack's careful measurements on beak depth in the three ground-finches: the large ground-finch (*Geospiza magnirostris*), the medium ground-finch (*G. fortis*) and the small ground-finch (*G. fuliginosa*). Over evolutionary time, natural selection has caused the beaks of the two species to diverge as the birds specialised on different foods. Character displacement is distinct from ecological compression. Ecological compression takes place on a much shorter time-scale and does not involve heritable change.

INTERSPECIFIC COMPETITION IN NATURAL COMMUNITIES

Over the last 15–20 years ecologists have come to change their views on the importance of competition between different species (interspecific competition) as a force shaping natural communities and driving evolutionary change. After all, natural selection is fundamentally about competition between the members of a species (intraspecific competition). Botanists, in particular, have long been aware that intraspecific competition is typically stronger than interspecific competition. This does not mean, however, that competition between species never occurs. As so often is the case in ecology, what is at issue is the relative importance of an effect. Competition between species for shared resources undoubtedly occurs, but this does not necessarily mean that it is very important. Furthermore, it is perfectly possible that interspecific competition is more important in some communities than in others.

An early review of the evidence for and against the theory that interspecific competition is important in natural communities was provided by Connell. Connell pointed out that three general methods have been used to detect interspecific competition under natural conditions. The first is to look and see whether nature conforms to the predictions that follow from this theory. MacArthur's observations on the *Dendroica* warblers fit into this category. He showed that the feeding and nesting habits of the warblers were consistent with the hypothesis that interspecific competition was important in the structuring of that community.

The second approach described by Connell is to look for 'natural experiments'. For instance, the decline in red squirrel numbers in Britain as grey squirrel numbers increased is consistent with the theory's predictions.

The third approach and the one which Connell favours, is to conduct controlled experiments in the field. This alone, Connell argues, allows one to ensure that everything varies in the same way between treatment and control except for the factor being tested. Now, insistence on controlled field experiments is a very strict requirement. After all, we do not have controlled field experiments to test the hypothesis that hunting by humans has led to a reduction in the numbers of African elephants, but no one seriously doubts this. Nevertheless, Connell's insistence on careful experimentation is helpful. It might, for example, be the case that red squirrel numbers have declined in Britain due to the effects of disease rather than competition by grey squirrels.

It is perhaps unsurprising that the most convincing evidence, using Connell's criteria, for the importance of interspecific competition comes from plants. This is because experiments can be conducted on plants more easily than on animals. For instance, many early experiments showed the importance of competition between the roots of canopy trees and those in the understorey in temperate woods. Trenches were dug around plots containing understorey trees, cutting the roots of neighbouring adult trees and so reducing competition for water and nutrients. In every case the smaller trees inside the trenched plots survived and grew much better than those in nearby control plots. It should be mentioned that many of these studies did not distinguish between intraspecific and interspecific competition.

From Connell's review it looks as though many species seldom reach the population densities necessary for interspecific competition to be important. It seems that their numbers are kept down by intraspecific factors, by physical conditions or by interspecific factors, such as predation and parasitism, that are not normally included within the definition of interspecific competition.

It has been concluded that intraspecific competition is not of great importance in structuring natural communities. 'As a future trend we can expect a further depreciation of competition, both intra and interspecific, as being a major force in ecology and evolution. The manifold influences of weather, climate and other physical factors will assume greater importance and the claim of universality of competition will probably be replaced by a gradual re-evaluation of the role of predation'.

Some ecologists feel that den Boer has gone too far but there have certainly been some ecologists who have changed their minds and now feel that interspecific competition is less important than they previously thought. For instance, Sinclair had extended the work of Bell on grazing in the Serengeti and had concluded that interspecific competition was the dominant structuring process in the Serengeti community, producing the observed separation and coexistence of herbivores. However, Sinclair, as a result of 'natural experiments' including variations in the numbers of wildebeest, concluded that predation appeared to play as much of a role as interspecific competition in structuring the community. In particular, most of the larger herbivores are found nearer to the wildebeest than expected from a random distribution. This is contrary to the 'interspecific competition hypothesis', but agrees with the 'risk of predation hypothesis'.

One example where interspecific feeding competition is still believed to be important in community structure is provided by Gilbert. Gilbert and his co-workers studied the ten most common hoverflies in an ancient woodland near Cambridge. They found that the ratios of the proboscis lengths of these species were significantly more constant than expected by chance. This agrees with Hutchinson's suggestion that closely related species should reduce interspecific feeding competition by having dissimilar feeding apparatuses. There is also strong evidence for character displacement in hyaenas and other carnivores as a result of interspecific competition.

As ecologists have demanded better evidence for the importance of interspecific competition, so a number of the cases of apparent character displacement have been shown to be due to other causes. However, it still looks as though Lack's example of character displacement in the ground-finches *Geospiza fortis* and *G. fuliginosa* is valid. It could be argued, as an alternative hypothesis to that of character displacement, that the intermediate beak depths seen on the islands where only one of the ground-finches is found.

DO PLANTS NEED NICHES?

Temperate chalk grasslands, like the one in Fig. 14.16, may contain as many as 40 species of vascular plants in each square meter. Over 150 tree species can be found in a single hectare of some tropical forests. Yet most plants require much the same basic resources: light, water, carbon dioxide, oxygen, nitrate, phosphate, a dozen or so minerals in solution and space. Can it really be that different plant species have different niches? Silvertown reviewed advances in plant community ecology and questioned whether Gause's competitive exclusion principle held for plants.

Fig. 14.16. A species-rich temperate short grassland on a shallow chalky soil.

Rather surprisingly, Silvertown and Law argue that competition between species may be less important in plants than in animals. They maintain that terrestrial plants are often surrounded by other members of the same species and that this means that competition between different plant species is unimportant. This argument is surprising because the problem with traditional niche theory in plants is that each plant seems to be competing with individuals from many other species, as, for example, in chalk grasslands and tropical forests.

A more convincing argument for the maintenance of high species richness in many plant communities is that occasional episodes of density-independent mortality can delay and greatly reduce the effects of interspecific competition. These non-equilibrium theories date back at least as far as Andrewartha and Birch who argued that many insect species were affected far more by such occasional density-independent disasters than by density-dependent mortality. The crucial point about some of these non-equilibrium models is that calculations suggest that the average time required for interspecific competition to eliminate species may be longer than the average time between speciation events. In other words, by the time one species has become extinct through interspecific competition, another one or two may have evolved. The larger the community, the more important this effect becomes. Tropical rainforests are places where such non-equilibrium coexistence may be particularly important.

Predators and parasites may also stop a single species from increasing in numbers to a level where it can out-compete other species. Janzen has argued that in the tropics, whenever a single plant species begins to increase substantially in numbers, host-specific herbivores congregate. These stop and even reverse the growth in the numbers of that particular plant species and prevent it from seriously competing with other species.

Some botanists have maintained that plant species may conform to the predictions of traditional niche theory and occupy distinct niches. Grubb introduced the notion of the regeneration niche. He argued that different plant species often need specific conditions for their seeds to germinate and for their seedlings to survive and grow. Grime has further examined the huge variety of ways in which plants regenerate. The fact that adult plants may not occupy distinct niches does not mean that this must be the case for all stages of the life cycle. The niche of an organism needs to be considered over the entire life cycle of the organism.

COMMUNITY STRUCTURE OF FISH ON CORAL REEFS

Individual coral reefs may hold an astonishing diversity of fish species. The debate in recent years about what allows the coexistence of so many species has mirrored the same debate about plant communities. Essentially, two reasons have been suggested as to why coral reefs have so many fish species. First, that coral reefs are communities at equilibrium showing precise resource partitioning in response to the competition between the various fish species. Second, that the number of fish species on coral reefs is kept high largely by stochastic processes. Stochastic processes are unpredictable and operate in a relatively density-independent fashion. This is the opposite of the traditional, equilibrium hypothesis which emphasises density-dependent competition between species. Advocates of the stochastic hypothesis even refer to 'assemblages' of reef fish, denying that the fish really coexist as a community where the actions of one member affects other members.

Although there are not yet enough data to rule out either of these two explanations, experiments on the recruitment of new fish to the reefs may provide the answer. The traditional equilibrium model assumes that the probability of an individual fish larvae surviving to reproduce is limited in a density dependent manner by the abundance of the adult fish. The alternative stochastic model predicts that recruitment to the adult phase is independent of the density of the adults.

15

Trophic Levels

INTRODUCTION

There are many approaches to studying the ecology of organisms in their natural surroundings. We have already seen several of these i.e., changes in population size, the occupation of niches and so on. Another way is for ecologists to study organisms from the point of view of their feeding relationships—what they feed on and what feeds on them.

There are a number of reasons why this is a good approach. First, one of the major problems faced by all organisms is how to obtain enough energy and nutrients to survive, grow and reproduce. Second, if the feeding relationships of a group of organisms can be unravelled, then we get a clearer understanding of how those organisms interact. Finally, there simply is not enough time, nor sufficient funds, to study every species in detail. Ecologists therefore often cannot always look at each and every species, one by one. Rather, they study the autotrophs, the herbivores, the carnivores, the decomposers and so on, lumping together the species that have a similar role in the community.

In this chapter we will look at the arrangement of organisms into these community groups or trophic levels. Trophic literally means feeding, so trophic levels are the levels or positions at which species feed. Examples of trophic levels include 'herbivores' and 'decomposers'. In this chapter we will look at the characteristics of organisms in each of the trophic levels. We will then examine approaches to the study of the feeding relationships of organisms. Finally, we will see whether there are any rules governing the trophic levels and feeding relationships observed in nature.

AUTOTROPHS

Autotrophs are organisms that literally 'feed themselves'. Unlike heterotrophs, they do not require organic compounds as their source of energy. Autotrophs can be divided into two groups, the photoautotrophs and the chemoautotrophs and we will look at each of these groups in turn. The autotrophs are sometimes also referred to as producers. This is because they manufacture or produce the organic molecules on which all other organisms depend for their source of energy.

Photoautotrophs

The autotrophs with which we are most familiar are the photoautotrophs or photosynthetic organisms. As their name suggests, photoautotrophs obtain their energy from light—the Sun. The organisms best known to us as photoautotrophs are, of course, plants. Most ecologists accept a five kingdom classification of life. Under this classification, from which viruses are excluded, organisms are classified into the following five kingdoms: Prokaryotae (unicellular), Protoctista (eukaryotes that do not fit into the other kingdoms: often unicellular), Plantae (multicellular eukaryotes with photosynthetic nutrition), Fungi (multicellular eukaryotes with hyphae and living saprotrophically or parasitically) and Animalia (multicellular eukaryotes lacking cell walls and living heterotrophically).

Of these five kingdoms, three show photoautotrophic nutrition. Almost all species of plants are photoautotrophs, the exceptions being parasitic plants such as broomrapes (*Orobanche* spp.) and dodders (*Cuscuta* spp.).

The other eukaryotes that show photoautotrophic nutrition are some protoctists including *Euglena*, green algae, red algae and brown algae. These organisms are able to photosynthesise in essentially the same way as plants.

The third kingdom with members that show photosynthesis is the Prokaryotae. This group is divided into the Cyanobacteria and the Bacteria. All cyanobacteria show photosynthesis. Indeed, they look rather like chloroplasts and there is now very strong evidence that the chloroplasts possessed by eukaryotes once were cyanobacteria which somehow became permanent parts of eukaryotic cells. This is part of the endosymbiotic theory of the origin of the eukaryotic cell.

While all cyanobacteria show photosynthesis, only some bacteria do. Indeed, bacteria show just about every sort of nutrition known and there is little doubt that they should be divided into at least two kingdoms, each as distinct from the other at the biochemical level as animals are from plants or fungi. There is still a great deal that remains to be known about the biochemistry and ecology of the photosynthetic bacteria, but the number of different ways in which they solve the problem of obtaining energy are fascinating.

They differ fundamentally from other photoautotrophs in being anaerobic. Nor do they carry out the photolysis of water with the synthesis of NADPH and release of oxygen. Instead of water being the source of electrons used along the electron transport chain in photosynthesis, various inorganic or organic compounds act as electron donors.

The particular electron donors used are characteristic of different groups of bacteria. In green sulphur bacteria, sulphide is the electron donor. The purple sulphur bacteria can use sulphide and sulphur itself. The non-sulphur purple bacteria are even more complicated.

Under anaerobic conditions they are photoautotrophs using either hydrogen or sulphide as their electron donor. Under aerobic conditions, though, they will, in the presence of light, use a simple organic compound such as lactate ($CH_3CH(OH)COO^-$) as their source both of electrons and of carbon. In the dark they can survive and grow as heterotrophs.

In all the bacteria so far discussed, sunlight is trapped by bacteriochlorophyll which absorbs light at a characteristically high wavelength, 870 nm, compared to the 680–700 nm of chlorophyll 'a' which is found in eukaryotes. There is, however, an extraordinary bacterium called *Halobacterium halobium*. This species is unable to grow at salt concentrations less than 3 M, that is five times as salty as sea water! Accordingly, the bacterium lives in drying saltpans and in salted fish. *Halobacterium* is photosynthetic, but lacks bacterial chlorophyll. Instead its membrane contains the protein bacteriorhodopsin. At the center of this protein is a molecule of the visual pigment retinaldehyde. Just as in the rods of the human eye, this substance is bleached by light. The bacterium uses the energy associated with this process to eject protons from its cell. These protons are then allowed to flow back into the cell via an ATPase, thus synthesising ATP just as chloroplasts do.

Photosynthetic bacteria have various ways of obtaining their carbon. In the green sulphur bacteria, the only route of carbon dioxide fixation is by the reductive carboxylic acid cycle. This involves a reversal of the tricarboxylic acid (Krebs) cycle together with the reactions acetyl-CoA → pyruvate → phosphoenol pyruvate → oxaloacetate. The purple sulphur bacteria use both the Calvin cycle and this reductive carboxylic acid cycle.

Chemoautotrophs

Different authors use the term chemoautotroph in different ways. The easiest is to use it to refer to organisms in which energy is obtained solely from the oxidation of inorganic electron donors without the use of light. Organisms covered by this definition are clearly autotrophic yet non-photosynthetic. However, it is not the case that non-photosynthetic organisms which can obtain energy from the oxidation of inorganic electron donors are necessarily chemoautotrophic. For instance, some iron-oxidising bacteria can oxidise Fe^{2+} to Fe^{3+} but are heterotrophic, obtaining their energy principally from the breakdown of organic compounds.

Unlike photosynthetic organisms chemoautotrophs or chemosynthetic organisms as they are often called, are restricted to just one kingdom—the Prokaryotae. From an ecological perspective, some of the most important chemosynthetic organisms are involved in the nitrogen cycle. These include *Nitrosomonas* and *Nitrococcus*.

At a depth of 2500 m the scientists found great concentrations of chemosynthetic bacteria feeding on the hydrogen sulphide (H_2S) thrown out by the hot vents. Across most of the ocean floor animals are dispersed very thinly. Of course, no light can penetrate to such depths, so most animals rely on a slow 'rain' of organic matter from above. Around these volcanic vents, however, are found great concentrations of animals. There are unique tube worms 1.5 m in length and remarkable clams 30 cm long, both of which feed on the bacteria. Because the vents are hot, they set up convection currents which draw in organic fragments from the surrounding area. Hitherto unknown fish and blind white crabs act as scavengers which congregate at the vents and benefit from these convection currents.

The Mariana Trough snail has a well-developed radula in addition to these bacteria. This probably means that it does not depend solely on the bacteria for its nutrition. On the other hand, the tube worm (*Riftia pachyptila*) and the clam (*Calyptogena magnifica*) at the Galapagos Islands vents probably rely on their chemosynthetic bacteria for all of their nutrition. The tube worms lack a gut and the clam has a gut which is so small that it may be useless.

There is now a tremendous interest in the ecology and evolution of these chemoautotrophic communities. It has been argued that lift on Earth evolved some 4200 million years ago in such an environment.

DECOMPOSERS

After the producers, the decomposers are the next most important organisms in a community. Without producers, there could be no decomposers, no herbivores and no carnivores. Without decomposers, the remains of the other organisms in a community would simply accumulate. Eventually the world would run out of carbon dioxide or nitrate or phosphate or other inorganic substances essential for life. The decomposers break down the organic waste products and dead remains of organisms into the inorganic substances needed by the producers.

The simplest self-perpetuating community is therefore one consisting just of producers (whether photo or chemoautotrophic) and decomposers. Such a community is illustrated in Fig. 15.1 which shows a community based on photoautotrophs and decomposers. The photoautotrophs obtain their energy from sunlight and their minerals and water from a reservoir such as soil or a lake. Photosynthetic organisms also require carbon dioxide for photosynthesis and oxygen for respiration; these are obtained from the atmosphere, though they may be obtained from the atmosphere via the reservoir, as when submerged aquatics obtain their oxygen and carbon dioxide from the water in which they live. Photoautotrophs produce carbon dioxide in respiration, oxygen in photosynthesis and organic compounds in anabolism. Photoautotrophs cannot take up organic compounds. These organic compounds must therefore either accumulate in a sink or be utilised by the decomposers. The decomposers in turn usually require oxygen, unless they are obligate anaerobes and produce carbon dioxide and minerals such as ammonium (NH_4^+) which they return to the atmosphere and reservoir. If the producers are chemoautotrophs, the box labelled 'sun' and the arrow labelled 'sunlight' need to be deleted from Fig. 15.1.

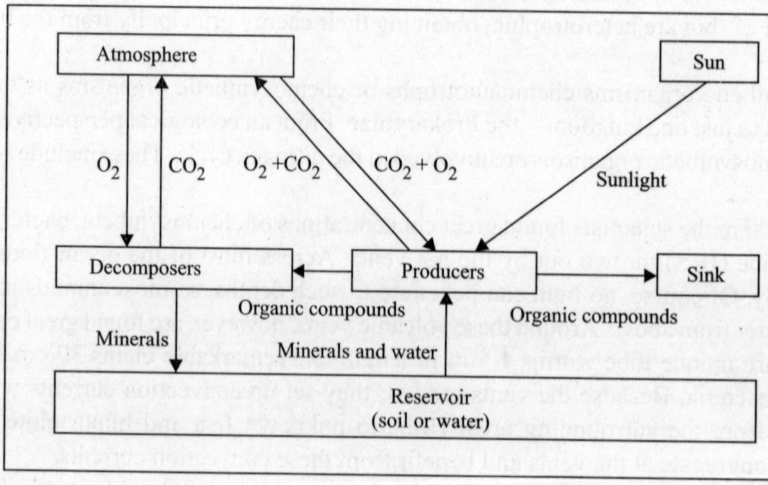

Fig. 15.1. The simplest self-sustaining community needs both producers and decomposers. This diagram shows a community based on photoautotrophs and decomposers.

Studies of natural communities often pay little attention to decomposers. Maybe this is because decomposers are often micro-organisms. Whatever the reasons, decomposers play a vital role in nature and certainly deserve careful investigation. Decomposers are, of course, heterotrophs. Organisms that ingest small particles of organic matter derived from the dead remains of plants or animals (detritus) are known as detritivores. Four of the five kingdoms contain decomposers — Prokaryotae, Fungi, Protoctista and Animalia. Accurate generalisations are difficult, but most decomposition is probably carried out by saprotrophic (as opposed to parasitic) fungi, by bacteria and by invertebrates. The best way to see the significance of decomposers is to look at an example where their role has been carefully investigated. As more is known about terrestrial than aquatic decomposers, the example given below is a terrestrial one.

Decomposition on The Forest Floor

In some communities dead organic matter accumulates, but in most forest environments the rate at which litter is added to the soil equals the rate at which it is removed by decomposition. The amount of dead organic matter measured in forests at different latitudes is greatest towards the poles and least at the equator. As the total biomass per unit area in a forest is greatest at the equator, this means that almost all of the organic matter in a tropical rainforest is tied up in the living organisms. This is also the case with the minerals. Tropical rainforests therefore rely on rapid nutrient cycling for their maintenance.

The breakdown of leaf litter is due to physical and biological factors. Breakdown in the tropics is fast for a variety of reasons. These forests are hot and humid. The lowland climate is probably less seasonal and the weather less erratic than at higher latitudes. These physical factors in themselves speed up organic breakdown. Additionally, they provide an ideal environment for many decomposers.

Decomposition of Dead Plant Matter

When biological materials decay they undergo humification. Humus is the organic matter resulting from the partial decomposition of organic material in the soil. It is typically dark in colour. Functionally, it is of great importance: it acts as a reservoir, holding minerals and water until they are absorbed by plants. Structurally, humus is extremely complicated. Apart from certain inorganic deposits, such as those found in invertebrate shells and in vertebrate teeth and bones, the compounds in humus most resistant to decomposition are the lignins of vascular plants. Lignins, along with cellulose, hemicelluloses and pectins, are found in plant cell walls. Chemically they are diverse and their biochemistry is still not fully understood. They are composed mainly of phenolic polymers which eventually give rise to humic acid.

The main aryl aldehydes in the polymers in lignin. These substances are extremely resistant to microbial breakdown which why lignin is so resistant to decomposition. In soils, certain groups of fungi and bacteria are the main organisms responsible for breakdown of lignin. The reactions produce the dark pigment melanin which accounts for the characterstic colour of humus.

Careful examination of the process of decay dead plant tissue reveals characteristic patterns in the order in which different decomposers colonise the material. Once the grass has been cut and stored, there is a rapid decrease in the micro-organisms found on the leaves when the plant was still alive. Simultaneously, there is a build-up of species normally found on stored material. The build-up of these species depends on several factors, especially the water content of the hay. As every farmer and gardener with a compost heap knows, growth of micro-organisms in damp hay is accompanied by the release of heat. This heat is due to respiration by the micro-organisms. Providing the hay is not too dry, temperatures of 60°C may commonly be reached. Under certain conditions, the temperature may even be such as to lead to spontaneous combustion!

HERBIVORES AND CARNIVORES

Although the distinction between herbivores and carnivores is clear, it has its limitations. Some ecologists consider both plant and animal eaters as predators or parasites. Herbivores can be labelled as predators if they are larger than their prey and consume most or all of a plant in a relatively short amount of time; they are parasites if they live in or on their hosts for long periods of time. According to this definition, most mammalian herbivores are predators; most insect herbivores are parasites. However, locusts are predatory whereas deer browsing on the lower leaves of a tree are parasitic. It might be thought that arguing over terms like this serves little purpose. However, it can help an ecologist to see trophic interactions from a different perspective.

There are some important differences between most plants and most animals from the perspective of the organisms that feed on them. Generally speaking, plants are immobile. Because they cannot run away from animals, they use a variety of strategies to escape from their herbivores. They may have defence mechanisms, producing spines or tough leaves or manufacturing toxins.

Oak trees, for example, produce unpalatable tannins in their leaves. Plants may try to hide from herbivores. Many short-lived plants exist as seeds in the seed bank and produce a sudden abundance of growing individuals when environmental conditions are appropriate. This flood of seedlings may be more than the herbivores can eat and increases a plant's chances of reproducing before it is eaten. Such seed dormancy is an example of a plant hiding both in space (buried in the soil or leaf litter) and in time.

In some respects carnivores face even greater problems than herbivores in the acquisition of food. In particular, their food is often very mobile. On the other hand, although herbivores may not need to be very mobile to obtain their food, they may need great mobility to escape from carnivores! An advantage of being a carnivore is that food is much less likely to be toxic. It is also easier to digest and tends to contain more protein.

OMNIVORES

An omnivore is an organism that eats both plants and animals. (Many of the technical terms in ecology were coined when organisms were, by and large, classified into two groups—animals and plants. Nowadays, most ecologists would classify a species that fed on fungi and animals, for instance, as an omnivore). Omnivores are, by definition, generalists. That is, they are not specialists, but rather obtain their food from a great variety of sources. Humans are omnivores par excellence. Most humans eat both animals and plants, while fungi are also popular with many. Further, the range of animals and plants we eat is huge.

At first it might be thought that omnivores have the best of both worlds. When there are lots of animals available, they can feed on the animals and if animals are scarce they can switch to plants. However, nothing in life is free. If omnivory has benefits, it must also have costs. The major cost is that omnivores, being generalists, are less efficient at using any single food source than are specialists. For instance, humans lack the gut bacteria capable of digesting cellulose that are found in many specialist herbivores. So we need plant matter with more protein and starch in it than do ruminants and equids. We are also less efficient at getting the last scrap of nutrition from an animal carcase than are dogs (specialist carnivores), for instance. Because we are omnivorous, we need an alimentary canal suited for both animal and plant food. We need teeth suitable for tearing meat and for grinding plant matter. We need enzymes for attacking animal food and those for attacking some of the chemicals made only by plants—starch, for instance.

FOOD CHAINS

Food chains are attempts to demonstrate the fate of individual organisms in a particular habitat. Consider a clover plant in a mixed deciduous oak wood in Europe. Suppose it is eaten by a snail, and the snail is killed and eaten by a thrush, which in turn is caught by a sparrowhawk. Traditionally, food chains are drawn with the producer at the bottom. Successive trophic levels are connected by arrows which lead from the organism being eaten to the organism eating it. Sometimes the Sun, as the ultimate source of almost all food chains, is placed at the bottom. Although decomposers play an important role in natural communities, they are usually not included. A generalised food chain is shown in Fig. 15.2. Herbivores are also called primary consumers; first-level carnivores secondary consumers; second-level carnivores tertiary consumers; third-level carnivores quaternary consumers; and so on. It is rare for a food chain to have more than five levels.

A quite different hypothesis to explain why food chains are relatively short has been put forward by Pimm and Lawton. They argue, on theoretical grounds, that the longer a food chain is, the less stable it is in the face of environmental fluctuations. After all, our hypothetical tenth-level carnivore may become extinct if anyone of the trophic levels beneath it is greatly reduced in number, for whatever reason. At the moment it is difficult to distinguish between Pimm and Lawton's explanation and the energy limitation one. This is partly because communities with considerable environmental fluctuations often also have lower levels of primary productivity and partly because the sorts of experimental interventions needed to test between the explanations have not been carried out.

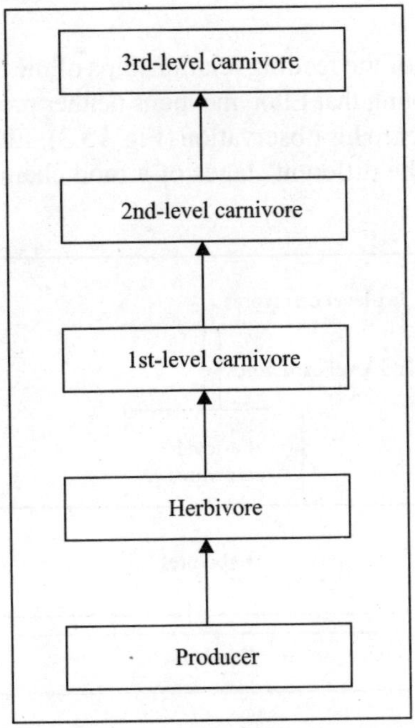

Fig. 15.2. A generalised food chain showing the passage of food from producers to third-level carnivores.

Perhaps the most useful way to think of a food chain is as a history of the fate of particular atoms. After all, a snail is very unlikely to eat an entire clover plant, though it might nibble a few leaves. As a story, however, of what happens to individual atoms, a food chain is a valid scientific record. On the other hand, a single food chain obviously cannot tell us much about what is happening to an entire community. For example, a number of herbivore species will eat clover and the sparrowhawk will also take other birds, some small mammals and even beetles. Consequently, a different way of representing such relationships, known as a food web, is often drawn.

FOOD WEBS

A food web is really a collection or matrix of food chains. It shows the pattern of energy or nutrient flow throughout a community. Food webs do have the advantage that they are more realistic than food chains; for instance omnivores can be shown on a food web but not on a food chain. A disadvantage with food webs is that unless you are already familiar with most of the organisms in it, it is difficult to learn much simply from looking at one. Another disadvantage of food webs is that they usually fail to give any indication of the relative importances of the different food chains—one line may represent 90 per cent of an animal's diet, another 10 per cent, yet this is rarely made clear.

A final disadvantage with them is that because they usually contain so many species, in their final published form they may tell us as much about the interests of the people who determined them as about the feeding relationships of the organisms themselves.

Various techniques are used to determine food webs (and food chains). Direct observation may allow many of the feeding relationships to be identified. Analyses of faecal pellets, stomach contents, regurgitated owl pellets and so on may add more information. Radio-active isotopes such as ^{32}P have also been used to trace the fate of atoms in a community.

PYRAMIDS OF NUMBERS

Food webs are useful descriptions of the feeding relationships of the organisms in a community, but they are non-quantitative. It is worth noting that Elton mentions neither producers nor decomposers. He coined the term pyramid of numbers to refer to his observation (Fig. 15.3). Elton noted that it is hard to obtain the actual numbers of individuals in the different stages of a food chain and the situation has not changed much since then.

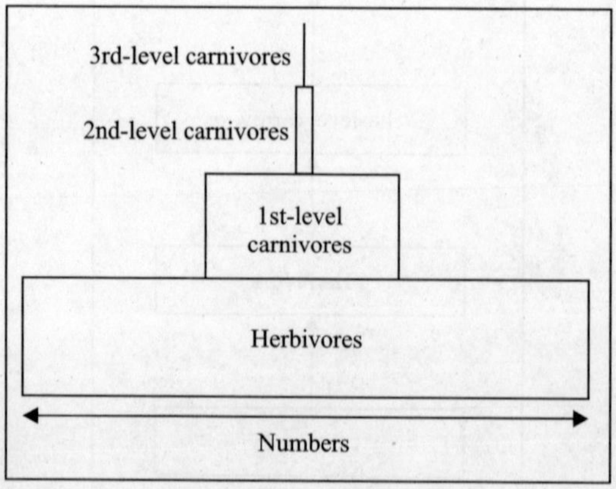

Fig. 15.3. Typical pyramid of numbers. Producers have been omitted. The width of the boxes indicates the relative numbers of the organisms at each of the trophic levels. The most convenient horizontal scale is probably a logarithmic one. The highest carnivore(s), in this case the third-level carnivore, is/are sometimes referred to as the 'top carnivore(s)'.

Although Elton thought that 'The general existence of this pyramid of numbers hardly requires proving, since it is a matter of common observation in the field', the existence of parasites frequently leads to inverted pyramids of numbers (Fig. 15.4a). If producers are included, inverted pyramids of numbers may again result as when, for instance, a single tree supports many much smaller organisms. Figure 15.4b shows the inverted pyramid of numbers in the oak wood at Wytham near Oxford.

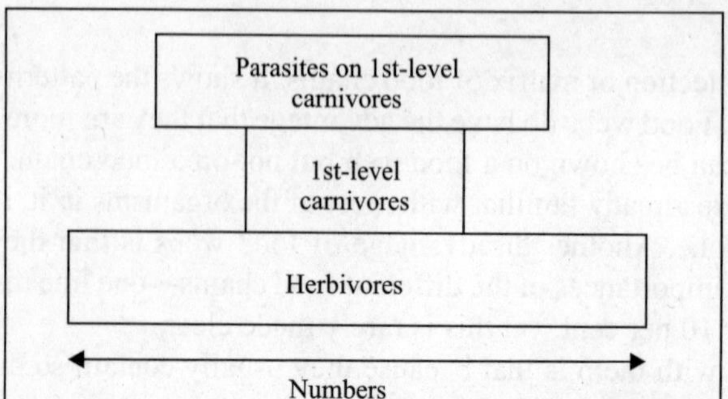

Fig. 15.4a. Hypothetical inverted pyramid of numbers showing more first-level carnivore parasites than first-level carnivores.

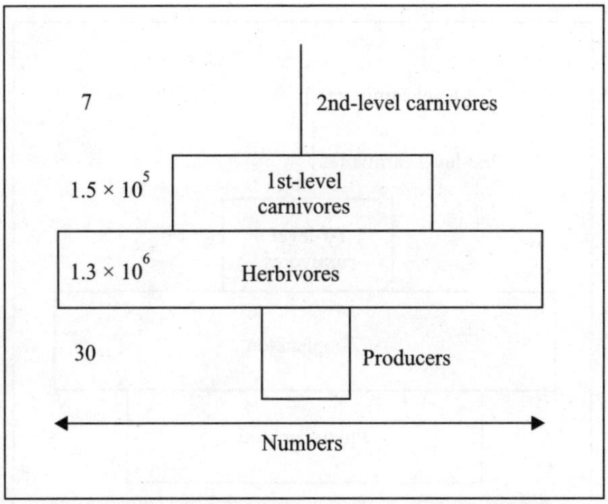

Fig. 15.4b. Inverted pyramid of numbers in Wytham oak wood, Oxford. The horizontal scale is logarithmic. The numbers at the right-hand side refer to the numbers of the organisms at each trophic level per hectare. Only oaks were counted as producers.

PYRAMIDS OF BIOMASS

Because some pyramids of numbers are inverted due to the very large size of some producers or the very small size of parasites, ecologists started to determine pyramids of biomass. As the term suggests, pyramids of biomass are calculated by determining, for a given unit area, the biomass of the producers, the biomass of the herbivores, the biomass of the first-level carnivores and so on.

It is true that very few pyramids of biomass have ever been determined, as they are even more time-consuming to determine than are pyramids of numbers, but it does seem likely that most are indeed pyramidal in shape and not inverted. The great game communities on the savannahs of Africa appear so abundant compared to the vegetation that it looks as if a pyramid of biomass would be inverted, with the herbivores weighing more than the grass and other vegetation. However, even a sparsely covered savannah has a great many roots and the human eye perhaps overestimates the density of the large herbivores.

There is no theoretical reason why pyramids of biomass should never be inverted. Indeed, at least one instance is known where they are. In the oceans a given mass of phytoplankton at certain times in the year supports a larger mass of zooplankton (Fig. 15.5). The reason this is possible is largely because the rate of reproduction of the phytoplankton is much greater due to their much smaller size. Consequently, the phytoplankton always produce enough for the zooplankton to eat. Of course, this means that the turnover of the phytoplankton is much more rapid than the turnover of the zooplankton.

A close analogy would be that the mass of food in most people's homes is much less than the mass of the people that live there. Again, this is because the turnover of the food is far quicker than the turnover of the people.

Ecology textbooks sometimes dismiss pyramids of numbers and pyramids of biomass before going on to consider pyramids of energy. Here, though, it is worth emphasising that pyramids of energy are not 'better' than pyramids of numbers or biomass. Each gives an ecologist a different sort of information about the relationships between organisms in a community. Which one prefers will depend on what data one wants to collect and why.

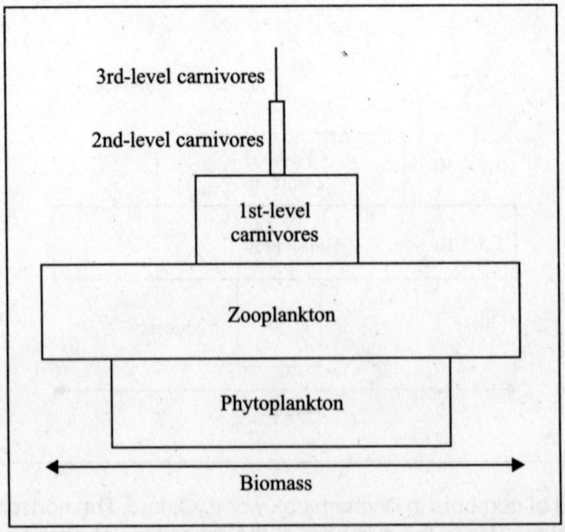

Fig. 15.5. Inverted pyramid of biomass of zooplankton feeding on phytoplankton.

SECTION IV

Biodiversity

16

Biological Diversity

INTRODUCTION

Although protection of biological diversity is central to conservation biology, the phrase 'biological diversity' has different meanings to different people. The definition given by the U.S. Office of Technology Assessment is 'the variety and variability among living organisms and the ecological complexes in which they occur'. This concept can be subdivided into three levels. Biological diversity at its most basic level includes the full range of species on Earth, from species such as bacteria, viruses and protists through the multicellular kingdoms of plants, animals and fungi. At finer levels of organisation, biological diversity includes the genetic variation within species, both among geographically separated populations and among individuals within single populations. On a wider scale, biological diversity includes variations in the biological communities in which species live, the ecosystems in which communities exist and the interactions among these levels (Fig. 16.1).

All levels of biological diversity are necessary for the continued survival of species and natural communities and all are important for the well-being of humans. Genetic diversity is needed by any species in order to maintain reproductive vitality, resistance to disease and the ability to adapt to changing conditions. Genetic diversity within domestic plants and animals is of particular value in the breeding programmes necessary to sustain modern agricultural species. Species diversity represents the range of evolutionary and ecological adaptations of species to particular environments. The diversity of species provides people with resources and resource alternatives; for example, a tropical rain forest with many species produces a wide variety of plant and animal products that can be used as food, shelter and medicine. Community-level diversity represents the collective response of species to different environmental conditions. Biological communities such as deserts, grasslands, wetlands and forests support the continuity of proper ecosystem functioning, providing beneficial services to people—services such as flood control, protection from soil erosion and air and water filtering.

Biological diversity includes the entire range of species found on Earth. A species is generally defined in one of two ways:

1. A group of individuals that is morphologically, physiologically or biochemically distinct from other groups in some characteristic (the morphological definition of species).
2. A group of individuals that can potentially breed among themselves and do not breed with individuals of other groups (the biological definition of species).

The morphological definition of species is the one most commonly used by taxonomists, biologists who specialise in the identification of unknown specimens and the classification of species. The biological definition of a species is the one most commonly used by evolutionary biologists because it is based on measurable genetic relationships rather than on somewhat subjective physical features. In practice, however, the biological definition of a species is hard to use because it requires knowledge of which individuals are actually capable of breeding among themselves—information that is rarely available.

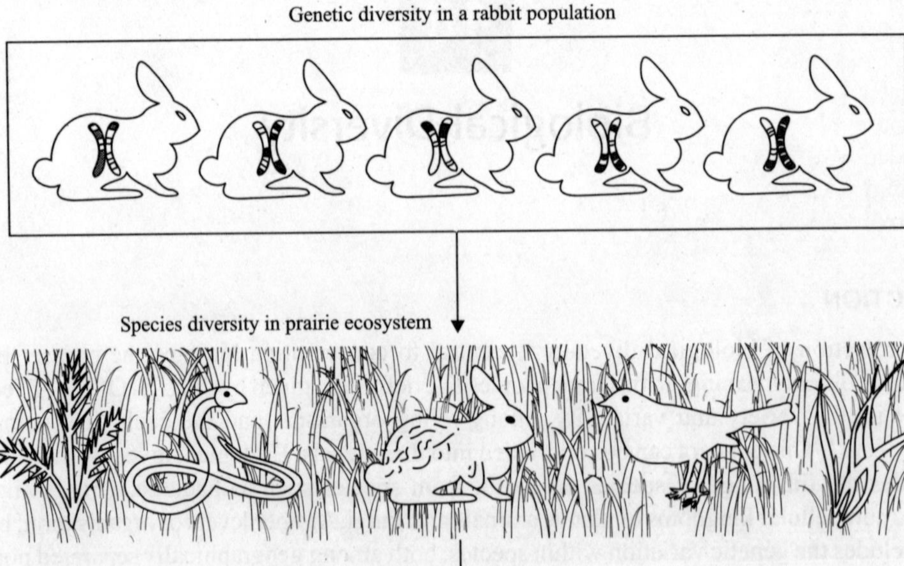

Fig. 16.1. Biological diversity includes genetic diversity (the genetic variation found within each species), species diversity (the range of species in a given ecosystem) and community/ecosystem diversity (the variety of habitat types and ecosystem processes extending over a given region).

Problems in distinguishing and identifying species are more common than many people realise. Differences between species are not always as clear-cut as either of the definitions would imply. For example, a single species may have several varieties that have observable morphological differences, yet the varieties are similar enough that they are still considered to be a single biological species (Fig. 16.2).

Different breeds of dogs, such as German shepherds, collies, dachshunds and beagles, all belong to one species and readily interbreed despite the conspicuous differences among them. Alternatively, there are closely related 'sibling' species that appear very similar in morphology or physiology, yet are still considered biologically separate and do not interbreed. In practice, biologists often have difficulty distinguishing variation within a single species from variation between closely related species. Further complicating matters is the fact that what are otherwise distinct species may occasionally hybridise, producing intermediate forms that blur the distinction between species. Such hybridisation is particularly common among plant species in disturbed habitats. Finally, for many groups of species, the taxonomic studies needed to determine species and identify specimens have not yet been done. The inability to clearly distinguish one species from another, whether due to similarities of characteristics or to confusion over the correct scientific name, often slows down efforts at species protection.

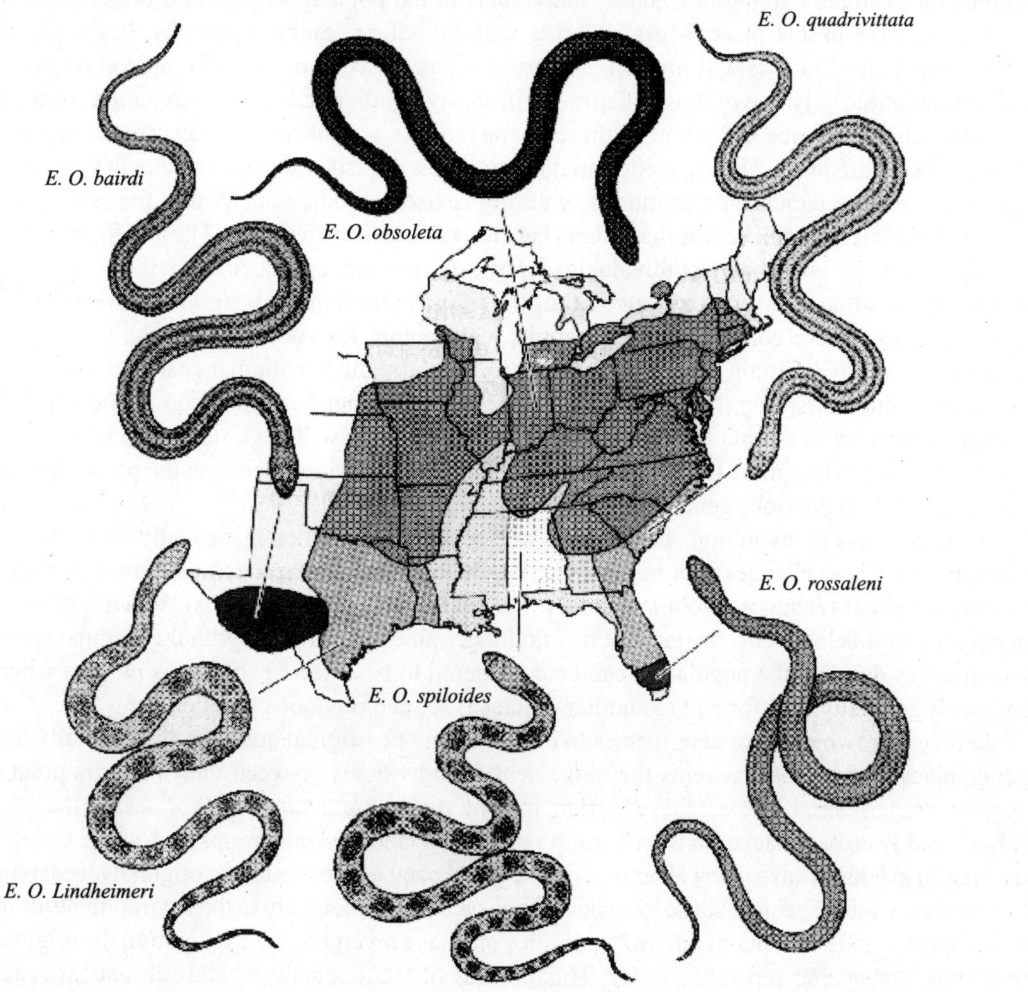

Fig. 16.2. The rat snake *Elaphe obsoleta* is a variable species that has many distinct subspecies. Each subspecies has its own geographical range and particular patterns of colour and banding. The subspecies interbreed where their ranges overlap.

ORIGIN OF NEW SPECIES

The biochemical similarity of all living species indicates that life on Earth probably originated only once, about 3.5 billion years ago. From one original species came the millions of species found on Earth today. The process of new species formation, known as speciation is continuing today and will most likely continue in the future.

The process whereby one original species evolved into new, distinct species was first described by Charles Darwin and Alfred Russel Wallace more than 100 years ago. This theory of the origin of new species is widely accepted today in the scientific community. The wealth of new information provided by the fossil record and by modern molecular biology research has continued to support this theory.

The theory of evolution is both simple and elegant. Imagine a population of a species—mountain rabbits, for example—living in Canada. Individuals in the population tend to produce more offspring than can survive in that place. Most offspring will die before reaching maturity. In the population of rabbits, one pair of rabbits will produce numerous litters of six or more offspring, yet on average, in a stable population, only two of those offspring will survive. Individuals in the population show variations in certain characteristics and some of these characteristics are inherited—they are passed genetically from parents to offspring. These genetic variations are caused by spontaneous changes in the chromosomes and by the rearrangement of chromosomes during sexual reproduction. Within the rabbit population, some individuals have thicker fur than others because of genetic differences. These differences in genetic characteristics will enable some individuals to grow, survive and reproduce better than other individuals, an idea that is often referred to as 'survival of the fittest'. Our hypothetical thick-furred rabbits will be more likely to survive cold winters than rabbits with thinner fur. As a result of the improved survival ability provided by a certain genetic characteristic, the individuals with that characteristic will be more likely to produce offspring than the others; over time, the genetic composition of the population will change. After a series of cold winters, more thick-furred rabbits will have survived and produced thick-furred offspring, while more thin-furred rabbits will have died. More rabbits in the population now have thicker fur than in previous generation.

In the process of evolution, populations are constantly adapting genetically to changes in their environment. These changes may be biological (new food sources, new competitors, new prey) as well as environmental (changes in climate, water availability, soil characteristics). When a population has undergone so much genetic change that it is no longer able to interbreed with the original species from which it was derived, the population can be considered to be a new species; this process whereby one species is gradually transformed to another species is termed phyletic evolution.

In order for two or more new species to evolve from one original ancestor, there usually has to be a geographical barrier that prevents the movement of individuals between the various populations of a species. For terrestrial species, these barriers may be rivers, mountain ranges or oceans that the species cannot readily cross. Speciation is particularly rapid on islands. Island groups such as the Galapagos and the Hawaiian Islands have many examples of insect and plant groups that were originally local populations of a single invading species. These local populations adapted genetically to the environments of particular unoccupied islands, mountains and isolated valley and have diverged sufficiently from the original species to now be considered separate species. This process of local adaptation and subsequent speciation is known as adaptive radiation. One of the best-known examples of adaptive radiation is that of the Hawaiian honeycreepers, a group of specialised bird species that apparently derives from a single pair of birds that arrived by chance in the Hawaiian Islands tens thousands of years ago.

The origin of new species is normally a slow process, taking place over hundreds, if not thousands, of generations. The evolution of new genera and families is an even slower process, lasting hundreds of thousands or even millions of years. However, there are mechanisms whereby new species can arise in just one generation. Particularly in plants, there may be unusual, unequal divisions of chromosome sets at the time of reproduction that result in offspring with extra sets of chromosomes; these offspring are known as polyploids. Polyploid individuals may be morphologically and physiologically different from their parents and if they are well suited to the environment, may form a new species.

Even though new species are arising all the time, the present rate of species extinction is probably more than 1000 times faster than the rate of speciation. The situation is actually worse than this grim statistic suggests. First, the rate of speciation may actually be slowing down because so much of the Earth's surface has been taken over for human use and no longer supports evolving biological communities. There are now fewer populations of each species and thus fewer opportunities for evolution. Many of the existing protected areas and national parks may be too small in area to allow the process of speciation to occur (Fig. 16.3). Second, many of the species threatened with extinction are the sole remaining representatatives of their genus or family; examples include the coelocanth fish of the Indian Ocean and the African cheetah. The extinction of these taxonomically unique species representing ancient lineages is not balanced by the origin of new species.

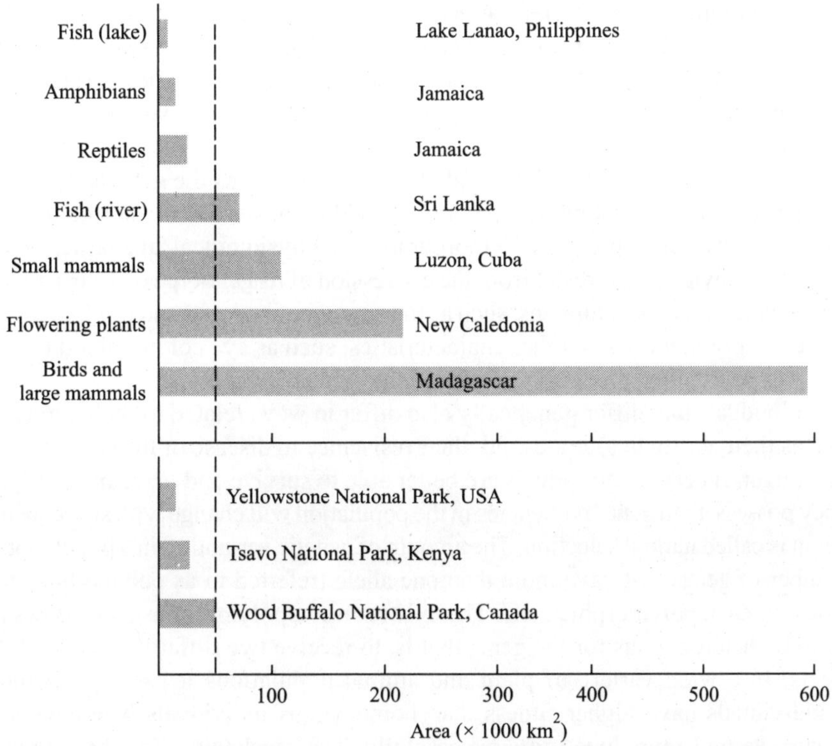

Fig. 16.3. Certain groups of organisms apparently need a minimum area in order to undergo the process of speciation. For example, for small mammals, the smallest islands (Cuba and Luzon) on which a single species is known to have given rise to two species are 1,00,000 km^2. Even the largest national parks are probably too small to allow for the evolution of new species of flowering plants, birds or mammals, although they might be large enough for the continued evolution of fishes, amphibians and reptiles.

GENETIC DIVERSITY

At each level of biological diversity—genetic, species and community—biologists study the mechanisms that alter or maintain diversity. Genetic diversity within a species is often affected by the reproductive behaviour of individuals within populations. A population is a group of individuals that mate with one another and produce offspring; a species may include one or more separate populations. A population may consist of only a few individuals or millions of individuals, provided that the individuals actually produce offspring. A single individual of a sexual species would not constitute a population. Neither does a group of individuals that cannot reproduce constitute a population; for example, a group of 10 dusky seaside sparrows would not constitute a population if all of them were males.

Individuals within a population are often genetically different from one another. This genetic variation arises because individuals have slightly different forms of their genes, the units of the chromosomes that code for specific proteins. These slightly different forms of a gene are known as alleles and the differences can arise through mutations—changes that occur in the deoxyribonucleic acid (DNA) that constitute an individual's chromosomes. The various alleles of a gene may produce forms of a protein that differ in structure and function, in turn affecting the development and physiology of the individual organism.

Genetic variation increases when offspring receive unique combinations of genes and chromosomes from their parents via the recombination of genes that occurs during sexual reproduction. Genes are exchanged between chromosomes and rearranged during the crossing over that occurs during meiotic cell division and new combinations are created when gametes from two parents unite to form a genetically unique offspring. Although mutations provide the basic material for genetic variation, the ability of sexually reproducing species to randomly rearrange alleles in different combinations dramatically increases the potential for genetic variation.

The total array of genes and alleles in a population is referred to as the gene pool of the population, while the particular combination of alleles that any individual possesses is referred to as its genotype. The phenotype of an individual represents the morphological, physiological, anatomical and biochemical characteristics of the individual that result from the expression of its genotype in a particular environment (Fig. 16.4). Some characteristics of humans, such as the amount of body fat and tooth decay, are strikingly influenced by the environment, while other characteristics, such as eye colour, blood type and forms of certain enzymes are determined predominantly by an individual's genotype.

Sometimes individuals that differ genetically also differ in ways related to their survival or ability to reproduce, such as their ability to tolerate cold, their resistance to disease or the speed at which they can run away from danger. If certain individuals are better able to survive and leave more offspring because of the alleles they possess, then gene frequencies in the population will change with subsequent generations. This phenomenon is called natural selection. The amount of genetic variability in a population is determined by both the number of genes that have more than one allele (referred to as polymorphic genes) and the number of alleles for each polymorphic gene. The existence of a polymorphic gene allows individuals in the population to be heterozygous for the gene, that is, to receive two different alleles of the gene from their two parents. In a wide variety of plant and animal populations it has been demonstrated that heterozygous individuals have higher fitness than homozygous individuals heterozygous individuals tend to grow, survive and reproduce more successfully. The mechanism for this appears to be that: (i) having two different forms of, say, an enzyme gives the individual greater flexibility in dealing with life's challenges; and (ii) nonfunctional or harmful alleles received from one parent are masked by the functioning alleles received from the other parent. This phenomenon of increased fitness in highly heterozygous individuals, also referred to as hybrid vigor, is widely known in domestic animals.

Thus, genetic variability within a population can be measured as:

1. The number (and percentage) of genes in the population that are polymorphic (have more than one allele).
2. The number of alleles for each polymorphic gene.
3. The number (and percentage) of genes per individual that are polymorphic.

Genetic variability can affect diversity at the species level. A species consists of one or more populations and although most mating occurs within populations, individuals occasionally move from one population to another, allowing the transfer of new alleles and genetic combinations between populations. This genetic transfer is referred to as gene flow.

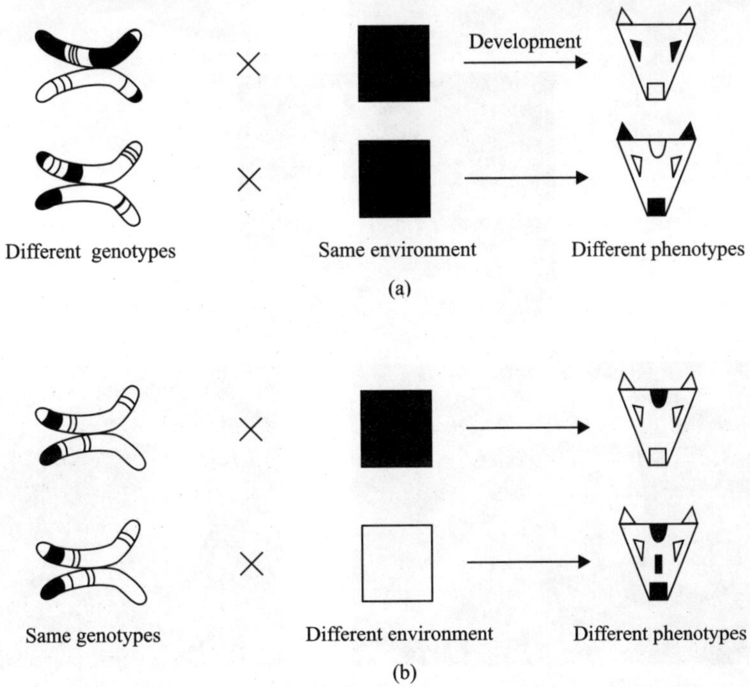

Fig. 16.4. The physical physiological and biochemical characteristics of an individual—its phenotype—are determined by its genotype and the environment (e.g., hot vs. cold climate; abundant vs. scarce food) in which the individual lives.

Genetic variation also occurs within domesticated plants and animals. In traditional societies, people preserved any new forms they found that were well-suited to their needs. Through generations of this process of artificial selection by humans, varieties of species were developed that were productive and useful to local conditions of soil, climate and crop pests. This process has continued in modern agriculture with scientific breeding programmes that manipulate genetic variation to meet human needs. Thousands of varieties of widespread crops, such as rice, potatoes and corn, have been incorporated into the breeding programmes of modern agriculture. Among animals, the huge number of breeds of domestic dogs, cats, rabbits, chickens, cattle and pigs are evidence of the potential of artificial selection to alter gene pools. Figure 16.5 shows an example of some results of artificial selection. Genetic variation is also maintained in specialised collections of species used in scientific research, such as *Drosophila* fly stocks used in genetic research, tiny *Arabidopsis* plants used in plant science research and mice used in physiological and medical research.

Fig. 16.5. Pigeon hobbyists have bred and preserved unusual forms of domestic pigeons, all originally descended from the rock dove.

BIOLOGICAL COMMUNITY

A biological community is defined by the species that occupy a particular locality and the interactions between those species. A biological community together with its associated physical environment is termed an ecosystem. Within an ecosystem, water evaporates from biological communities and the Earth's surface, to fall again as rain or snow and replenish terrestrial and aquatic environments. Soil is built up out of parent rock material and decaying organic matter. Photosynthetic plants absorb light energy and this energy is released back into the atmosphere as heat during the lives of plants and animals, as well as after they die and decompose. Plants absorb carbon dioxide and release oxygen during respiration, while animals and fungi absorb oxygen and release carbon dioxide. Mineral nutrients, such as nitrogen and phosphorus, cycle between the living and the non-living compartments of the ecosystem.

The physical environment affects the structure and characteristics of a biological community; but the biological community can also have effects on the physical characteristics of the ecosystem. In a terrestrial ecosystem, for example, wind speed, humidity and temperature in a given location can all be affected by the plants and animals present there. In aquatic ecosystems, such characteristics as water turbulence and clarity, water chemistry and water depth affect the characteristics of the biological communities, but in turn biological communities such as kelp forests and coral reefs can affect the physical environment.

Within a biological community each species utilises a unique set of resources that constitute its niche. The niche for a plant species might consist of the type of soil on which it is found, the amount of sunlight it receives, the amount of soil moisture it requires, the type of pollination system it has and its mechanism of seed dispersal. The niche for an animal might include the type of habitat it occupies, the types of food it eats over the year and its requirements for overwintering dens. The niche of each species is the space where that species exists in nature and is not outcompeted by other species.

The niche of a species often includes the stage in succession that the species occupies; succession is the gradual process of change in species composition, community structure and physical characteristics that occur following natural and human-caused disturbance to a biological community. Particular species are often found at particular successional stages. For example, sun-loving butterflies and annual plants are found most commonly early in succession, in the months immediately after a hurricane or a logging operation has destroyed an old-growth forest. At this stage, the ground is receiving high levels of sunlight, with high temperatures and low humidity during the day. Over the course of years and decades, the forest is gradually re-established. Different species, including shade-tolerant wildflowers and birds that nest in holes in dead trees, are found in these late-successional stages of mature trees and old-growth forest. Similar patterns of succession-stage specific organisms are found in other biological communities, such as grasslands, lakeshores and the intertidal zones of oceans.

The composition of communities is often strongly affected by competition and predation. Predators often dramatically reduce the numbers of populations of prey species and may eliminate some prey species from certain habitats. Predators may indirectly increase biological diversity in a community by keeping the densities of other species so low that competition for resources does not occur. A good example of this is the marine intertidal ecosystem in which a large sea star (starfish), *Pisaster*, feeds on fifteen species of mollusks that cling to the rock. As long as the predatory sea star is present, the mollusk species do not compete with each other for space on the rocks, since they are eaten too fast to allow them to achieve high population densities. Under these circumstances, all fifteen species are able to occupy the niche. If the sea star is removed, however, the mollusks increase in abundance and start competing for space on the rocks. In the absence of predation, competition between species reduces the number of species; eventually only a few of the original fifteen species remain, with some rocks taken over by just one species. In plant communities as well, species diversity is often higher when intense grazing by animals prevents competition among plants than when grazers are absent.

The number of individuals of a particular species that the resources of an environment can support is termed the carrying capacity. A population's numbers are often well below the carrying capacity when it is held in check by a predator population. If the predator is removed, the population may increase to a point where it reaches the carrying capacity or may even increase beyond the carrying capacity to a point at which the environment is damaged and the population crashes.

The population size of a species may also be controlled by other species that compete with it for the same resources; for example, two heron species may eat the same type of fish. Population size may decline if a competing species becomes abundant or may increase if the competitor becomes scarce or is eliminated from the community.

Community composition is also affected by mutualistic relationships, in which two species benefit each other. Mutualistic species reach higher densities when they occur together than when only one of the species is present. Common examples of such mutualisms are fruit-eating birds and plants with fleshy fruits; flower-pollinating insects and flowering plants; the fungi and algae that together form lichens; and corals and the algae that live inside them. Another example involves plants that produce special chambers that are inhabited by ants. The ants protect the plants and the ants' wastes are used by the plants as a source of nutrients; in turn, the ants have a secure home to occupy. At the extreme of mutualism, species may form a symbiotic relationship, in which two species are always found together and apparently cannot survive without each other. For example, the death of certain types of coral-inhabiting algae due to unusually high water temperatures in tropical areas may be followed by the weakening and subsequent death of their associated coral species.

TROPHIC LEVELS

Biological communities are organised into feeding levels called trophic levels that represent ways in which energy is obtained from the environment (Fig. 16.6):

1. Photosynthetic species (also known as primary producers) obtain their energy directly from the Sun. In terrestrial environments, higher plants such as flowering plants, gymnosperms and ferns are responsible for most photosynthesis, while in aquatic environments, seaweeds, single-celled algae and cyanobacteria (blue-green algae) are the most important. All of these species use solar energy to build the organic molecules they need to live and grow. Only a tiny percentage of the light energy falling on the Earth's surface is actually captured by the photosynthetic process; the great majority of light energy is reflected back into space or absorbed by the air, ground and water and radiated back as heat. And in fact most of the energy captured by each subsequent trophic level is used up in the basic metabolic processes of keeping the organisms alive and is released into the environment as heat.
2. Herbivores (also known as primary consumers) eat photosynthetic species. For example, in terrestrial environments gazelles and grasshoppers eat grass, while in aquatic environments crustaceans and fish eat algae. Because much plant material, such as cellulose and lignin, is indigestible to many animal species or is simply not eaten, only a small percentage of the energy captured by photosynthetic species is actually transferred to the herbivore level.
3. Carnivores (also known as secondary consumers or predators) eat other animals. Primary carnivores (such as foxes) eat herbivores (such as rabbits), while secondary carnivores (such as bass) eat other carnivores (such as frogs). Since carnivores do not catch all of their potential prey, and since many body parts of the prey are indigestible, again only a small percentage of the energy of the herbivore trophic level is transferred to carnivores. Carnivores usually are predators, though some combine direct predation with scavenging behaviour and others, known as omnivores, have a substantial proportion of plant foods in their diets. In general, predators are larger and stronger than the species they prey on, but they usually occur in lower densities than their prey.

   Parasites form an important subclass of predators. Parasites are typically small in size and do not kill their prey immediately. The effects of parasites may range from imperceptibly weakening their prey to totally debilitating or even killing their prey over time. Parasites include a wide range of species, such as disease-causing bacteria and protists, intestinal worms, parasitic fungi, beetle larvae that feed on the inner bark of trees and plants such as mistletoe and dodder that parasitise other plants. Parasites are often important in controlling the density of prey species. When their prey populations are at a high density, parasites can readily spread from one host individual to the next, causing an intense local infestation of the parasite and a subsequent decline in prey density. When prey densities are low, parasites are less able to move from one host to another and the effects of parasites on the prey population are correspondingly weak.

4. Detritivores (also known as decomposers) are species that feed on dead plant and animal tissues and wastes, breaking down complex tissues and organic molecules. Detritivores release minerals such as nitrogen and phosphorus back into the environment, where they can be taken up again by plants and algae. The most important detritivores are fungi and bacteria, but a wide range of other species play a role in breaking down organic materials for example, vultures and other scavengers tear apart and feed on dead animals, dung beetles bury and feed on animal dung and worms break down fallen leaves and other organic matter.

Principles of Community Organisation

As a consequence of less energy being transferred to each successive trophic level in biological communities, there tends to be a lower biomass (total weight of living material) and fewer individuals at each level. The greatest biomass in an ecosystem will be that of the primary producers. And as a general rule in any community, there will be more individual herbivores than primary carnivores and more primary carnivores than secondary carnivores. For example, a forest community generally contains more individual insects and insect biomass than insectivorous birds and more insectivorous birds than raptorial birds (birds such as hawks that feed on other birds).

Although species can be organised into these general trophic levels, their actual requirements or feeding habitats within the trophic levels may be quite restricted. For example, a certain aphid species may feed only on one type of plant and a certain lady beetle species may feed on only one type of aphid. These specific feeding relationships have been termed food chains. The more common situation in many biological communities, however, is for a species to feed on several items at the lower trophic level, to compete for food with several species at its own trophic level and in turn to be preyed upon by several species at the next higher trophic level. Consequently, a more accurate description of the organisation of biological communities is a food web in which species are linked together through complex feeding relationships (Fig. 16.7). Species at the same trophic level that use approximately the same resources in the environment are considered to be in a guild of competing species.

Species-specific ecological requirements are an important reason for the inability of many species to increase in abundance within a community. For example, the diets of most herbivorous insects are restricted to just a few plant species and many insects can only eat certain plant parts. The reason for this is partly that their mouthparts and body forms are only suited for certain feeding behaviors, but more important, their digestive systems are often surprisingly limited in their ability to extract nutrients and tolerate the toxic defensive chemicals produced by many plants. So even though a forest may be full of vigorously growing green plants, some insect species may be unable to complete their development and reproduce because they cannot get the specific food they require.

To survive, individuals of a species require a variety of resources, such as food, water, minerals and a safe resting place; the continuous availability of these resources is an important component in the ability of a species to persist in an environment. A low supply of anyone resource may limit the size of a population; the necessary resource that is in least supply may be the limiting resource on population size. In many communities, there are occasional episodes when one or several resources become limited and vulnerable species are eliminated from the community. For example, although the water is normally not a limiting resource to rain forest organisms, episodes of drought that last for weeks and even months do occasionally occur, even in the wettest rain forests. At such times, animal and plant species that need a constant supply of water may be eliminated from the community. In another example, bird species that are specialised to feed on insects may be unable to eat or feed their young during days or weeks when unusually cold, dry or windy weather prevents insects from flying; in this situation, the insects are the limiting resource for the bird population.

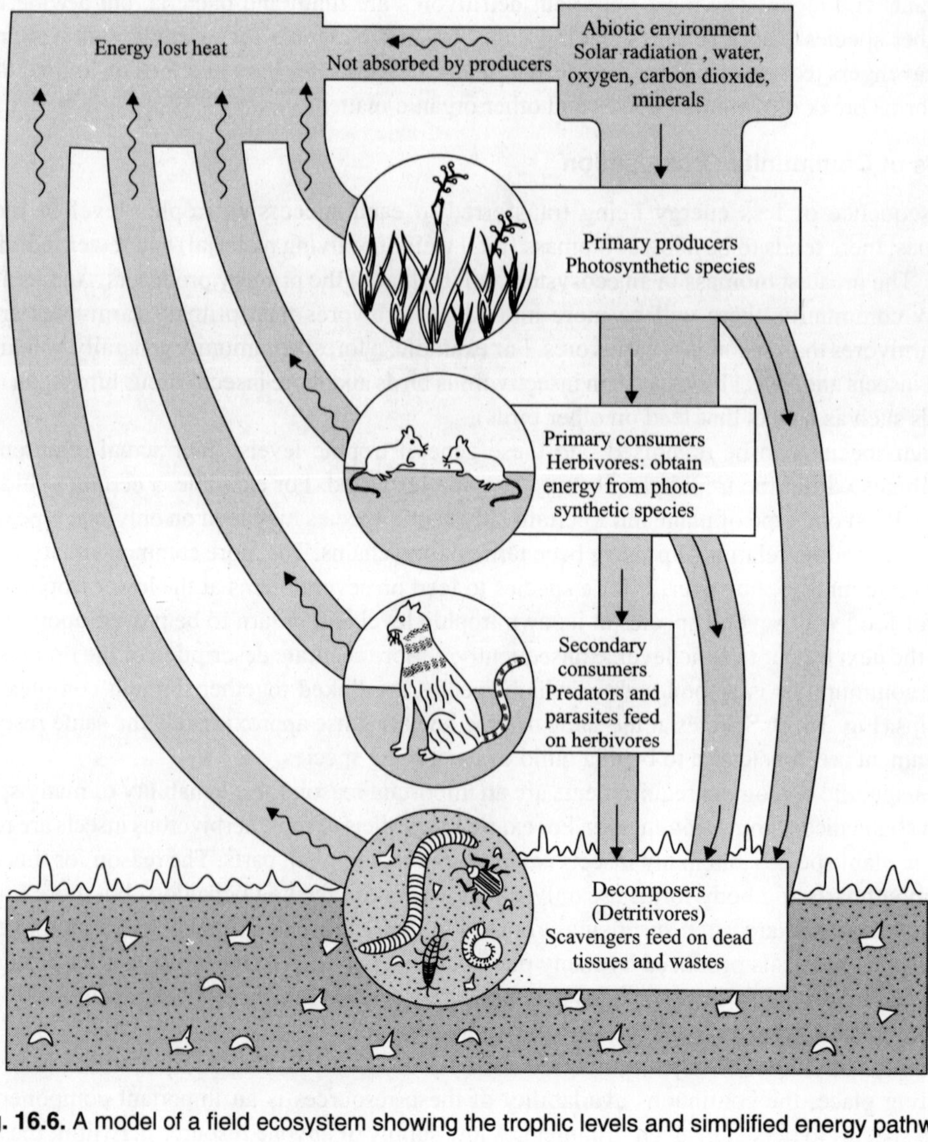

Fig. 16.6. A model of a field ecosystem showing the trophic levels and simplified energy pathways.

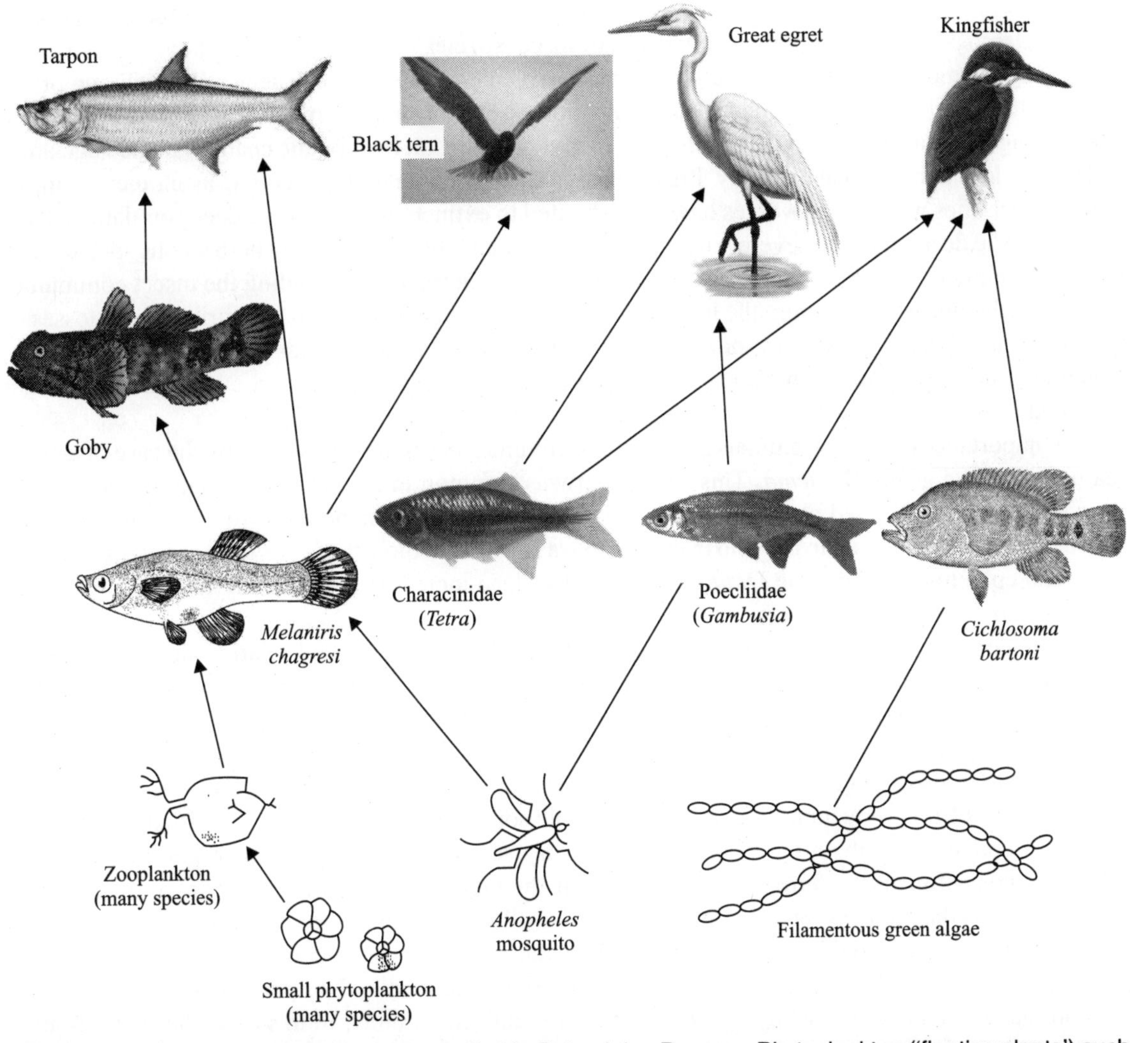

Fig. 16.7. Diagram of an actual food web studied in Gatun lake, Panama. Phytoplankton ('floating plants') such as green algae are the primary producers at the base of the web. Zooplankton are tiny, often microscopic, floating animals; they are primary consumers, not photosynthesisers, but they, along with insects and algae, are crucial food sources for fish in aquatic ecosystems.

Keystone Species

Within biological communities, certain species may be important in determining the ability of large numbers of other species to persist in the community. These crucial species have been termed keystone species. Protecting keystone species is a priority for conservation efforts, because if a keystone species is lost from a conservation area, numerous other species might be lost as well. Because many tropical insect species appear to be highly specialised in their feeding behaviour, subsisting on just one or a few related plant species, Raven has argued that the extinction of each tropical plant species potentially results in an extinction cascade, with an additional loss of 10 to 30 insect species. With the recent evidence of a greater diversity of insect species than previously suspected, it is even possible that the extinction of a single plant species could eliminate hundreds of insect species.

Among the most obvious keystone species are top predators, since these species are often important in controlling herbivore populations. The elimination of even a small number of individual predators, constituting a minute amount of the community biomass, can result in dramatic changes in the vegetation and a great loss in biological diversity. Presented one such example in the sea otter; in another example, in many localities where gray wolves have been hunted to extinction by humans, deer populations have exploded. The deer have then severely overgrazed the habitat, eliminating many herbaceous species. The loss of these plants was detrimental to the deer and to other herbivores, including the insect community that fed on the plants. In addition, the loss of plant cover has led to soil erosion, contributing to losses of the species that inhabit the soil. In short, the elimination of a keystone species—in this case, the wolves, which kept deer populations in check—can create a series of linked extinction events that result in a degraded ecosystem with much lower biological diversity at all trophic levels.

The importance of grazing animals in patterning communities is also illustrated by the case of a black sea urchin of the genus *Diadema*. This sea urchin was common in coral reefs throughout the tropical West Atlantic, where it fed on large, fleshy algae that were fairly uncommon. In 1981 following the unusually warm weather of an El Nino event, there was a massive die-off of *Diadema*, apparently caused by a viral epidemic. Without the *Diadema*, the fleshy algae increased dramatically in abundance, over growing and damaging coral reefs.

The importance of a keystone species may hinge on highly specialised relationships between the keystone species and other organisms. In many tropical forests, fig trees and fig vines (*Ficus* spp.) appear to be keystone species in the functioning of vertebrate communities. Fig flowers are pollinated by small, highly specialised fig wasps, which mature inside the developing fig fruit. Mature fig trees produce continuous fig crops and generations of wasps are continually coming to maturity. As a consequence of this continuous fruit production, figs provide a reliable source of fruit to primates, birds and other fruit-eating vertebrates throughout the year, even during dry seasons. While fig fruits do not have the high energy content of many preferred lipid-rich fruits or the high protein content of an insect diet, during drought periods the fig fruits serve as 'famine food' that allows vertebrates to survive until their preferred foods are once more available. Even though fig trees and vines may be uncommon in the forest and the fruit may constitute only a small percentage of the total vertebrate diet their persistence is necessary to the continued functioning of many species in the vertebrate community. In this case, the fig trees are a keystone species because so many other species rely on them for food, but the health of the *Ficus* population rests on the health of the wasp population; the relationship between the trees and the wasps forms the foundation of the entire community's health.

Many keystone species play less obvious roles that are nevertheless essential to maintaining biological diversity. Some inconspicuous detritivorous species play a significant role in the functioning of a community. For example, dung beetles exist at low density in tropical forests and constitute only a fraction of the biomass. Yet the beetles are crucial to the community because they bury dung and carrion as a food source for their larvae. These buried materials break down rapidly, making nutrients available for plant growth. Seeds contained in the dung of fruit-eating animals are also buried, which may facilitate seed germination and the establishment of new plants. In addition, by burying and feeding on dung the beetles kill the parasites of vertebrates contained in the dung, helping to keep the vertebrate populations healthy.

The identification of keystone species has several important implications for conservation biology. First, as we have seen, the elimination of a keystone species from a community may precipitate the loss of many other species. Second, in order to protect a species of particular interest, such as a monkey, it may be necessary to protect the keystone species on which it depends either directly or indirectly, such as fig trees and dung beetles. Third, if the few keystone species of a community can be identified, these could be carefully protected or even encouraged if the area is being affected by human activity. For example, during selective logging, figs and other important fruit trees should be protected, while common trees that are not keystone species could possibly be reduced in abundance with little loss of biological diversity.

Keystone Resources

Nature reserves are typically compared and valued in terms of their size in hectares, because, in general, larger reserves contain more species than smaller reserves. However, area by itself may not be as significant as the range of habitats and resources that reserve contains. Particular habitats may contain critical limiting resources—'keystone resources' —that occupy only a small area of the habitat yet are crucial to many species in the community. For example:

1. Salt-licks and mineral pools provide essential minerals for wildlife, particularly in inland areas with heavy rainfall. The distribution of salt licks can determine the abundance and distribution of vertebrates in an area.
2. Deep pools in streams and springs may be the only refuge for fish and other aquatic species during the dry season when water levels drop. These water sources may be the only sources of drinking water for a considerable distance for terrestrial animals.
3. Elevational gradients may be a crucial feature of tropical conservation areas. Many fruit- and nectar-feeding vertebrates and insects require a continuous supply of food and one way they can accomplish this is by moving among communities searching for new food sources. A steep elevational gradient is typically occupied by a series of different plant communities, so migrating up or down a mountain slope is a common and efficient behaviour for an animal in search of new food sources.
4. Mangroves are communities of woody plants that occupy the narrow intertidal band in many tropical and subtropical regions of the world. Mangroves, complex communities themselves, are also important as nurseries for the juvenile stages of many marine species and as feeding grounds during high tide for many estuarine fishes and crustaceans.

Keystone resources may occupy only a small proportion of a conservation area yet be of crucial importance in maintaining many animal populations. The loss of a keystone resource could mean a rapid loss of animal species. For instance, removal of dead standing trees and fallen trees from forests may eliminate nesting and overwintering sites for many birds and other vertebrate species. When vertebrate species are lost, there may be linked extinctions of plant species that depend on the animals for pollination and seed dispersal.

MEASURING BIOLOGICAL DIVERSITY

In addition to the definition of biological diversity generally accepted by conservation biologists, there are many other definitions of biological diversity that have been developed in the ecological literature as a means of comparing the overall diversity of different communities. These concepts were used to test the theory that increasing levels of diversity leads to increasing levels of community stability. However, this line of research eventually led to the conclusion that there was no simple relationship between diversity and stability. The indices of biological diversity developed in the course of this research are primarily useful for comparing particular groups of species among communities. As Kimmins says 'Diversity can refer to all organisms in the community, but it is more frequently used to refer to one type or group of organisms. Thus we can talk about the diversity of vascular plants, of birds, of mammals and of the soil fauna'.

At its simplest level, diversity can be defined as the number of species found in a community, a measure known as species richness; but most definitions also include some measure of how evenly the total number (or abundance) of individuals is divided among species. For example, if there are 10 different bird species in a community of 60 individual birds, an even abundance would be 6 birds per species, while an uneven abundance would be 2 birds per species in 5 species and 50 birds in the sixth species. In the first case, no species would be considered to be dominant, while in the second case the community would be dominated by the sixth species.

In writing about diversity, Pielou maintained that diversity is 'a single statistic in which the number of species and evenness are compounded.' Many methods of calculating diversity have been proposed that combine these two types of information and while these methods generally give similar results, some are more strongly affected by sample size and species rarity than others. Also, it is problematic to evaluate communities in which individuals are of very different sizes, such as plant communities. If a community consists of a few large beech trees and many small painted trilliums, should the trilliums be considered dominant because there are more of them or the beech trees because they contain most of the biomass of the community? There is no easy answer to these and many other questions about mathematical diversity indices, which makes such indices useful primarily for ecological comparisons and less useful for applied conservation biology. As the Committee of Scientists concluded, the U.S. Congress intended the term diversity 'to refer to biological variety rather than any of the quantitative expressions now found in the biological literature.' This is echoed in a later conference on the management implications of biological diversity: 'Diversity indices must be used as an analytical tool and not used to define diversity'. In their present form, diversity indices only capture part of the broad concept of biological diversity.

Mathematical indices of biodiversity have also been developed to connote species diversity at different geographical scales. The number of species in a single community is usually described as alpha diversity. Alpha diversity comes closest to the popular concept of species richness and can be used to compare the number of species in different ecosystem types. The term beta diversity refers to the degree to which species composition changes along an environmental gradient. Beta diversity is high, for example, if the species composition of moss communities changes at successively higher elevations on a mountain slope, but is low if the same species occupy the whole mountainside. Gamma diversity applies to larger geographical scales; it is defined as 'the rate at which additional species are encountered as geographical replacements within a habitat type in different localities. Thus gamma diversity is a species turnover rate with distance between sites of similar habitat or with expanding geographic areas'. In practice, these indices will often be highly correlated. The plant communities of the Amazon, for instance, show high levels of diversity at the alpha, beta and gamma scales.

To Sum up

1. The Earth's biological diversity includes the entire range of living species. Biological diversity also includes the genetic variation that occurs among individuals within a species. At a higher level, biological diversity includes the biological communities in which species live and their interactions with the physical and chemical environment.

2. Genetic variation within species arises through the mutation of genes and the recombination of genes during sexual reproduction. This genetic variation allows species to adapt to a changing environment through the process of natural selection. In some cases this process leads to the evolution of new species. Genetic variation within species also allows scientists to breed domestic plants and animals suitable for human use.
3. Species interact within biological communities through such processes as competition, predation and mutualism. Species occupy distinct trophic or feeding, levels within communities that represent the ways in which they obtain energy. Individual species often have specific feeding relationships with other species that can be represented as food chains and food webs.
4. Certain keystone species appear to be important in determining the ability of other species to persist in a community. These keystone species are often top carnivores but may also be inconspicuous species. The loss of a keystone species from a community can result in a cascade of extinctions of other species. Certain keystone resources, such as water holes and salt licks, may occupy only a small fraction of a habitat, but may be crucial to the persistence of many species in an area.
5. Mathematical indices of biological diversity have been developed to examine and compare patterns of species distribution at local and regional levels. Such mathematical indices are chiefly useful for examining particular groups of species rather than the full range of species and interactions found in nature.

17

Creation of Biodiversity

INTRODUCTION

Patterns of biodiversity are created by ecological and evolutionary processes. Lessons from ecology and evolution are important for conservation. This chapter covers: (i) origins and remits of ecology and evolutionary science; (ii) ecological patterns and processes; and (iii) evolution and the diversification of species.

Biodiversity is the outcome of evolutionary and ecological processes. The purpose of this chapter is to provide a secure foundation from which you can explore these themes without feeling like you are swimming through blancmange.

The ecological domains are defined by the population, community or ecosystem and their interactions with the environment. The phenotype is central. Relevant ecological processes range from the autecology of individuals through to whole ecosystem function. The patterns and processes have an impact on genetic diversity but as a consequence. Ecological patterns and processes within communities and ecosystems may ignore the genetic domain which is manifested by species. The evolutionary domain is defined by a focus on genetic processes, patterns and consequences and the variation they create. The genotype is central. Relevant evolutionary processes range from adaptation (as genetic selection in response to the environment) through microevolutionary changes within a species to macroevolution, speciation and extinction. These changes may be driven by the environment but the genetic diversity remains the theme, linked to the species. Some genetic processes (e.g., mechanisms for mutation and variation) could function without any environmental influence.

EVOLUTION, ECOLOGY AND THE ORIGINS OF CONFUSION

Origins of Ecology and Evolutionary Science

A fascination with the origins, patterns and causes of nature's variety can be traced back at least as far as Greek texts, most famously Aristotle's (fourth century BC) *Historia Animalium* listing over 500 species followed by Theophratus (fourth to third century BC) with a *Historia Plantarium* describing plants and their uses. From the start attitudes to nature split between a vision of purposeful, harmonious systems versus nature as nothing more than individualistic bits linked like a machine. These classical writers may be an artificial horizon, a sort of Cambrian explosion of surviving texts. Ancient cave paintings (e.g., Vallon-Pont-d'Arc), unintelligible but haunting murals of humans and animals, hint at an awareness of the interactions of nature in humanity's archaic history. Medieval and pre-Enlightenment scholars maintained a robust interest in nature's variety, interactions and importance. The thirteenth century *Summa Theologie* of St Thomas Aquinas analyses the orderliness of nature, including the sentiment that could be drawn from the most recent biodiversity texts on ecosystem function: 'It is better to have a multiplicity of species than a multiplicity of one species'. Other insights were less useful—biodiversity included many dubious mythical wonders.

The European Enlightenment revolutionised attempts to classify the natural world. New classifications used the physical resemblance of animals and plants to catalogue and to create an order, in keeping with the growing intellectual confidence of the age. Older magical, symbolic, sometimes medical categorisations were lost. Many Enlightenment systems were firmly grounded on a vision of divine creation with species as immutable types, any new species at best filling up preordained pigeon holes. Most famous is the work of Carolus Linnaeus, a Swedish biologist, who devised the system for applying classifying and naming species still in use nowadays. The themes of other natural historians are familiar in present-day biodiversity research. In the 1700s John Harris published *Lexicon Technicum*, describing natural history as 'a description of any of the natural products of the earth, water or air, such as beasts, birds, fishes, metals, minerals, fossils, together with such phenomena as at any time appear in the material world, such as meteorites'. In 1749, Georges, Comte de Buffon, Keeper of the Royal Zoological Garden in Paris, started publication of his *Histoire Naturelle*, intended as an inventory of the Earth and its wildlife, highlighting patterns of geographical distribution, adaptation and migration, a global biodiversity assessment of its day.

Modern Ecology: Illegitimate Origins and Confused Childhood

The disparate heritage of ecology caused problems defining the legitimate territory of this science. The term ecology (then spelt oecology) was coined by the German zoologist Ernst Haeckel in 1866, in part to categorise the topic as a mere aspect of physiology and distinguish it from biology. Ecology was an awkward combination of the traditions of natural history (everything from grotesque menageries through to the ideas of Darwin) and the other major thread of nineteenth-century natural biology, physiology. At various times both traditions subsumed or abhorred the newcomer. In 1893 Sir John Scott Burdon-Sanderson, President of the British Association for the Advancement of Science, distinguished ecology as one of the three divisions of biology (physiology and morphology were the other two), with ecology defined as 'a philosophy of nature'.

Ecology is not a radiation of ideas from a secure central core nor a synthesis of ideas into a greater whole. It is an umbrella of concepts. Three blights (no secure pedigree, multiple parents and a subject matter unloved by the great biological themes of the late nineteenth century) got the subject off to shaky start.

A fourth source of confusion, in part a reaction against this sense of illegitimacy, soon developed. During the early twentieth century there was an emphasis on making field studies rigorously scientific to bolster ecology's status, to understand nature all the better to master and improve upon it. There was a desire to make ecology coherent by developing a quantitative, mechanistic, reductionist approach, as if nature could be stripped down like a big machine, echoing the rigours of physics and chemistry. Ecology had transferred its interest from explanations dominated by history to explanations that forgot time. Ecology as a science had no clear past and now ignored historical time as a process. G.E. Hutchinson's famous paper of 1959, 'Homage to Santa Rosalia or why are there so many kinds of animals', is a revealing example of the shift in emphasis. Specifically addressing global animal species' richness, Hutchinson lists factors to consider for a theory to predict the total—food chains, plant-animal interactions, complexity and stability of ecosystems, productivity, size, environmental rigour, competition and the environmental mosaic. The paper omits historical ecology.

Ecology was riven by rival schools of thought—population/community ecology concentrating on species and population versus ecosystem ecology analysing the cycles and flow of energy and resources with a sense of regulation and balance. Alternative visions included communities as integrated super-organisms versus individualistic assemblages. This contest between ecological systems as benignly self-regulating or driven by selfish contest echoed the Enlightenment, even the Greek, debates between a vision of nature driven nature by purpose versus trial and error. Then, just when ecology was trying to look grown up by talking in numbers, growing public awareness of environmental problems in the 1960s saw the term ecology purloined as a general social and political creed. Many scientists were wary of or annoyed by the blurring of their science with environmental politics. The emergence of a science of biodiversity has to some extent healed this rift.

ECOLOGY OF BIODIVERSITY

Search for Patterns and Processes

This section faces the very same challenge to produce a coherent review of the many ecological creating diversity and controlling its distribution: first, a hierarchy of factors differentiating three broad types of ecological factors, primary influences of time and space, second, templates such as physical habitat structure or patterns common to food webs and third, intimate processes depending on individual species actions and attributes.

Primary factors, geographical and physical

Primary ecological factors are large-scale global or regional physical influences. They determine the general environment within which the other ecological factors create local, intimate detail or disruption:

1. History and age: biodiversity is generally greatest in oldest ecosystems.
2. Gradients: biodiversity changes across environmental gradients (e.g., latitude, altitude, depth, aridity and salinity).
3. Area: biodiversity increases with increasing area.
4. Isolation or islandness: biodiversity decreases with increasing isolation.

Ecological Templates

Biodiversity at regional or smaller scale suggests patterns or structures to which species diversity adheres or is driven to fill up. Structures may be literal (e.g., the physical architecture of a habitat) or apparent (e.g., rules governing how species link together in food webs).

1. Productivity—available energy: globally biodiversity increases with productivity but locally increased or decreased productivity typically lowers biodiversity.
2. Scaleless patterns—ecological factors such as area or productivity have a measurable scale (e.g., diversity per hectare or relative to solar energy). Scale-less patterns include the consistent patterns found in food webs, patterns of species richness relative to population abundance and body size.
3. Habitat heterogeneity and patchiness—the greater the variety of types of habitat, the greater the diversity of species.
4. Habitat architecture—diversity increases with increasing architectural complexity of the physical habitat.

Interactive, intimate factors

Primary factors and ecological templates typically promote or reduce biodiversity in a predictable way—biodiversity increases with increasing area or decreases with decreasing habitat complexity. The third set of factors can enhance or suppress diversity depending on the spatial and time scales. They are characterised by the local, intimate scale at which they operate, directly dependent on the attributes, behaviours, interactions and susceptibilities of the species involved:

1. Succession: successional processes comprise the arrival, establishment and replacement of species through time. Changes may be driven by environmental agents and species altering habitats.
2. Interactions between species: interactions between species influence short-term, local diversity and power evolutionary diversification as winners and losers, exploited and exploiters struggle for dominance.
3. Disturbance: local physical destruction, as against regional, climatic variations linked to gradients, may destroy diversity. Equally, disturbance may be frequent enough to prevent a limited gang of species monopolising a habitat but not so frequent that everything is wiped out.
4. Dispersal and colonisation: dispersal (active or passive) and colonisation (arrival plus survival) depend on the attributes of individual species and represent the species interface with the primary factor of isolation.

Primary Factors

History and age

The biodiversity of an ecosystem will depend on the time it has existed and previous history. History as an explanation for present-day patterns of biodiversity was the factor that twentieth century ecology forgot. In part this was because history seemed little more than a *Just So story*; species are where they are because that is where they were, the patterns untestable without the use of a time machine. Modern biodiversity is inevitably contingent on past events. Increased understanding of the Earth's history has sharpened our awareness of history as a factor underlying existing biodiversity.

The distribution of related taxa across widely separated continents, especially in the southern hemisphere, had proven a challenge to nineteenth century biologists. The scatter was explained by lineages jumping across oceans. The revelation of continental drift during the 1960s permitted an alternative view. Present-day distributions are the result of two historic processes: the origination of lineages on a united super-continent and the subsequent split and continental drift of fragments. This physical splitting is vicariation and the resulting diaspora called vicariance. Vicariance patterns depend on where the common origin range was, how the now separate fragments once joined and where splits arose. Large-scale dispersal of lineages, which might include some jump dispersal, is responsible for range expansions, which then might suffer vicariation. In addition to continental drift a gradual expansion of the Earth resulting in a stretched crust may be a historical factor forming vicariant patterns (see Fig. 17.1).

A second major historical influence are refugia, discrete areas acting as sanctuaries during times of global stress. They form the shrunken fragments of larger biomes and expand, reforming into wider swathes during benign periods. The idea was developed from the fate of tropical rainforests during the Pleistocene glaciations of the northern hemisphere, though could apply to many global biomes. Not only do the refugia conserve existing biodiversity, but also the fragmentation may increase net diversity due to speciation, the new species surviving later reunion of refugias. The great age of tropical rainforests has been used as an explanation that seems to contradict the refugia hypothesis, the stability-time hypothesis. Tropical rainforests are ancient, stable habitats, allowing prolonged accumulation of species. Rainforests have existed for tens of millions of years. Pleistocene glaciations are much more recent. Both historic processes have added to their riot of biodiversity.

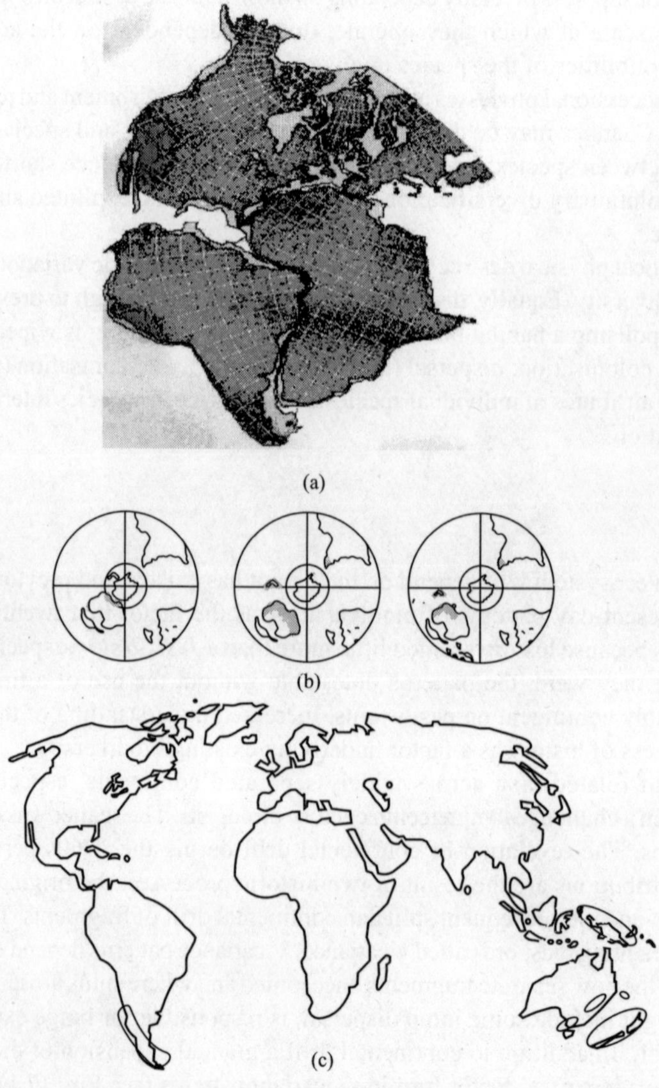

(a) (b) (c)

Fig. 17.1. Distribution of the Southern Beech, genus Nothofagus, a classic example of vicariance: (a) the modern continents when joined as Gondwanaland; (b) the subsequent separation of southern continents. The shaded areas show areas with fossil evidence for the presence of *Nothofagus*; and (c) modern distribution of *Nothofagus*, apparently very disjunct but in regions historically close together.

Age and history are important: massive physical processes such as continental drift and glaciation linked to dispersal of lineages and evolution explain the broad patterns of biodiversity distribution and endemism. Large-scale biodiversity differences between continents are contingent on the past. We would not have the unusual faunas of Australia and Madagascar if it were not for their history.

Gradients

Biodiversity does not litter the planet at random but typically changes across environmental gradients (see Fig. 17.2). The most famous gradient is the increase in biodiversity from high latitudes, the poles, to low latitudes, the tropics. Trends of light, temperature, climate and seasonality facilitate diversity. Increased diversity towards the tropics occurs in land mammals, birds, reptiles, amphibians, insects, snails, clams, trees and plankton. Marine fish diversity increases towards the tropics but some marine trends are sharply stepped, linked to distinct fronts between massive water fronts. Exceptions to the trend include some marine groups, e.g., penguins, perhaps the result of their south polar origins, although the diversity of ecologically similar Auks in high northern hemisphere water hints at other influences. Terrestrial exceptions include conifers, which are most diverse in temperate latitudes. The decrease of diversity with altitude is a response to similar environmental conditions, with similar exceptions of conifers most diverse at mid-altitudes of mountain ranges.

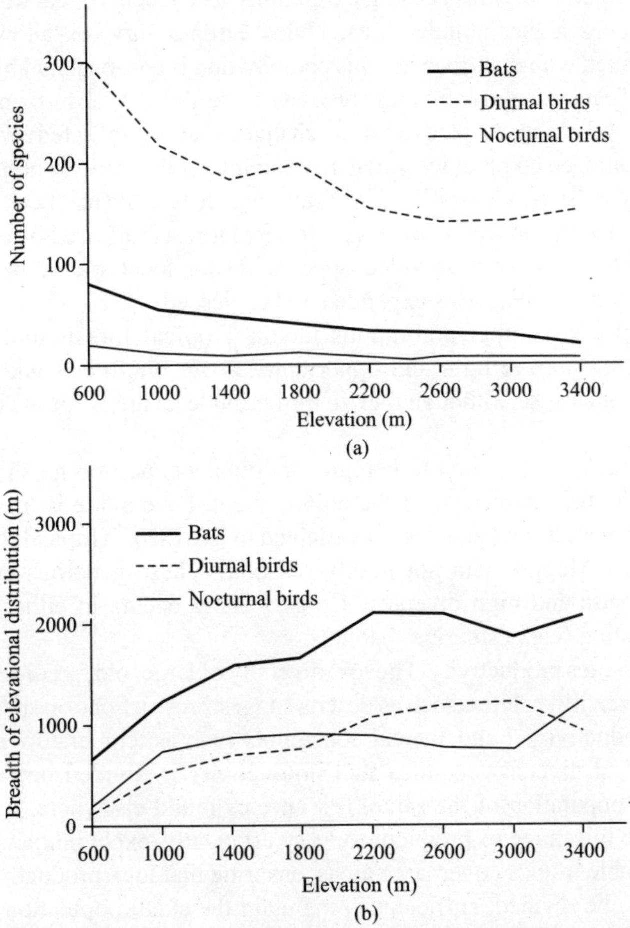

Fig. 17.2. Numbers of bat and bird species recorded over an altitude gradient in the Peruvian Andes. Numbers of species decline with increasing elevation. As altitude increases bat species show wider ranges, perhaps having to roam further to survive, in turn restricting numbers that may live together. Oddly birds do not show this trend, so different taxa may be affected by gradient in different ways.

Biodiversity changes with depth underwater but the deep sea remains largely unexplored, so generalisations are risky. Diversity seems to decrease with depth, certainly in high diversity habitats such as coral reefs. Other marine systems show greatest species richness at intermediate depths. Samples of deep-sea floor life increasingly hint at a rich variety. Depth is not simply an underwater analogue of altitude but a response to the synergy of light, temperature, pressure and dissolved substances such as oxygen. Geographical gradients also include aridity and salinity. Diversity is lower in arid regions compared to wet zones of similar area, altitude and distance to the sea. Aquatic diversity is highest in either sea or freshwater but decreases sharply at brackish concentrations in between.

Gradients can be divided between regulation gradients where the controlling factor is not used directly (e.g., altitude) or resource gradients where it is (e.g., nutrients), altering availability. Although biodiversity gradients are so striking, explaining how they arise is difficult.

Most attention has focused on the polar-tropical gradient. Time may be a factor. Pleistocene glaciations smothered polar and temperate regions causing extinctions. The recent retreat of the ice has not allowed time for species to recolonise higher latitudes. Meanwhile the tropics survived, allowing time for speciation. So time as a factor is linked with disturbance. This combination is encapsulated by the idea of ecosystem stability. The stability of ecosystems (combining resistance, the ability to absorb impacts without suffering change and resilience, the ability to recover from changes that are inflicted) was seen as a dominant influence. Stability encouraged complexity which in turn reinforced stability. These ideas are encapsulated as the stability-time hypothesis. Contrarily disruption and instability has been cited as promoters of tropical diversity. Contraction and fragmentation of tropical forests followed by expansion and fusion in response to glacial cycles may have provided opportunity for local speciation in the fragments, the diversification surviving once the forests expanded and coalesced.

Habitat heterogeneity is another ambiguous factor. Tropical forests and coral reefs are more heterogeneous, providing more refuges and opportunities. But rainforests and coral structures are a response to gradients, not a cause, although they in turn provide an arena for the diversification of other life dependent on them.

Tropical diversity seems closely tied to benign conditions neither too harsh or unpredictable. This link has been termed the favourableness of the environment. Favourable is tricky to define, not least because a harsh environment is not harsh for taxa adapted to live there. Tropical forests are warm but not too hot, wet but not waterlogged and not highly seasonal. These generous conditions permit high productivity, high biomass and high diversity. Low diversity occurs in either constant but extreme (e.g., deserts) or fluctuating (e.g., estuarine) habitats.

Favourableness promotes productivity. The low diversity of large, old, predictable (all characteristics supposed to promote diversity) habitats such as deserts suggests restrictions on speciation or colonisation, perhaps due to low productivity. Fundamental constraints such as temperature and light act as severe limitations. High tropical diversity requires that productivity is divided between more species not monopolised by larger population of the same, few species found elsewhere. A diversity of specialist species may exploit the full range of productivity very efficiently, excluding a more limited variety of generalists. The favourable tropics cover large areas, ensuring that total productivity is high, effectively a bigger energy cake to be divided, sufficient to maintain the small population typical of many rarer species, without undue risk of accidental extinction associated with tiny population or ranges. This possibility is termed the species-energy hypothesis.

The lack of physical constraints in the benign tropics may permit more interactions between species, more intense, more specialised, more likely to reach a conclusion as arms races spiral and competitors outwit one another. The interaction hypothesis is awkward to demonstrate and circular—diversity begets diversity. Nonetheless, interactions once accelerated in favourable conditions may snowball.

An alternative to this tangle of favourableness, productivity and interaction is that tropical high diversity is nothing more than a variation on the theme of another primary factor, area. The tropics harbour greater diversity because they cover a greater area. Not only is their extent absolutely larger than other global biomes, but also northern and southern hemisphere tropics join, unlike the separation of temperate or Arctic zones.

Area

Biodiversity increases with increasing area. This rule of thumb remains an awkward ecological pattern to explain. A large area of polar habitat will contain many fewer species than a smaller area of rainforest; glib comparisons can be dangerous. The pattern may be little more than a sampling artefact since bigger areas act much like bigger nets, collecting up more species without any especially ecological mechanisms at work.

The link between area and biodiversity is enshrined in species—area relationships. Numbers of species increase with increasing area. Data depicting this pattern are commonly plotted to give a species-area curve, the increase of species with area typically slowing as the pool of species still to find is used. Species number and area are often plotted transformed into their logarithms, resulting in a plot approximating to a straight line rather than a curve. Straight lines are much easier to model as equations and many species—area relationships can be described by the equation $S = cA^Z$. S is the number of species and A the area; c and z describe how species numbers change with area; c is the number of species per unit area and then z adjusts the effects of area which may vary between different habitat types. The hunt for consistent patterns has focused on z. Very similar ranges of z, between 0.25 and 0.3, for different habitats have been cited as evidence supporting some fundamental ecological rule governing species diversity in relation to area but z can be much more varied (see Fig. 17.3).

Explanations for the effect of area on diversity vary from suggestions that the patterns are little more than statistical artefacts through to area as the predominant motor of ecological diversity. If all species were distributed at random, a bigger area would contain more, but since they do not occur at random, except perhaps at a local scale, this simplest explanation does not hold. Other non-random distribution patterns may generate species-area curves. Many ecosystems support total numbers of species and population of each that follow a very particular pattern. Many species occur at low or moderate densities, a few at very low and a very few at very high abundances. Plot the numbers of species with population in each category and the distribution of species across population categories fits a pattern called the log-normal. Different sized areas drawing their species from a total pool that matches a log-normal pattern can create a classic species-area curve. The important ecological processes are those that produce the log-normal pattern in the first place, not the area.

Larger areas contain a greater variety of habitats—habitat heterogeneity. Not only will larger areas contain more types but also the types will include increasingly different habitats. Combine the increased variety of habitats as area increases with the variations in detail at different scales and area's effect in determining species number may be the result of habitat heterogeneity.

The extent of an area will affect evolutionary processes, the opportunity for speciation and risk of extinction. Biodiversity differences between provinces and continents of varying size echo their history. The species-time hypothesis recognises the role of time, larger areas harbouring more species due to more opportunity to speciate and less risk of extinction.

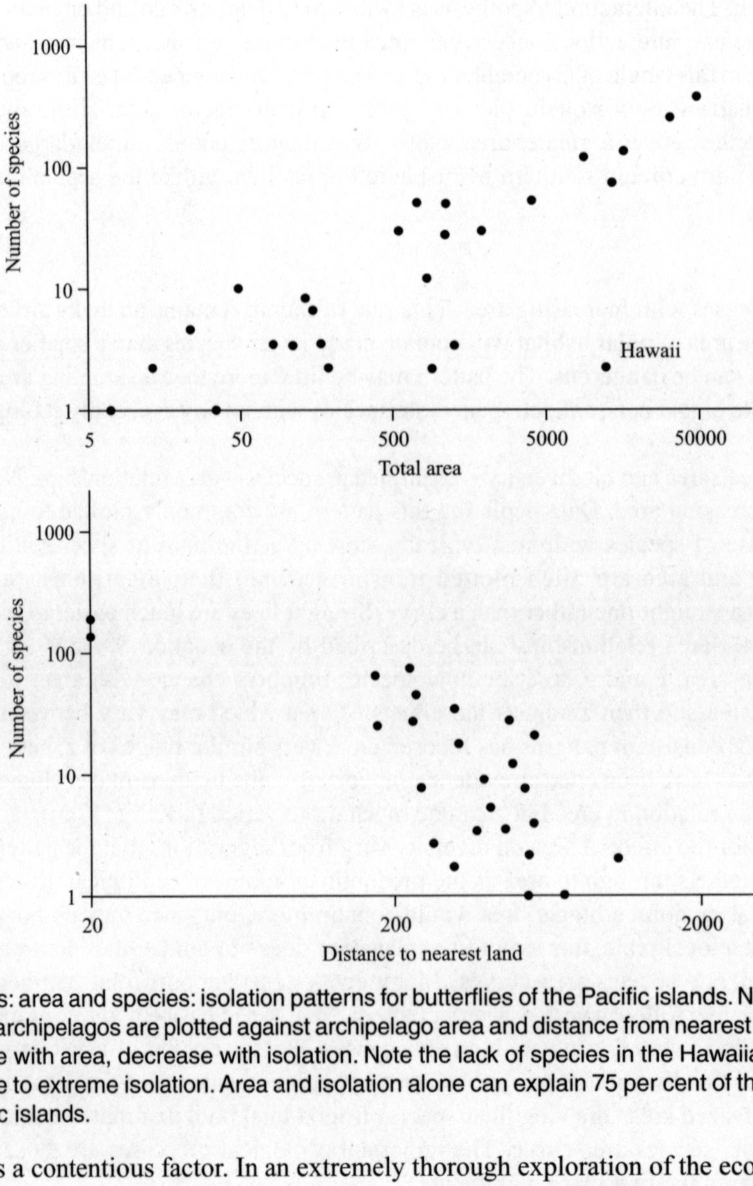

Fig. 17.3. Species: area and species: isolation patterns for butterflies of the Pacific islands. Numbers of species found in different archipelagos are plotted against archipelago area and distance from nearest mainland. Species numbers increase with area, decrease with isolation. Note the lack of species in the Hawaiian islands, despite their total size due to extreme isolation. Area and isolation alone can explain 75 per cent of the butterfly species diversity of Pacific islands.

Area remains a contentious factor. In an extremely thorough exploration of the ecology of diversity, Rosenzweig entitles a summary section 'A large piece of the puzzle: area effects', combining statistical, habitat heterogeneity and evolutionary processes. Other authors demean area as little more than a handy surrogate for the underlying mechanisms.

Isolation

More isolated habitats harbour a reduced biodiversity. Isolation is not simply the distance between a source of colonists and an island but the interaction of distance, species dispersal mechanisms and any special hazards disrupting access. Rosenzweig distinguishes mainland (essentially area) from island (isolation) patterns. Islands include not only literal oceanic isles but also separate patches of the same habitat scattered through the landscape. Genuine islands may be self-contained with species originating entirely from immigration and none depend on continual topping up by immigrants for survival. Mainlands may be populated by taxa that originated from speciation within the area. By this definition Hawaii is a mainland.

Island biogeography, epitomised by MacArthur and Wilson's Theory of Island Biogeography, depends on two processes: immigration, which varies with isolation from colonisation source to island and extinction, which depends on island area. As more species colonise, the pool of potential colonists is used up. In the mean time extinctions increase, perhaps simply because every species has some chance of accidental extinction or maybe because of interactions between the increasingly cramped arrivals. An equilibrium diversity is reached once immigration and extinction cancel each other out. This may be a dynamic equilibrium, as species colonise, go extinct and recolonise. A turnover of species is common, though hard to measure since a species may be found, then go extinct and re-establish before the next survey, so-called cryptoturnover.

Island biogeography predicts that should an island be perturbed so that it harbours more or less species than the equilibrium and assuming the perturbation does not alter the island habitat, then the equilibrium will re-establish. Experiments following the fate of islands intentionally stripped of fauna suggest that not only does the equilibrium restore but also other features such as the general patterns of predator and prey species, the trophic structure, are rebuilt to mirror the pre-disruption status quo. However the actual species involved may differ. Such results suggest there are places to be filled, limits and opportunities for diversity.

Isolation and area create source and sink habitats. Sinks are patches in which a species is unable to maintain a viable population without topping up by immigrants. Sources are areas where a species' reproduction is sufficient for survival. The arrival immigrants to a sink patch is called the rescue effect.

Isolation and inter-patch processes have been developed as metapopulation ecology. A metapopulation is a set of population in separate patches but linked by immigration (see Fig. 17.4).

Ecological Templates

Productivity

On a global scale biodiversity increases with rising productivity, one factor influencing planetary biodiversity gradients. At a local scale productivity may limit or promote species diversity. At low productivities diversity is low. Some species simply cannot survive on the limited resources. This may be a particular limitation to specialists. At very high local productivity, diversity, declines as a few, competitively dominant species, freed from resource constraints, monopolise the habitat. Other species are ousted, unable to resist the competition or perhaps simply unable to utilise the riches. The increased nutrients may also promote dynamic instability in food webs, the resulting species interactions leading to extinctions as population fluctuate dangerously. At intermediate productivity levels there are sufficient resources to allow many species to survive but none may take over; this is the intermediate productivity hypothesis. Intermediate productivity may maintain diversity as the variety of life generates local habitat heterogeneity, both of physical structures (e.g., plant growth forms) and of resources such as nutrients or moisture. So productivity links to the resource needs of individual species and competitive interactions. Productivity has also been linked to disturbance in the dynamic equilibrium model of species diversity. In this model diversity is determined by the synergy of production (and its effects on population growth and competition) and disturbance (varying in frequency and intensity).

Productivity may also tie in with scaleless patterns of species diversity such as species—abundance patterns. If species diversity fits species—abundance patterns which are not purely ecological but much more general rules governing collections of things and the productivity pie is sliced up accordingly, then the more pie (the greater the productivity), the more species will get a share, so occur in a habitat.

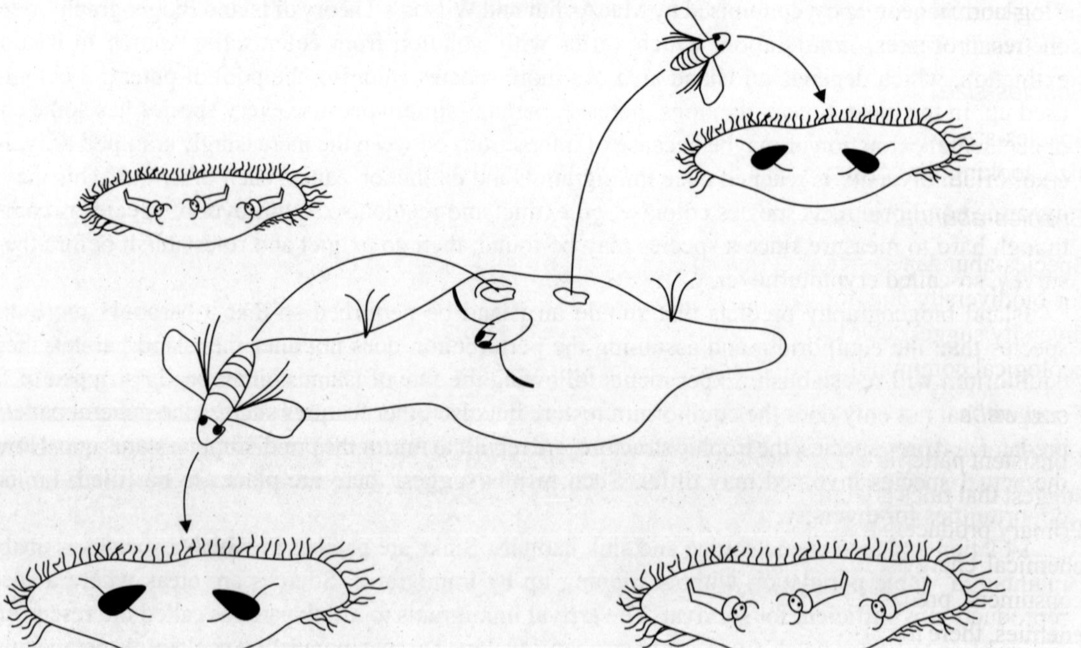

Fig. 17.4. A metapopulation of greater waterboatmen, *Notonecta* sp. The boatmen survive in the permanent central (source) pond. Peripheral ponds may be colonised by flying adults, which then breed. Ponds not colonised contain prey species that are wiped out in the presence of the boatmen. The peripheral ponds dry out each year so that the ecological stage is reset. Note how habitat patchiness and disturbance, dispersal and predation all affect local diversity.

Local declines in diversity at increased productivity have been shown in terrestrial and aquatic habitats, especially where nutrient enrichment occurs. Enriched aquatic systems lose diversity as algae, able to respond quickly to increased nutrients, bloom and smother larger plants. Enriched grassland swards also lose species as a few coarse grass species take over. In both cases removal of nutrients can reverse the trend and local diversity recovers.

Scaleless patterns

Scaleless patterns include apparent correlations between body size, population abundance and number of species, plus rules for the structure of food webs. The relationships between species numbers, population abundance and body size have been a focus of recent work. Plot the numbers of species in a community against their population and some consistent species-abundance patterns emerge. Models to explain these patterns focus on how species divide up available resources (resource apportionment) or statistical patterns that apply to collections of things in general be it ecological or otherwise. For example the 'geometric series' assumes a first species to establish monopolises a share of the resources, the second species in takes the same percentage of the remaining resources, the third the same percentage of what is left and so on until all resources are used up. Other models have no specifically ecological roots and make no assumptions about what the species do. Such models can just as well describe patterns of diversity in collections of any objects, whether insects, kitchen cutlery or chocolate bars. For example the log-normal distribution pattern arises when the abundance of items (in this case, population of species) is the result of many random, independent influences.

Species-size

Species-size patterns take a similar approach, plotting the number of species against categories of body size, looking for consistent patterns and trying to explain these with general models.

Species-abundance-size

Species-abundance-size patterns have proved frustrating. There are some potentially important lessons for biodiversity and conservation, in particular if any consistent patterns linking abundance to size or diversity suggest that there are some population sizes simply too small to survive. It is possible to find ecological communities to fit each model but none are generally applicable.

Food webs

Consistent patterns in food webs (species linked as those higher up the web exploit those below for food) suggest that rules govern web structure. A species' position in a web is defined by how it captures energy. Primary producers trap the raw energy of sunlight, sometimes other sources and convert this into stored chemical energy. Primary consumers including herbivores eat the primary producers and secondary consumers, predators and parasites, prey on the primary consumers. Since predators have their own enemies, there may be several levels of secondary consumers. Each of these categories is a trophic level.

Food webs show consistent patterns. First most contain no more than three or four levels. Longer webs can be found, more out of a love of the unusual. For example, some marine webs may go up to eight, assuming tortuous series of links from tiniest plankton to largest sharks, by way of four or five levels of secondary consumers eating one another. The consistent rule stays sound, most food webs reach three or four trophic levels. The second pattern is that omnivores, defined as a species that takes at least 5 per cent of its food from another trophic level than its main supply, are rare. Third ratios of numbers of predator to victim species, links per species and links between different levels are also strikingly consistent in real webs. As diversity of one increases so does diversity of the others.

Limits to food web length were originally ascribed to energy transfer from one level to the next. Only 10–20 per cent of energy eaten is assimilated, the bulk being lost via excretion or metabolism, so available energy rapidly peters out up the chain, leaving too little to support any more levels. If available energy was the limiting factor, webs from more productive habitats should be longer but they are not. Only in the most unproductive habitats (e.g., terrestrial Arctic) is there evidence for truncation due to low energy. An alternative to energy limits is dynamic stability of webs as they become ever longer. Model food webs can be built to mimic realistic producer, predator and prey interactions, the fate of population tracked over many generations. The longer the model food webs, the more unstable they become. As population rise and fall, interactions ripple through the web and feedback lurches through the system. Long webs lose levels, collapsing back to more stable two, three or four levels. Dynamic models generate realistic web length limits, realistic predator-victim ratios, numbers of links and rare omnivory.

Energy flow and dynamics may be linked. In low productivity systems stability increases with an increase in productivity. In high productivity systems any further increase in productivity promotes instability. Taken together productivity may limit web length in low productivity habitats, instability in high productivity systems.

Realistic links and ratios can be created by a third, strikingly simple, theory, the Cascade model. Ignoring population dynamics and energy the Cascade model suggests that species are arranged in a hierarchy, perhaps by body size, so that each feeds only on those below and is fed on by those above. The total number of links per species is fixed, but how species link is randomised. The Cascade model predicts web length limits poorly.

Habitat heterogeneity

A varied physical environment harbours a greater biodiversity. Heterogeneity refers to the kinds, variety and distribution of different types of habitat. All environmental factors vary across time and space. The more variation, of size, age, type and amount of patch, the greater the diversity of life. Patches can be physically discrete areas or variations in the distribution of resources. This patchiness, coupled with changes over time, mitigates against a few species monopolising the environment.

No species can be good at exploiting every environment. There is a trade-off, strong evolutionary pressures to optimise strengths whilst losing less successful attributes. The process will be self-reinforcing, with less and less to be gained from trying to exploit patches that other species can utilise more effectively. Even an apparently homogenous environment will become varied as different species colonise, altering their local environment. The result is that different patches will vary in their usefulness to species (see Figure 17.5).

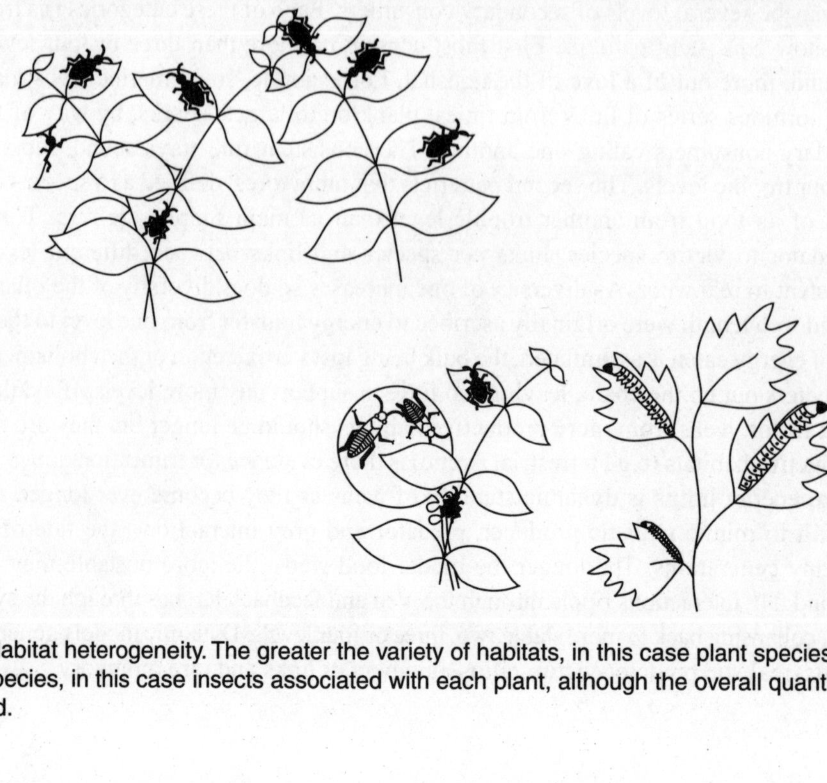

Fig. 17.5. Habitat heterogeneity. The greater the variety of habitats, in this case plant species, the greater the variety of species, in this case insects associated with each plant, although the overall quantity of habitat has not changed.

Patchiness mitigates against monopoly by just a few species. Changes in one patch (losing out to a competitor) may be compensated for by a different fate in another patch (the victorious competitor wiped out by its predator). Local variation between patches results in regional coexistence of species. This is especially so if the ecology of separate patches is out of synch. For example, if disease strikes at a particular stage in a life history, population will survive in some patches that are not at that vulnerable stage. Habitat heterogeneity has been shown to promote coexistence of predators and prey, e.g., many insect predators such as ladybirds can wipe out an individual patch of prey but cannot find all prey colonies hidden in different patches and of competitors, e.g., many plant species which differ so little in requirements that competition would be severe coexist regionally scattered across separate patches. So the fate of individual patches links with isolation and area to create regional diversity. Temporal variation will also maintain diversity. Seasonal changes can upset the otherwise inevitable outcomes of interactions between species, resetting the ecological stage. Different species may suffer resource bottle necks at different times of year, even if they are using the same resource.

Habitat diversity (heterogeneity and complexity) and species diversity co-evolve. The richness of habitats in a region depends in part on the species to recognise their existence. Species discriminate the finer grain of habitats (be it patches or complexity) because selection forces them to. Natural selection finds the habitat diversity. Some regions with apparently very limited environmental diversity, e.g., the Fynbos of South Africa or Rift Valley Lakes, can support a surprising diversity of species (plants and fish respectively) that show a very fine-grained division of the environment.

Habitat complexity

The more complex the architecture of a habitat the greater the diversity of species. Complexity refers to the structural architecture of each habitat. The amount and shape of structural complexity in the environment will create opportunities for species. A structurally complex ecosystem (e.g., the three-dimensional architecture of a forest) provides more physical habitats for species than a simple one (e.g., an equivalent area of tidal mud). Habitat complexity sounds intuitively obvious but is difficult to untangle from other factors. For example in the forest-mudflat comparison above, the diversity of plants in an area of wood provides many more sources of food for herbivores (and in turn their enemies) than a mudflat (see Fig. 17.6).

An important insight into complexity is that there is more of it at smaller scales. Many natural phenomena, whether soil structures or vegetation, are fractal. They do not have a finite size but measurements of dimensions such as area or edge increase the finer the scale at which they are measured. To a one meter tall grazing deer a small herb like a buttercup is scarcely a mouthful; to a 5 mm long insect a buttercup is a complex climbing frame, with leaves, shoots, flowers and buds, each offering different resources. So the availability of microhabitats increases at smaller and smaller scales. As a result species biodiversity increases at smaller scales. There is more room for organisms of smaller body sizes. The biodiversity of terrestrial insects may in part be the result of this pattern, linked to the complexity and scale of the physical habitat that plants create. Many habitats are exactly the right scale for the diversification of insect-sized life.

Intimate Processes

Succession

Diversity changes as species establish, interact and alter the environment. The biodiversity of a site increases over ecological time, immigration, establishment and loss of species creating a successional pattern. Succession is the non-seasonal turnover of species over time. The turnover is predictable with recognisable early colonists and late survivors..The progress of succession is directional, the sequence of early and late successional communities repeated across separate patches of the same habitat type in the locale. The directional, replicated changes may converge on a final community. Successional change comprises a sense of direction, of species turnover and of physical change as the biodiversity creates its own structures and alters the environment.

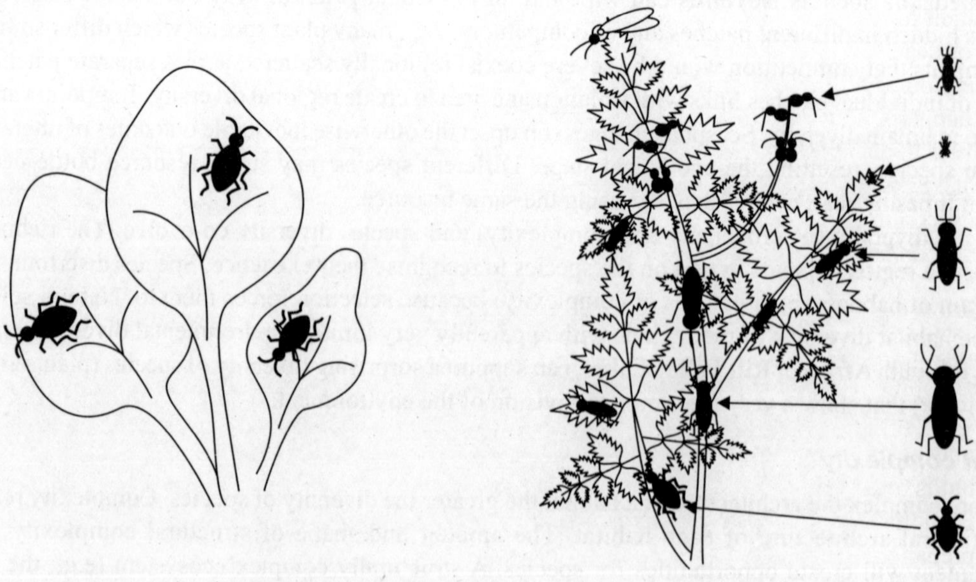

Fig. 17.6. The greater the habitat complexity the greater the diversity of species. In this case the physically complex plant hosts more species. The variety of different sized and shaped spaces on the complex plant creates opportunities for more species, which vary in body size.

Early twentieth century ecology was mesmerised by the hint of rules and progress hinted at in successional patterns. Theories of successional change were developed primarily for plant communities. The sequence of plants is a major factor in creating landscape, so succession may be important in determining plant biodiversity and in turn, the diversity of wildlife exploiting the plants and the physical habitat they provide. Original theories saw successional change as a predicable drive towards a final ecological superorganism, the species of plants so integrated that characteristics of each community were greater than the sum of the parts, the result an inevitable climax community, perhaps fine-tuned to maximise efficient energy transfer. This vision faltered faced with successions leading to multiple outcomes, none conspicuously a climax, successions that appeared to reverse and no evidence for increased energy efficiency. Succession is now widely regarded as the result of individual species' life histories and attributes, each living its own life and the consistent patterns arising because of the same main environmental pressures acting on the same basic set of life forms. The attributes of plants, interactions with and changes to their environment and interactions with herbivores and diseases that exploit the plants in turn drive the process. Succession arises from the intimate ecology of physiology and life histories of species interacting with their local environment. Five processes interact: disturbance that creates an opening, immigration, establishment, competition and reaction as species alter their habitat.

Important species attributes include dynamics (maximum growth rates, size, longevity and establishment) and physiology (requirements for or tolerance of light, water and nutrients).

Succession is a dynamic process and diversity can increase or decrease. Succession on virgin habitat such as a volcanic island or newly exposed glacial morraine is called primary succession. Secondary succession occurs on terrain stripped of plants but retaining previous biological influences such as soil conditioning and seed banks. Succession can be categorised by the driving forces of change. Change in response to external influences (e.g., climate change) is allogenic. If driven by the organisms themselves, altering their habitat succession is autogenic. Degradative succession occurs on organic debris such as a cow pat or dead leaf, the end result being the habitat's destruction.

Succession may be yet another ecological process that is nothing more than a statistical phenomenon that just happens to involve wildlife. Successional patterns can be modelled using nothing more than the probabilities that a species will arrive or be lost, regardless of the previous history of the site or present inhabitants. Species come and go at random.

Huston comments: 'Succession is the key to understanding the regulation of nearly all aspects of biodiversity or ecological time series'.

Interactions between species

Interactions between species can promote or destroy local diversity in the short term but are a major force for evolutionary diversification. Interactions between species fall into four broad categories:

1. Competition—species contest a resource. Where the resource is limited inferior competitors may be driven to extinction. Competition is often depicted as detrimental to all species involved, but is often asymmetrical, with one suffering, the other unaffected.
2. Exploitation—one species exploits another as a food source. Exploitative interactions include predation, parasitism, disease and herbivory. The victim loses, the exploiter gains. Diversity is decreased if exploiters wipe out victims or promoted if the monopoly of a few dominant competitors is prevented.
3. Mutualisms—both species benefit from the presence of the other and their interaction. Such relationships are often tightly forged, by coevolution, into a mutual dependency. Mutualisms promote diversity by opening up new ecological opportunities.
4. Engineering—interactions due to the activities of one species benefiting others but with no reciprocal gain and typically not tightly coupled by coevolution. Ecological engineering may be physical (e.g., beaver dams) or functional (e.g., nutrient cycling).

Competition

Competition between species occurs when they use a resource in such short supply that there is not enough to sustain their potential reproduction, growth, abundance or distribution. Competition was regarded as the major process regulating local diversity. Species exactly overlapping in their exploitation of a resource could not coexist indefinitely. One wins, others lose. Competition could exclude species, lowering local diversity. However, combined with disturbance and patchiness, a mosaic of competitive interactions could result in a variety of outcomes, maintaining diversity. Longer term evolutionary pressures would select for species adapted to avoid competition by niche differentiation (living in a sufficiently different way to avoid competition), character displacement (anatomical differences), niche shifts (short-term modification of niche to avoid competition) and complementarity (species sharing a resource would use it at different times or places). A species cannot be good at everything and there is a strong trade-off between improvement of core abilities and evolutionary abandonment of what is done poorly. Competition provides a selection pressure with species diversifying by refinement of different abilities simply to stay in the evolutionary game. Competition theory suggested that there were limits to how similar species could be and therefore a limit to how many species could pack into an ecosystem but golden rules proved elusive. Existing niche differentiation might show that competition had been an important evolutionary pressure in the past but was not active nowadays, the species and diversity patterns ghosts of competition past. Re-evaluation of competition's role has recognised its importance but as only one of several interactions (see Fig. 17.7).

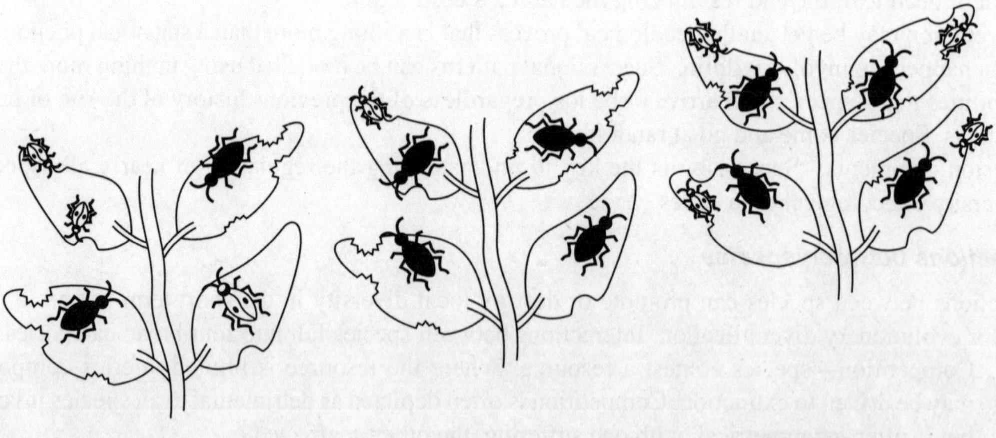

Fig. 17.7. Competition. Two beetle species feed on the same host plant. Competition almost wipes out one in the short term, threatening local diversity, but evolutionary diversification due to separation of feeding sites consolidates diversity in the longer term.

Competition in evolutionary time is also a mixed force. Avoidance of competition by differentiation, complementary rather than overlapping exploitation of resources and exploitation of entirely new resources are powerful forces for speciation. However, the eclipse of some higher taxonomic groupings (e.g., Palaeozoic squid) ousted as dominant marine predators by the rise of the fish represents a loss of diversity.

Exploitation

Competition's fall, from pre-eminence as the process determining local diversity was in part due to increased understanding of the importance of exploitative interactions, especially predation. Predation and competition were often cited as if mutually exclusive rivals but they have very similar effects. Exploitation can increase or decrease local diversity and is a force for evolutionary diversification. The difference lies in the structure of the interaction. Competition for food involves species feeding at the same trophic level, exploitative interactions involve links between different levels (eating your competitor could count as either). The impact of exploiters varies with the intensity of interaction, their behaviours and synergy with competitive processes. Generalist predators maintain diversity by eating anything they find in proportion to its availability or by homing in on the most abundant prey by switching their hunting activity to concentrate on it and massing where the prey is concentrated. Either strategy inflicts most damage on prey species that potentially monopolise a habitat by out-competing other species. Specialist predators may have the same effect if they specialise on would-be monopolisers but can reduce diversity if their target is a rare taxa. Exploitation works like physical disturbance. Too little allows a few species to take over, too much reduces all prey. Intermediate levels generally promote local diversity (the intermediate predation hypothesis). The role of predators prompted the idea of keystone species, which by their presence or absence define local community diversity. The keystone idea is now extended to include other roles such as keystone engineers or mutualists.

Mutualisms

Natural historians recognised the importance of mutualisms, perhaps as evidence of a divinely created, harmonious nature, until overthrown by post-Darwinian evolutionary theory with its imagery of contest and struggle. Mutualism became the forgotten triplet of competition and exploitation but the importance of mutually beneficial links is fundamental to ecological diversity and planetary health. Mutualisms are defined by benefit to both species, though the benefits could be seen as reciprocal selfishness rather than benign altruism. Mutualisms are typically cemented by specific behavioural or physical ties forged by co-evolution between the participants. These ties can be so specialised that neither partner can survive without the other and no alternative species can step in should one partner become extinct. Mutualisms fall into four categories. Trophic mutualisms are built on the exchange of food. Most often the two partners provide nutrients that the other cannot obtain itself or in sufficient quantities. Examples include coral polyp/algae, forest trees/fungal mycorrhizae and plants/nitrogen fixing bacteria. Pollination mutualisms and seed dispersal mutualisms involve plants and animals. In return for bringing pollen to a flower or transporting away seeds, the animals are rewarded with food such as nectar or pollen. Protection mutualisms comprise interactions in which at least one partner benefits from reduced exploitation or competition, e.g., ants tending aphids repel parasites and predators, in return receiving honey dew, while the cleaner fish ridding other fish of parasites receives a food reward.

The most biodiverse ecosystems on Earth, tropical rainforests and coral reefs, are founded on trophic mutualisms. Mutualisms open up new niches. Even species without direct mutualists have indirect benefactors. In a food chain linking plant, herbivore, predator and parasite, the plant and predator are mutualists, with the plant hosting the predator's food, the predator preventing its prey from overexploiting the plant. Similarly the herbivore and predator's parasite are mutualists. Mutualists are an additional resource prompting competition as well as cheat strategies. The image of nature conjured up by the importance and variety of mutualisms has been described as green in root and flower.

Engineering

The importance of ecological engineers is a recent insight, born of the increasing integration of ecosystem ecology (concerned with energy and material fluxes) and community ecology (founded on species). Living species drive abiotic processes and engineer their environment. The activities of some ecological engineers could be described as mutualistic but are different because of the lack of close evolutionary ties. In a classic example of engineering, the beaver, constructing a pond for its own benefit, inadvertently provides a habitat for many other species. The beaver's actions are vital but not co-evolved with the beneficiaries. Engineering therefore includes many interactions where one participant benefits, the other is unaffected. Engineer's impacts work by the impact on either biogeochemical cycles or the habitat structure. Some structural engineers create physical structures out of their own bodies (e.g., coral and trees), so-called autogenic engineering. Others create alter existing structures (e.g., beavers and woodpeckers), so-called allogenic engineering. Biogeochemical engineers alter chemical cycles and nutrient availability, for example bioturbation of sediments altering nutrient (resource) and oxygen (regulation) gradients.

Reawakening interest in mutualisms and the importance of engineers driving ecosystem function have altered our understanding of the role of species interactions as a source of biodiversity. Competitive and exploitative interactions are very important, creating local variations and evolutionary escalation. Besides the direct impact of exploiters on victims and competitors on each other, indirect effects ripple through ecosystems. Many competitors or victims may escape single, strong interactions but are buffeted by many diffuse conflicts. The parasites of a herbivore are the plant's mutualists. Two competitors that share a predator magnify predation on each other since their enemy has a larger total pool of prey. Two competitors with different predators benefit from the activities of each other's enemies. Meanwhile ecological engineers riddle their habitats with geochemical variation and physical structure. Evolution multiplies these intimate links, particularly host-parasite systems as host speciation spawns parasite co-speciation. This intimate tangle, populated by mutualists and driven by ecological engineers, has challenged the mechanical, reductionist ecology of biodiversity. Ecosystems may be examples of complex adaptive systems, defined by their diversity of components, complexity derived from simple rules generating varied patterns, with energy and materials in constant flux and the ability to adapt and evolve. The sense of patterns emerging that are more than or at least different from the sum of their parts shows the wheel of scientific fortune changing. Species interactions are vital to the generation and maintenance of biodiversity. So, biodiversity begets biodiversity after all (see Fig. 17.8).

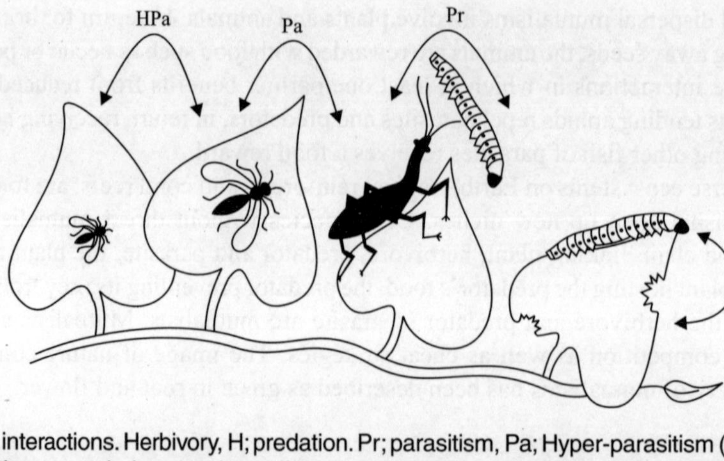

Fig. 17.8. Trophic interactions. Herbivory, H; predation. Pr; parasitism, Pa; Hyper-parasitism (parasite of a parasite), Hpa. In addition the assassin bug predator of the caterpillar is a mutualist of the plant, killing herbivores. The hyper-parasite a mutualist of the assassin bug, killing parasites so restricting their future impact on the bug and therefore is also a mutualist of the plant.

Disturbance

Natural disturbance is an important factor creating or destroying local diversity. Disturbance takes many forms and acts at many scales. As an intimate ecological factor, disturbance includes physical abiotic impacts such as fire, flood, storm, landslide and treefall. Disturbance is defined as a discrete event, a sudden physical destruction, causing mortality faster than population or cover can be replaced. The loss frees up resources which can be exploited by survivors and colonists. Disturbance is different from continuous stress which slows growth or rare, massive catastrophes such as meteor impacts. Local physical disturbance is the ecological process outlined here, though predation and disease are also a form of disturbance, but the result of biotic interactions.

The impact of any disruption varies with five factors: disturbance frequency (number and predictability), intensity, synchrony (one large disturbance versus many, smaller out of synch patches plus timing relative to victim's life histories), size of gap that results and effects on resources. The disturbance alters biodiversity by immediate deaths, changes to resource availability (amount and type), physical alteration of the habitat and evolutionary adaptation.

The effect of disturbance is inextricably linked to successional, competitive interactions and habitat patchiness. Disturbance disrupts inevitable outcomes, particularly monopolisation of habitat and resources by a few superior competitors. The disturbance may keep population so low that resources are never in short supply and competition does not occur. Or disturbance can reset and redirect the struggle, creating diverse outcomes, especially if separate patches are perturbed at different times and intensities, resulting in a mosaic of communities. The disturbance may alter a patch's environment so much that it is no longer hospitable for previous occupants. Disturbance plays a vital role in conspiring with deterministic forces, such as competition and physical structure, such as heterogeneity, to create dynamic patterns of biodiversity.

Disturbance can also decrease biodiversity, if too frequent, severe or extensive to cause extinctions and inhibit recovery. Diversity is greatest at intermediate levels of disturbance, a general result encapsulated in the intermediate disturbance hypothesis. Too much disturbance wipes out species, too little allows monopoly. Defining intermediate is awkward and dangerously circular: intermediate disturbance promotes diversity because it is a level of disturbance at which diversity is maximised. It is better to consider disturbance and competitive population growth rate processes simultaneously. The two have been formally linked in the dynamic equilibrium model of local diversity. The link highlights intermediate disturbance promoting diversity when growth rates and competition are high. High growth rates as a result of productivity promote diversity by aiding recovery from disturbance. The model suggests disturbance effects vary with trophic level. High level taxa have smaller population, which are more vulnerable to extinction and require greater energy to recover. An increase in disturbance will endanger higher level diversity more than that of lower trophic levels, especially when productivity is low. At high productivities but low disturbance, biotic impacts of keystone predators and herbivores become important as a means of breaking up low diversity monopolies. At low productivities but high disturbance, keystone resource species accelerating resource supply are important. At low productivities and disturbance, keystone mutualists are important (see Fig. 17.9).

Dispersal

Diversity varies with species' abilities to reach a site which in turn depends on attributes and behaviour. Dispersal, the movement or transport of individuals, can be very conspicuous whether it is dandelion seeds on a summer's breeze or herds of migrating wildebeest. Because dispersal is awkward to measure or model and the fate of those that do not complete their journey often unknown, its role was poorly researched. Biodiversity deals in variety, population, communities and ecosystems but it is individuals that disperse. Individual based models, taking into account that all individuals and their fates are different, have advanced our understanding of dispersal's consequences for local diversity. At the opposite extreme, large-scale supply side ecology, the rate and timing of recruitment, particularly of dispersing juvenile stages of marine life swirling in gigantic oceanic currents, helps to explain regional and local diversity.

Dispersal is commonly divided between migration, the active, mass movements of whole population and dispersal, the active or passive transport of individuals. The individuals in question may be adults, juveniles, eggs, spores or bits capable of vegetative growth. The precise mechanisms, timing and rates are very species specific. Marine plankton may disperse at the mercy of the tides. Some plant seeds disperse in tightly coevolved mutualisms with one species of animal carrier. All species have some power of dispersal, even those in uniform, stable habitats. Different types of dispersal create different patterns. Jump dispersal over great distances and barriers may found regional population. Metapopulation species rely on dispersal to maintain their scatter of discrete colonies. Diffusion, the gradual movement and expansion of range, will mesh with habitat heterogeneity, resulting in an unevenly spread population. Even in a homogenous terrain, diffusion can create patterns. The spread of population, oscillating through peaks and troughs of abundance typical of many species, will meet like waves on the surface of water. As high and low population of interacting species ripple across the landscape, complex patterns emerge.

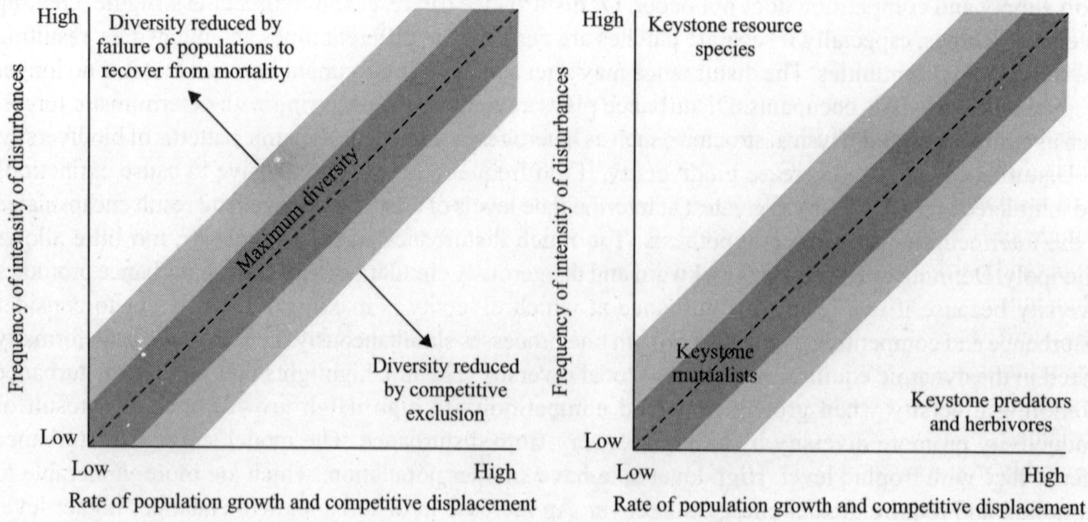

Fig. 17.9. Huston's 1994 synthesis of disturbance, population growth and interactions as determinants of local diversity. The left graph shows diversity reduced if either growth and competition or disturbance is high, without any compensating influence. The right graph shows the importance of different keystone interactions under three disturbance/population growth regimes.

Supply side ecology, the rate and pattern of arrival, can be a dominant influence. Inconsistencies in the diversity of many coral reefs and marine benthic communities have shown that variable recruitment, driven by the rate and timing of dispersal, causes local population occasionally to be deluged by arrivals, whilst other areas receive none. These occasional events swamp local interactions that regulate diversity in between. There are lessons for conservation. The plagues of the crown of thorns starfish (*Ancanthaster planci*), famed for the havoc they wreck on corals, may outbreak due to occasional unpredictable mass arrival of larvae. Ultimately the dispersal of individuals builds the ranges within which evolutionary events (such as speciation) and ecological primary factors (such as vicariation) work. From time and space to individuals and back again, the ecological factors creating biodiversity are an integrated loop rather than a reductionist shopping list.

ECOLOGICAL RENAISSANCE: EMERGING GENERAL RULES

Ecology has at times appeared an intractable science. Atoms are so small and galaxies so large that they allow mathematical generalisations but ecosystems and their inhabitants appeared to be messy, contingent and unpredictable. Whilst ecologists could do experiments and explain in detail what happened to one species of water bug on one marsh or tussocks of grass up on a mountain, it was difficult to extend these lessons to anywhere else. However the period since the mid-1990s has seen significant developments which hint at unifying principles and general ecological rules.

The relationship between size and metabolic rate has been a key area. Metabolic rate reflects many key aspects of a species' life such as growth and reproduction, from bacteria to blue whales. Metabolic rate varies consistently with body size so that smaller organisms use more energy per gram of body weight than bigger species. Work on forest trees has shown consistent relationships between size and abundance so that for every tenfold increase in tree trunk diameter, the number of trees decrease by a hundredfold, but the horde of smaller saplings uses the same total energy as the smaller number of larger trees. Each size class uses the same amount. Similar patterns occur in animal communities. These consistent patterns suggest a template for the distribution of resources and because of the relationship between metabolism and body size, the diversity of life.

A second approach is macroecology, which recognises that the local species richness (those bugs in the pond or tussocks on the hill) is influenced by distributions and abundance at larger spatial and temporal scales. Macroecology has highlighted several key patterns where the influence of the large scale on the local is apparent. First, there are the familiar relationships between species richness and area, isolation latitude and altitude. Second, consistent range size patterns such as species at higher latitudes have wider ranges and species with wide ranges occur at more local sites than species with narrow ranges geographical. Abundance patterns include individually abundant species having wider ranges than rare species and consistent patterns of a few abundant and many rare species in most communities. Finally there are body size relationships, such as the greater number and higher population of smaller species compared to larger ones. These large scale phenomena, broadly consistent across taxa, habitats and time, have been explained based on models using the relationship between energy and biomass and the constraints these inevitably impose on species.

A major ecological conundrum may also be near resolution—that complexity (e.g., more species, more interactions) fuels instability (e.g., extinctions). The revelation that complexity begat instability came from models of food webs in the 1970s and did much to shatter the early twentieth century view that greater ecological complexity reinforced stability and that natural systems existed in some sort of equilibrium. However, the result is also at odds with the conspicuous complexity of many ecosystems such as tropical forests and coral reefs, which do not spontaneously collapse. New models have built in more realistic species dynamics and results suggest that many weak interactions result in greater stability. Once again, all those species matter.

The final and boldest development has been a unified theory of biodiversity and biogeography (biodiversity in this case defined strictly as the richness and abundance of species). This model unifies the competing visions of ecological communities as regulated either by strong interactions between species fighting to occupy a limited number of niches or as looser assemblages thrown together by dispersal and extinctions. The theory makes the assumption that all individuals in a community with the same broad trophic role, e.g., all predators or all herbivores, are equal in terms of competitive ability and chance of survival, although different species vary, for example in their likelihood of colonising. The model then assumes that there is a limited space/resource to fill up, that new individuals can establish only if others are lost and that individuals survive or are lost at random as space becomes available. The underlying maths is fearsome but the model produces strikingly realistic predictions of species richness and abundance across a range of ecosystems and a single parameter, based on two variables (total number of individuals in the regional community in question and the rate of speciation), is sufficient to define the outcome for each community.

EVOLUTION OF BIODIVERSITY

The ecology describes patterns of and provides explanations for the biodiversity of extant ecosystems. Ecological processes also have evolutionary consequences. They interact with genetic diversity via adaptation, microevolution and speciation. The day-to-day dramas played out on the ecological stage are the driving force for the multiplication of species in evolutionary time. The environment provides continual pressures to diversify via adaptation, innovation and exploitation of new ways of life. Taxonomic biodiversity is the most conspicuous result of ecological and evolutionary processes driving the multiplication of species.

Multiplication of Species

Diversification of genes: environmental forces

New species arise from two sources. An individual lineage may change over time so that subsequent generations are sufficiently different from ancestral types that they are classified as a separate species. This is anagenesis or chronospeciation. There is no net gain of species. Speciation that spawns a net gain in numbers is cladogenesis, whether it be splitting of the ancestral line into two (or more) new species or budding of new species from an ancestral lineage that also survives. These processes can be gradual, the accumulation of inherited genetic differences selected by ecological processes or sudden, perhaps random, with no clear adaptational benefit—Raup's survival of the luckiest.

Taxonomic biodiversity is increased by cladogenesis so it is important to understand why lineages speciate. The genetic variation in population that underpins all of biodiversity and is central to evolutionary processes is almost unavoidable. Genetic diversity within a species is created both by the environment, as local population adapt to variations between local conditions and by processes within the genome that would occur even in a homogenous environment, e.g. mutation. The drive to genetic diversity is irresistible.

Some genetic diversity can be categorised as adaptational. Genetic differences reflect variations in local conditions with selection of genotypes beneficial in the prevailing conditions and loss of the sub-optimal. The genetic differences may reflect tolerances to different conditions or create a functional diversity so that individuals in a population can exploit varied opportunities, but perfect adaptation is probably impossible, not least because environments constantly alter. Adaptational diversity may create key innovations, allowing exploitation of lifestyles new to that taxa or novel to any life. Key innovations are one evolutionary factor unleashing adaptive radiations, resulting in geologically rapid diversification of a lineage. Innovations may be utterly new. They also include pre-adaptations, a character allowing ecological or evolutionary exploitation of a new lifestyle whilst essentially carrying out its original function and exaptations, characteristics again allowing ecological or evolutionary expansion, but the character is used in a different way from its original purpose, fortuitously effective in a new role.

Major periods of innovation, defined by their global extent, diversity of innovations and new ecosystems may spring from revolutions in global biogeochemical cycles. The two main innovation revolutions since the Cambrian (one the Cambrian marine diversification, the other the Mesozoic marine modernisation) coincide with massive biosphere changes such as volcanism, warming and sea level changes which could alter nutrient supply. Increased supplies of nutrients and energy opened up opportunities and once evolution and speciation accelerated, a positive feedback of arms races, competition and mutualisms would snowball.

Genetic and resultant evolutionary biodiversity can also arise without any adaptational context. The exchange of genetic material through an interbreeding population results in genetic drift. Genetic drift can be a potent force in small population where the original genetic diversity may be limited to start with, a biased sample of the total variety in a larger population. In a small population a few matings can result in drift rippling through subsequent generations. Small population prone to such effects can be created in many ways: founder population of a few individuals colonising new habitat; isolated remnants marooned in refugia; bottlenecks created when population crash; metapopulation where the total population may be high but is scattered between separate patches that do not readily interbreed.

Diversification of genes: genetic systems

Just as the environment forges genetic variation within population, so processes within the genome itself create and maintain diversity. Mutations are any changes in the genome, whether one nucleotide or larger sequences. Types of mutation are unpredictable, uncontrolled and independent of the natural environment, though rates may vary. Mutations include errors in DNA (deoxyribonucleic acid) replication, insertion of transposed sections and rearrangement of chromosomes via fusions and inversions. The rate of changes to Eukaryote nucleotides has been estimated as 4×10^{-9} per base pair per generation. Since eukaryotes typically harbour billions of base pairs, every individual carries some mutations. We are all mutants. So mutation will create genetic variation even within the least genetically diverse taxon, a source of hope for genetic recovery for some endangered species. Genetic drift creates diversity within a freely mating population. Although the frequency with which individual alleles are transmitted to offspring may be broadly predictable (based on their frequency in the parent population and the rules which govern their sharing out between gametes), only a minute fraction of gametes actually contribute to the next generation at fertilisation. The alleles in the billions that do not make it are winnowed. The impact of drift depends on population size and are magnified in small population. Drift effects can be substantial in large population, for example species with breeding opportunities monopolised by a few individuals such as red deer stags coveting harems. The limited number of individuals contributing to the genetic diversity creates a small effective population size.

The distribution of individual alleles on chromosomes can also be rearranged by recombination. Eukaryote chromosomes exchange sections so that alleles that would otherwise stay linked together are combined into new sets. The total genetic diversity at the allele level does not increase but the permutations do. Recombination is most effective in outbreeding population. Self-fertile taxa end up recombining with themselves. Prokaryotes have mechanisms that achieve broadly similar remixing of alleles.

Selection and maintenance of genetic diversity

Environmental and genome processes combine to produce genetic diversity within population; the genetic foundation of each individual is tested by selection. Selection occurs within ecological time and forges a lineage through evolutionary time.

Selection acts directly on the phenotype. The genotype is affected indirectly. Directional selection favours individuals with an advantageous characteristic, selecting the genotype responsible and reducing genetic diversity by elimination of alternatives. Since many general phenotype characteristics are a synergy of complementary features not governed by one gene, directional selection effectively selects sets of genes. Stabilising selection favours intermediate characteristics and extremes are winnowed out. Stabilising selection fixes combinations of alleles that build the intermediate character rather than variations within each allele so again genetic diversity is lost. Disruptive selection favours different extremes; the intermediates are at a disadvantage and maintain the variation in alleles that cause the switch between extremes, not different alleles themselves. So selection tends to decrease genetic diversity. However, fluctuations through space and time alter selection pressures, even reversing them, providing an important brake to such losses. Genetic dispersal and gene flow act to homogenise genetic diversity within a population. Highly mobile species show less variation than the more sedentary. The dispersal characteristics of a species are important. Some species known to be poor dispersers nonetheless show little genetic variation throughout their range. This may be because they have only recently expanded into the range, the homogeneity an echo of their close-knit origins. Species living as a metapopulation can show marked losses of genome diversity if extinction of patches wipes out unique variations. The toll from selection and drift is countered by imports from flow. The outcome teeters, critically dependent on how far individuals move between birth and mating versus the strength of selection. There can be a critical distance below which flow can compensate for any effects of selection. The conservation of small, fragmented population often depends on understanding such pressures. In summary, mutation creates genetic diversity, whilst selection and drift destroy.

Breeding systems affect genetic diversity. Inbreeding allows disadvantageous, often recessive alleles expression. Enfeebled offspring show inbreeding depression. If they breed in turn, the degradation may snowball into a genetic meltdown. This is a particular threat if a once large population is suddenly reduced in size. Genetic meltdown becomes the final straw after the first strike that reduced abundance. Small founder population which may be homozygous to start with so that any deleterious alleles have been weeded out already are in less danger. Outbreeding generally suppresses the worst impacts of disadvantageous alleles. Outbreeding depression can occur via immigration of individuals poorly adapted to local conditions which mate with well-adapted locals. There is a balance: mate with those too close and risk inbreeding, mate with those from too far and risk outbreeding depression.

Given the environmental and genome promotion of genetic diversity, it is not surprising that there is no evidence of limits to local, small-scale selection for want of genetic variation. We do not know if there are limits to major evolutionary changes, for instance radical rearrangement of body plans. These may be prevented by lack of genetic foundation. Perhaps the genes do exist but developmental rules halt any such revolutions. Even without the capacity for unfettered creativity, the environment and genes have conspired to create a wealth of species.

Speciation

Speciation is the creation of new species. Whilst commonly pictured as a diversification or radiation of a lineage, speciation requires mechanisms both to create and then maintain the diversification into cohesive species. Only then can a genotype establish, its individuals isolated from interbreeding with other species. The familiar images of radiating, diversifying lines might be better thought of as collapsed spokes from a wider fan of individuals that once freely interbred.

The threshold from variation within a species (adaptation, microevolution) to speciation and macroevolution is a barrier. The very same processes creating genetic diversity within a species also maintain cohesion. Gene flow may smother variation; unusual genotypes may be less fertile or selected out by the environment or ignored by would-be mates. Selection may hone well-adapted species, marooned on peaks of near perfection, from which any evolutionary departure is difficult.

Mechanisms are needed to push genetic diversity over the speciation threshold. There are four main modes of speciation:

1. Allopatric: Geographical separation of a previously continuous population.
2. Peripatric: A variant of allopatric involving isolation of a peripheral population.
3. Parapatric: Speciation in adjacent population stretched over an environmental gradient.
4. Sympatric: No separation during speciation.

Allopatric speciation

This requires the separation of population either by creation of a physical barrier, extinction of the middle range of a population or jump dispersal across an existing barrier prior to speciation. The barrier prevents gene flow and the population's genomes diverge. The divergence can be adaptive if environments differ either side of the barrier or random due to mutations, recombinations and chance losses. If the diverging population rejoin, they may still interbreed and coalesce, maintain the divergence but with a hybrid zone in between or not interbreed at all, the latter representing full speciation. This reproductive isolation can be maintained by genetic incompatibility, discrimination in choice of mate or the physical impossibility of mating (see Fig. 17.10).

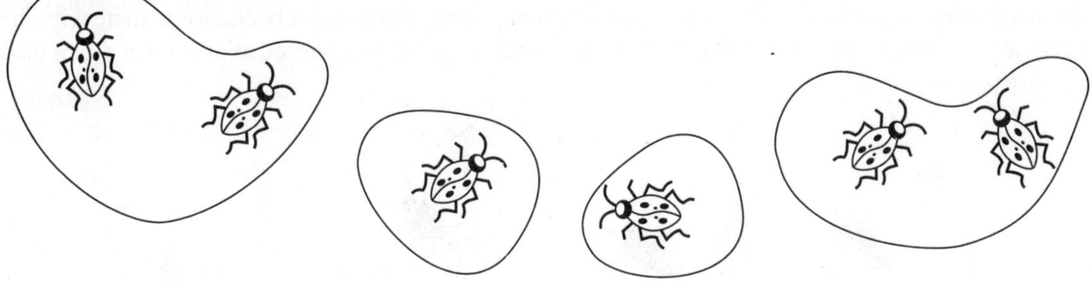

Fig. 17.10. Allopatric speciation. A united population is split, in this case by a physical barrier of rising sea levels. The evolutionary history of the isolated population differ, to the extent that once reunited they no longer successfully interbreed. Two new species exist where there had been one.

Peripatric speciation

This involves isolation of a small peripheral population from the main range. Otherwise the process of geographical isolation followed by divergence are the same as allopatric speciation. The small size of peripheral population creates specific features. Small numbers of founders inevitably contain only limited, biased subset of the original species' gene pool. This provides a head start to divergence, magnified by inbreeding and homozygosity, allowing expression of unusual characters previously suppressed. The cause of isolation may add impetus.

Founders colonising an island may have abundant resources and little competition. Alternatively peripatric speciation in collapsed remnants of larger population can face genetic bottlenecks. These founder effects emphasise the vagaries but importance of chance events and genetic loss. Alternatively, adaptations may flourish in new or empty niches. Old genetic complexes are shaken up, innovations and radiations are marked. This adaptive, creative result is called the founder flush.

Parapatric speciation

Variation throughout the range of a species and resulting local adaptation will create genetic gradients. Once such a cline starts, gene flow across the gradient may decrease, especially if dispersal is poor, there is selection against hybrids and increasingly pure types straying into the wrong side of the cline. The more pronounced that the differences become, the more the trend is reinforced. A true hybrid zone will develop and if reproductive isolation becomes established, eventually will disappear leaving two adjacent species. Distance over the range of the population is the creative force, since genetic variation driven by genome mechanisms alone can create clines even across a homogenous habitat. Geographical and ecological disruption can aid and abet progress.

Sympatric speciation

This involves no spatial separation of population. Separate, cohesive genotypes are established and maintained whilst in contact with one another. Models of sympatric speciation have relied heavily on selection.

Disruptive selection

Disruptive selection by selection of different phenotypes is controlled by one gene and elimination of intermediates. The selection of phenotypes may be due to ecological interactions such as predation. Once different phenotypes are established, natural selection will encourage reproductive isolation through habitat selection or positive assortive mating (different phenotypes positively choose to mate with their own kind). Sympatric speciation linked to habitat selection has been suggested as a major cause of insect diversity (see Fig. 17.11).

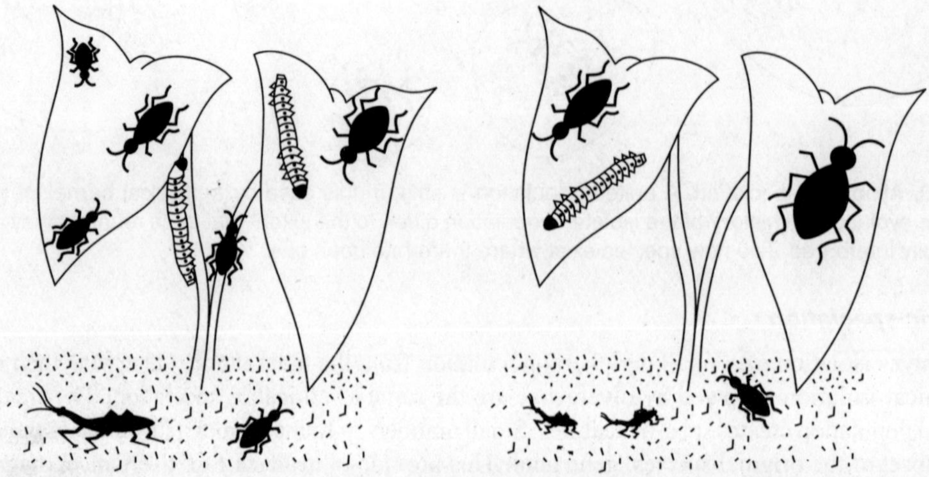

Fig. 17.11. Sympatric speciation by disruptive selection. A species of beetle, which shows wide variation in body size, shares part of its range (on the plant) with a predator. The predator can overpower small but not large beetles. Eventually large and small beetle population exist as a result of predator selection. If differences between the beetle population are consolidated so that they no longer interbreed successfully two new species exist where there had been one.

Competitive speciation

Competitive speciation is a variant of disruptive selection. Competition for resources within a species selects for different phenotypes able to avoid intense competition whilst intermediate types are lost.

Polyploidy

Polyploidy relies on the multiplication of whole sets of chromosomes so that offspring have double (or more) the normal complement. Gametes of polyploids typically contain more than the normal one-half set of chromosomes, so if fertilised (whether by another polyploid or not) the offspring has a different number of chromosomes. Polyploidy relies on abnormal cell division to create the overloaded gametes. Autopolyploidy occurs by fertilisation between polyploids within a species. The resulting offspring often suffer low fertility, their mass of chromosome literally tangled during cell division to create gametes but vegetative reproduction can be strong. The potato is a good example. Allopolyploidy relies on hybridisation between different species of polyploids. Offspring are usually fully fertile, gamete production smoothed as the chromosome contributions of each parent separate out with their own kind, avoiding tangles. Many wheat varieties are allopolyploids. Polyploids are often bigger and more productive than their forebears. Polyploidy is rare in animals but a major speciation mechanism in plants, with over 43 per cent of dicotyledon and 58 per cent of monocotyledon species polyploid.

Other potential sympatric mechanisms include chromosomal rearrangements that result in reduced fertility of heterozygotes. In the meantime homozygotes, potentially displaying distinctive features, can establish population which become cohesive, much like disruptive selection. This is called stasipatric speciation. Macrogenesis, the dramatic mutation to produce a hopeful monster, is a possibility, demonstrable in some laboratory population but with little support in nature.

The relative importance of all mechanisms is still hotly debated. Examples occur of all types, their precise role dependent on the taxa in question.

Speciation in Action

Speciation, especially the evolutionary consequences on a geological time scale, are a culturally familiar idea in the developed world, for example the diversity of the iconic Darwin's finches on the Galapagos Islands. However, are there examples of speciation taking place before our eyes to back up and test the different models and hypotheses? Yes, in particular amongst insects, which are both familiar and have fast enough generation times to study.

A neat example of peripatric events is shown by the population of the mosquito *Culex pipiens* in the tunnels of the London Underground. The subterranean population are genetically distinct from their above-ground relatives. Population within individual tunnels are also genetically distinctive: the limited genetic diversity within the tunnel population is suggestive of founder effect bottlenecks providing only a limited gene pool and that the foundation of new colonies is a rare event. Whilst mosquitoes from different underground population will still interbreed with one another if brought together, they very rarely will do so with above-ground mosquitoes. The London Underground system was begun only in 1863. This is incipient speciation within a few human lifetimes.

A similar example occurs in the natural tunnels left behind by lava flows in Hawaii. Within these caverns live troglodyte plant hoppers of the genus *Oliarus*. One species, *Oliarus polyphemus*, has been studied in detail. Plant hoppers from population in adjacent tunnels look the same but have distinctive courtship songs, whilst those within the same tunnel sing to the same music sheet. They are reproductively isolated not just physically, tunnel to tunnel, but also behaviourally, by song. *Oliarus polyphemus* is regarded as a complex of incipient species.

Insects also provide examples of sympatric speciation. The wild fruitflies *Drosophila melanogaster* of 'Evolution Canyon' on Mount Carmel in Israel are a fine example, not least because so much laboratory work has been carried out using this very species to examine genetics and evolutionary processes. The canyon has opposite slopes only 100–400 meters apart, well within the dispersal range of the flies. The south-facing slopes are hot and dry and conditions fluctuate creating a more stressful environment than the cooler, wetter north facing slopes. Flies from either slope show marked, genetically based adaptations to their respective slopes, for example for heat tolerance or drought stress, but these differences should be obliterated by dispersal of flies and mixing of genes between the population. However, flies from either slope show very strong selection for mates from their own slope. Again, incipient sexual isolation although the flies are not physically separated, unlike the underground's mosquitoes or Hawaii's *Oliarus* bugs.

BIODIVERSITY AND ECOSYSTEM FUNCTION

Initial studies tried to conceptualise the ways in which ecosystem function might vary with species richness. There are several models, all commonly describing how the effectiveness of ecosystem function varies as species richness declines.

1. *Linear models*: Ecosystem function degrades in proportion to loss of species. In effect every species matters, more or less equally.
2. *Redundancy models*: As species richness declines, the ecosystem function remains largely unimpaired as other species in the habitat take up the functional roles of those that are lost. Species can substitute for one another. There are in effect spare species in the habitat. Ecosystem function collapses sharply once a minimum threshold of species richness has been breached.
3. *Singular models*: These models regard all species as making a unique contribution. The rivet model uses the analogy of rivets popping out of a machine: lose a few and not much changes, but once too many are lost, a bit of the machine falls off. In natural systems the equivalent is that several species may be lost with little impact until enough are lost to cause a sudden loss of ecosystem function. Functional effectiveness then stabilises until the next time when enough remaining species have been lost. In the keystone model, ecosystem function depends on one vital species to drive the process. Once the keystone is lost, ecosystem function degrades markedly whilst losses of other species has little effect.
4. *Idiosyncratic models*: The complexity, contingency and context of ecological interactions might make the outcome of species losses unpredictable. Ecosystem function may degrade or improve, depending on which species are lost, in which combinations and when and where.

Whilst these representations of the relationship are helpful, they do not explain how the processes might work. Three main hypotheses have emerged:

1. *Sampling effect*: This explanation suggests that the more species there are in a habitat, out of those locally available, then it is more likely that the most functionally effective of the local species pool will be included. So, on average, the more species there are, then the more likely it is that the most productive or most effective nutrient conservers are present. However, a monoculture of the most effective species would be every bit as effective so this hypothesis simply recognises the results of sampling any array of items, rather than suggesting that diversity itself creates any special bonus.
2. *Species complementarity or niche differention models*: Species vary in their needs, the resources they use and their precise ways of life. The more species there are in an ecosystem, the wider the range of resources that will be used and processes that will be biologically driven.
3. *Positive interactions*: Species may interact in ways which help one another, either through specific mutualisms in which species are adapted to benefit one another or by altering the environment in ways which accidentally facilitate survival of other species.

Some Classic Studies

Despite the apparent simplicity of the key question—does biodiversity affect ecosystem function?—experiments to investigate the relationship have proved very demanding of resources and a minefield for interpretation. There are now several classic studies which have produced robust results.

Ecotron, UK

Pioneering experiments, the Ecotron, from 1993 onwards using sixteen climate-controlled rooms in which miniature ecosystems could be created and species dynamics and geochemical cycles carefully monitored. Experiments could combine multitrophic communities—plants, herbivores, predators and parasites—and include both above and below soil systems. Experiments could be run over several years. The Ecotron facility was used to test the effects of varying species richness on ecosystem functions within the microcosms. Lower species richness resulted in lower overall productivity, the first experimental demonstration that for effective ecosystem functions the number of species in a habitat matters.

Cedar Creek, USA

The Cedar Creek Biodiversity experiments use outdoor plots of prairie grassland plants to examine how species richness affects plant productivity and geochemical functions such as carbon and nitrogen cycling. Individual plots can be planted with varying numbers of species, between one and twenty-four, including a variety of functional groups (e.g., nitrogen fixers) and tracked over several years as the communities and functions respond to natural variations in climate; responses of insects, diseases and invasive plants can also be monitored. Key results are that both productivity and nutrient use increase with species richness. Productivity increases not only are linked to the presence of one species (so, not the sampling effect model) but also are long-term responses to different combinations of species, in particular the variety of functional groups. Both species richness and functional types matter.

Biodepth, Europe

The Biodepth project uses a similar approach to Cedar Creek, extended across geographically diverse context. Plots of plants, varying in richness, from one to thirty-two and in functional groups, have been set up in seven countries, from Ireland to Greece and Sweden to Portugal. Ecosystem processes such as productivity, decomposition and nutrient cycling have been monitored. The key results are that above-ground biomass of plants increased with species richness and variety of functional groups but there was also variation with the precise composition of species and between sites. Species richness and variety both matter, but there is evidence of idiosyncratic responses to combinations of species and site.

All these projects are ongoing, increasingly focused on how the systems respond to environmental change, especially to climate. Whilst the debate remains fierce and the species and habitats involved vary, the general message is unmistakable; the variety and richness of life on Earth is fundamental to Earth's natural economy. Living things drive the global environment, not into mythical, idyllic balance but quite the reverse, preventing the planet closing down.

ECOSYSTEM INTEGRITY AND ECOSYSTEM HEALTH

The significance of biodiversity for our well-being combined with the wide conceptual horizons of biodiversity, connecting the biological sciences to economics, politics and culture, have encouraged ecologists to develop more holistic paradigms of natural systems, combining the species, the ecosystem functions and how these affect humanity. One approach has been the idea of ecosystem integrity. Put simply this is the ability of the ecosystem to support and maintain the full range of species, communities and processes expected in the natural habitats of a region. Note how this combines four elements: first, the structure and organisation (e.g., numbers of species, relative abundance); second, the natural functions (or vigour), third, the ability to resist damage or the resilience to recover and fourth, the undegraded potential to develop and adapt to change. Degradation of any of these represents a loss of integrity. This approach alters how we think about the natural world. The habitats within a nature reserve may support a wealth of species and communities and be apparently pristine, but the ecosystem may have limited opportunities to change or adapt, perhaps because we manage it to meet our targets, stifling change or perhaps because there is not the room or resources to allow change. The notion that change is an important aspect of natural systems also challenges how we conserve systems. There is a movement to allow 'wilding' of conservation landscapes, where we do not intervene to meet chosen management priorities but let nature do whatever it will, even if the precise end result is not known.

The 1992 biodiversity convention talked of 'the need to safeguard the health and integrity of the world's ecosystems'. The very existence of something called ecosystem health has been controversial amongst scientists, but, like integrity, is an approach recognising the need to cross disciplines and boundaries to effectively conserve the natural world. The notion of ecosystem health recognises that ecosystems are degraded by the stresses we inflict overwhelming the natural resistance and resilience. The ecosystem becomes dysfunctional, just like we do when we are poorly and that an approach of both prevention and cure is needed. Ecosystem health emphasises the need to define and measure loss of function, to diagnose causes and detect early warnings. The idea is an effective analogy, helping to get across the impact we are having on the natural world. Ecosystem health echoes the holistic approach of ecosystem health, incorporating the vigour, organisation and ability of ecosystems to resist and recover from damage. The idea has attracted flak, sometimes conceptual such as the implication that ecosystem may be like superorganisms, sometimes pedantic such as the suggestion that the word 'health' cannot be used like this. The key ideas of integrity and health have been taken on board. In 2004 the UK government conservation umbrella the Joint Nature Conservation Committee (JNCC) published guidelines for site and habitat assessment which recognised the need to measure both the structure and function of habitats and use these to judge their condition against their desired state, rather than just taking a snapshot of their condition, their publication popularised explicitly using the idea of ecosystem health.

As the idea of biodiversity has encouraged the natural sciences to engage with the wider world, the real problem could be that many of these useful ideas become muddled together. Ecosystem integrity emphasises the role of natural ecological and evolutionary processes, wherever they may lead and the conservation of natural organisation and function. Ecosystem health emphasises the measurement of resilience, vigour and productivity and the need to detect and prevent their degradation. Both are allied to the concept of sustainable development which emphasises economic processes and development without compromising the future well-being of the planet, which is where the whole UN involvement in biodiversity and the Rio treaty began. They are all different in detail but all three recognise the same essentials for our future survival.

IMBALANCE OF NATURE

The balance of nature has a strong hold on our imagination and conveys a cosy vision of harmony. However, nature's creative imperative is anything but balanced. The diversity of life around us, at least the most familiar aspect of species richness, can be reduced to a very simple equation. Species diversity equals speciations minus extinctions. Since there are still very many species alive on Earth, nature is profoundly imbalanced, with speciation generating a surplus over and above the ravages of extinction. Quite how big that surplus is remains one of the most difficult questions facing biologists. The diversity of life on Earth is still surprisingly poorly known.

18

Types of Biodiversity

INTRODUCTION

Discovering and defining the richness of life is central to our knowledge of biodiversity. Biodiversity strikes a Pavlovian nerve, conjuring up images of rare mammals and rainforests but it embraces much more than these. There are three main components to biodiversity: genetic, organismal and ecological. The genetic and subcellular diversity includes all biodiversity expressed within individual cells plus non-cellular organisms such as viruses. The diversity of genetic information is central to this category but the variety of metabolic pathways and molecular biology of life also represent important diversity. Taxonomic diversity is dominated by our focus on species. Species are but one level at which organisms can be classified and the diversity of other categories such as families or phyla provides additional insights. Ecological diversity includes whole communities, habitats and ecosystems, including domestic stocks.

Biodiversity is defined as all hereditarily based variation across all levels of organisation, from the genes within a single local population or species, to the species composing all or part of a local community and finally to the communities themselves that compose the living parts of the multifarious ecosystems of the world.

TYPES OF BIODIVERSITY

Genetic and Subcellular

Genetic, molecular and metabolic diversity are often overlooked. They represent the founding diversity of life, not just as it is nowadays but with echoes of the ancient past and potential futures. All three generate diversity within and between species and ecosystems.

Genetic diversity provides the core differences that divide life into its major types, especially important when visible structures and shape provide no reliable guide. This genetic diversity is witness not only to the deepest divisions of life but also to some shared characteristics, stretching across huge taxonomic distances. Such genetic characteristics speak of the relatedness of all life forms one to another. Genetic diversity is the ultimate divisor and link. Genetic and metabolic diversity are especially important to understanding time, evolution and taxonomic characteristics of life. Genetic material is the raw material from which future biodiversity will be spawned.

Genetic diversity is made up of genes. A gene controls the expression and development of a particular feature of a living organism. The precise form of the character will vary (e.g., hair or eye colour in humans) and each of the variations of the gene is called an allele. So a gene in an organism will be one allele from a larger set. Genes are built of deoxyribonucleic acid (DNA), a linear molecule built like a ladder and twisted into the famous double helix. The rungs of the ladder consist of molecules called nucleotide bases, two joined to make each rung. There are four different bases, cytosine, guanine, adenine and thymine, which pair up to build the rungs as base pairs. Cytosine pairs with guanine, adenine with thymine. Each set of three base pairs makes up a triplet and each triplet is the code sequence for an amino acid. There are sixty-four permutations in which three consecutive base pairs can occur, more than enough to code for the twenty amino acids common to all organisms. So the diversity of the genetic code is a hierarchy starting with the sequence of individual rungs then triplets, the order and length of sets of triplets, the resulting alleles and quantity of genetic material in different organisms. In addition to the active genes there is extra DNA apparently doing nothing.

The result is that genetic diversity even within a single species is so vast that the information is much larger than the total number of individuals. The measurement of genetic diversity is a rapidly improving area.

The metabolic diversity of life, particularly amongst bacteria, is commonly overlooked, but represents a fundamental and ancient diversification. The metabolic abilities of bacteria can be directly useful to humans (e.g., fermentation) or indirectly as part of wider geochemical cycles that are ecosystems services maintaining the health of the environment. Metabolic diversity can be divided between utilisation of different energy sources, energy release (respiration) and nitrogen fixation.

Molecular biodiversity is less familiar but offers an important alternative approach to describing subcellular variety, which is important for food, pharmaceuticals cosmetics and industrial compounds (e.g., dyes). Molecular biodiversity can be described using molecular structures or functions (e.g., growth stimulant, anti-microbial, alarm) and audited by habitat, biogeographically region or species. For example, South American amphibians are characterised by their variety of steroids, temperate grasslands their morphine derivatives and northern conifer forests their terpenes. Although not such a familiar image of biodiversity as pandas and rainforest, the new analytical and informatics techniques are rapidly allowing molecular biodiversity to catch up as a significant approach to understanding the richness of life on Earth.

Taxonomic Diversity

The number of species currently alive on Earth is often thought of as synonymous with the term biodiversity. Work at the species level has practical advantages. Funding bodies (often the general public) recognise species. Conserving a species is a simpler concept than conserving a species' gene pool. Species projects are often easier to organise than conserving whole habitats. However, the species is but one category in a hierarchy of classification. Other categories are important measures of biodiversity sometimes providing different, contradictory messages to species-level analyses.

Systematics is the branch of biological science responsible for recognising, describing, classifying and naming organisms. Part of this work involves a taxonomic (taxa = category, nomic = name) hierarchy, including species and other levels at which organisms can be described. Linked to this are taxonomic nomenclatures (naming systems) which provide internationally agreed rules on how species are to be named. International codes exist for animals, plants and bacteria, plus there is a provisional system for viruses. The taxonomic classifications also portray our understanding of the evolution and relatedness of life, the phylogenetic classification. Humans have a natural tendency to classify things, biodiversity being no exception. Classification depends upon what we can observe and what we think it means (often influenced by social, political and religious attitudes). Taxonomic classifications of life have varied throughout time.

Table 18.1. The taxonomic hierarchy. Each level in the hierarchy is a category at which fundamental features are shared. Each category will consist of one or more of the next category down (e.g., a genus consists of one or more species). Sometimes subcategories are used (e.g., subclass, suborder). Recent analysis of genetic biodiversity has caused a rethink of what had seemed the ultimate, kingdom-level divides, a salutary lesson.

Taxonomic hierarchy	*Human beings*	*Mount kupe bush shrike*	*African elephant*
Kingdom or domain	Kingdom Animalia; Domain Eucarya	Kingdom Animalia; Domain Eucarya	Kingdom Animalia; Domain Eucarya
Phylum or division	Chordata	Chordata	Chordata
Class	Mammalia	Aves	Mammalia
Order	Primates	Passeriformes	Probosidea
Family	Hominidae	Corvidae	Elephantidae
Genus	*Homo*	*Telophorus*	*Loxodonta*
Species	*Sapiens*	*Telophorus kupeensis*	*Africana*

Recently systematics and taxonomy have not been popular fields with biologists or funding bodies but burgeoning biodiversity studies have focused attention on our woefully incomplete catalogue of life. An urgent task in biodiversity research is to rebuild skills in systematics. Table 18.1 gives the typical classification hierarchy, from species up to the fundamental divisions of life into kingdoms or domains. Note that there is marked variation in diversity even at very high levels of classification. Recognised numbers of phyla in five kingdom classification of life are: Prokaryote 16, Protista 27 (at least), Fungi 5, Animalia 32 and Plantae 9.

Defining species

The species remains the focus of most biodiversity research, but there is no precise, simple, universal definition of what a species is. Some alleged species fit all criteria of all definitions, others are hard to fit to any.

The word species literally means outward or visible form. Classifications up to the twentieth century relied on the physical, often outward, similarity of features to distinguish a species. The use of visible characters to define a species is the morphological species concept. This approach is widely used with collections of dead specimens or fossils which can provide no other information on breeding and ecology. Species with varied individuals, e.g., camouflaged, mimics or variable growth forms, risk being split into several species.

Insights from evolution and genetics prompted the biological (or isolation) species concept. A species consist of populations of interbreeding individuals, able to reproduce successfully with other populations. Central to this definition is the idea of reproductive isolation, be it a physical barrier or behavioural, physiological or genetic inability to mate and produce fertile young. The isolating mechanisms can be pre- or post-mating. Pre-mating includes separate habitat, time, behavioural and physiological mismatch, mechanical inability to mate and gamete inability to fertilise. Post-mating includes hybrid inviability (embryos or young die), hybrid sterility (young are born but cannot reproduce) and hybrid breakdown (young will mature and breed but their offspring are less viable). This definition does not work with asexual species. A surprising number of recognised species living side by side appear to mate and exchange genetic material without the species involved coalescing into an indistinguishable melange. Many plants show interbreeding. Tests of the American buffalo genome have revealed domestic cattle genes but no morphological blur. Species defined by breeding are also difficult for systematists. Not only are the breeding abilities of fossils impossible to fathom but also over evolutionary time species change, so much so that they might be morphologically distinct, but at every step interbreeding between immediate parent-offspring generations was possible. The morphologically distinct ancestor and descendent species are linked by a seamless reproductive thread.

Refinements to the biological species concept include the following: the recognition species concept, emphasising not the barriers to reproduction but shared characteristics permitting reproduction, the cohesion concept, a species defined by cohesive mechanisms that prevent interbreeding with other species and evolutionary diversification within the species and the ecological concept, species defined by occupation of a niche unique compared to other species in its range and evolving separately from populations beyond its range. All remain flawed by hybridisation and problems of practical measurement.

A third approach is the evolutionary species concept. This emphasises the phylogenetic concept with species defined as the smallest population (sexual) or lineage (asexual) diagnostically distinct from other such populations and with a discrete lineage. The diagnostic characters are often genetic. When applied to species recognised by morphological or biological concept criteria, the sheer genetic diversity within species often suggests they should be splintered into many separate lineages. The evolutionary concept defines a species by a single, distinct lineage with an identifiable evolutionary history and fate.

Ecological Diversity

Taxonomic diversity classifies types of organisms and their relatedness but organisms do not live in isolation from one another or the physical world. Humans have long recognised different ecosystems of apparently interdependent life. Ecological science focuses on these patterns and processes, hence ecological diversity is the inclusive term for this third category.

Ecological diversity covers a host of concepts; ecosystems, communities, assemblages, habitats, biomes and biogeographical regions. These are not one and the same, indeed some may not be biodiversity proper. The term ecosystem embraces the living organisms and non-living (abiotic) features such as climate and geology of a site. Some have argued that the inclusion of abiotic components excludes ecosystems from biodiversity. A community refers to organisms living together, essentially the live component of an ecosystem. Again this deceptively simple idea is problematic. The term implies a linkage, an interdependence of species that may not exist. Even tightly linked communities will harbour fleeting tourist species, moving through without necessarily playing any role whilst other communities may be very loosely tied assortments, often described as an assemblage. Habitat conjures up precise images such as the giant panda habitat, but the term may not mean anything if the species is not present; does giant panda habitat cease to exist should the giant panda become extinct? Biome is the term associated with global or continental scale, regional ecosystems defined by vegetation and fauna, in its turn largely determined by climate. A formation is a similar concept, relying solely on vegetation data. Even colloquially familiar biome terms such as rainforest or wetland become difficult to define precisely due to global variations or to give sharp boundaries where they grade into each other.

Ecological biodiversity also includes geographical foci such as high diversity hot spots, continental and oceanic islands plus regions of endemism. If ecological diversity (whether ecosystem, community or any other tricky concept) was nothing more than the sum of its parts (e.g., the species list) this category of biodiversity could be largely ignored, but ecological diversity has its own importance. Ecological patterns emerge at the ecosystem and community level which are more than the sum of the parts of the species present. Ecosystems can (albeit rather mechanically) be said to provide ecosystem services (or functions) such as nutrient and gas cycling. An ecosystem that loses a species may not be merely minus one species but may function in a very different way with knock-on consequences to other ecosystems and perhaps globally. So the ecological category of diversity is important.

Classifications of ecological diversity vary with scale, just like the genetic (nucleotide molecules up to differences between populations) and taxonomic (subspecies to domains). Three main scales are commonly for used ecosystems: global, regional and national. Global classifications rely on major vegetation types (biomes) typically defined by combinations of dominant vegetation, landscape and climate, sometimes plus biogeographical position. Regional classifications are often based on practical definitions. National classifications work at a finer scale, often based on precise species presence and abundance. Classifications of wetlands (global, regional or national) are good examples of ecosystem categorisation, the difficulties of definition and how resulting classifications can vary with the purpose of the systems used.

Wetland classification

Global classifications

At the gross, global level, wetland habitats are subsumed within the broader biomes. Wetlands are characterised by the presence of water at the surface or in the root zone, unique soil conditions and hydrophyte (water loving) plants. Distinguishing different types of wetlands is problematic. The duration and depth of inundation vary. Wetlands are often marginal habitats between truly aquatic and terrestrial systems. Size can vary from small ponds to huge regional expanses. Definitions also vary according to the aim of the classification scheme and cultural traditions, e.g., in the United States, swamp means forested, in Europe typically not so. At a global scale nine main wetland types are commonly distinguished using general biome and landscape characters:

1. Bogs: acidic, peat-building wetlands with water and nutrient input from precipitation, typically in wet, cool climates and characterised by *Sphagnum* mosses.
2. Fens: peat-building but with nutrients and hydrology influenced by surrounding landscape and geology, neutralising acidity. (Note that bogs and fens are often combined together as mires, a generic term for peat-based wetlands).
3. Swamps: so waterlogged that water usually covers surface; dominated by a few species e.g., reeds or characteristic trees, e.g., swamp cypress.
4. Marsh: inundation often seasonal: diverse, herbaceous vegetation with little or no peat accumulation.
5. Floodplains: periodic over-spill from lakes and rivers; very varied.
6. Shallow lakes: open water up to a few meters deep.
7. Salt marsh: tidal herbaceous coastal sward of temperate latitudes.
8. Mangroves: tidal coastal woodlands between 32°N and 30°S, characterised by mangrove trees; exclude saltmarsh from tropics by competition but limited to tropics by climate.
9. Anthropogenic: wetlands created by humans. e.g., paddies.

National classifications of wetlands

At a national level much more detailed classifications can be devised, often differing depending on the purpose for which they are used. In the United States there are over fifty wetland classifications systems in use for water regulation, recreation and wildlife conservation. Here is an example from the UK based on detailed botanical classification.

Classification by species and floristic composition: the National Vegetation Classification

The National Vegetation Classification (NVC) is a systematic and comprehensive classification of UK plant communities. Vegetation types have been distinguished by computer analysis of the presence and absence of individual species and their relative abundance. The classifications for four habitat types (woodland, mires, grassland and aquatic) are available. All four contain communities identified as wetland, a hundred in all, with coastal ecosystems still to be added. Each community type is described in detail, listing plant species, abundance, physical and chemical conditions in the habitat and distribution. Each type is given its own code and name, derived from the diagnotistic vegetation. For example, S9 *Carex rostrata* swamp is the ninth recognised swamp community from the aquatic habitats, characterised by the bottle sedge, *Carex rostrata*:

1. Woodland: Seven wet woodland communities.
2. Mires (bogs, wet heaths and fens): Thirty-eight wetland communities.
3. Grassland: Three inundation communities.
4. Aquatic: Twenty-four submerged and twenty-eight swamp/tall herb fen communities.

Domestic Biodiversity

Biodiversity is widely thought of as an entirely natural phenomenon but definitions now recognise the importance of domesticated species, ecosystems forged from human activity and some cultures intimately tied to their environment. Biodiversity plays a role in the ethical, religious and social values of societies. Differences between cultures affect our valuation of biodiversity, in turn affecting conservation. The fate of biodiversity in different cultures may provide important insights since humans are now the dominant influence on remaining species and ecosystems. Several indigenous cultures are (or were) founded on an ecological intimacy with biodiversity. The native plains tribes of North America relied on the buffalo herds, not only as food but also as an icon central to their beliefs; and customs. The Marsh Arabs of Iraq have a social and agricultural system dependent on their wetland home.

Domestic biodiversity is a tiny fraction of the whole but provides 90 per cent of human food supplies. Domestic stocks include life from all kingdoms. In addition landscapes created by humans, whether farmland, forests, gardens and urban landscapes, are new ecosystems. The extent of human interference with domestic biodiversity varies. Truly domestic species have been subject to marked artificial selection, so that many characteristics will be markedly different from wild ancestors such as cattle, sheep and pigs. Domesticated species (exploited captives) retain many similarities to wild brethren, e.g., reindeer, yak, camels. Feral populations are domesticates that have reverted to living wild. They can create problems, for example the feral cat predation of native marsupials in Australia, but are unique additions to wildlife. Domestication leads to a proliferation of forms, cultivars in plants and breeds in animals, which can be defined by heritable, distinct and uniform characteristics different from other such groups. Plant domestication spans five categories, first the wild ancestor stock. Weedy relatives often flourish in marginal habitats impacted by humans, a genetic bridge between wild relatives and the truly domestic. Primitive cultivars are genetically diverse stock still open to natural selection. Modern cultivars are the result of purely artificial selection of supermarket qualities. Advanced breeding lines are the recent additions, their genomes created by laboratory, manipulation, perhaps including genetic diversity impossible through natural processes.

Numbers of domesticated species are small. Of the 3,20,000 vascular plants, 500 are domesticated and 15–20 are major crop species. The 50000 vertebrates (excluding fish) have provided 30–40 domesticates, plus 60 semi-domesticated. Another 200 fish and invertebrates are commonly used in aquaculture and 4 insects (honey bee, various silkworm caterpillars and the cochineal bug). Fungi and microbes are grown commercially for food, brewing and biotechnological uses (e.g., pharmaceuticals). The lack of species is made up for by breeds. Estimates of total numbers of breeds vary, for example cattle 780–1090, sheep 860–1200, pigs 260–480 (if we cannot keep track of domestic diversity is it any wonder that numbers of wild species are so difficult to estimate). The core domestic farm mammals (cow, buffalo, horse, ass, pig and sheep) comprise 3237 breeds of which 474 are rare and an additional 617 have gone extinct since the end of the nineteenth century. The variety of breeds is a resource. Many have adaptations to local conditions, these adaptations can be exported to new ranges either by moving the animals or controlled breeding. The known traits of breeds allow easier control of artificial selection and discovery of new mutations. The variety is also insurance against the unforeseen. However, this richness has often been overlooked, the potential value of domestic variety suddenly revived in the face of major threats such as climate change.

The inventory of domestic diversity also includes more ominous items. Manipulations of genes in some animals and plants have created hybrids which would probably be impossible through natural selection. Incorporation of genes from one species into another is now commonplace, creating transgenic species. Transgenic sheep, pigs, cattle, horses and rodents have been created containing genes to enhance obvious features, for example mouse genes controlling fur structure in sheep enhance fleece productivity. Human genes have also been incorporated into some animal species to produce medical products such as growth hormones. Genes can even be combined between kingdoms with bacterial genes incorporated into some mammals. Such advances now include the ability to coalesce species, most famously the goat/sheep hybrid which is an example of a chimera species.

Chimeras and deadly viruses are biodiversity's stranger corner but our attitudes to nature have always been coloured by the culture of the day.

MEASURING BIODIVERSITY

Quantifying the variety and richness of life on Earth is central to studies of biodiversity. This section provides estimates of taxonomic and ecosystem diversity.

Numbers of Known Species

Inventories of the present-day diversity of species on Earth can use known species and estimates of likely numbers (including likely maxima and minima). The numbers known and reliability of estimates vary greatly with taxa. Birds and mammals are well documented, though there are still some surprises, e.g., the Udzungwa partridge (*Xenoperdix udzungwensis*), a bird species first identified in 1991, in a dinner.

Numbers of many invertebrates and micro-organisms remain darkly mysterious. Estimates of the rate at which new species are discovered can provide some insight into these voids.

Estimates of global diversity have changed markedly throughout history. It is important to separate these counts of known taxa from estimates of the actual numbers, which use various assumptions to include projections of likely numbers as well as the as yet undescribed.

The total described inventory for some taxa such as multicellular animals and green plants gives good minimum estimates, but there are problems even with these organisms we have described. For some there is a debate over the numbers of genuine species versus subspecies. In more diverse, typically less well-known taxa, including some of the numerically and ecologically dominant groups such as worms and molluscs (Phyla Annelida and Mollusca respectively), there is confusion because several nominal species (anything that has been described and had a scientific name attached to it) are actually specimens of the same species. Each of the names is merely a synonym for the same thing, perhaps arising because of variable shapes, names given to different stages in the life cycle, earlier names remaining unknown to later taxonomists or confusion due to reclassification. The numbers of known mollusc species varies between 45000 and 150000, with 70000 a compromise. Extensive, tangled synonyms need painstaking revision. Even without these problems, the literature of described species is confused and scattered.

Estimating the Actual Numbers of Species

If we cannot agree on the numbers of species we claim to know, the chances of estimating numbers of those we do not know seems rash. However, the actual total of species on Earth at the present time has become something of a Holy Grail in recent years. Many approaches have been used to estimate the actual total of species. Most involve using some data from a taxon, region or ecosystem and scaling up the patterns to global proportions. Each has strengths and weaknesses and usually different answers. The Global Biodiversity review settles for 12.5 million species (see Fig. 18.1).

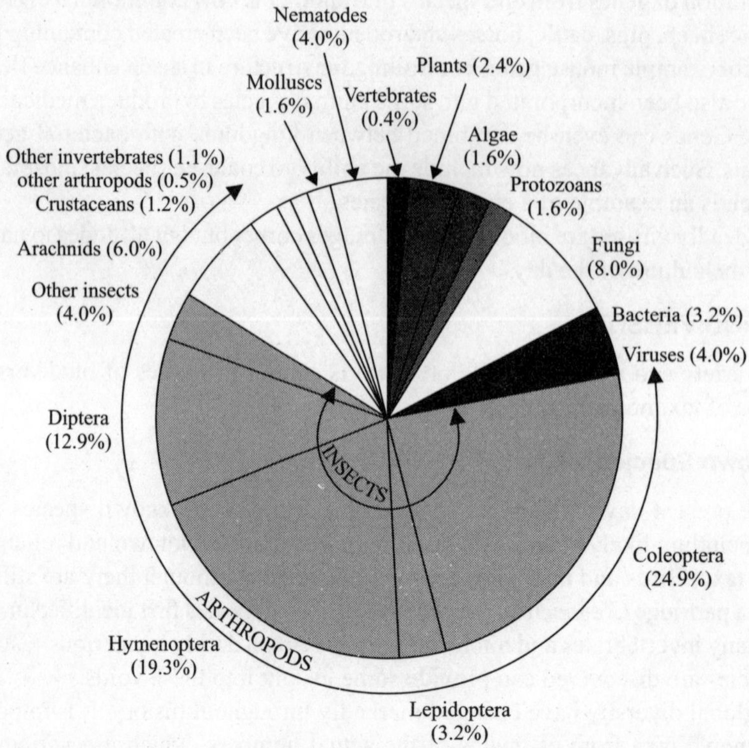

Fig. 18.1. Proportions of different taxa from conservative estimates of actual total numbers of species. The diagram includes taxa likely to contain 1,00,000 + species plus vertebrates

Time series

This technique uses the rates at which new species are being described and extrapolates the trends to look for an asymptote. Whilst we may be reaching an asymptote for a few taxa (e.g., birds and mammals) there have been recent bursts of finds for many invertebrate groups so that time series seem very unreliable.

Expert opinion

Specialists with intimate knowledge of particular taxa can be asked to provide their best guess of likely totals and the results summed. Expert opinion has varied greatly over time, with estimates being revised ever upward.

Empirical relationships

There is a host of ecological patterns, often described in detail for individual communities, that can be scaled up from the original local data to global proportion. These include the ratios of species number to body size, species to area and species to population abundance, plus patterns derived from food webs and plant-herbivore associations (see Fig. 18.2).

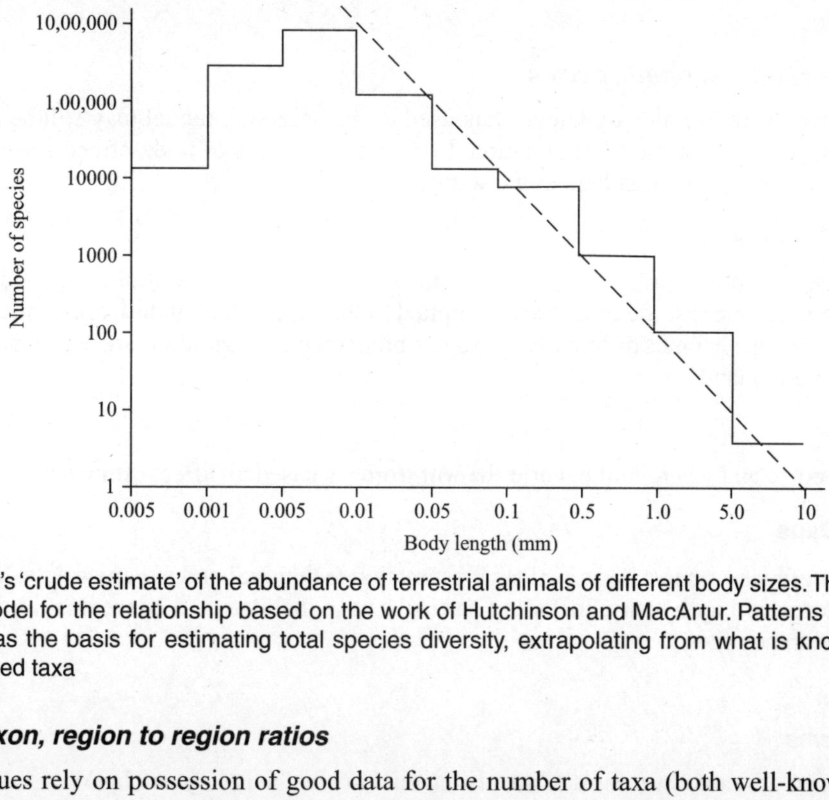

Fig. 18.2. May's 'crude estimate' of the abundance of terrestrial animals of different body sizes. The dashed line is a similar model for the relationship based on the work of Hutchinson and MacArtur. Patterns such as these can be used as the basis for estimating total species diversity, extrapolating from what is known to include poorly described taxa

Taxon to taxon, region to region ratios

Such techniques rely on possession of good data for the number of taxa (both well-known and more obscure groups) in a particular region. The ratio of species of a well-known taxa (e.g., butterflies) to total taxa (e.g., all other insects) for the reliably surveyed region can then be used to estimate total numbers by counting the numbers of the same well-known taxa in comparatively poorly surveyed regions, assuming the same ratio of well-known taxon species to all other species holds true and multiplying up to get an estimate of the total. A classic example has been the use of butterfly to insect ratios from the United Kingdom. British insect fauna is unusually well recorded, so the data are reliable. There are 67 species of extant butterfly to about 22000 insect species in total, a ratio of 1:328. Globally there are some 17500 species of butterflies. Assuming the UK butterfly to total insect ratio holds globally, this gives an estimate for insect species numbers of 5.75 million.

Problems with Estimation Techniques

Scaling up

Many of the methods used rely on taking detailed data from a local, thoroughly documented site and scaling up to global proportions. Very many ecological patterns and processes are scale dependent. The patterns shown in local studies might not scale up globally in a reliable way. Patterns and processes could be diminished or exaggerated.

Comparing taxa

These gross estimates commonly lump taxa together. Yet this assumes that patterns and processes that work well for one taxa might not fit another. Data for ants might not be reliably extrapolated to estimate beetle numbers.

The fraction of undescribed species

Estimates rely on what we already know. Data used as the basis of a model may still be missing many undescribed species. It is a catch–22 situation: to estimate numbers of undescribed species we have to know about the undescribed numbers in the source data we use.

Species definitions

The criteria used to define a good species vary with taxa. The minute separations used to distinguish sub and sibling species amongst the birds are not applied to the desperately underworked micro-organisms. These problems afflict habitats differently. Sub and sibling species separations are poorly drawn in marine studies compared to land.

Workforce

The size, expertise and where in the world the workforce is based all affect estimates.

The Major Gaps

The problems of bias amongst described taxa, estimation techniques and expertise have left major gaps in our knowledge of biodiversity, including particular taxa, habitats and conceptual topics, super-rich groups and reference sites.

Ecosystems

Marine systems

Although marine systems cover most of the planet, our knowledge is very patchy. Whilst the biodiversity associated with coral reefs has been compared to that of tropical rainforests, other habitats such as the surface waters of open oceans resemble deserts. Sampling of deep sea sediments coupled with greater interest in micro-organisms has provided evidence of very high diversity. The diversity of deep sea life may be a result of the great extent of this habitat, a macro-ecological phenomenon if local diversity is sustained by the extended regional species pool. The open ocean can spring surprises. In a new genus of bacterium, *Prochlorococcus*, was identified; it may be the most abundant cellular life on Earth.

Tropical rainforests

Although rainforests appear to support the greatest biodiversity on Earth, research effort is still overshadowed by the sheer size of the task. Particular habitats within the forest have yielded tantalisingly diverse communities, but whether such results can be extrapolated reliably remains unknown. For example, one explanation for the variety of herbivorous tropical forest insects is that each is very specialised, the forest environment in effect parcelled up into very small niches, each species perhaps feeding on only one host plant. Recent results suggest that these insects may be less specialised, with perhaps only 2–5 beetles unique to each tree species, rather than the 160+ used to estimate Earth's arthropod richness at 30 million.

Intraterrestrials

Up until it was believed that life was restricted to the upper meter or two of soil or seabed. Revelatory discoveries have now detected life in rocks as deep as 3.2 km down in South African gold mines, with some microbiologists suggesting we live on top of a 6 km deep, hot biosphere. These crust-dwelling microbes have been named the intra terrestrials. Some appear to rely on energy sources derived from the rock itself, for example water-basalt reactions that release hydrogen, although other communities may be relying on percolation of dissolved organics. DNA sequences from marine and terrestrial samples suggest many are related to the microbes of deep sea hydrothermal vents. The rocks may also be old. Intraterrstrials have been detected in 14 million year old marine sediments. Intraterrestrial life appears to go very slowly, effectively dormant for long periods and is lived at low densities. These extraordinary communities remain elusive.

Taxa

Parasites

Given that most organisms seem to suffer parasitism, it is tempting to imagine parasite diversity as vast as the range of potential hosts. This is especially so in the case of those insects that lay their eggs in or on living hosts (typically other arthropods), the host being killed only once the parasite matures and hatches. Such insects are specifically termed parasitoids; this distinction from other parasites reflects the almost invariably fatal impact on the host, unlike other parasites. Given that insects make up the overwhelming majority of species, if every species had its own parasitoid (and parasitoids are in their turn attacked by their own super-parasitoids) this automatically increases total species massively. However, evidence of parasite and parasitoid biodiversity is sketchy.

Micro-organisms and fungi

Diversities of microbial eukaryotes, Archea and Bacteria, Fungi and viruses are poorly known, despite the significant, perhaps dominant, roles they play in the Earth's geochemical cycles. Terrestrial subsurface microbial biomass may rival that of all land and marine plants whilst oceanic viruses are the most abundant of all life forms. Difficulties of sampling, identification and defining exactly what is a species plague the work with these groups. Free-living microbial eukaryotes (often more familiar as 'Protozoa') may be relatively species poor, with estimates of between 10000 and 20000 species. Most species have global distributions and an abundant seed bank of cysts or spores allows species to reappear when environmental conditions suit. They show no geographically driven diversity, no known endemics. The diversity of prokaryotes, Archea and Bacteria is difficult to estimate because clonal reproduction and the frequency of gene transfer undermine the use of definitions of species familiar to animals and plants: 4500 have been described but analysis of prokaryote genetic diversity in soil and water samples suggests this is wholly unrealistic. DNA-based techniques define species as genomes with 70 per cent or more homology of DNA. Using this criterion, soils and water samples of between 30 and 100 cubic centimeters contain between 3500 and 11400 genomic species. There could be millions of prokaryotic species. The visible wildlife we hold dear, the giant pandas and blue whales, are no more than surface decoration.

Nematodes

Nematodes are a phylum of worms, capitalising on a very uniform body plan but extremely abundant in very many habitats. Problems from lack of expertise and identification have hindered investigation of this phylum but recent evidence hints at high species diversity. About 15000 species are known but some nematode experts have suggested that there may be millions of species, basing this claim on the abundance of these worms. As the apocryphal story puts it, they are numerous and found throughout all habitats and organisms so that if all other organic life was to be stripped away a ghostly outline would remain sculpted in nematodes.

Mites

These are members of the Phylum Arthropoda. Like the nematodes, problems of lack of expertise and identification have held back work but there are suggestions of high species diversity.

Insects

Despite the considerable work devoted to the class Insecta there remain very poorly known groups and habitats. Many of the estimates of total global species richness are based on insect or arthropod richness.

Conceptual gaps

Super diverse groups

Some taxa have been described as super (or hyper or ultra) diverse (or speciose), reflecting an inordinate variety of species. This accolade suggests that the patterns and underlying processes may be unusually magnified or even different to the typical determinants of diversity. Understanding such differences is not only important but also difficult and may require separate models to estimate total numbers.

Special species and areas

All diversity is not equal. An individual species, perhaps just one comprising an unusual lineage, might be rated as special, more worthy of protection than another species with many related species in a diverse genus. How to measure the phylogenetic isolation and uniqueness of a species is a new field. Some taxa may be very useful indicators of the overall diversity of a site, their diversity accurately reflecting the variety of other species which cannot be measured for want of time and techniques. Similarly focal taxa are groups combining concepts of the special and representative and could be useful signals of important habitats. Essential areas and reference sites are also concepts only recently developed. The idea of centers of diversity has been developed, reflecting the evolutionary heartland (or perhaps last refuge) for a taxon. Linked to this, complementarity has been developed as one criterion to assess regional diversity, especially for plants. All the species of a taxon make up the total complement. The single most important site (or country) for this taxon is that with the highest proportion of this complement; the site with the highest proportion of the remaining species becomes the next most valuable. Conversely centers of diversity for birds have concentrated on aggregations of endemic species. Regions of high diversity or endemicity for one taxon are not necessarily so for another. All these ideas, their usefulness and identification are comparatively new, debatable and poorly researched.

GLOBAL BIODIVERSITY PATTERNS

Taxonomic Biodiversity: Global Patterns of Taxonomic Diversity

Global patterns of species diversity are well known for only a few groups. There are marked patterns, in particular concentrations of species within biodiversity hot spots, biogeographical variations between continents and trends such as the increase of species in most taxa towards the tropics. Birds and plants have been studied in detail, reflecting a bias in expertise and available data, but already producing useful insights for conservation planning.

Bird species number nearly 9700, the most speciose of terrestrial vertebrates compared to over 4000 amphibians, 6550 reptiles and 4327 mammals. Birds are one of the most thoroughly recorded taxa: their distribution is unevenly spread across the main biogeographic realms, with 3083 species in South America, 2280 in Asia, 1900 Africa, 950 in the Palaearctic, 900 in Australiasia and 800 in the Nearctic. Within continents the uneven pattern continues. Table 18.2 outlines species, family and endemic diversity from the top five South American countries plus Galapagos Islands (see Fig. 18.3).

Table 18.2. Bird taxonomic diversity of top five South American countries plus Galapagos Islands. Note Ecuador's high diversity despite small size and Galapagos' proportion of endemic species (Year 2004).

Country	*Area, km^2*	*Species total*	*Endemic species*	*Family total*	*Endemic families*	*Vascular plants, species number*
Colombia	11,38,914	1700+	67	79	20	51,220
Peru	12,85,216	1538	112	86	23	17,144
Brazil	85,11,965	1492	185	84	23	56,215
Ecuador	2,83,561	1388	37	82	21	19,362
Venezuela	10,98,581	1340	40	79	20	21,073
Galapagos Islands	7,845	136	25	38	0	

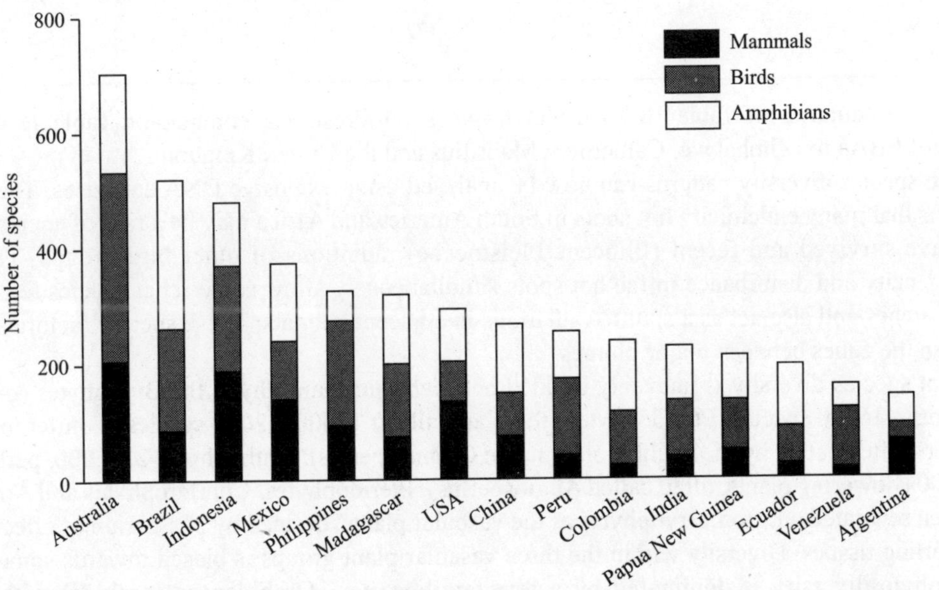

Fig. 18.3. Top fifteen countries for endemic mammals, birds and amphibians (Year 2004).

Birds have been used as a focus to identify priority areas for conservation, primarily using endemics. Birds are potential good indicators of biodiversity. Birds are found throughout the world and have diversified in all terrestrial regions and habitats. Their taxonomy and species distributions are well known; there is also evidence for correlation between their diversity and that of other terrestrial vertebrates and vascular plants. The International Council for Bird Preservation has run an extensive global programme to identify Endemic Bird Areas (EBAs). The survey relies on endemics or more exactly restricted range species breeding over a total range of no more than 50,000 sq. km. Any areas with two or more restricted range species are cited as EBAs; 2609 birds fitted the range criterion, plus 59 known to have gone extinct since 1800. The analyses suggested 221 EBAs, 70 per cent in the tropics, embracing 2480 species, 77 per cent of known threatened bird species. EBAs are ranked into three categories: Priority I is most important/threatened while Priorities II and III are less important/threatened. These categories include estimation of biological importance (restricted range species relative to area, taxonomic uniqueness and known significance to other plant and animal taxa) and degree of threat (threat to birds, proportion of area within IUCN recognised protected areas). Table 18.3 gives numbers of EBAs, bird species and numbers in priority categories for the five countries containing the largest number of EBAs.

Table 18.3. Five countries with largest numbers of endemic bird areas (Year 2004).

Country	*EBAs (inclusive of those shared with other countries)*	*Restricted ranges species confined to countries' EBAs*	*Total restricted range species, confined plus shared*	*Priority I EBAs*	*Priority II EBAs*	*Priority III EBAs*
Indonesia	24	339	411	16	7	1
Brazil	19	122	201	8	8	3
Peru	18	106	216	4	7	7
Colombia	14	61	189	4	9	1
Papua New Guinea	12	82	172	3	63	3

All five countries of Table 18.3 harbour tropical rainforest. For comparison Table 18.4 provides details of EBAs for Zimbabwe, Cameroon, Mauritius and the United Kingdom.

Bird species diversity patterns can now be analysed using extensive DNA databases. The evidence suggests that many endemicity hot spots in South America and Africa may be a mix of ancient lineages that have survived and recent (Pliocene/Pleistocene) radiations of other bird groups. The habitat heterogeneity and disturbance in the hot spots simultaneously allow some relict species to cling on in some patches and also act as a spur to radiations in adjacent habitats. These species factories are often found in the zones between major biomes.

Plant species diversity is unevenly divided between four main Phyla: the Bryophytes (mosses and liverworts) 16000 species; Pteridophytes (ferns and allies) 10000–12000 species; Coniferophytes 700 (conifers, often called by the slightly older name Gymnosperms); Anthophytes 2,50,000, perhaps up to 7,50,000 (flowering plants, often called Angiosperms). Pteridophytes, Coniferophytes and Anthophytes are often separated out from Bryophytes as the vascular plants, defined by possession of effective water transporting tissues. Diversity within the three vascular plant groups is biased towards some families. Anthophyte diversity is dominated by a few families, e.g., Orchidaceae (orchids) with between

25000 and 35000 species, Composites (thistles, dandelions) 20000 and legumes (peas and beans) 14200. The Coniferophytes contain 500 conifers, 100 cycads then some numerically tiny but remarkable species, notably the fossil relict Gingko tree (one species) and the slow-growing, two-leafed *Welwitschia* of southern African deserts. The Pteridophytes consist of 9000 fern species and again, small numbers of allied groups including horsetails and Lycopods, plants that were once the dominant vegetation of Palaeozoic forests growing as trees. Plants are unevenly distributed between the continents: Latin America 85000 species, Africa 40000–45000, Asia 50000, Australia 15000, North America 17000 and Europe 12500. Eighteen plant diversity hot spots support 50000 species, 20 per cent of the world's vascular flora, but between them cover only 5 per cent of the land's area.

Table 18.4. Endemic bird areas of Zimbabwe, Cameroon, Mauritius and UK (Year 2004).

Country and name of EBA	*Area of EBA, km²*	*Restricted range species confined to EBA (+ shared with other EBA*	*Per cent of EBA area protected*	*Priority category*
Zimbabwe East Zimbabwe Mountains	4900	2 (+ 2)	5–10	III
Cameroon Cameroon Mountains	7300	26 (+ 2)	5–10	I
Cameroon Cameroon/ Gabon Lowlands	40000	5 (+ 2)	20–30	III
Mauritius Whole island system	1900	6 (+ 4)	0–5	I
UK No EBAs	–	1	–	–

Species richness within the sometimes unnatural confines of political boundaries reflects size, topographic heterogeneity and complexity plus climate. Detailed studies of tree and vine inventories from South American tropical forests have compared species richness to climatic factors such as total rainfall and seasonality, plus soil nutrients. Diversity showed clear increases with total rainfall and length of rainy season. Although climate alone was a very effective predictor of species richness, soil factors could be important, especially increased species richness with increased nitrogen availability at higher altitudes. Such results can be useful to focus attention on known high total rainfall and long rainy season sites as of particular importance for conservation, even if good species inventories are not yet available. The greatest species richness is in the tropical forests but many oceanic islands have very high endemicity. Tiny St Helena in the Atlantic, Napoleon's final exile, has seventy-four endemics out of eighty-nine species. Table 18.5 gives known vascular plant numbers and endemicity for the top five species-rich countries and other countries for comparison.

Revelations of plant species richness continue at a smaller scale. A study of a one hectare plot in the Ecuador *terre firme* tropical forest in the late 1980s found 473 species (with stem diameter at standard breast height of greater than 5 cm). Given the estimated 3000 tree species in Ecuador, this represented 16 per cent of the country's tree flora but 49 per cent of species were represented by only one individual. Similar inventory work from Colombian forest, working in even smaller plots of 0.1 hectare area but including trees and herbs, found up to 313 species per plot.

Table 18.5. Flowering plant diversity, selected countries.

Country	*Flowering plant species*	*Conifers and allies*	*Ferns and allies*	*Number of endemics (% of species)*	
Brazil	55000	–	–	–	
Colombia	35000	–	–	1500	(4.3)
China	30000	200	2000	18000	(56)
Mexico	20000–30000	71	1000	3624	(13.9)
USSR	22000	74	207	–	
Madagascar	8000–10000	5	500	Up to 8500	(68)
UK	1550	3	70	16	(1)
Zimbabwe	4200	6	234	95	(2.1)

Plant species diversity has also been the target of the IUCN Plant Conservation Programme to identify Centers of Plant Diversity (CPDs). The rationale for CPDs is that their identification and subsequent protection will conserve the majority of wild plants; 241 CPDs have been identified so far. The main criteria are species richness and numbers of endemics. The IUCN programme also takes into account value of the gene pool to humans, diversity of habitat types, sites containing significant proportion of species adapted to specific conditions found therein and imminent large-scale threat.

Ecological Biodiversity

Describing the extent of global ecosystems

Ecological biodiversity shows global patterns akin to taxonomic trends. Global classifications depict polar to equatorial biomes, largely created and defined by geography and climate. Some habitats are restricted by latitude (e.g., tropical rainforest, coral reefs) but others occur across the planet, though their precise regional form and complement of species varies.

Global biogeography has long recognised differences between the major continents and their allied peripheries (see Fig. 18.4). The six main divisions are Nearctic (North America), Neotropical (Central and South America), Palaearctic (Europe, Russia, central and eastern Asia), African (Africa including Madagascar), Oriental (India, Indo-China and South East Asian islands down to Borneo) and Australasia (Australia, New Zealand, New Guinea). Increasingly detailed global classifications of biomes have been devised, increasingly using remote sensing from satellites. These attempt to classify the main terrestrial ecosystem types produce schemes that broadly reflect global climate systems and sometimes other natural or human factors, e.g., soils or landuse. Udvardy's combination of the use of dominant vegetation type and biogeographic is still widely used and has been used as a base against which to measure human impacts on biomes at a global scale (see Fig. 18.5).

Table 18.6. Example vegetation types and extent from Udvardy's system

Realm	*Biome and province*	*Area, km²*
Afrotropical	Tropical Humid Forest	
	Congo rainforest	21,95,019
	Guinean rainforest	7,09,112
	Malagasy rainforest	1,47,862

(Contd...)

Realm	*Biome and province*	*Area, km²*
	Evergreen sclerophyllous vegetation	
	Cape Sclerophyll	99,663
Neotropical	Tropical Humid Forest	
	Amazonian	28,64,623
	Campechean	2,79,695
	Colombian coastal	2,73,266
	Guyanan	10,90,396
	Mudieran	19,88,840
	Panamanian	1,28,872
	Atlantic forest	2,23,944
	Evergreen sclerophyllous vegetation	
	Chilean sclerophyll	47,988

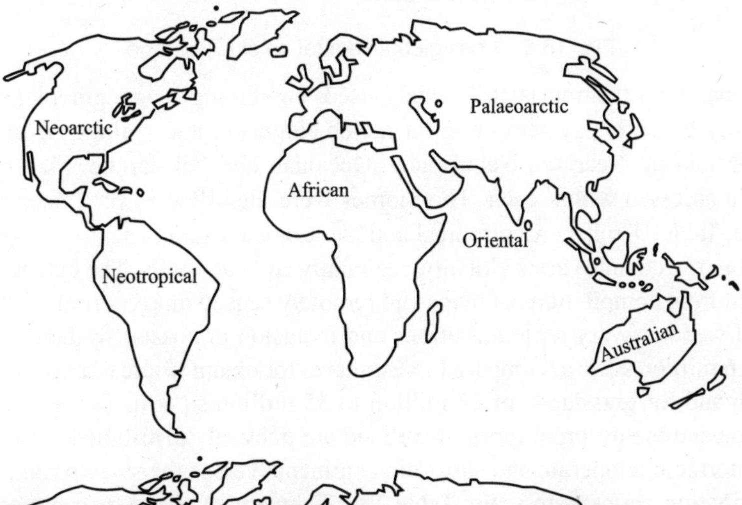

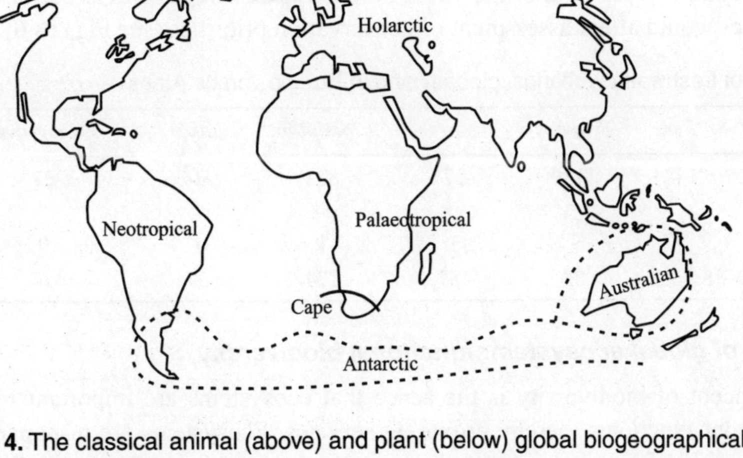

Fig. 18 4. The classical animal (above) and plant (below) global biogeographical realms.

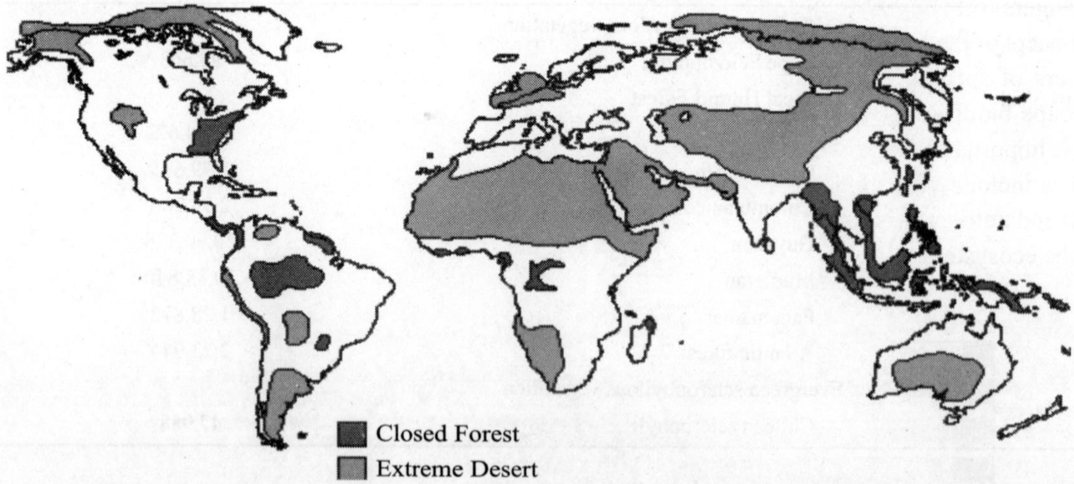

Fig. 18.5. Gross global vegetation classification.

Udvardy's global classification is still widely used, combining biogeographical and biome-based categories. Udvardy based his system on eight major biogeographic realms: Afrotropical, Antarctic, Australian, Indo-Malayan, Nearctic, Neotropical, Oceanian and Palaearctic. Major vegetation types, biomes, were then assessed within each. The biomes were identified as provinces reflecting regional variations in form. Table 18.6 lists Afrotropical and Neotropical examples.

The distribution of wetlands varies globally, regionally and nationally. The extent of global wetlands has been estimated from compilations of maps and remotely sensed images from satellites. Estimates of the global area of wetlands vary with definitions and inclusion of coastal wetlands but range between 5.3 million and 8.6 million sq. km, compared to estimates for extant tropical rainforest of 9.4 million to 12 million sq. km and for grasslands of 24 million to 35 million sq. km. The total extent of wetlands varies with latitude and the different types of wetland are unevenly distributed with vast tracts of bogs dominant across northern temperate and subarctic continents versus the swamps and floodplains of the tropics. The distribution varies with scale. Table 18.7 compares wetland distributions globally and in Europe. Differences would affect assessment of conservation priorities (see Fig. 18.6).

Table 18.7. Extent of freshwater wetlands, globally and in Europe versus Africa.

	Bogs	*Fens*	*Swamps*	*Marsh*	*Floodplain*	*Lakes*	*Mangrove*	*Anthropogenic*
Global	1867	1483	1130	274	823	114	27	1300 (rice paddies)
Europe	54	93	1	4	1	1	0	Minor
Africa	0–38		85	57	174	39	6	46

The importance of global ecosystem: functional biodiversity

Central to the concept of biodiversity is the sense that ecosystems are important for what they do. Ecosystems carry out functions, thereby providing services. Ecosystems are responsible for fluxes of energy and materials. The importance of ecosystems as providers of services adds to our awareness of the dangers from degradation and loss and of the value of biodiversity. Functional biodiversity meshes the concept of the ecosystem, dominated by ideas of flow and flux of resources with the community, the numbers of species. Ecosystems are a synergy of genetic, population, community, ecosystems and landscape biodiversity. Ecosystem function is the sum total of their activity, apparent even when the precise importance of individual species as keystones, driving the processes remain unclear. The array of services include vital global life support through atmospheric quality, climatic control, protection of coasts and nutrient cycling. The importance of functional biodiversity is best revealed by an example of just one ecosystem.

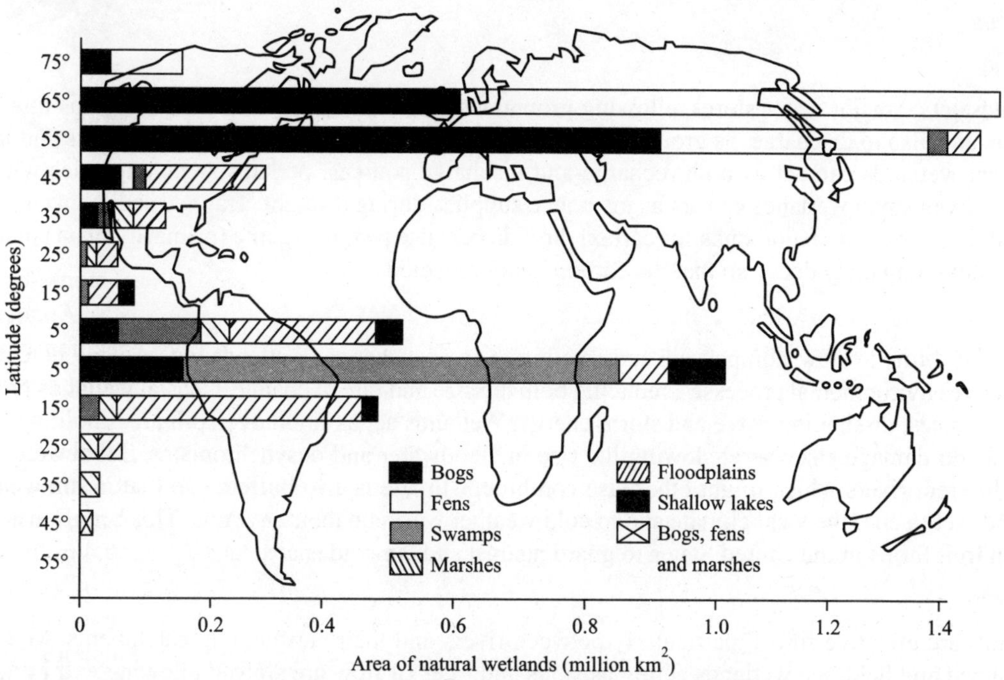

Fig. 18.6. Latitudinal extent of different wetland types.

Importance of wetlands

Wetlands throughout the world support their own unique wildlife. By its very existence this biodiversity has an intrinsic value. Tangible, highly valuable benefits also derive from wetland ecosystems. These can be broadly divided into production (in ecological terms the living components, in economic terms the stock), the ecosystem functions (in economics, the services) and the intrinsic cultural and biological diversity in its own right (attributes to economists). The astonishing variety of benefits, local to global, provided by wetlands is outlined next.

Production

Wetlands are the most productive natural ecosystems on Earth.

The following examples are all sustainable uses of wetlands, although in some cases more intensive versions of the same exploitation also exist, destroying the original habitat:

1. Consumptive: Wetlands yield fish, shellfish and prawns. Plant products include wild fruit and vegetables, gardened crops and small-scale arable. In addition plants provide grazing plus harvested fodder, building poles, thatch, craft materials, latex, resin, tanning products, beer and medicines. Note that these products are not the preserve of subsistence economies. High vale crops such as thatch, rushes, sedges and wild cranberries are harvested in developed countries. Animals are hunted for fur and skins, as well as sport. Energy is provided by fuelwood and peat.
2. Non-consumptive: Tourism and recreation rely on the wetland landscape and wildlife without being direct consumers.

Services

Storage

Wetlands act as major water stores, allowing groundwater to recharge as water slowly seeps into deep aquifers and also to discharge, as groundwater tops up surface aquatic habitats such as rivers and lakes. Note that wetlands can act as both recharge and discharge sources, perhaps switching roles with the seasons. Even small wetlands can act as insurance supplies during drought. Trapped sediment can build up land. Peat bogs are major sinks for carbon dioxide because peat is organic (primarily plant) material that cannot completely decay so that the carbon is not recycled.

Buffering

This is the ability to slow, compensate and ameliorate against change. Wetlands buffer many potentially destructive environmental processes, reducing both the size and rate of change. Coastal wetlands provide shore defences, dissipating wave and storm energy. Wetlands act as flood overspill areas holding water that may do damage elsewhere, slowing the rise of floodwater and desynchronising floodwater peaks from different rivers which might otherwise combine. Climate is also buffered so that in hot weather, wetlands act to cool the local climate and in cold weather warm up their environs. This benefit has been used in fruit farms in the United States to guard against sudden cold snaps.

Cleaning

Wetlands are effective filters, particularly between rivers and their surrounding catchments. Materials are trapped and held, the wetlands acting as sinks and rates of flow are slowed allowing extra time for natural processes to neutralise potentially harmful inputs. Wetlands are known to trap and clean sediment run-off, organic detritus, sewage, excess nutrients (especially phosphorous and nitrogen compounds), acidic inputs, metals, pesticides and pathogens. These effluents can be from a readily identifiable point source such as a mine-working or sewage treatment plant. In addition wetlands are particularly useful to control diffuse, non-point source inputs that plague so many catchments, e.g., nutrient run-off from farmland, which cannot easily be collected and cleaned by technical intervention.

Pathways

The waterways ramifying throughout many wetlands provide pathways for natural fluxes such as nutrient recycling and alluvial deposition, increasing fertility. Open water also provides literal routes for movement by animals and humans.

Attributes

Cultures utilising wetlands have developed customs, architecture and landscapes as a result of this interdependence. In some cases the entire society is intimately tied to the wetland system, creating unique cultural systems, e.g., the Marsh Arabs. So often dismissed as miasmal wastes in urgent need of draining, wetlands are actually a vital asset.

Islands and hot Spots: the Earth's Extra-special Places

Biodiversity is not evenly distributed. Within the general patterns of taxonomic and ecosystem distribution, there are striking concentrations of species, associated with certain ecosystems. Identification of these special places is important for practical conservation and to understand the underlying causes that drive diversification. These foci can be special by virtue of several, sometimes contrasting, factors:

1. Very high total species numbers.
2. Endemicity, whether of common or unusual lineages.
3. Unusual combinations, characteristics of communities.
3. Super speciose taxa.

Continental hot spots

Biodiversity hot spots defined by endemism and the uneven distribution of species have been cited for birds and plants in earlier sections. A few countries, primarily tropical, have been described as megadiversity countries, unusually rich in all forms of biodiversity, although data for such categorisation rely on higher vertebrates, plants and a few insect groups. The megadiversity countries are Mexico, Colombia, Ecuador, Peru, Brazil, Democratic Republic of the Congo (formerly Zaire), Madagascar, China. India, Malaysia, Indonesia and Australia. However, political boundaries are an inappropriate guide and generalisations do not take into account the uniqueness and biodisparity found in other countries. Eighteen global hot spots identified by endemicity of many taxa have been identified. Table 18.8 summarises data for four.

Table 18.8. Example of continental biodiversity hotspots (Year 2004).

	Higher plant species	*Endemic mammal species*	*Endemic reptile species*	*Endemic amphibian species*	*Endemic swallowtail butterffy species*
Cape region, South Africa	6000	15	43	23	0
Colombian Choco	2500	8	137	111	0
Western Ghats, India	1600	7	91	84	5
South West Australia	2830	10	25	22	0

Globally significant continental hotspots are often regions of habitat heterogeneity caused by habitat change. The resultant patchwork allows older taxa to survive and radiation of new species (see Figs. 18.7 and 18.8).

Islands

Island diversity centers are very different from the continental hot spots, often poor in species but sheltering unusual relicts, strange combinations and extraordinary evolutionary radiations. Islands also have a historic role in our awareness and understanding of biodiversity, be it the evolutionary inspiration of the Galapagos or that icon of extinction, the dodo, from Mauritius.

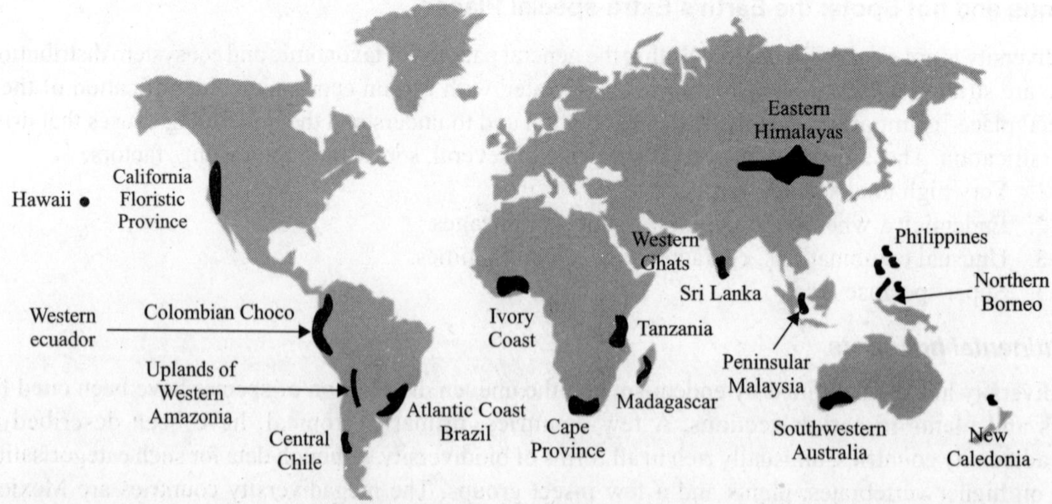

Fig. 18.7. Global biodiversity hotspots, defined by endemicity and also threat from human pressures.

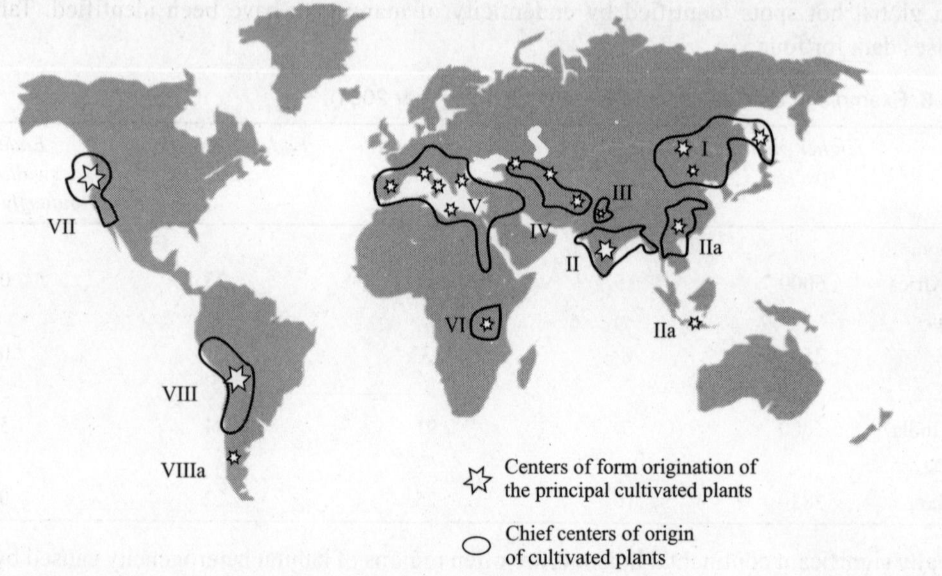

Fig. 18.8. Centers of origin of cultivated plants, based on the work of Vavilov. Stars indicate the centers of origination for the form of crop plant used, the outlined surrounding areas the broad region from within which the plant was domesticated.

Islands have distinctive ecological communities. Historical contingency (what was living there before islands split from mainlands, when the separation occurred) and the vagaries of colonisation create unusual combinations of wildlife. Relict faunas, long extinct on main continents, survive and in some cases flourish. *In situ* speciation creates spectacular radiations, such as the very speciose fruit flies of Hawaii. Animals take up unexpected roles; on Hawaii one moth has a caterpillar that does not chew leaves but is a deadly ambush predator, snatching other insects. A few taxa have become particularly characteristic of islands, e.g., pigeons, rails and tortoises, all capable of erratic but long-distance dispersal, yet not so mobile that once established, new arrivals dilute the evolutionary aftermath. The same traits develop amongst these taxa: gigantism in birds and reptiles, dwarfism in mammals, flightlessness in birds. The wildlife of islands is also distinguished by being extinction prone.

There are two main types of high diversity islands: continental islands such as New Zealand, Australia and Madagascar and tropical oceanic islands such as Hawaii, Galapagos and Mauritius.

Continental islands

Table 18.9 gives comparative biodiversity data for Australia, New Zealand, Madagascar and Kenya as a contrasting continental country.

Table 18.9. Diversity, threat and uniqueness of three continental islands, with Kenya as a comparison. Note that counts of extinct species exclude those lost from very small islands around the mainlands.

Description	*Australia*		*Madagascar*		*New Zealand*		*Kenya*	
	Total	*Endemic*	*Total*	*Endemic*	*Total*	*Endemic*	*Total*	*Endemic*
Flowers			8000					
			10000		2160		6000	
Gymnosperms	15000	80%	5	68%	22	82%	6	4%
Ferns			500		189		600	
Mammals	282	210	105	67	–	3	309	10
Birds	571	351	250	97	285	74	1067	7
Reptiles	700	616	252	231	40	40	187	15
Amphibians	180	169	144	142	3	3	88	10
Fish	113	110	40	38	30	27		
IUCN Plant Centers of Diversity	8		1		3		1	
Endemic Bird Centers	7		5		2		2	
Threatened species								
Plants	2024		194		232		144	
Mammals	38		50		1		17	
Birds	39		28		26		18	
Reptiles	9		10		1		2	
Amphibians	3		0		3		0	
Fish	16		0		2		0	
Animals known to have gone extinct	18		3		14		0	
Extinct endemic plants	173		0		5		0	
Unusual wildlife	Marsupial and reptile diversity		Lemurs; extinct flightless Elephant Bird		Flightless birds			

Oceanic islands

Oceanic islands supporting special diversity are typically very isolated, volcanic (the altitude creating diverse habitats) and tropical. Low-lying tropical atolls lack habitat diversity whilst temperate islands are often too exposed to extremes of weather (see Table 18.10). Island wildlife has proven particularly susceptible to extinction, due to several factors:

Evolutionary innocence

Evolutionary innocence island species have proven very vulnerable to introduced competitors, predators, parasites and diseases. Many islands lack taxa such as large mammalian predators and ants, the very groups that are dominant ecological keystone taxa on mainlands. Relict fauna and particular traits such as flightlessness further increase the threat posed by alien arrivals.

Table 18.10. Diversity, threat and uniqueness of three oceanic island systems.

Description	*Hawaiian Islands (Pacific)*	*Galapagos Islands (Pacific)*	*Mauritius (Indian ocean)*
Vascular plants	900 of which 850 are endemic	540 of which 170 are endemic	878 of which 329 are endemic
Threatened native plants	343	82	269
Introduced plant species	4000	195	–
IUCN Plant centers of diversity	1, whole archipelago	1, whole archipelago	1, whole archipelago
Endemic bird centers	1, whole archipelago	1, whole archipelago	1, whole archipelago
Animals known to have gone extinct	86	5	41
Extinct endemic plants	108	2	24
Species of land snails	c.1000, all endemic	90 of which 66 are endemic	109 of which 77 are endemic
Extinct snail species	29	1	25
Unusual wildlife	High endemism, super-species rich taxa, e.g., fruit flies	High endemism. unusual species, e.g., marine iguana	High endemism; once home to the dodo

Small populations

Small populations in many island species occur as small, isolated populations specialised to live in a narrow habitat range and with no pool from which recolonisation can boost numbers. Genetic diversity may be enfeebled by inbreeding and bottlenecks due to population size. In addition co-evolution of island taxa one with another may result in extinctions cascading through the island once one or two species are lost and ecological links are broken.

Lack of disturbance

Physical disturbance to island habitats can be increased once humans arrive, for example direct action such as fire to clear land or indirectly through impacts of introduced animals such as pigs grubbing up land.

Human exploitation

Direct human harvesting has wiped out species such as the dodo and moas and endangered others, e.g., giant tortoises. Habitat loss as humans clear land is just as great a threat, especially given the small habitat areas and species ranges of many island taxa.

Islands remain under threat, even the most famous. The Galapagos Islands have become a focus of concern, in part due to increased tourism but also because of protests from local people angry at restrictions from conservation laws. Protests have culminated in intentional habitat destruction and threats to conservationists. The focus on hot spots has also received more theoretical criticism, suggesting that our fascination with these small areas risks not doing much to prevent the damage to the rest of the world, the cold spots, which provide the majority of the ecosystem functions upon which we rely.

MORE THINGS IN HEAVEN AND EARTH

Fragments of ancient continents, far-flung islands and impenetrable jungle–our fascination with biodiversity is in part the mystery and unknown. Sad to say, despite recent expeditions, there is unlikely to be a population of Brontosaurs lurking up the River Congo but there are other surprises. Mauritius, now recognised as an oceanic hot spot, was the home of the dodo (*Raphus cucullatus*), discovered in 1598 and extinct by 1670. The dodo was the epitome of the absurd, a fat, waddling, flightless pigeon permanently clad in juvenile plumage, too stupid to avoid being eaten into extinction. Analyses of fossil bone structures reveal an athletic, leggy creature, confirmed by the earliest illustrations. The dodo was an innovative design but fatally maladapted to human interference.

Continental hot spots also hold surprises. Several virulent diseases such as Lassa fever and Ebola virus seem to flare from African rainforest hot spots. Ebola outbreaks do not only kill humans. A forty-strong chimpanzee (*Pan troglodytes*) clan in the Tai Forest of Cote d' Ivoire lost twelve to Ebola. Ebola virus replication is error prone; the frequent mutations that result permit infection of a variety of species. The Tai chimps were proficient hunters of small mammals and perhaps picked up the disease from the rodents that have boomed since when Liberian refugees fleeing civil war crossed into the area. The dodo and the Tai chimps may seem idiosyncratic but prompt important questions. How do human pressures, whether hungry sailors or refugees, impact natural systems?

19

Soils

INTRODUCTION

Soil is one of the most important ecological factors. Plants depend for their nutrients, water supply and anchorage upon the soil. Even for the free-floating aquatic plants which derive their nutrients dissolved in the water medium around them, soil (mud) is important as chief storage of all the nutrients, which are made available to the water medium. Soil system is indeed very complex and dynamic, undergoing continuous change and the rates of such changes being influenced by a number of other factors of the environment. Marsden-Jones after growing plants in different types of soil concluded that soil may affect plants by affecting seed germination, size and erectness of the plant, vigour of the vegetative organs, woodiness of stem, depth of the root system, susceptibility to drought, frost and parasites, number of flowers per plant and the time of flowering etc. Warming emphasised the importance of soil characteristics and proposed five ecological groups of plants, as: (i) oxylophytes—usually found on acid soils; (ii) halophytes—on saline soils; (iii) psammophytes—on sand; (iv) lithophytes—on rock surfaces; and (v) chasmophytes—in rock crevices.

DEFINITION AND COMPOSITION OF SOIL

Soil is thus usually defined as 'any part of earth's crust in which plants root'. Muddy bottoms of ponds, porous rock surfaces, ravines or glacial deposits, bottoms of lakes, peats etc., all are thus soils. But this is a limited definition of soil, as we know that soil is actually formed as a result of long-term process of complex interactions leading to the production of a mineral matrix in close association with interstitial organic matter—living as well as dead. In soil formation, modification of parent mineral matter takes fairly a long time. Such a modification is actually the result of interactions between climatic, topographic and biological effects.

Soil is thus not merely a group of mineral particles. It has also a biological system of living organisms as well as some other components. It is thus preferred to call it a soil complex, which has the following five categories of components:

1. Mineral matter: A matrix of mineral particles derived by varying degrees of breakdown of the parent material-rock.
2. Soil organic matter or humus: An organic component derived from long and short-term addition of material from organisms growing above and below ground i.e., plants, animals, micro-oroganisms.
3. Soil water/Soil solution: All water contained in soil together with its dissolved solids, liquids and gases. Soil water is held by capillary and absorptive forces both between and at the surface

of soil particles. Soil water in reality is a dilute solution of many organic and inorganic compounds, which is the source of plant mineral nutrients.

4. Soil atmosphere: It occupies the pore space between soil particles, which, at any time, is not water-filled. Its composition differs from the above-ground atmosphere in the sense that it is normally lower in oxygen and higher in carbon dioxide content.
5. Biological system: To the above, there may also be added, the biological system, as each soil has a distinctive flora as well as fauna of bacteria, fungi, algae, protozoa, rotifers, nematodes oligochaetes, molluscs and arthropods.

A perusal of the above shows that soil is not merely mineral matter, but a complex of several other types of component. Thus, biologically the soil may be said as 'the weathered superficial layer of the earth's crust in which the living organisms grow and also release the products of their activities, death and decay'. According to Dokuchayev, a famous Russian pedologist, 'the soil is a result of the actions and reciprocal influences of parent rocks, climate, topography, plants, animals and age of the land'. It can be represented by the following formula:

$$S = (g.e.b.)\Delta t$$

where, s-soil; g-geology; e-environment; b-biological influences; t-time.

Before we deal with the individual components of the soil complex in detail, it would be worth to understand first, how soil is formed in nature?

FORMATION (ORIGIN) OF SOIL

The whole process of soil formation is generally divided into two stages: (i) weathering—breakdown of bigger rocks into fine, smaller mineral particles; and (ii) soil development or pedogenesis—modification of the mineral matter through interaction between biological, topographic and climatic effects, which ultimately lead to the development of any of a great variety of potential soil types.

Weathering Process

Bare rock surfaces are exposed to various types of physical, chemical and biological processes which lead to physical and chemical disruption of their components. Physical processes of weathering include action of water, temperature, glaciers, gravity etc., which cause weathering of rocks through such processes as wetting-drying, heating-cooling, freezing, glaciation, solution and sand blast etc. The chemical processes of weathering include hydration, hydrolysis, oxidation-reduction, carbonation, chelation etc., which are due to chemical composition of rocks, chemicals in the atmosphere, as well as those produced as a result of activity of living organisms, such as lichens, fungi, bacteria, blue-green algae, bryophytes etc.

In biological process, lichens are able to extract nutrients from bare rocks. Lichens, fungi and bacteria on rock surface retain the water for a long period during which the chemical processes can proceed, splitting the rock alumino-silicates by hydrolysis and carbonation into the simpler clay alumino-silicates. Algal partner of lichen through photosynthesis, increases the amount of available organic matter at rock surface.

Various exudates from these organisms and the respiratory carbon dioxide accelerate the process of weathering, as some of the lichen acids (organic acids) generally dissolve mineral components. Rock weathering is, therefore, for a short time, a physico-chemical process but soon it becomes biogenic, increasing in its rate. The weathering processes are physical as well as chemical.

Physical

1. Wetting-drying: It is the disruption of layer lattice minerals which swell on wetting.
2. Heating-cooling: It is disruption of heterogeneous crystalline rocks in which inclusions have differential coefficients of thermal expansion. Particularly in dry climates, due to sun heating, large boulders flake at surfaces.
3. Freezing: This is the disruption of porous, lamellar or vesicular rocks by frost shatter due to expansion of water during freezing.
4. Glaciation: Larger masses of snow and ice-glaciers, while falling may cause physical erosion of rocks through grinding process.
5. Solution: Some more mobile components of rocks, such as calcium chlorides, sulphates etc., are simply removed by agents like water.
6. Sand blast: In arid, desert conditions the rocks are disrupted by physical action of wind, sand etc.

Chemical

1. Hydration: As a result of taking water, due to reversible change of haematite to limonite $Fe_2O_3 \rightleftharpoons Fe_2O_3 \cdot 3H_2O$), the rock swells. This swelling causes the disruption of sandstones etc.
2. Hydrolysis: In this process, components like alumino-silicates of rock breakdown, during which elements such as potassium and surplus silicon are washed out which give rise to simpler mineral matter like clay alumino-silicates. For example, hydrolysis of orthoclase ($K_2Al_2Si_6O_{16}$) to kaolinite ($Al_2O_3 \cdot 2SiO_2 \cdot 2H_2O$).
3. Oxidation-reduction: Some oxidation-reduction chemical reactions, such as reversible change of Fe^{3+} to Fe^{2+}, cause disruptin of rocks, because Fe^{2+} is more soluble than Fe^{3+}.
4. Carbonation: Some chemicals, produced in the atmosphere and those during the metabolism of micro-organisms bring about carbonation. As for example reversible change of $CaCO_3$ to $Ca(HCO_3)_2$ leads to solution loss of limestone or disruption of $CaCO_3$ cemented rocks as the hydrogen carbonate is more soluble than the carbonate.
5. Chelation: Some chemical exudates, produced through biochemical activity of micro-organisms like lichens, bacteria, etc., are able to dissolve out mineral components of the rocks. These metals dissolved with organic products of microbial activity are known as chelates. For example, acids produced by lichens and bacteria have strong chelating properties.

Pedogenesis

During weathering, the rocks are broken down into smaller particles. But this is not the true soil and plants cannot grow in this matter. The weathered material undergoes further a number of changes, which is a complex process, known as pedogenesis or soil development. Whereas in weathering, mostly physical and chemical factors are involved, pedogenesis is largely a biological phenomenon. During this phenomenon, living organisms such as lichens, bacteria, fungi, algae, micro-arthropods, molluscs etc., as a result of secretion of organic acids, enzymes, CO_2 production, addition of organic matter after their death, bring about geochemical, biochemical and biophysical processes. Due to all this the crusts of weathered rock debris are converted to true soils consisting of a complex mineral matrix in association with a variety of organic compounds and a rich micro-organism population.

Thus during pedogenesis, there are added various organic compounds, dead organic matter, living organisms etc., to the mineral matter. As a result of mineralisation of dead organic matter, the minerals are then gradually added to different layers of developing soil. This soil, when fully developed can be seen having a number of layers-horizons of a soil, known as soil profile, which will be described later in this chapter.

FACTORS AFFECTING SOIL FORMATION

We have seen that during the whole sequence of soil formation—weathering and pedogenesis, there are involved many factors. The nature and type of the soil, forming at a particular place is largely governed by the nature of the parent material and its interaction with climate, topography, organisms etc. The subdivided soil-forming factors are as follows:

Passive Factors

1. Parent material: Its physical constitution influences the soil aeration, leaching rate and texture, whereas the chemical composition influences the chemical characteristics of soil.
2. Topography: It influences soil formation through drainage and retention of water. Soil aeration is also governed by topography which affects the solifluction and insolation of chemicals.
3. Time.

Active Factors

1. Rainfall: It determines the direction of solute translocation according to precipitation/evaporation (P/E) ratio. In combination with other factors rainfall also determines depth to water table, which would affect the level of capillary water and aeration.
2. Temperature: Through its interaction with precipitation it governs P/E ratio. Through its influence on physico-chemical processes, it controls the rate of organic turnover in soil. Also; it affects growth rate of vegetation and other organisms.
3. Humidity/Evaporation: Their interaction influences rainfall and the amount of water movement.
4. Biosphere effects: Living organisms are very important in pedogenesis as they speed up and modify the physcio-chemical processes:-
   (a) Phytosphere: (i) direct plant activities such as secretion of organic acids, enzymes, respiratory CO_2; (ii) input of organic matter after death; and (iii) micrometerological influence of vegetation cover.
   (b) Zoosphere: (i) direct interaction between primary production and consumption by animals with consequent effect on input of organic matter to the soil; and (ii) soil-dwelling micro-arthropods, molluscs, lumbricids, etc., have a direct influence on organic matter turnover and incorporation.
   (c) Micro-organisms: Bacteria and fungi play an important role in the geochemical, biochemical and biophysical processes of pedogenesis.

SOIL PROFILE

As pointed out earlier, due to interaction of various factors influencing the, process of pedogenesis, there develop a variety of soil types. It depends upon the nature of parent matter and other factors i.e., climatic, topographic and biological etc., as to which type of the soil will develop in any area. Different types of soil may be defined by the nature of the mineral matrix, the vertical distribution of organic matter and the movement and redeposition of various inorganic components. Soils are described and identified by reference to their profiles. Soil profile is 'the sequence and nature of the horizons (layers) superimposed one above the other and exposed in a pit-section dug through the soil mantle'. The smallest three-dimensional volume of a soil needed to give full representation of horizontal variability of soil is termed a pedon. A soil horizon is defined as 'a layer which is approximately parallel to the soil surface and that has properties produced by soil forming processes but that are unlike those of adjoining layers'. Horizons may, usually be identified visually in the field but they also have chemical and physical characteristics which can be diagnosed in the laboratory.

Although, profiles of different types of soil differ markedly in respect of their physico-chemical and biological properties, Fig. 19.1 shows a hypothetical soil profile with its principal horizons. It is not always true that all these horizons are always present in each profile.

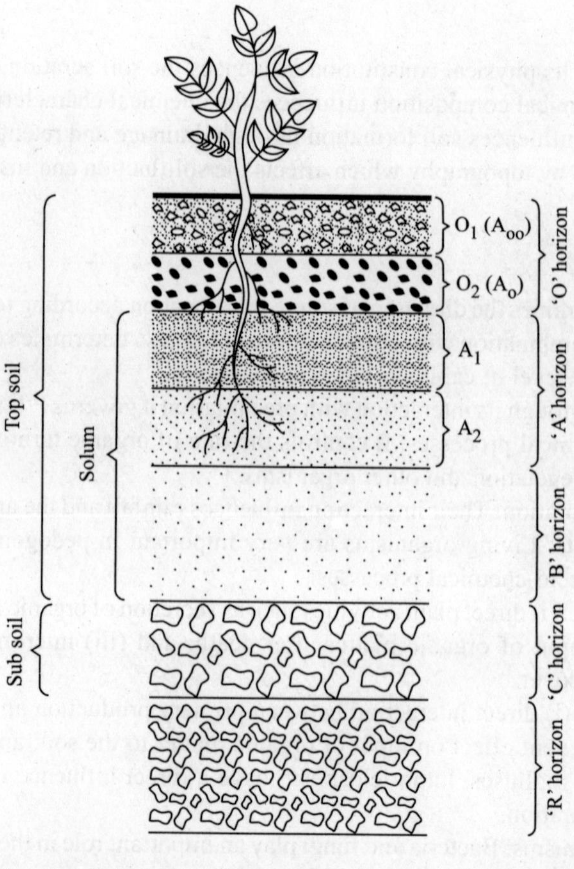

Fig. 19.1. Hypothetical diagram of the soil profile to show principal horizons.

The different horizons of a soil profile historically in Russian terminology, were classified into ABC terminology. However, according to the present one in general use, the soil profile (Fig. 19.1) consists of the following five main horizons.

The 'O' Horizons

These are the organic horizons forming above the surface of the mineral matrix, mainly composed of fresh or partially decomposed organic matter. This horizon is well developed in forests and may be completely absent in grasslands.

This horizon is divided into following two sub-layers:

1. O_1 (A_{00}) region: This is the uppermost layer consisting of freshly fallen dead organic matter as dead leaves, branches, flowers and fruits, dead parts of animal etc. These do not show evident breakdown.
2. O_2 (A_0) region: It is just below the O_1 region, in which decomposition has begun. Thus organic matter is found under different stages of decomposition and micro-organisms like bacteria, fungi, actinomycetes are frequently found. Upper layers contain detritus in initial stage of decomposition, in which material can be faintly recognised, whereas the lower layers contain fairly decomposed matter, the duff.

The 'A' Horizons

These are the mineral horizons formed either at or adjacent to the surface. These are rich in organic matter and/or show downward loss (eluviation) of soluble salts, clay, iron or aluminium, being consequently rich in silica or other resistant minerals. This is thus also known as zone of eluviation-downward loss or leaching.

This horizon is divided into following two sub-layers:

1. A_1 region: It is dark and rich in organic matter. The amorphous, finely divided organic matter here becomes mixed with the mineral matter, which is now known as humus, which is dark brown or black coloured. This region having a mixture of finely divided organic matter and the mineral matter is also called humic or melanized region. In forest soils this region is less deeper than those of the grasslands.
2. A_2 region: This region is of light colour in which the mineral particles of large size as sand are more, with little amount of organic matter. Chiefly in areas with heavy rainfall, the mineral elements and organic chemicals are rapidly lost downwards in this region, making it light-coloured. This is thus also known as podsolic or eluvial zone or zone of leaching.

The 'B' Horizons

These are the mineral horizons forming below the surface in which one or more of the following features can be present: (i) enrichment with inwashed clay (lessivation), iron, aluminium, manganese or organic matter; (ii) residual enrichment with sesquioxides or silicate clays which has occurred other than by the removal of carbonates or readily soluble salts; (iii) sesquioxide coatings of mineral grains sufficient to give a more intense colour than horizons above or below; and (iv) alteration of the original rock material to give silicate clays or oxides in conditions where (i), (ii) and (iii) do not apply.

This is just below the 'A' horizon and can also be divided into B_1 (A_3), B_2 and B_3 regions, depending upon the stages of soil development in the area. B_1 (A_3), if present in the process of successional development, the pesence of which depends upon the extent of development of horizons, above and below it.

The 'B' horizon is dark-coloured and coarse-textured due to the presence of silica-rich clay organic compounds, hydrated oxides of aluminium, iron etc. Since, the chemicals leached from A_2 region, become collected in this horizon, it is also known as zone of illuviation or illuvial zone. This zone is poorly developed in dry area.

A_1, A_2 and B collectively are also known as mineral soil or solum.

The 'C' Horizons

These are the mineral horizons below the 'B' but excluding true bedrock and without any characteristics of' 'A' or 'B' horizons. It consists of incompletely weathered, large mases of rocks.

The 'R' Horizons

This is the parent, unweathered bedrock, upon which there is collected water.

SOME PROCESSES IN SOIL FORMATION, CHARACTERISTIC TO THE CLIMATE TYPE

It is clear form the above account that climate and other factors largely affect the soil formation. The type of soil thus depends upon the nature of interactions between such factors. The characteristics of the soil formed may be analysed through study of their profiles.

Some of the following processes are intimately related with soil formation as influenced by climate and other factors. These are characteristic to a particular type of climate. Some of them which are much pronounced, are as follows:

Laterisation

One of the passive factors—time, required for soil formation is much variable, since it is strongly influenced by climatic and parent material factors. Under tropical conditions, soil formation may occupy a much longer time-scale. Mohr and van Baren wrote: 'The process of laterisation in the tropics...must be regarded as a geological phenomenon. This is because the time factor, next to climate, is of the greatest importance in determining the final stage in the cycle of weathering.' Under high temperature regimes and high rainfall in the tropics, the silicate minerals are very unstable and a large loss of silica, probably as colloidal silicic acid occurs. The ratio of silica/sesquioxide becomes very low. The aluminium and iron sesquioxides of the parent minerals are resistant to decomposition and some are synthesised *in situ* from ionic Al and Fe liberated during alumino-silicate breakdown. This process known as laterisation is most common on base-rich parent matters and it leaves a residue of primary laterite containing little else but iron and aluminium sesquioxides and a few resistant parent minerals.

Melanisation

The process is very common in regions of low humidity where humus formed from the organic matter, along with water becomes mixed in the 'A' horizon of the soil. Due to melanisation, the 'A' horizon becomes dark-coloured.

Podsolisation

In temperate climates, particularly with moderate rain coupled with a nutrient-deficient, well-drained parent material, podsolisation is very common. The chief vegetations are the genera of Ericaceae and conifers, which produce a litter with high lignin and low nutrient (particularly calcium) contents and rich in phenolic compounds which inhibit microbial activities. The litter layer is thus acidic. The water percolating through such litter, being acidic dissoloves out with it minerals and humus content from 'A' horizon. These leached materials reach to the lower horizons, being collected in the form of a hard, distinct layer in the 'B' horizon. Due to loss of chemicals the 'A' horizon turns to a light ash colour. This process is known as podsolisation and soils as podsols.

Gleisation

In humid climates, high ground water level or the retention of surface water in the soil, water-logged and consequently reducing conditions are fairly common. Under such reduced conditions firstly due to ferrous compounds, the soil colour becomes blue-grey or grey and secondly the rates of decomposition of organic matter are very slow. These together result into the accumulation of a sticky, compact layer of blue-grey or grey colour at the bottom of 'B' horizon. This process is known as gleisation and the soils as gleys.

SOIL CLASSIFICATION

There are several systems of soil classification. On the basis of the mode of their formation, particularly the nature of the origin of mineral matter, soils are sometimes classified as: (i) residual soils; and (ii) transported soils.

Residual soils are those in which the whole process of soil formation i.e., weathering and pedogenesis occurs at the same place. Thus in these soils the soil formation occurs at a place where their parent matter (rock) is present. Transported soils are those where the weathered material is taken away at other places. This is at these different places, where, through pedogenesis, soil formation is completed. The weathered matter may be taken away to other places by several agents. Depending upon the nature of these transporting agents, the transported soils may be:

1. Colluvial (by gravity).
2. Alluvial (by running water).
3. Glacial (by glaciers-large masses of snow, ice).
4. Eolian (by wind).

The earliest attempts of a detailed classification of soils were made in Russia. This classification is based upon a combination of climatic/vegetation data and soil morphology. Dokuchayev's final classification of Russian soils is given in Table 19.1.

Table 19.1. Dokuchayev's classification of soils.

		Class A. Normal or Zonal soils		
Zones	1. Boreal	2. Taiga	3. Forest-steppe	4. Steppe
Soil types	Tundra (dark brown soils)	Light grey podsolised soils	Grey and dark grey soils	Chemozem
Zones	5. Desert steppe	6. Desert		7. Subtropical and tropical forest
Soil types	Chestnut and brown soils White soils	Aerial soils Yellow soils		Laterite or red soils
		Class B. Transitional soils		
8. Dry land moor soils or moor-meadow soils		9. Carbonate containing soils (rendzinas)		10. Secondary alkali soils
		Class C. Abnormal soils		
11. Moor soils		12. Alluvial soils		13. Aeolian soils

The US Department of Agriculture Soil Survey has been developing continuously several systems of soil classification from time to time. The US System in its revised form has been widely used in mapping of American soils. This classification is chiefly based upon the various factors, such as nature of parent matter and climatic factors etc., which influence the soil development. The soils are classified into three major types: (i) zonal soils; (ii) intrazonal soils; and (iii) azonal soils. Each type has been further sub-divided into suborders. An outline of this classification is given in Table 19.2.

Table 19.2. USDA revised system of classification of soils.

Order	*Suborder*	*Great soil groups*
Zonal soils	Soils of the cold zone	Tundra soils.
	Light-coloured soils of arid regions	Desert soils. Red desert soils. Sierozem. Brown soils. Reddish-brown soils
	Dark-coloured soils of semi-arid, subhumid and humid grasslands	Chestnut soils. Reddish Chestnut soils. Chemozem soils. Prairie soils. Reddish Prairie soils
	Soils of the forestgrassland transition	Degraded Chemozem. Noncalcic Brown or Shnatung Brown soils
	Light coloured podsolised soils of the timbered regions	Podsol soils. Gredy wooded or Grey Podsolic soils. Brown Podsolic soils. Grey-brown Podsolic soils. Grey-brown Pedosolic soils. Red-yellow Podsolic soils
	Lateritic soils of forested warm-temperate and tropical regions	Reddish-brown Lateritic soils Yellowish-brown Lateritic soils. Laterite soils
Intrazonal soils	Halomorphic (saline and alkali) soils of imperfectly drained arid regions and littoral deposits	Solonchak or Saline soils.• Solonetz soils. Soloth soils
	Hydromorphic soils of marshes, swamps, seep area and flats	Humic Gley soils (includes Wiesenboden). Alpine Meadow soils. Bog soils. Half-Bog soils. Low-Humic Gley soils. Planosols. Groundwater Podsol soils. Groundwater Laterite soils
	Calcimorphic soils	Brown Forest soils. (Braunerde. Rendzina soils. Lithosols. Azonal soils Regosols (includes Dry Sands). Alluvial soils

The most recent classification has been given by USDA., 7th Approximation, which recognised ten orders. Each order is identified on the basis of characters of its horizons i.e., sequence, morphology, colour, constitution etc. Further sub-division of each order is made into sub-orders, great groups, subgroups, families and series. The various orders of this system and their possible equivalence to those in Thorp and Smith's system are given in Table 19.3.

Table 19.3. Various orders of soils in the USDA.

Order	*Approximate equivalents in Thorp and Smith*
Entisols	Azonal soils and some Low Humic Gley soils
Vertisols	Grumusols
Inceptisols	Ando., Sol Brun Acide, some Brown Forest, Low-humic Gley and Humic Gley soils

(Contd...)

Order	Approximate equivalents in Thorp and Smith
Aridisols	Desert, Reddish Desert, Sierosem, Solonchak, some Brown and Reddish Brown soils and associated Solonetz
Mollisols	Chestnut, Chemozem, Brunizem (prairie), Rendzinas, some Brown, Brown Forest and associated Solonetz and Humic Gley soils
Spodosols	Pod sols, Brown Podsolic soils and Ground water Podsols
Alfisols	Grey-brown Podsolic Grey Wooded soils, Noncalcic Brown soils. Degraded Chemozem and associated Planosols and some Half-Bog soils
Ultisols	Red-Yellow Podsolic soils, Reddish-Brown Lateritic soils of the US and associated Planosols and Half-Bog soils
Oxisols	Laterite soils, Latosols
Histosols	Bog soils

SOIL COMPLEX–COMPONENTS AND PROPERTIES

As pointed out in the beginning of this chapter, the soil is not simply one thing but actually made up of several components which all together make soil complex. The soil is, in fact a three-phase system of solid, liquid and gaseous components, each of which has its own physical and chemical properties, in an equilibrium or transient-state, relationship with the others. In addition to these components, each soil has a distinctive flora and fauna of bacteria, fungi, algae, blue-green algae, protozoa, rotifers, nematodes, oligochaetes, molluscs, arthropods etc. These soil organisms make the biotic component of the soil. The liquid and gaseous phases are, in small volumes, fairly homogeneous, whereas the solid phase is heterogeneous, consisting of a range of different-sized inorganic particles of silica, silicate clay, metal oxides and other minor components. These all are present in varying degrees of association with different types of organic matter.

The various components of the soil complex are: (i) mineral matter; (ii) soil air; (iii) soil water; (iv) soil solution; (v) soil organic matter or humus; and (vi) soil organisms. We shall now discuss in brief the nature of each of these components of soil complex.

Mineral Matter

As a result of varying degrees of weathering of parent material-rock, the mineral particles of different sizes are formed. In a given sample of soil, there may be present, different-sized particles in different proportions. Depending upon their size (diameter basis), the International Society of Soil Science has given different names to these mineral particles, which are as follows:

Name of the particle	Diameter range (mm)
Clay	Less than 0.002
Silt	0.002–0.02
Fine sand	0.02–0.20
Coarse sand	0.20–2.0
Stones and gravel	Above 2.0

Soil texture

It is determined by 'the relative proportion of mineral particles of different sizes present in the soil'. Stones and gravels are excluded from the textural classes. To separate the particles of different sizes, the organic matter of soil is oxidised and inorganic cementation is removed to breakdown structural aggregates. Fine fractions, as clay and silt are removed from coarser material by sedimentation technique and the coarse fraction is dried and graded by sieving. Stones and gravels are retained by a 2.0 mm perforated plate sieve, the coarse sand retained on a 70 mesh wire sieve (0.23 mm per opening) while the fine sand passes through the 70 mesh. Silt and clay fractions are graded according to sedimentation velocity in a liquid column.

On the basis of the proportion of different-sized particles, soils are classified into different textural groups. Figure 19.2 shows an arbitrary textural classification of natural soils which may be used with laboratory particle size analyses, explained above or crudely with 'finger tests' in the field. As shown in Fig. 19.2, some of the various groups of soil are shown in Table 19.4.

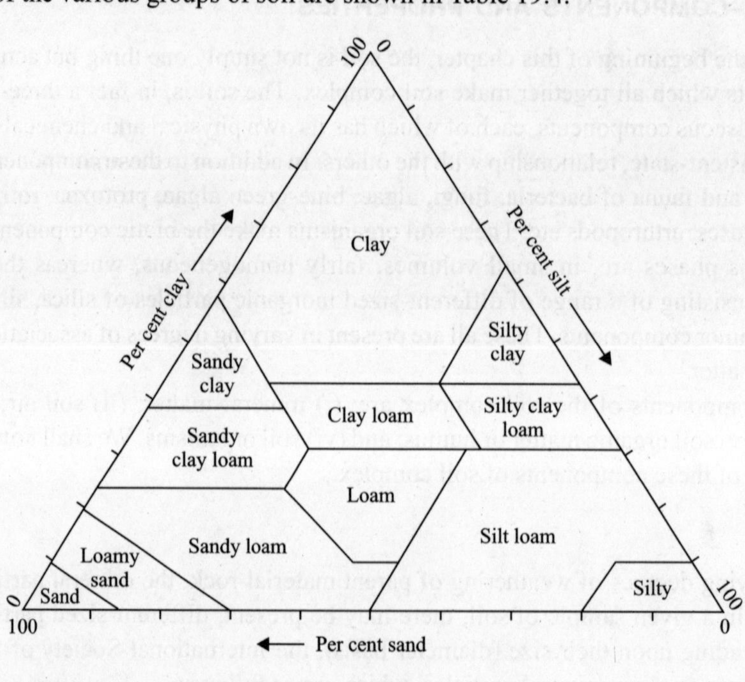

Fig. 19.2. Diagram to show classification of soil texture i.e., different textural groups of soil.

Table 19.4. Various groups of soil.

Textural group	*Relative proportion of different-sized mineral particles*
Sandy soil	85% Sand + 15% Clay or Silt or both
Loamy sand	70% Sand + 30% Clay or Silt or both
Loam soil	50% Sand + 50% Clay or Silt or both
Silt	90% Silt + 10% Sand

Soil texture directly influences soil-water relationships, aeration and root penetration through its relationship with interparticle pore space. Indirectly it also affects the nutritional status of soil, the clay fraction being main source of many plant nutrients and of cation exchange activity. Sandy soils are nutrient-deficient due to high porosity. The relative proportion of the mineral and organic solids to the liquid or gas-filled pore space are determined by the particle size distribution of the mineral matter (soil texture) and the binding of these fundamental particles into larger units or aggregates.

Soil structure and porosity

Soils in nature are generally found in the form of aggregates. In fact the groups of mineral matrix (as shown in Fig. 19.3a) are aggregated together to form larger units or aggregates (crumbs). The aggregated condition, depending upon the mode of arrangement of the groups of mineral matrix, gives a definite pattern to the soil, referred to as soil structure. The aggregates of soil are more or less well-defined structural units which are generally much larger than the textural particles, from a few millimeters to several centimeters in diameter.

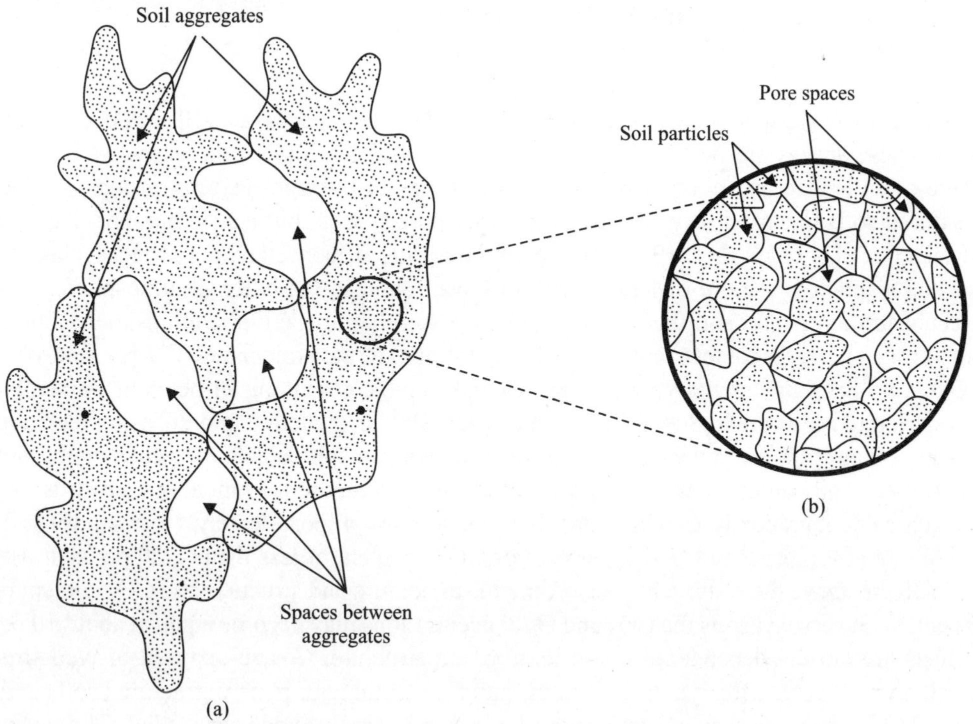

Fig. 19.3. (a) Diagram shows soil aggregates and spaces between them; and (b) A part of soil aggregate to show soil particles and pore spaces between them.

On the basis of the final shape of these aggregates, following common types of soil structures are generally recognised in nature:

1. Prism-like: Where the vertical axis is more pronounced, with its flattened sides, appearing like a pillar. The top of the pillar may be either round (columnar) or flat (prismatic).
2. Platy: Where the horizontal dimensions are more pronounced, than the vertical ones. These appear as flattened, compressed pad-like structures, which may be thick (platy) or thin (laminar).

3. Block-like: Where all the three dimensions are of more or less similar size and equally developed, thus making the aggregates as cubical structure, with round or flat faces.
4. Spheroidal: Where all the axes of the aggregates are of more or less same length with their faces curved or irregular. These are further distinguished as granular, where the aggregates are less porous and crumby, where the granules are more porous. Spheroidal soils are supposed best for plant's growth.

There are spaces between the mineral particles (Fig. 19.3b) as well as between the larger aggregates (Fig. 19.3a). The spaces, however, between the aggregates are much larger than those between the textural particles. Sandy soils have larger pore spaces (but total space area being lesser) than those in between textural particles of a clay or silt soil.

Detailed structure of soil is investigated by: (i) separating and size grading stable aggregates by passing the sample through a nest of suitable sized sieves with a standardised washing procedure; and (ii) by measuring pore space of undisturbed sample. Earthworms and similar animals are important in creating pore space and in stability of aggregates as they affect aeration of soil. Thus structure of a soil is of considerable importance in relation to root penetration.

Soil Air

As noted above, there are spaces in between mineral particles (pore spaces) as well as between aggregates of the soil. These spaces are occupied, in addition to various liquids, by the gases, which make the atmosphere of soil. Size and number of such spaces differ in different textural groups of soil. Thus coarse-textured soils like sand (Fig. 19.4b) have larger pore spaces (but less in number) than the fine-textured soils like clay or silt (Fig. 19.4a). Soil aeration is thus influenced by pore spaces. Gases present in the soil atmosphere are more or less similar to those found in the external atmosphere. However, oxygen concentration is less and that of CO_2 higher in soil than those in the atmosphere. For aerobic soils, the nitrogen + argon content remains close to that of external atmosphere (79 per cent v/v) while O_2 and CO_2 vary in complementary proportions to make up the remaining 21 per cent. Anaerobic soils such as waterlogged soils are highly deficient in oxygen. By using suction and diffusion techniques for sampling the soil atmosphere in the fields, it has been shown that appreciable oxygen contents are found in the majority of soils unless they are almost entirely water saturated. From above towards the water table, oxygen content gradually falls and the CO_2 content rises proportionately. Normal, dry soil rarely has a CO_2 content of more than 0.5–1.0 per cent or an O_2 content of less than C, 20 per cent. Between about 15 to 30 cm above the water table, according to soil texture and structure, the CO_2 content reaches C, 5 per cent. With very wet soils the CO_2 and O_2 concentrations may become equal at about 10 per cent. These effects are closely dependent on soil texture and structure. Coarse-textured or well-structured soils have higher gaseous diffusive transfer rates than fine-textured or poorly structured soils under wet conditions. This is because they contain many large pores which remain gas-filled. The water-filled pores of the fine-textured soils form potent barriers to gaseous diffusion; O_2 diffuses ten thousand times more slowly in water than in gas. In dry soil, the situation is very often reversed, as the fine-textured soils have a greater total pore space and provide a larger gas-filled cross-sectional area for diffusion. The major part of the gaseous transfer in the soil is diffusive.

In this way, soil aeration regimes are very important in growth of roots, seed germination and microbial activities. Poor soil aeration suppresses root hair development and may reduce rates of absorption of water and nutrients.

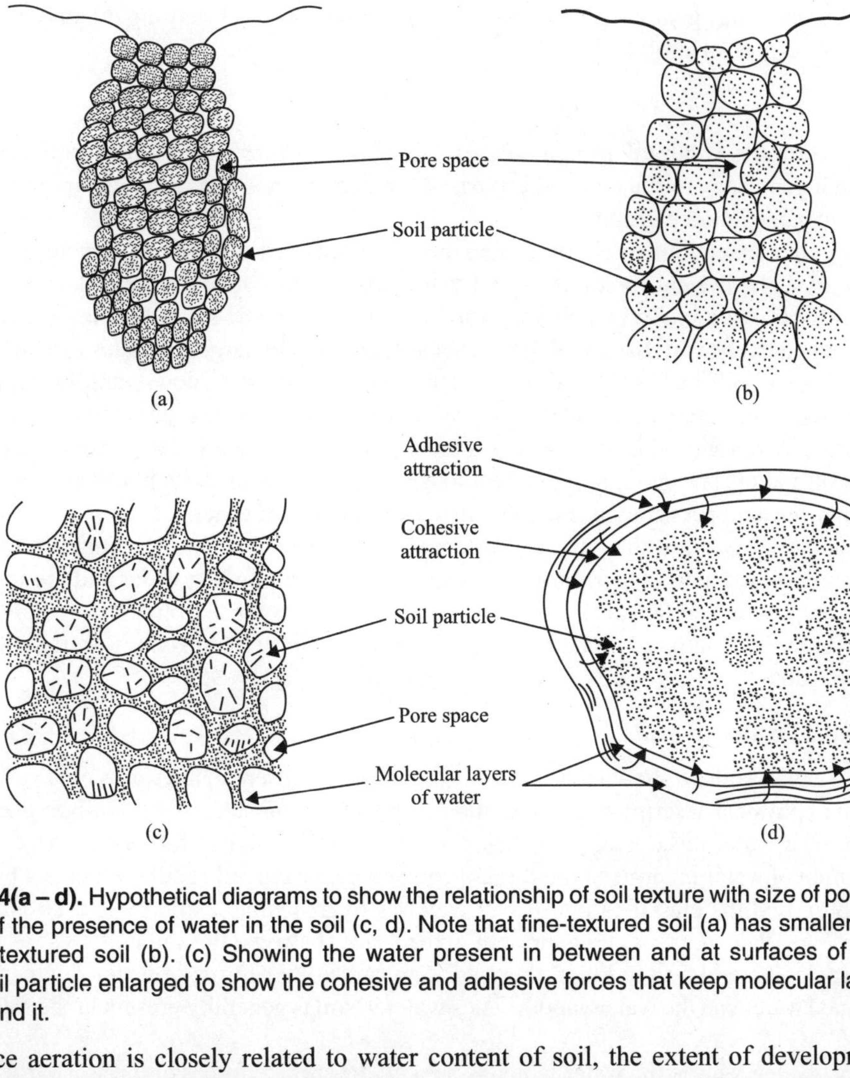

Fig. 19.4(a – d). Hypothetical diagrams to show the relationship of soil texture with size of pore space (a, b) and mode of the presence of water in the soil (c, d). Note that fine-textured soil (a) has smaller pore spaces than coarse-textured soil (b). (c) Showing the water present in between and at surfaces of the soil particles. (d) a soil particle enlarged to show the cohesive and adhesive forces that keep molecular layers of water held up around it.

Since aeration is closely related to water content of soil, the extent of development of air-space tissues differs in different kinds of plants. Under normal soil water conditions, oxygen entering the root from the soil will probably satisfy the oxygen demand of the cortex. As soil water content increases, the liquid path (dissolved O_2) to the root will increase and a point reached where the external supply will be insufficient to meet the root respiratory demand. Under fully waterlogged conditions there will be no external O_2 supply at all. In all plants, however, there is alternative route of gas exchange in the ground tissue intercellular space which connects ultimately with the atmosphere, through the leaves and stomata.

In normal mesophytes the porosity of ground tissue is so low (2–7 per cent by volume) that it is unlikely to be an adequate diffusion path for O_2 transport to the underground parts. In wet land species up to 60 per cent of the plant body is pore space and consequently diffusion resistance is much lower and internal ventilation improved. Plants vary in their ability, to form aerenchyma, from species to species and degree of soil anaerobiosis. Plants which cannot form aerenchyma appear to be intolerant of wet conditions. Thus hydrophytes under suboptimal aeration develop morphological, anatomical and physiological adaptations. They generally have shallow root systems, special aerating organs and tissues and have an ability to respire anaerobically.

Soil Water

Principal source of soil water is rainfall (precipitation). Some of the water which falls during rainfall is lost as run-off, which goes back to the source i.e., sea, ocean and some is lost during evaporation. The remaining water percolates into the ground.

Like soil aeration, the water content of soil is also much related to its texture and structure and also with the magnitude of such physical forces as capillary attraction, cohesion and adhesion.

Thus size of the mineral particles and their shape and number of the pore spaces are important in the amount of water retained by the soil. Sand which are coarse-textured with larger particles can hold water only loosely due to bigger pore spaces. The water in such soils generally runs down rapidly reaching to deeper layers. Thus sandy soils are well drained. Clay soils, with much proportion of colloidal fractions as humus etc., can retain much water. Silts also like clays are able to retain much water. Loams, which are a mixture of sand and silt and/or clay are supposed as best soils for plant's growth, as such soils are very fertile, being rich in nutrients, have proper aeration and hold fairly large amount of water.

Soil aeration and soil water are thus greatly influenced by the texture and structure of soils. As indicated above, sandy soils due to bigger pore spaces are unable to hold much water and are thus well drained. Such soils are also called physically dry soils. In some soils as those of halophytic conditions, although water is present in abundance, but due to high degree of salinity (much concentration of salts), it is not available to plants. Such soils are referred to as physiologically dry soils.

Behaviour (forms) of water in soil

Different forms of water in soil are shown in Fig. 19.5. All the water present in the soil is not available to plants. The problem of physical description of the status and movement of water in the non-homogeneous, three-phase matrix of a water-unsaturated soil has exercised soil physicists for many years. Briggs visualised the retention of water in terms of the forces provided by the curved capillary menisci between soil particles and proposed the classification of soil water as hygroscopic, capillary and gravitational water. As shown in Fig. 19.4c, d, the water is present in the space (between the soil particles) as well as around the soil particles. In addition to these, there are also some other forms of water retained in the soil, such as combined water and the water vapour. Thus water in soil is generally present in the following five forms (states):

1. Hygroscopic water: This is the water which is held by the surface forces of the soil particles. As shown in Fig. 19.4d soil particle is surrounded by a film of molecular layers of water. The water is held tightly around the particle as a result of cohesive and adhesive forces and it cannot be easily removed by the plants.
2. Capillary water: This water in soil is mainly held between spaces (pore spaces) of the soil particles and angles between them, forming a system of capillaries (Fig. 19.4c). This water, is held by the interparticle meniscus effects and removed by the gravity, is known as capillary water. This is the most important form of water to plants.
3. Gravitational water: This is the water that moves downwards through a moist soil in response to gravity. It is thus removed from the soil by gravity alone. In fact gravitational water is surplus to the water retaining capacity (due to surface forces and interparticle meniscus effects) of soil and thus drains from it reaching to deep saturated zone of the earth-ground water, the upper surfaces of which is water table.

4. Combined water: This is the form of water present as hydrated oxides of aluminium, iron, silicon etc., in the soil. The above said four forms of water together with their possible relationships with the soil and plant water status, are shown in Fig. 19.5.
5. Water vapour: Some water in soil is also present in its vapour state in the pore spaces between soil particles.

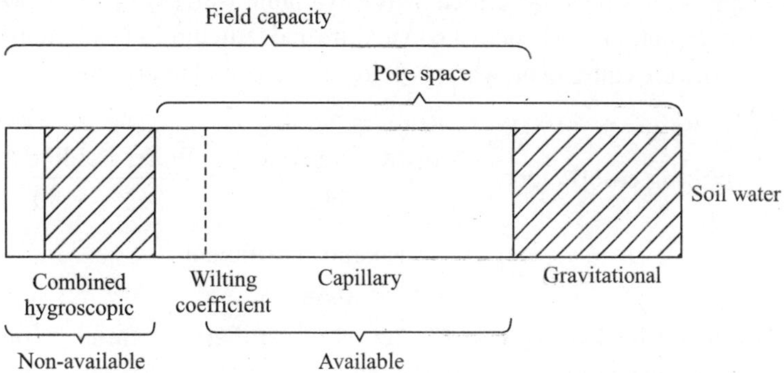

Fig. 19.5. Generalised diagram showing the different forms of soil water and their possible relationship with soil and plant water status.

After rainfall, the water enters the soil and it fills the space between soil particles and the volume that can be filled is known as the pore space (Figs. 19.3b, 19.4a, b), which varies with the texture and structure of a soil. As stated earlier, all the water present in a soil is not available to plants. The total amount of water present in the soil is called holard. Of this, the amount of water that can be absorbed from soil by plants is known as chresard or available water and the rest, which is not absorbed by plants is called echard or non-available. Thus the amount of water between the full field-capacity as a maximum and the wilting coefficient as a minimum represents the available water (Fig. 19.5).

Terms used to specify soil water status

Following are some of the terms that reflect the water status of soil and which are generally used for comparative studies on different soils.

1. Soil water potential: This is an expression of the total reduction of water potential in the soil, due to matrix, soluble, external pressure and gravitational effects.
2. Field capacity: When a soil holds all the water it can, but no gravitational water, it is said to be at its field capacity. This is generally defined as 'the water content of an undisturbed soil (per cent oven-dry weight) after it is saturated by rainfall and drainage of gravitational water has completely stopped'. In practice for determining field capacity, soil is thus sampled about 48 hours after rain. This value for a soil may be taken as 'constant'. Thus field capacity of soil may be taken as the total amount of capillary, hygroscopic and combined water plus water vapour.
3. Moisture equivalent: It is loosely defined as 'the water content (per cent oven dry weight) retained by the undisturbed soil, when about 30 gram of it are allowed centrifuge-drainage at 1000 gram for 30 minutes'.
4. Water-holding capacity (Storage capacity): This is the extent to which a soil can hold capillary water against gravity. In other words, it equals to field capacity less hygroscopic water. It is defined as 'the amount of water (per cent) retained by unit weight of dry soil when immersed in water under standardised condition'.

5. Hygroscopic co-efficient: It is defined as 'the water vapour (per cent): absorbed by unit weight of dry soil, when placed in an atmosphere completely saturated with water vapours'.
6. Permanent wilting percentage (Wilting co-efficient): It is 'the amount of water (per cent overn-dry weight) that remains in a soil when permanent wilting beings in plants growing in soil'. The water which may be extracted from soil between the field capacity percentage and the permanent wilting percentage has been described as the available water (Fig. 19.5) which is a function of the pore size distribution and hence of soil texture and structure. The relationship between wilting co-efficient, moisture equivalent, water holding capacity and hygroscopic co-efficient is generally expressed as:

$$\text{Wilting co-efficient} = \frac{\text{Moisture equivalent}}{1.84} = \frac{\text{Water holding capacity}}{2.90}$$

$$= \frac{\text{Hygroscopic co-efficient}}{0.68}$$

The status of soil water is also expressed in terms of capillary potential (= metric potential). The relationship between soil water content and water potential is usually ascertained by the use of Tensiometer, a pressure membrane apparatus.

Soil Solution

From the point of view of plant nutrients in soil, there are two important chemical components—organic matter and the chemicals in solution. There exists a weak solution of various salts, alongwith other liquids and gases in the soil mass. This soil solution contains almost all the essential minerals. Complex mixtures of minerals as carbonates, sulphates, nitrates, chlorides and organic salts of Ca, Mg, Na, K etc., are found as dissolved in water. The chemical nature of soil solution depends on the nature of the parent matter and the climatic and other factors (including chemical nature of organic matter) involved in pedogenesis. Thus, soil solutions of rivers, lakes, ponds, forests and grassland differ widely from each other. These are of much ecological significance.

The soil solution is the primary source of inorganic nutrients for plant roots and in the case of the more important cations, the concentration of this solution represents an equilibrium with the cations absorbed on the exchange complex. The cation exchanging property of soil is mainly due to its clay fraction and this property is caused by the presence of unsatisfied negative charges at the surface of clay and organic particles; metal cations and hydrogen ions are bound at these sites, sufficiently loosely to show exchange with ions in the bathing solution. This property of exchanging cation is called cation exchange capacity (CEC) of soils. The CEC of sandy soils is associated mainly with their organic content but that of finer textured soils with their clay content. The presence of ion exchangers in soil is of much importance both in pedogenesis and in the soil-plant nutritional relationship. Most metallic elements which are taken up by plants are absorbed as cations but they exists in three forms in the soil: (i) as poorly soluble components of mineral or organic matter; (ii) absorbed onto the cation exchange complex; and (iii) in small quantities in soil solution. Soils with more or less optimal concentrations of various nutrient solutes in them are called eutrophic, whereas those with suboptimal concentration of these nutrient salts are called oligotrophic.

Soil acidity: Soil acidity is associated with the presence of hydrogen and aluminium ions on the exchange complex and the existence of an equilibrium solution of hydrogen ions in the interstitial water of soil. It is expressed in terms of pH, which is defined as the negative longarithm of hydrogen ion activity where activity is understood to mean effective concentration. pH values of soil show much correlations with the soil type, vegetation type, profile horizon, thus affecting plant's growth, lime requirement and mineral nutrition.

The pH values of natural soils show much variation, ranging from 3.0 to 8.4. However, extreme values also do occur. In some unusually alkali soils with high sodium carbonate content, pH values reach up to 10.0–10.5, while drained gleys have 2.0 or even less.

Pearsall suggested some boundary pH values which are ecologically significant. Plants regarded as calcicoles usually occur in soil with pH 6.5, whereas calcifuges in those with pH value below 3.8–4.0. Soils above pH 6.5 are generally cation-saturated (those containing free $CaCO_3$ called calcareous soils), while soils below pH 3.8–4.0 are strongly desaturated and have a considerable content of exchangeable hydrogen. These pH limits are also reflected in the nature of the soil organic matter: raw humus or mor is associated with soils of below pH 3.8, while mull is characteristic of the more cation-saturated soils of pH 4.8–5.0 and above.

Soil pH strongly affects the microbial activities, as below pH 5.0 bacterial as well as fungal activities are reduced. Some soil-borne diseases, like root-rot of cotton, root-rot of tobacco, potato scab etc., may be controlled by lowering the pH of the soil, whereas others like club-root of crucifers and *Rhizcotonia* root rot by increasing the soil pH.

Some plants grow on soils with high salt concentrations. Such plants are known as halophytes as some mosses, ferns, *Distichlis stricta*, *Artemisia tridentata*, *Rhizophora*, *Ceriops* etc. In arid regions (with low rainfall, where evaporation rates exceed precipitation), soils become generally saline. Salinity brings, about marked physiological effects on plants, affecting osmotic concentration of cells. Most of the halophytes show xerophytic tendencies and vivipary i.e., germination of seeds before the fruit is shed from the parent plant.

Soil Organic Matter or Humus

Organic matter is the chief source of mineral's return to soil. It has also a much water holding capacity. It can hold minerals as ions by absorption and binds clay particles into aggregates. It makes clay soils porous thus increasing aeration of soil and also causes easy percolation of water in soil.

Nature of soil organic matter

Soil organic matter (humus) represents the equilibrium between input, originating from primary photosynthetic production and the degradative and resynthetic processes associated with soil organisms. The input is largely of carbohydrates, lignin compounds, fats and proteins with smaller quantities of free amino acids, alkanes, terpenoides, carotenoids, flavonoids, alkaloids, polyphenols, resins etc. Most of these, as carbohydrates, cellulose etc., are easily decomposed and have short residence time in soil. Soluble sugars, lipids, proteins, amino acids, organic acids etc., are minor constituents of organic matter. However the major constituents of humus are humic complexes, which make the greatest proportion of it in most soils. They comprise a mixture of diverse phenolic polymers like lignih, resins etc.

Sources and rates of production of organic matter

Organic matter to soil is contributed by surface-living plants and animals (dead leaves, twigs, branches of plants and bodies of animals) as well as by micro and macro-organisms living in the soil (a variety of micro-organisms and animals and dead roots of plants etc.). In thick forests, in each season there accumulates thousands of tonnes of litter on the soil surface. All dead, fresh organic matter fallen recently to the ground is called litter. In deciduous forests, where all leaves fall at one time, large amounts of litter are produced as a crop each year, whereas in evergreen forests the leaf fall is gradual with simultaneous decomposition. Thus, litter here is less pronounced. In grassland, however, there is little accumulation of litter as compared with the forests. This is due to marked differences in the types of vegetations in two habitats. Just beneath the fresh litter there may be present material derived from the litter of previous season in which decomposition might have progressed. This partially decomposed litter is called duff. Thus initial input of organic matter is in the form of surface litter and underground dead root matter. But as a result of various soil processes, chiefly activities of micro-organisms in the soil, this organic matter (litter) is grossly modified, which is now known as humus or soil organic matter in current practice. Humus is dark, finely divided, amorphous organic matter.

Formation of humus

In humus formation the first step is the comminution and transport of the original plant fragments by the soil fauna. The details of the various kinds of these animals will be given later under 'soil organisms'. After this initial breakdown, commenced by the soil fauna, micro-organism's, attack (bacteria, fungi, actinomycetes) increases rapidly and the original material is soon cleared of its labile components like soluble sugar, polysaccharides, proteins and fats leaving a residue of more resistant lignin and lignin-derivatives. Thus in many soils, a large proportion of the humus is derived from the joint activity of the soil fauna and micro-organisms.

The nature of the organic matter is governed both by input and by 'soil metabolism' , consequently vegetation, climate, parent material and topography all have a strong influence. The composition of the organic matter of soil is thus determined not only by litter chemistry but also influenced by the nature of microbiological resynthesis. The rate of humus formation and its accumulation are thus governed by nature of micro-organisms in decomposition, temperature, moisture, aeration and pH of soil.

Form of humus

In temperate climates the range of organic matter types reflects the gross differences between soil groups. Depending upon the nature of parent matter, vegetation and climatic condition, there develop two common types of humus: mor humus and mull humus. These two terms were originally used by Muller. These are as follows:

Mor humus

Mor humus also known as raw humus, mor is characteristic of true podsol (a type of soil) and formed at a low pH (below 3.8–4.0). Moderate rain alongwith a nutrient-deficient, well drained parent matter due to podsolisation produce a mor humus. Earthworms are absent. Thus litter does not become mixed with mineral soil. It decays slowly *in situ* to form O_2 layer of true mor. This is a compact, rather tough, black or brown peaty material in which the individual fragments are easily recognised microscopically as plant material. The most prominent feature of the podsol is the intense eluviation of the surface horizons with mobilisation of organic matter and its redeposition at lower levels. The fully developed podsol is usually very acidic and its cation exchange complex is strongly desaturated. Thus, organic matter in acidic conditions does not decompose rapidly due to inhibition of microbial activities. Organic matter thus accumulates as peat or muck.

Mull humus

In contrast with mor humus, mull is characteristic of brown forest soils and some other mollisols of temperate climates. Fine-textured, nutrient-rich rock with broad-leaved forest cover develop mull humus forms. It is mirobiologically rich than mor, supporting a greater bacterial and fungal population. It is associated with soil pH values above 5.0, an abundance of divalent cations and well-developed earthworm fauna. As a grey, brown-grey or blackish material, it is diffusely incorporated amongst the soil mineral particles by biological mixing. Since microbial decomposition is rapid, no plant or microbial remains are recognisable (*cf* mor) in mull. It is amorphous and colloidal in nature.

Soil Organisms

In addition to its solid, liquid and gaseous components, each soil has its own distinctive flora and fauna (bacteria, fungi, actinomycetes, algae, blue-green algae, protozoans, rotifers, nematodes, earth-worms, molluscs, arthropods etc.), which make the 'biological system' of soil complex. Among these organisms, some help in maintenance of soil fertility through nitrogen fixation and others are responsible for return of essential elements back to soil by decomposition of dead organic matter. Following are the possible ways in which these organisms play an important role in each soil:

1. Nitrogen fixation: Bacteria, like *Azotobacter*, *Clostridium* and *Rhizobium* and blue-green algae, like *Nostoc*, *Anabaena*, *Tolypothrix*, *Oscillatoria*, etc., fix atmospheric nitrogen, thus involved in soil fertility.
2. Root growth: Some bacteria and fungi in the rhizosphere zone of soil are known to secrete growth hormones, like IAA, thus affecting root growth.
3. Antibiosis: In some soils, some micro-organisms produce metabolic products which may inhibit the growth or cause killing of other micro-organisms: This phenomenon of antagonistic effects through metabolic products is known as antibiosis.
4. Soil-borne diseases: Some fungi and bacteria, as well as animals like nematodes etc., are serious pathogens of roots of plants. This causes considerable loss to the annual yield of plants.
5. Soil formation: Some blue-green algae and bacteria produce mucilaginous substances which help in binding of soil particles into larger aggregates. Burrowing worms, as earthworms and other animals besides helping in humus formation, also improve aeration of the soil. This may affect root development and penetration.
6. Decomposition of organic matter: Besides, what is said above, the most important role of a variety of soil organisms is that they act upon dead organic matter, changing it to humus and finally to minerals, which are returned to the soil. A variety of bacterial, fungal, actinomycetes and soil animals are involved in this process.

In humus formation, earthworms, millipedes, isopods, dipterous larvae, slugs and snails, play important role in comminution of the litter. These animals also carry fungal and bacterial spores, which are then inoculated into the whole mass of a litter and these provide a way for succession to various microarthropods (spring tails, mites etc.) and other organisms which feed within the decaying tissues on fungal hyphae and other micro-organisms. Other agents in clearing of microbial colonies are protozoa, eelworms (nematodes) and enchytraeid worms.

SOIL EROSION

The top layer of the soil is the vital component, since all the nutrients required by plants are present in this layer. This top layer of soil is, therefore, said to be the feeding zone of plants. This fertile top soil is most valuable natural resource and it usually lies at most places at a depth of 15–20 cm, over the face of the land. Soil, as most people think, is not a dead inert matter of minerals. But a healthy soil is indeed alive and dynamic consisting of micro-organisms as bacteria, fungi, algae, protozoa, worms and insects.

Although, soil erosion is a natural process and is as old as the Earth itself, yet today soil erosion problems far exceed natural formation of soil. It takes approximately 500 to 1000 years for an inch of the top layer to build up. There are several ways by which this fertile, top soil is lost and wasted. This loss of top soil or disturbance of the soil structure is known as soil erosion. Odum included soil erosion as a part of soil pollution, while other authors called soil erosion as creeping death of the soil. The problem of soil erosion is throughout the world. In USA over 77 million acres of land has become seriously eroded. Soon after the rain begins the cream of soil is skimmed off with every spell of showers.

Of all potentially arable land only about 44 per cent is under cultivation. The rest i.e., 56 per cent is unsuitable for farming due to inherent soil problems and man-induced problems. Only about 2.5 million sq. km arable land is irrigated at great cost and with many side-effects. Massive irrigation is harmful to fertility due to salination. In a short time, approximately 600 million hectares of potential farmland will be lost to soil erosion, salinisation and water-logging. Meanwhile the world population will reach from 4 billion to 6.25 billion. Even if another 300 million hectares of presently unused land is brought under farming by the year 2000, the net result will be that the area of farmland per person will dwindle from 0.31 to 0.15 hectares. Soil loss is maximum in region with high population densities. Continuous cultivation of same crop also adds to soil loss. The present rate of soil erosion is over 2500 million tonnes per year i.e., over half a tonne of soil every man, woman and child on the planet. Can we afford such a loss? Soil erosion is one of the most difficult problems facing the world, particularly 10 country as ours. The Indian subcontinent is faced with severe silting problems in Bhakra, Rihand and other multipurpose dams. The life of Bhakra dam is feared to be reduced by 250 years due to lack of proper attention towards silt prevention at Govind Sagar Lake, the main reservoir of the dam. Due to alarming rate at which silt is accumulating, it may not last for more than 150 years (designed originally to last for 400 years). In Pakistan also, Tarbela Dam is facing the same problem due to silt being washed from the deforested and eroded Himalayas. The Border Road Organisation while constructing roads along the Sutlej also could not take adequate steps to prevent debris and soil from entering the river. The Spiti river discharges a considerable amount of silt in the Sutlej. Since there is no vegetation in Spiti valley, soil erosion is inevitable and the rocks from new mountains crumble under the flow of water. In the eastern hills of Nepal about 38 per cent of the land area has fields whose top soil has been washed away.

Sediment loads from major rivers of the world (Table 19.4) show that while Yellow River in China carried 1.6 billion tonnes of soil to ocean each year, the Ganges follows close with about 1.5 billion tonnes and Mississippi, the largest river of the USA, carried just about 0.36 billion tonnes into the gulf of Mexico. The Ganges and Mississippi have almost same discharge.

Table 19.4. Sediment load of some major rivers.

River	*Country*	*Annual sediment load (million metric tons)*
Yellow	China	16000
Ganges	India	14550
Amazon	Several	3630
Mississippi	USA	3000
Irrawaddy	Burma	2990
Kosi	India	1720
Mekong	Several	1700
Nile	Several	1110

The loss of cropland soil is maximum in India (Table 19.5), being 18.5 per cent of the total soil loss at the global level. This is much serious since India has only 2.4 per cent of the land area of the world. In USSR it is 9.80 per cent whereas in USA only 6.70 per cent.

Table 19.5. Estimates of excessive erosion of top soil from world cropland.

Country	*Total cropland (million acres)*	*Excessive soil loss (million tonnes)*
USA	4210	17000
USSR	6200	25000
India	3460	47000
China	2450	43000
Total	16320	132000
Rest of world	15060	122000
Grand total	31380	254000

Kinds of Soil Erosion

Various types of soil erosion are recognised. However, based on the rate at which soil loss takes place, there are two main types of soil erosion:

Normal or geologic erosion

It occurs under normal, natural conditions by itself without any interference of man. As natural, it is very slow process and equilibrium between loss and build up is hardly lost, unless there is some major disturbance by a foreign agent.

Accelerated soil erosion

This type of removal of soil is very rapid and never keeps pace with the soil formation. This is the most serious type of loss, generally caused by an interference of an agency like man and other animals.

Agents of Soil Erosion

Based on the various agents that bring about soil erosion and the form in which the soil is lost during erosion, there are following types of soil erosion in nature:

Water erosion

It is caused by the action of water, which removes the soil by falling on as rain drops, as well as by its surface flow action. Depending upon the form of the lost soil, it may be:

Sheet erosion

Here the removed soil is like a thin covering from large area. This sheet is lost more or less uniformly.

Rill erosion

If sheet erosion occurs with full force, the run off water moves rapidly over the soil surface cutting well-defined finger-shaped groove-like structures, appearing as thin channels or streams. These are known as rill erosions.

Gully erosion

This results due to the convergence of several rills (thin channels formed during rill erosion) towards the steep slope, which form together wider channels (grooves) of water, known as gullies. These gullies may become still wider and deeper due to further rains.

Wind erosion

Soil erosion by wind is common in dry (arid) regions where soil is chiefly sandy and the vegetation is very poor or even absent. In our country erosion by wind affects approximately 50 million hectares of land, most of which is in Rajasthan. As in water erosion, wind erosion also is triggered by the destruction of natural vegetation cover of land by overfelling and overgrazing. Once, the top soil is laid bare to the fury of strong gales it gets blown off in the form of dust storm and sand storm. The high velocity winds blow away the soil particles. Depending upon the whole mechanism of the soil removal, this may be of the following types:

Saltation

In such arid regions where rainfall is low, drainage is poor and high temperatures prevail, water evaporates quickly leaving behind the salts. Salt accumulation occurs mainly in lowlands around the oceans. The salts are normally chlorides, sulphates, carbonates and nitrates of potassium, magnesium and sodium and chlorides and nitrates of calcium. The major portion of such salty soil is carried by wind in the form of small leaps which is caused by direct pressure of wind on small particles of soil.

Suspension

The wind throws away smallest soil particles into air, which, move as fine dust with the wind. By this way soils are transported to fairly long distances.

Surface creep

The heavier particles of soil that are not easily thrown up by wind, are simply pushed or spread along the surface by wind.

Landslides or slip erosion

The hydraulic pressure caused by heavy rains increases the weight of the rocks at cliffs which come under the gravitational force and finally slip or fall off. Sometimes entire hillock may slide down. In this type of erosion, water and gravity act together.

Stream bank erosion

The rivers during floods splash their water against the banks and thus cuts through them. Particularly at curves, water strikes with great speed and the bank caves in alongside. This type of erosion is also known as riparian erosion.

Deforestation and overgrazing

Effects of deforestation and overgrazing are well known on soil loss. In India, annual loss of soil nutrients in this way is of the order of 5.37 million tonnes of NPK. Besides this, the expenses of soil removal which gets silted in irrigation tanks, reservoirs, sea and riverbeds are also quite high. Overgrazing is a major hazard affllicting pastures, forests and mountains. There are huge semi-arid zones in India used for grazing. But grazing destroys the little cover and enhances wind and water erosion. There must be developed crops for arid regions. For example, there are some plants which yield hydrocarbons (substitute for oil) and grow well in arid conditions. These include jojoba, a plant which yields a sap like diesel oil and milk-weed. The Jodhpur Arid Zone Research Institute is already experimenting with such varieties. In overgrazed areas there should be grown fodder trees like Ku-babul, which in irrigated conditions gives enough green fodder per acre to maintain six cows. Most villages should be asked to grow such plants instead of grass on panchayati pastures. Goats and sheep should be kept in enclosures and fed with tree fodder, through which many more calories will be converted into meat and wool. These animals waste huge quantities of calories roaming on pastures. Ku-babul can supply the high quality feed.

According to the report of National Committee on Environmental Planning and Coordination (NCEPC), mismanagement of India's land and water resources has led to an environmental threat of alarming proportions. In other countries, environmentalism is frequently an exercise in aesthetics. But in India, it is an economic task of the utmost importance.

The total land area of 304 million hectares of India, about 175 million hectares are exposed to serious environmental threat. This area is suffering from degradation of various kinds. The main causes are water and wind erosion and water-logging. Water and wind erosion alone has affected 150 million hectares of fertile land. Our entire cultivated area is only 160 million hectares. Thus our degraded land is more than all our arable land put together.

In a primarily agricultural country like ours, land is by far the most valuable asset we have, with a million times more than all our steel plants and power stations put together. And yet we remain spectators to steady destruction of our most valuable resource. According to the report of NCEPC, India was losing more than 6000 million tonnes of top soil per year in 1972 which in terms of major nutrients—N, P, K alone. Today the loss is many times more. Our first task must be to identify the areas of maximum concern and the means to check damage in these areas.

Every year in India, water erosion alone takes away more than 6000 tonnes of top soil, containing worth of nutrients. This erosion causes the rapid siltation of tanks and reservoirs. Silt is choking our estuaries and harbours. Eroded soil is deposited on riverbeds, raising their levels and leading to devastating floods, which cause Rs. 2000 crores of damage each year. Denuded land cannot soak up water like wooded land and an increasing amount of rain water is thus running waste into the sea, causing silting and flooding in the process. Hence there is much less water stored in underground aquifers, which are vital for maintaining river flows in the dry season and providing water for tubewell irrigation.

Rains have been eroding Kerala coast. The bad effect of monsoon in Kerala is erosion of its coast line by the augmented rough sea water. During monsoon there occurs extensive damage due to sea erosion and thus coconut trees are uprooted and several huts destroyed, rendering mainly fishermen homeless.

The coastal line suffers erosion during monsoon when the Arabian Sea is very rough. Of the 560 km coast line of Kerala about 320 km consisting of sandy beaches is subject to severe sea erosion.

The only permanent method to check this erosion is construction of sea walls and groynes. Sea walls are already constructed in some vulnerable areas of coast line.

Uncontrolled wind erosion due to deforestation leads to decline in the rainfall also. One of the possible reasons why moisture-laden clouds pass over western Rajasthan without precipitating, is the fine dust suspended in the air over the desert. This dust is, of course, the direct result of wind erosion caused by deforestation. In the Himalayan ranges the annual rainfall has declined by 3 to 4 per cent.

SOIL CONSERVATION

We have discussed above the soil erosion and its consequences. Unless this loss of soil is checked, it would amount to a great loss for mankind. Thus man has made use of his ecological training in the preservation of this one of the most important resources.

Principles of Soil Conservation

As already pointed out the chief agents of soil erosion are water and wind. The actual art of soil conservation is based on certain basic principles, which include: (i) protection of soil from impact of rain drops; (ii) to prevent water from concentrating and moving down the slope in a narrow path; (iii) to slow down the water movement when it flows along the slope; (iv) to encourage more water to enter the soil; (v) to increase the size of soil particle; (vi) reduction in the wind velocity near the ground by growing vegetation cover, ridging the land etc.; and (vii) to grow the strips of stubble or other vegetation cover which might catch and hold the moving particles of soil.

Keeping in view the above said principles, ecologists have devised several methods, given below, to prevent the loss of soil during its erosion.

Methods of Soil Conservation

The various methods may be broadly arranged into the following types:

1. Conservation achieved by the use of plant vegetation cover (Biological methods):-
   (a) In areas with normal farming, where vegetation itself is used for soil protection (Agronomic practices).
   (b) In areas with low and moderate rainfall, where normal farming is not practically possible (Dry farming).
   (c) In areas suitable for successful growth of grasses, which are used as soil binders to check soil erosion (Agrostological methods).
2. Conservation achieved by supplementing the biological methods so as to increase the time of concentration of water, to reduce the velocity of water or afford protection against damage due to run off (Mechanical methods):-
   (a) Construction of small basins along the contours (Basin listing).
   (b) Construction of small basins along the slopes to intercept and divert the runoff water (Contour terracing).
3. Conservation achieved by purely mechanical methods including construction (Other methods):-
   (a) Formation or widening of gullies (Gully control)
   (b) Growing vegetation alongside, construction of drains, stone pitchings etc., (Stream bank protection).
   (c) To check the velocity of wind by tree plantations (windbreaks) (Afforestation).

Biological methods

By the use of plant or vegetation cover.

Agronomic practices

Natural protection by growing vegetation in a manner that reduces soil loss. These are:

1. Contour farming: An oldest method, useful in areas with low rainfall is the preparation of the field with alternate furrows and ridges. Ridges at the same level are known as 'contours'. The

water is caught and held in furrows and stored, which reduces run off and erosion. On slopes, however, this type of farming is coupled with terracing.

2. Mulching: It is effective against wind as well as water erosion. Some such plants as maize stalks, cotton stalks, tobacco stalks, potato tops etc., are used a mulch (a protective layer formed by the stubble, i.e., the basal parts of herbaceous plants, especially cereals attached to the soil after harvest). Mulches (2–3" thick) reduce soil moisture evaporation and increase amount of soil moisture by addition of organic matter to soil.
3. Crop rotation: It decreases soil loss and preserves the productivity of land. The same crop year after year depletes the soil mineral. This is overcome by culting legumes. A typical rotation consists of one year of fallow, followed by winter heat.
4. Strip cropping: It involves the planting of crop in rows or strips to check flow of water. It may be contour strip cropping (strips planted along the contour at 90° to the direction of slope), field strip cropping (strip planted parallel to each other) or wind strip cropping (strip planted in straight parallel rows at 90° to the direction of prevailing wind).

Dry farming

This practice is useful for croplands grown in low and moderate rainfall areas, where ordinary farming is at risk. Crop production, animal husbandry and growing grazing fields are the only possibilities of checking erosion. Methods employed differ in different areas. Some of them are following the land, strip cropping, crop rotation, contour farming etc.

Agrostological methods

Grasses such as *Cynodon dactylon* are utilised as erosion-resisting plants. They are grown in strips between the crops. They act as stabilisers when grown in gully and sodding. Such methods include:

1. Lay farming: This aims to grow grasses in rotation with the fields crops, which helps in building up the structure of soil, preventing soil erosion and improving its fertility.
2. Retiring lands to grass: It involves to grow grasses on such lands where major proportion of the top soil has been eroded. Generally grasses are allowed to grazing under suitable climate conditions.

Mechanical methods

These methods are used as supplements to biological, methods. These are:

Basin listing

Basin listing, means to construct small basin along the contours to retain water which also reduces its velocity.

Contour terracing

Contour terracing constructs a channel along the slope to intercept and divert the run off water.

This may be:

1. Channel terrace digs channels at suitable intervals and the excavated soil deposited as a wide, low ridge along the lower edge of the channel.
2. Broad based ridge terrace constructs ridge along both the sides of the channel.
3. Bench terrace constructs a number of platforms along contours or suitable graded lines across the slope.

Other methods

Gully control

To check the formation or widening of gullies by constructing bunds, dams, drains or diversions through which excess runoff water is channelled.

Stream bank protection

To grow vegetation alongside the river bank, to construct drains, concrete or stone pitching etc., for checking the cutting and caving of river banks.

Afforestation

Trees as windbreaks are planted in deserts which check the velocity of wind. Windbreaks are planted across the area at 90° to the prevailing wind. They check the spread of sand dunes or desert conditions or blowing away of the fertile top soil. Windbreaks may be planted in several rows. Afforestation is applied to Indian deserts, where such plants as *Lowsonia alba*, *Agave americana*, *Thevetia neriifolia*, *Calotropis gigantea*, *Ricinus communis*, *Zizyphus jujuba*, *Acacia catechu*, *Cassia*, *Acacia nelotica*, *Dalbergia sissoo*, *Mangifera indica* and *Tamarindus indica* serve as useful windbreaks.

20

Succession

INTRODUCTION

Vegetation is not static and unchanging—it can be altered in many ways. For example, wetland ecosystems are often drained by human. When this happens the species in the community which are adapted to living in the waterlogged habitats die out and are replaced by other species which are more characteristic of drier conditions. Drained land can be ploughed and planted with crops or used for other agricultural purposes. If such drained land is abandoned it is quickly invaded by other species and turns into grassland, scrub or given time, woodland. Often vegetation changes occur which are not caused by human intervention. Swamp cypress needed a brief phase of drying out to regenerate and how, if the area remained under water for hundreds of years, swamp could turn into shallow lake. The reverse can also occur—a shallow lake which dries out occasionally can be invaded by *Taxodium* and become swampland.

These are examples of vegetation changes, but not all can be called succession. Succession is a natural change in the structure and species composition of a community. Changes caused by the direct influence of humans in clearing and replanting land have never been included in the definition of succession. If human influence declines the changes which then take place are usually classified as succession.

Succession is often taken to mean changes in vegetation alone. Other organisms associated with the vegetation types also change, but it is the plants which dominate and most influence the community structure. When applied to vegetation, succession implies a directional change from one community to another. The original concept of vegetation succession, as proposed by Clements, was that succession represented a unidirectional series of changes which could not be reversed.

Clements' ideas on succession and community classification are now thought to be too precise and rigidly organised to reflect the realities of ecosystem change. We have already seen that, under the right conditions, swamps can become lakes and lakes can become swamps—so reversals in succession may occur by the alteration of an environmental factor, in this case flooding. Because of criticisms of Clements' views, definitions of succession were broadened to include all community changes including regeneration of the same vegetation type and minor fluctuations in community structure. This broader view was perhaps rather too all-encompassing to be helpful for the understanding of vegetation changes. The term 'succession', as used by ecologists today, falls somewhat in between these two earlier views. It is not simply the one-way changes in vegetation in a developing or maturing habitat nor is it now taken to include the minor fluctuations or changes found in communities due to seasonality or other cyclical changes. Succession is thought of as a series of changes in community structure and species composition.

At its most extreme, this sequence of succession starts on bare, uncolonised ground which has never had any vegetation growing on it before or in newly formed lakes. This is known as primary succession.

It is uncommon to find the early stages of primary succession today. Bare areas are rare, but they may occur on sand dunes, the lava flows of volcanos, new volcanic islands, landslips and glacial debris left by melting glacier ice. The remains of glacial tills left by retreating ice sheets and lakes forming in hollows and blocked glacial valleys were very common at the end of the last ice age. At that time, much of Canada and north-eastern Europe would have been exposed and subject to primary succession.

Most bare areas of land today are exposed because existing vegetation has been destroyed by fire or covered with flood silt or volcanic ash. The vegetation recolonising the area will be affected by conditions not available in primary successions. This includes the remnants of soil, organic matter, seeds and even some plants which have survived the changes. The sequences of vegetation developing on these previously vegetated and disturbed areas are called secondary successions. The changes on abandoned cultivated land are also secondary successions.

The sequence of vegetation types which occur in primary successions is often called a sere. Hence a hydrosere refers to the series of aquatic communities found in a lake, for example and a xerosere is the sequence on dry land. The type of xerosere which develops on rock is often called a lithosere. Typical examples of a xerosere and a hydrosere are described later in the chapter.

CAUSES OF CHANGE

We have already seen how a certain community will develop in particular environmental conditions. If successional change occurs in the community this suggests that the environmental conditions have altered. The vegetation change can be caused by effects of the plants themselves, autogenic succession, or from external environmental influences, that is by allogenic succession. A large number of factors can influence vegetation change. Many of these are summarised in Fig. 20.1. Some of these are due to human interference, catastrophic events or the effects of seasonality and are not part of succession, although they may have considerable effects on the communities. Both human influence and catastrophes can change or destroy vegetation so that succession can occur, but the changes they create are not strictly themselves part of the succession. Other autogenic and allogenic factors are important in succession and are listed on the lefthand side of the diagram.

Autogenic succession can be brought about by changes in the soil caused by the organisms growing there. These changes include accumulation of organic matter in the litter or humic layers, alteration of soil nutrient status or change in pH caused by the plants growing there. The structure of the plants themselves can also alter the community. For example, as larger species like trees mature, they cast shade on to the developing woodland floor which tends to exclude low-growing light-demanding species. Shade-tolerant species will invade the area instead.

Allogenic changes are caused by external environmental influences, not by the vegetation. For example, soil changes due to erosion, leaching or the deposition of silts and clays can alter the nutrient content and water relationships in the ecosystem. Animals can also be important as pollinators, seed dispersers and herbivores. They can also increase the nutrient content of the soil in certain areas or shift soil about (as termites ants and moles do) creating patchiness in the habitat. This may create regeneration sites which favour certain species.

Climatic factors may be very important, but on a much longer time-scale than any others. Changes in temperature and rainfall patterns will promote changes in communities. Evidence for the great effect of climate on vegetation is preserved in the fossil record of the last 5,00,000 years or so. Fossil pollen, preserved in peat and infilled lakes, show that the huge oscillations in climate which caused the icecaps to advance and retreat also greatly affected the vegetation of most of the world. In cold phases, tropical forest was succeeded by savannah and deciduous forest became treeless tundra. As the climate warmed at the end of each ice age, great successional changes took place. In Britain the tundra vegetation of the south of Britain and bare glacial till deposits in the north underwent succession to mixed deciduous forest. The greenhouse effect, caused by the release of massive amounts of carbon dioxide into the atmosphere, is now increasing global temperatures quite rapidly. In the next century we may be able to watch allogenic succession caused by changing climate to a degree never before seen by ecologists.

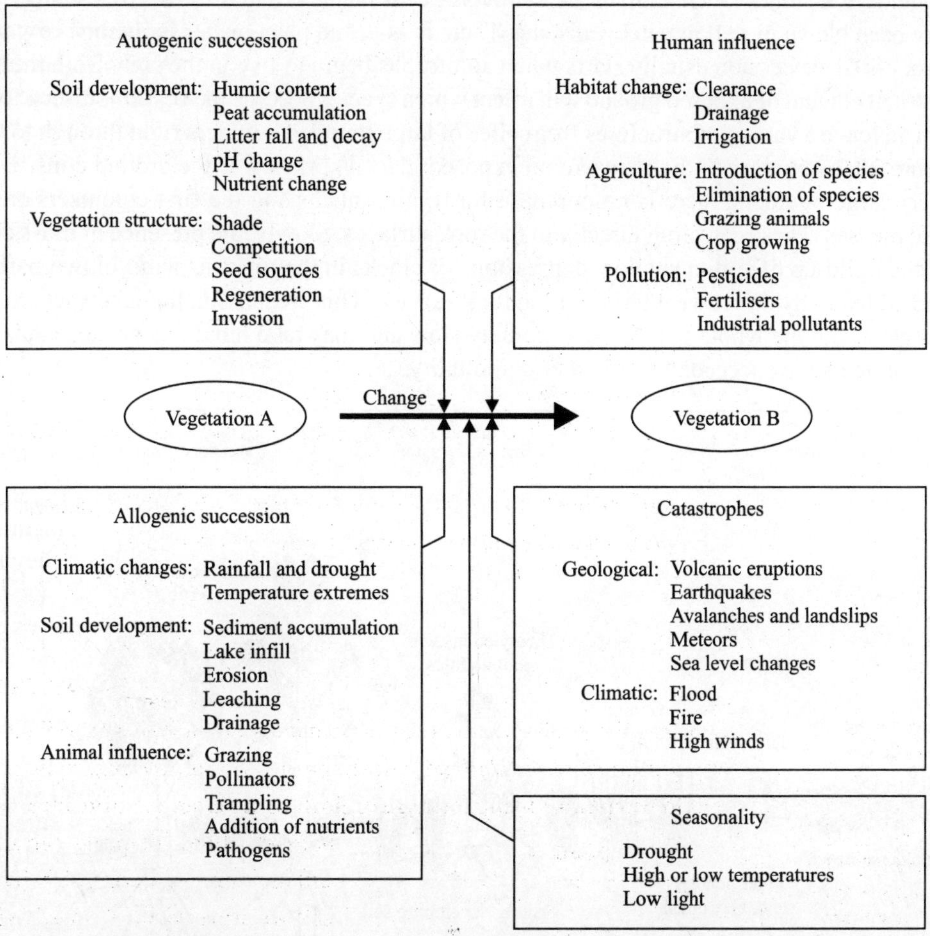

Fig. 20.1. The various factors which influence vegetation change. Those on the left are the main promoters of succession. Those on the right tend to produce vegetation change not considered as succession.

EXAMPLES OF PRIMARY SERES

Xeroseres

Where bare ground is colonised by organisms the subsequent primary succession is called a xerosere. About 13000 years ago, the melting of the ice sheets from the land in the northern hemisphere must have created vast expanses where primary xeroseres could occur. Primary succession can still occur on landslips and where valley glaciers retreat to provide ground for such seres. Our understanding of the primary colonisation and later sequence of succession comes from two sources. The study of the early phases of seres can be made by direct observation of recently exposed sites. The later stages of a sere can also be investigated by analysing fossil records from regions exposed at the end of the last glacial period and by examining recent community structure.

Bare rocky areas are often invaded first by animals such as spiders which can hide in the cracks between boulders or stones. Such animals are carnivores or scavengers and they live by feeding on insects which have been blown in or flown in by mistake. There is as yet no plant matter for herbivores to survive.

The process of succession usually starts when autotrophs begin to live in the area. Both the structure and the nutrient content of exposed ground will affect which species become the first colonisers. Retreating ice sheets can leave a variety of structures from piles of large rounded boulders right through to very fine clays. Figure 20.2 summarises the xerosere on exposed glacial surfaces. If the ground consists of solid rock or very large boulders, there is no purchase for rooting plants and the first colonisers are usually lichens and mosses. These can cling directly to the rock surfaces. Slowly, the presence of moss or lichens promotes the build-up of soil in shallow depressions or cracks in the rocks as wind-blown particles are caught and added to by direct weathering of the rock surface. This creates small pockets where vascular plants can grow, but the whole process is extremely slow and may take hundreds or thousands of years before the pioneers are succeeded by another community.

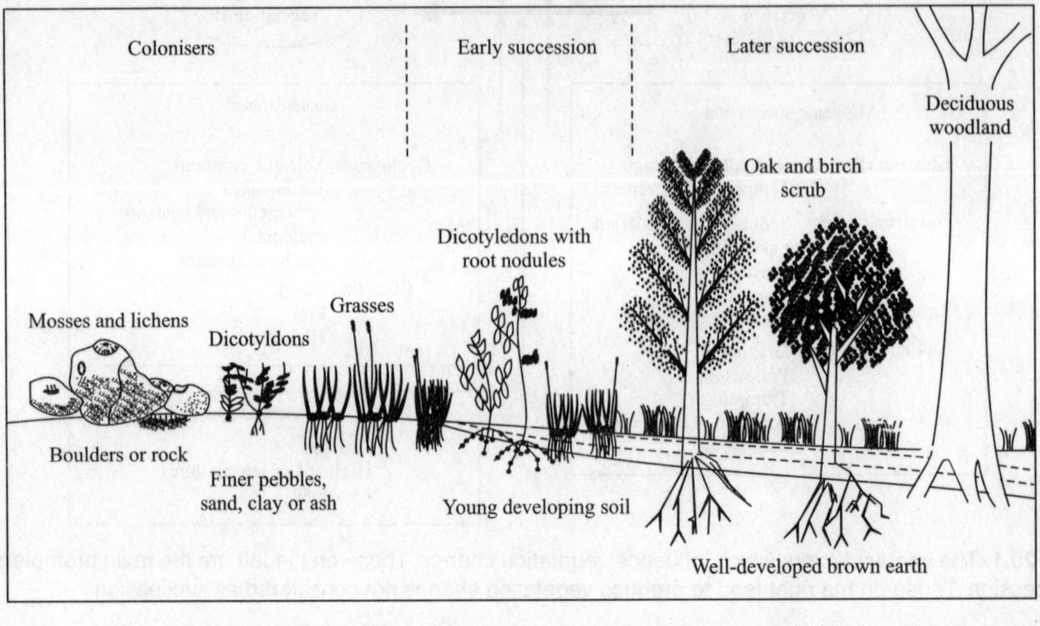

Fig. 20.2. A summary of the early stages in a xerosere, which depend on substrate type and the subsequent successional changes which occur after colonisation.

If finer particles, such as silts or clays, are present in quantity in the substrate, then the initial pattern of colonisation is very different from that on rock. Such areas already have more nutrients and a more balanced water regime than bare rock. The substrate can be penetrated by roots, so that vasular plants can colonise immediately. Such sites tend to be invaded by dicotyledonous herbs or if they are quite nutrient rich, by grasses. Even clay and silt substrates are poorly structured and nutrient poor compared with mature soils, so it is not surprising that the second community in these sites is rich in plants with nitrogen-fixing bacteria held in root nodules. Such communities include some horsetails (*Equisetum* spp.), legumes like *Astragalus*, *Lupinus* and *Trifolium*, the mountain avens (*Dryas* spp.) and in areas where trees can grow, alders (*Alnus* spp.).

In some cases the nitrogen-fixing species are the primary colonisers, as is the case on the ash falls of the Mount Saint Helens eruption where *Equisetum* and *Lupinus* were the dominant species. It is difficult to understand why nitrogen fixers are not always the first species to colonise a site. It has been suggested that such species have relatively large seeds because they need an initial store of nitrogen to produce their first root nodules. The heavy seed is a poor dispersal unit, so that these plants are unlikely to spread far into open areas. Most colonisers have lightweight seeds which are ideal for long-distance dispersal. It may also be that plants with nitrogen-fixing bacteria are only at an advantage where the roots of plants are crowded close together and competition for available nitrogen is great. *Equisetum* is not a seed plant: it is dispersed as spores, so this may explain its abundance on the slopes of Mount Saint Helens. The primary coloniser on volcanic ash on the Mexican volcano El Chicon was also a sporebearing plant, in this case a fern, *Pityrogramma calomelanos*.

Later in the xerosere, the grass and herb mixture is invaded by scrub. Early invasion of scrub is slow, but once a few bushes have become established, they form an ideal site for roosting birds. Once birds are common, bush invasion speeds up as the birds deposit many seeds on the site in their droppings which are also rich in nutrients. In northern Europe, most mid-successional species like hawthorn (*Crataegus monogyna*), privet (*Ligustrum vulgare*), blackthorn (*Prunus spinosa*), wayfaring tree (*Viburnum lantana*) and spindle (*Euonymus europaeus*) have berries which are eaten by birds as a means of seed dispersal. Colonising trees such as the wind-dispersed birches (*Betula* spp.) and bird-dispersed oaks (*Quercus* spp.) may also establish at this stage. Birch and oak are both species which require high light intensities for their seedlings to establish. They are pioneer trees. The birches are usually short lived and fail to regenerate once the oaks mature and cast too much shade. Once woodland is established it may be invaded by more shadetolerant species such as limes (*Tilia* spp.), ash (*Fraxinus excelsior*) and hornbeam (*Carpinus betulus*).

Hydroseres

Hydroseres are primary succession sequences which develop in aquatic environments such as in lakes. After the glaciers of the last ice age retreated, the new lakes collected inorganic particles such as sand and clay washed in from the bare hillsides around. Once primary xeroseres developed on the bare land, sediment in wash decreased—as the vegetation reduced erosion, so the infilling of the lakes slowed. The early changes in plant succession are allogenic and very slow as a deep lake has to become shallow enough for rooted aquatics to grow. This may take hundreds to many thousands of years depending on lake depth, erosion and drainage patterns in the catchment area.

The open water of lakes may be colonised by free floating aquatic plants like the duckweeds (*Lemna* spp.) or the alga *Chara*. At the edges of the lake where the light penetrates to the floor submerged aquatics, with their roots in the sediment, grow. These often include genera such as *Potamogeton* and *Myriophyllum*. Once plants like this root in a lake, the biomass is increased and productivity increases too. More organic matter falls to the lake floor creating a fine mud.

Once submerged aquatics colonise a lake, the successional changes become more rapid and are mainly autogenic as organic matter accumulates. Inorganic sediment is still entering the lake and this is trapped more quickly by the net of plant roots and rhizomes growing on the lake floor. Areas of the lake soon become sufficiently shallow for floating-leaved aquatics to grow. These plants are rooted in the mud, but some or all of their leaves float on the water surface. These include water lilies (*Nuphar* and *Nymphaea*). The large surface leaves cast deep shade which reduces the submerged aquatic species. The plants decay to form organic muds which, as the lake becomes shallower, are invaded by emergent vegetation, such as *Phragmites* and *Typha*, to form reed swamp. These plants have creeping rhizomes which knit the mud together and produce large quantities of leaf litter as they die back each autumn. This litter is resistant to decomposition and reed peat rapidly builds up, accelerating the autogenic change. If the peat becomes raised above the water surface at dry times of the year, the community changes again to one dominated by sedges (*Carex* spp.). Sedge peat accumulates above the water level and so is no longer totally waterlogged. It can then be invaded by trees—usually alders (*Alnus*) and willows (*Salix*). The development of wet woodland (called fen carr) is often seen as the last stage in the hydrosere before the area dries out completely and typical deciduous oak woodland develops.

Often many of these communities can be seen together at a single lake, the lake has an edge with emergent rooted aquatics in the foreground; in the centre are emergent reed beds with irises. Bushes are invading this area and the trees in the background are on dry land. The vegetation types in the succession are mainly determined by water depth. In the lake center it is still too deep for rooted aquatics to grow, but towards the edges the lake is shallower so that rooted aquatics can grow there.

As a lake fills in with sediment, the open area decreases in size and the vegetation types move inwards as the water becomes shallower. Figure 20.3 shows a cross-section of a lake. You can see how the different vegetation types are associated with different water depths and also how the sediments build up and fill the lake. Notice how the different plant types produce different sediments beneath them. Because different vegetation types produce different sediments a record of what has grown in the lake is fixed in its infill deposits. Old lakes, which have completely filled in, can be cored to give a long tube of sediment which goes right down to the bottom of the original lake. Analysis of the sediment shows the successional changes which have occurred in the lake during its history. The right-hand side of Fig. 20.3 shows what a core taken from such a lake might show. Such a permanent record of successional changes cannot be obtained from terrestrial seres as the top surface of the soil is changing and being reworked rather than building up in a sequence. So hydroseres provide us with more accurate information about succession than do xeroseres. The patterns revealed by this record are discussed below.

PATTERNS OF SUCCESSION

Variation in Seres

As already mentioned in early concepts of succession described seres as unidirectional vegetation changes. The presentation of 'typical seres' like those described still suggests that seres are predictable sequences of change along a linear path from bare rock to forest or from open water to sedge fen. But is this in fact the case?

Succession is typically far too slow for ecologists to observe, so that direct study of vegetation changes on one spot is impossible. Usually our understanding of succession is built up from study of several sites where more than one stage of the sere occur. This situation occurs in lakes or along the valleys of retreating glaciers. Such studies are not completely satisfactory—they show the rough direction of the seral changes, but not the actual history of change. What is needed is a record of vegetation at one locality for thousands of years. This is only possible where vegetation build-up is sequential and records plant types. This is where the infilled lake cores mentioned in are so valuable. Figure 20.3 shows how the different sediments record the vegetation changes from a core which can be retrieved for analysis. Radiocarbon dating of the organic matter the core provides evidence of sediment accumulation rates and the length of time any one vegetation type in the succession lasted.

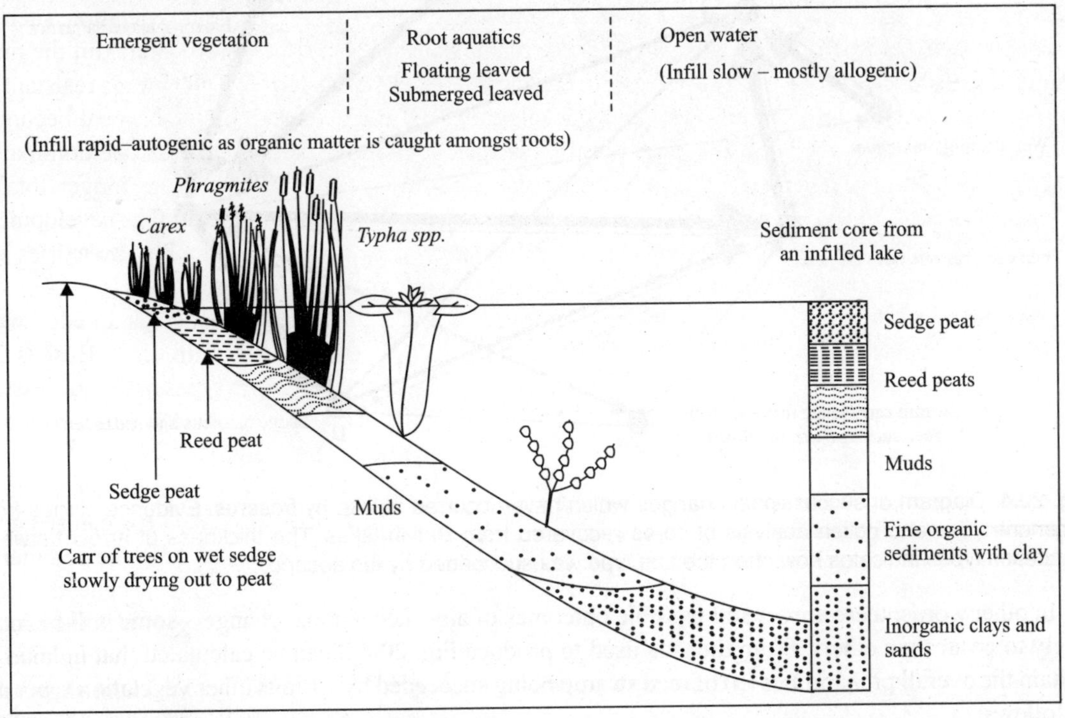

Fig. 20.3. The vegetation zones and sediment types usually found in an infilling lake. If an old, infilled lake is cored to take out a column of sediment (illustrated on the right), a record of the lake's history can be discovered.

Analysis of many such cores from British wetland sites shows that what actually occurs in hydroseres is a fascinating complexity of succession. The pattern of succession found in the lake hydroseres investigated is shown in Fig. 20.4. The typical sequence as already described, which is often considered as the most likely course of succession, would follow the arrowed lines running round the outside of the diagram as A → B → C → D → F → oak woodland. As you can see from Fig. 20.4 this is by no means the only sequence of vegetation change recorded in lake cores. Submerged aquatic vegetation (A) is more likely to be succeeded by reed swamp (C) than by floating-leaved aquatics (B). Although B is usually succeeded by C it can revert to A, presumably due to an increase in water depth in the lake. Similarly reed swamp (C) can be followed directly by fen carr (F) without the intermediate stage of sedge fen (D).

The study shows that seral succession is not a fixed one-way sequence of vegetation change, but a much more complex process where each vegetation type may be invaded by one of a number of different communities. So reed swamp vegetation will not always develop into sedge fen but may be invaded by trees or by Sphagnum bog.

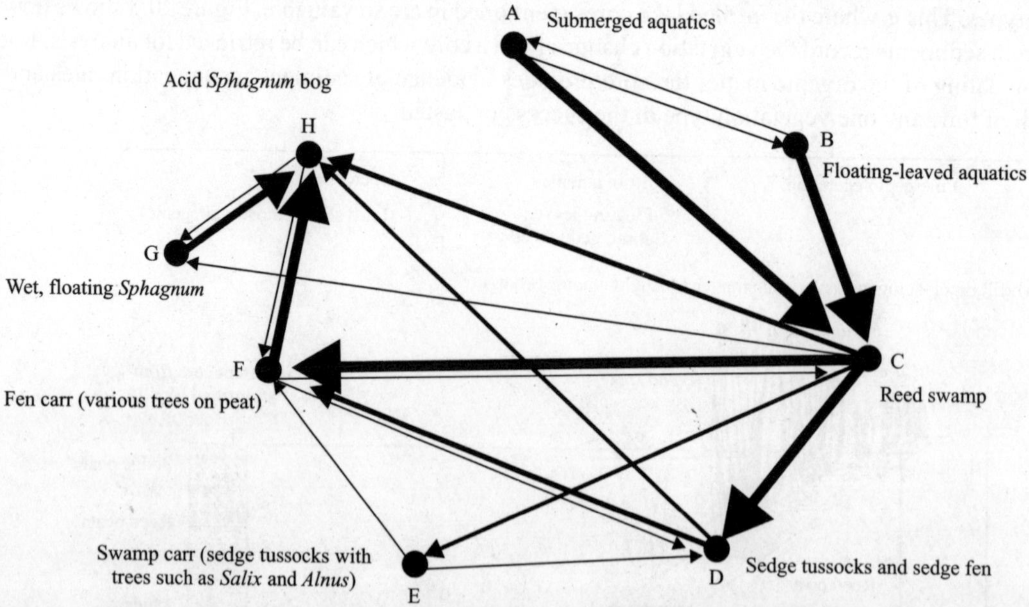

Fig. 20.4. Diagram of successional changes which have occurred in lake hydroseres. Evidence comes from sediment types and pollen analysis of cores recovered from British lakes. The thickness of arrow between vegetation types indicates how often the first type was succeeded by the second.

In other words, there are several possible outcomes of any successional change—some will be more likely to occur than others. From the data used to produce Fig. 20.4 it can be calculated that in lakes in Britain the overall probabilities (p) of reed swamp being succeeded by various other vegetation types are as follows:

succession by: floating-leaved aquatics p = 0;
sedge fen p = 0.26;
swamp carr p = 0.16;
fen carr p = 0.26;
wet floating *Sphagnum* p = 0.1;
Sphagnum bog p = 0.21.

At any site, the probability of a vegetation type being succeeded by another specified type will depend on the environment at that site. Another point to notice from Fig. 20.4 is that fen carr is not the final stage in the hydrosere succession. It appears that carr usually gives way to *Sphagnum* bog. In fact *Sphagnum* communities can develop directly from several seral stages including reed swamp, sedge fen and carr. *Sphagnum* growth is promoted by several conditions. If the water in the hydrosere becomes very nutrient poor, for example where reeds fringe the lake and act as a nutrient sink by catching the richer runoff water from the lake edge, the *Sphagnum* will develop inside the bank of reeds towards the center of the lake. If fen peats become oxidised and acidification occurs, because there is only infrequent flooding, then *Sphagnum* can invade in these conditions too. In carr, the shade cast by the trees tends to exclude many shade-intolerant hydrosere communities and again *Sphagnum* is a coloniser of such pools. Where the input of rain, which is nutrient poor and acidic because of carbon dioxide dissolved in it as carbonic acid, is high (over about 1000 mm a year) *Sphagnum* can even envelop the slopes around the lake hydrosere to form blanket bog.

End of the Succession

The crucial question is whether the vegetation changes of seres progress to an end point of a final community type. In other words, does succession cease when a particular vegetation type develops or does it have no end, progressing so far, then changing back to an earlier community?

In the early days of the concept, the answer was thought to be obvious. Clements believed that succession progressed steadily from communities with small simple plants, like mosses, through to a final stable community with large dominant advanced plants like trees. He called this final community type the climax vegetation. The nature of the final community was understood to be well suited to the environment in the area, in particular the climate and was often called the climatic climax vegetation. We have already met the kinds of communities which Clements would have called climatic climax as representatives of the major vegetation zones. They include evergreen broad-leaved tropical forest, temperate deciduous forest and boreal conifer forest.

A problem with the concept of climatic climax came with the realisation that the climate is not stable. Oscillating climatic changes of glacial cold phases and warm interglacials occur. Even within our apparently uniform interglacial, oxygen isotopes show a gradual cooling since about 3000 BC. Historic records also show shorter periods of climatic change such as the little ice age of the seventeenth century and now the greenhouse effect may well be reversing the cooling trend. The realisation that climate varied substantially led many ecologists to reject the concept of the climatic climax.

This rejection does not alter the fact that we can recognise vegetation types which are usually associated with climatic zones. These are mature communities which seem to develop at the end of primary or secondary seres. The term late successional vegetation, as a replacement for the discarded term climatic climax, has been suggested to describe them. In some respects this phrase is no better. If vegetation changes are cyclical or at least very variable, for hydroseres, then what is 'late' in the succession when one community can be replaced by another apparently earlier in the sere?

Can we classify vegetation on the basis of its stability? A particular phase in the sere, such as the rock with lichens stage in a xerosere or open water with floating aquatics in a hydrosere, may last for thousands of years. In this sense it is very stable, but in the end allogenic factors alter the environment and change occurs. For a succession to come to an end, the final vegetation type must be able to regenerate on the site. It is not really clear if this is the case for communities such as forest or *Sphagnum* bog. Certainly trees are very long lived and therefore, within our time-scale, seem to form a highly stable vegetation type. But do they regenerate? Many trees seem unable to establish directly underneath adults of the same species. This creates a shifting patchwork of regeneration. Often major phases of regeneration only occur after some catastrophe has destroyed part of the forest. In such cases, which species regenerate may depend on the type of catastrophe. For example, fire may promote pines (*Pinus* spp.) or *Eucalyptus*, while wind damage favours broad-leaved species rather than evergreens. So regeneration of damaged communities may not always end in the same species structure.

The second requirement for stability is no autogenic or allogenic effects. Is this true for tree communities? The individual tree species can, in a relatively short time, promote changes in the soil profile and consequently in ground flora and soil organisms. So autogenic change is possible under trees. In other words, examples exist which show that neither of the requirements for a totally stable community are met consistently by what might be recognised as climax or late successional forest vegetation.

Perhaps succession is still continuing in tree communities, but on a time-scale which we have difficulty recognising and so we see such communities as stable. One problem is the apparent similarity between forest ecosystems of all types. All have canopies, trunk space and a shaded ground flora. Wet tropical forest looks pretty much the same whatever the tree species and ground flora composition. These obvious similarities may simply mask the variation between communities in forest systems. As an analogy, reed and sedge fen are easily distinguished because we can readily distinguish *Phragmites* from *Carex*, but if we were the size of a small beetle crawling through the stems, the succession from one to the other might be harder to identify!

Diverted Seres

So far we have looked at the network of successions which may (or may not!) lead to a stable final vegetation type such as deciduous forest or *Sphagnum* bog. Sometimes the natural sequence of succession is halted or diverted by some external factor which influences the vegetation structure or community composition. At some stage in the sere the autogenic changes, which would usually occur to promote the development of the natural succession, are prevented. Autogenic change is usually prevented in two ways—by the removal of vegetable matter which would normally build up and alter the substrate; and by the exclusion of invading plants of the next seral stage. Both these environmental constraints are applied by grazing animals and by fire. Quite often diverted seres end in a stable grassland community maintained by heavy grazing pressure or regular fires. If the environmental constraint ceases, then the grassland is rapidly invaded by species of other communities.

The effect of intensive grazing on succession has been dramatically illustrated in Europe and Australia in the last 50 years. In the early twentieth century rabbits (*Oryctolagus cuniculus*) were extremely common in Europe, especially in Britain where they are thought to have been introduced by the Normans in the eleventh or twelfth century. In 1896 a viral disease of rabbits was discovered in South America and called myxomatosis. It was introduced into France in June 1952 and reached Britain in October 1953. In only two years the disease had spread and reduced the British rabbit population to about 1 per cent of its former level.

The effect on chalk grassland was rapid. Before 1954 rabbits had heavily grazed such areas and were responsible for keeping the grass very short and thus maintaining the downland communities. Once rabbits had died out in an area, the previously short-cropped grasses grew taller and more diverse as species which had been eaten by the rabbits became established. Many plants flowered in profusion including *Primula* species and orchids. Soon, however, the diversity disappeared again as the more vigorous grasses increased in number and size. Many more seedlings survived, especially those of woody plants such as heathers (*Calluna, Erica*) and gorse (*Ulex*) and scrub composed of hawthorn (*Crataegus*), juniper (*Juniperus*) and oak (*Quercus*) grew up. In other words, the diverted sere of grassland began to change into the typical scrub-woodland succession. The effects on the chalk heath of the invasion of woody species and subsequent autogenic acidification of the soil. Such changes to the soil mean that even if the bushes are subsequently cleared, the soil is too acidic for the original chalk heath vegetation to re-establish. Autogenic changes also occur under hawthorn which seems to enrich the nutrient content of the soil. If it is cleared, weeds like nettle (*Urtica dioica*) and goose grass (*Galium aparine*) grow instead of the grassland flora.

The changes in the vegetation caused by the collapse in rabbit numbers affected various animals. Many insect species increased in numbers, but yellow ants (*Lasius flavus*) decreased as their ant hills were overshadowed. Perhaps the most serious effect of myxomatosis on a single species, other than the rabbit, was to the large blue butterfly (*Maculinea arion*). This species requires the presence of the ant *Myrmica sabuleti* to complete its life cycle. The ant only occurs in turf less than 1 cm tall and the large blue declined and finally became extinct in Britain in 1979. The sudden decrease in rabbit numbers had a catastrophic effect on their predators too. Without this ready supply of food, many carnivores such as foxes, stoats, weasels and birds of prey failed to breed and declined in numbers. This decrease in predators, plus the growth of ground cover, led to an overabundance of small mammals such as voles (*Microtus*). Weasels, owls and other carnivores soon made use of this increase in small mammals and recovered to their earlier population sizes. More details of the amazingly complex interplay of flora and fauna revealed when a single species is removed from an apparently stable ecosystem can be found in Sumption and Flowerdew.

Grazing by domestic animals such as cattle and sheep can maintain short grassland and prevent the natural succession towards scrub and woodland. Frequent fire also has an environmental effect which can produce diverted seres. Bushes and trees, especially in their young stages, are susceptible to fire and are killed by it. Grasses, on the other hand, can survive fires and regenerate from basal meristems just below ground. Tall grass, which dies back in the dry season, produces a dry, highly flammable layer of surface litter. Such grasslands will catch fire and burn quite regularly and in fact require frequent burns to maintain them and prevent succession to woodland.

Sometimes grazing can have the opposite effect from diverting succession to grassland and instead promotes scrub growth. In the African savannahs, overgrazing of areas can prevent the fires which would kill young woody species as there is no dry grass to burn. In such conditions dense unpalatable thorny thickets (*Capparis* spp.) can grow up. This sometimes occurs near rivers where hippopotamuses (*Hippopotamus amphibius*) live. Their close cropping of the vegetation prevents fires and may allow fire-susceptible woody plants to establish.

HUMAN INFLUENCE ON SUCCESSION

This chapter has concentrated so far on the natural autogenic and allogenic changes which lead to succession in vegetation or to diverted seres. But a large amount of the terrestrial surface of the Earth carries vegetation which is no longer in natural succession. Human interference for agricultural production has halted vegetation change or completely destroyed the natural vegetation.

The most obvious effects of farming practices are found on arable land where the natural vegetation has been cleared and monocultures of crops are grown instead. Such plants are usually only grown for a few months before they are harvested, the soil turned over and a new crop planted. The ground is maintained permanently at the colonisation stage of a secondary sere. The only wild plants able to survive such practices are quick-growing annual weeds which can flower and set seed before the crop is harvested. Another major managed vegetation type is grassland. Grazing by sheep, cattle or other domestic animals or mowing, either for short turf or once a year for hay, all maintain grassland and prevent succession to scrub.

The effects of human land management on preventing or diverting succession is obvious all around us. Such land use has been widespread in Europe for at least 6000 years. Within the last 500 years agricultural expansion on other continents has increased dramatically. This has resulted in the clearance of large areas of natural vegetation. Most controversial, perhaps, is the clearance of tropical rainforest for the creation of grassland for cattle. Such large-scale clearance seriously alters the ecosystems involved in a way which small-scale shifting cultivation does not. There is nutrient loss from the system via export of wood, burning and leaching. This is often followed by erosion of the topsoil, further damaging the environment. It is uncertain what would happen if a secondary sere were allowed to develop on such land. Even quite small amounts of tropical forest which have been cleared and allowed to regenerate are usually invaded by fast-growing coloniser trees. Some areas in Brazil, known to have been cleared over 800 years ago, are still clearly recognisable as they do not have the native forest growing there.

21

Biomes

INTRODUCTION

Many ecologists have tried to classify the principal vegetation types into which the world is divided. The world vegetation types have come to be known as biomes and aquatic biomes have also been recognised. It is impossible in a single chapter to look at all the biomes in any detail. Instead, we will try to give a flavour of what life must be like for some of the organisms found there. We will also consider why particular biomes are found where they are.

No two ecologists seem to agree on how many biomes there are. This is hardly surprising as a biome is not a natural unit. If individual ecosystems are the 'species' of ecology, then biomes are the phyla or divisions. No two taxonomists seem to agree on a system of classification, so it is hardly surprising that a definitive list of biomes cannot be produced. This does not mean, however, that the concept of the biome has no use. Biomes provide a convenient shorthand for describing the world's flora and fauna. Traditionally biomes have been defined mainly in terms of their vegetation. However, in our descriptions below we will look at the animals in them (and sometimes the fungi, protoctists and prokaryotes too) as well as at the plants.

WORLD'S TERRESTRIAL BIOMES

The distribution of the world's major terrestrial biomes is shown in Fig. 21.1.

Tropical Rainforest

Tropical rainforests occur near the equator where rainfall is abundant and occurs throughout the year. They are found in South and Central America, West and Equatorial Africa, South-east Asia, Indonesia and North-east Australia. This continual combination of warmth and moisture allows continuous plant growth to occur. Even today we still know desperately little about the ecology of these tremendously impressive and productive ecosystems. We do not even know to within a factor of two how many different species they contain. It has been said that if all the soil invertebrates found in a cubic meter of tropical rainforest soil are collected, there will probably be at least one species there which has never been named or described by scientists. With the possible exception of the ocean benthos, there is little doubt that tropical rainforests contain the greatest diversity of life of any of the world's biomes, though the reasons for this remain unclear.

Contrary to popular belief, undisturbed tropical rainforest is not impenetrable. This is because so little light is able to get through to the forest floor that relatively few plants can grow there. It is up in the canopy, some 30–50 meter high, that most photosynthesis occurs. It is only recently that ecologists have begun to study these canopy trees. One technique is straightforward in principle: climb the trees. This method is increasingly being used, though it has its dangers: some people have fallen and been killed in the process.

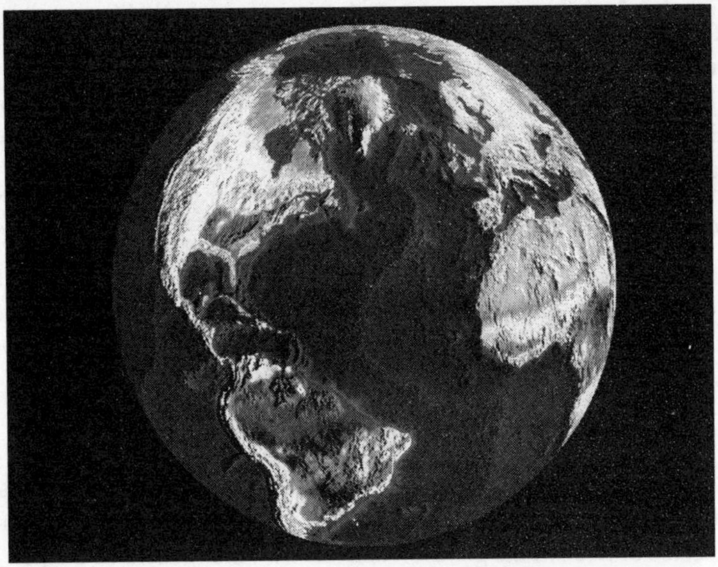

Fig. 21.1. The world's major terrestrial vegetation types.

Many trees in the canopy are covered by epiphytes plants growing on other plants. Whole communities are found up here. Indeed, nutrient cycling within the canopy is probably important and many trees produce aerial roots which absorb nutrients just as roots in soil do. Much of the animal life too is confined for most of the time to the canopy. Fruits are found throughout the year and specialised fruit-eaters have evolved among the insects, the birds and the primates.

Other animals concentrate on leaves. All species of sloth (*Choloepus* and *Bradypus*), for instance, have extremely large multi-compartmented stomachs which hold cellulose-digesting bacteria. A full stomach may account for 30 per cent of the body weight of a sloth and meals may be digested there for more than a month before passing into the relatively short intestine. Some of the adaptations shown by sloths to this indigestible diet are remarkable. Their body temperature is low, about 30–34°C and variable, falling at night, during wet weather and when the animals are inactive. This helps conserve energy. One apparent disadvantage with their slow life style is that they cannot run from predators. Camouflage, provided by two species of cyanobacteria which live in their fur and turn it green, helps reduce predation. At any event, they seem quite successful as sloths are the most abundant large mammals in tropical South American forest.

Elfinwood

To some extent elfinwoods resemble tropical rainforests in miniature. They get their name from elves, the mythical little people recorded in many European fairy stories. Elfinwoods occur high up on tropical mountains in Africa, South America and New Guinea. Here it is cold, but the climate is nonseasonal. The cold leads the trees to be small and stunted, with contorted branches and a low canopy of broad evergreen leaves. The branches are covered with curtains of lichens, mosses and filmy ferns. These epiphytes occur because the relative humidity is typically very high and elfinwoods are sometimes called cloud forests.

Tropical Seasonal Forest

Tropical seasonal forests occur in humid tropical climates with a clear dry season during which trees may lose their leaves. They are found in India, Southeast Asia, West and East Africa, South and Central America, the West Indies and Northern Australia. It is their seasonality that probably accounts for their being less diverse than the rainforests. Tropical seasonal forests are often found where monsoons occur as these provide seasonal rain.

Tropical Broad-Leaved Woodland

Tropical broad-leaved woodlands contain small trees and replace tropical seasonal forests when the climate gets drier and the soils poorer. The canopy is typically only 3–10 m high and consists of trees or shrubs with twisted branches and thick, fire-adapted bark. Tropical broad-leaved woodland occurs in northern South America, the West Indies, southern Africa and Burma.

Thornwood

Thornwoods occur in the same parts of the world as tropical broad-leaved woodlands and in other parts of Asia and in Madagascar. The climate, though, is more arid. The name of the biome comes from the presence of spiny species of trees in the genus *Acacia* and in other genera of the pea family.

Temperate Rainforest

Temperate rainforests occur along the Pacific Coast of North America and in New Zealand, Australia and Chile. Their climate is cool and maritime, lacking great variation in temperature and with abundant summer rain and much cloudiness and fog. In common with tropical rainforest, they have rain throughout the year, though at sometimes of the year the 'rain' is condensed fog. This, fog comes from moisture brought in by winds from the sea.

The trees in temperate rainforest are the tallest in the world. In Australia, the dominant tree of these forests is the mountain ash (*Eucalyptus regnans*) which can grow to over 90 m in height. In North America, the dominant tree is the redwood (*Sequoia sempervirens*) which may reach 100 m. The New Zealand forests include trees in the genus *Podocarpus*. Unfortunately, the flora of the New Zealand forests is being severely damaged by introduced deer. Indeed, the slogan 'Save the rainforests' applies even more urgently to temperate than to tropical rainforest. Canada's rainforests are disappearing at an alarming rate, mainly due to large-scale logging. According to the World Resources Institute, two-thirds of British Columbia's temperate rainforest has been degraded by logging or other development. These rainforests are home to the rare and remarkably beautiful 'white Spirit Bear'—a white form of the grizzly bear (*Ursus arctos*) which is threatened with extinction.

Temperate Deciduous Forest

Temperate deciduous forests grow in continental climates with summer rainfall and severe winters. They are dominated by broad-leaved deciduous trees and are found in Europe.

Temperate Evergreen Forest

Temperate evergreen forests occur in a variety of climates. Those in California, the Mediterranean and Southern Australia are dominated by trees with tough, evergreen and broad, but relatively small, leaves. North of California in Western USA, conifers predominate, including Douglas fir (*Pseudotsuga menziesii*) and sitka spruce (*Picea sitchensis*). In New Zealand, various trees are found including southern beech (*Nothofagus*), which also occurs in the temperate evergreen forests of Chile.

The New Zealand forests include some remarkable birds. Perhaps the most unusual are the flightless kiwis (*Apteryx* spp.). New Zealand has no native mammalian ground predators and so many of the indigenous birds have lost the ability to fly. Although kiwis have poor eyesight and cannot fly, their sense of smell is excellent and they use this to hunt out worms on which they feed. Also in New Zealand is the kakapo (*Strigops habroptilus*), a very rare bird that is almost unable to fly. It was long thought extinct, but a few are known to survive, feeding on leaves, young shoots, berries and moss. A third unusual bird of these New Zealand forests is the kea (*Nestor notabilis*). This is one of the few New Zealand species to have benefitted from introduced species. Previously vegetarian, the kea has learned to attack sheep and is now the world's only carnivorous parrot.

Across much of Europe a form of temperate evergreen forest is found on mountains where the growing season is too short for deciduous trees. The trees include silver fir (*Abies abies*), Norway spruce (*Picea abies*) and several species of pine (*Pinus*). The forest has the structure of the mountain form of the boreal forest. Mammals include the herbivorous red squirrel (*Sciurus vulgaris*) and chamois (*Rupicapra rupicapra*) and the carnivorous pine martin (*Martes martes*) and lynx (*Lynx lynx*). Among the birds are the nutcracker (*Nucifraga carypcatactes*) and crested tit (*Parus cristatus*).

Temperate Woodland

In Whittaker's classification, temperate woodland occurs when the climate is too dry for true forest. It is dominated by small trees, but these often provide only an incomplete and open canopy, so that the appearance may be of grassland with occasional scattered trees. Temperate woodland therefore lies on a continuum between forest and shrubland, grassland and semi-desert. An example of temperate woodland is pygmy conifer woodland in the western United States. Here, open woodland dominated by pines (*Pinus* spp.) and junipers (*Juniperus* spp.) is found with shrubs and grass beneath.

Temperate Shrubland

Around the shores of the Mediterranean Sea, the annual rainfall is between about 300 and 800 mm and during the summer there is usually no rain for about four months. In these months the mean temperature is 20–25°C. In the coldest months the mean temperature is about 10°C and frosts are only sporadic. Very similar climates occur in southern California, parts of Chile, the Cape of Good Hope in southern Africa and South-western Australia. The vegetation is very similar in these separated regions, even though the individual plant species differ greatly between the regions.

The shrubs are between 1 and 5 m high and have small thick drought-resistant leaves. Mediterranean maquis and Californian chaparral are among the best known examples of this biome. Heathland is also included in this classification, although it can only just be classified as shrubland. All of these types of shrubland are greatly affected by fire. These fires burn off the above-ground stems. Regrowth then occurs from root systems that survive the fire and from buried seeds.

Californian chaparral contains shrubs of *Quercus*, *Ceanothus, Arctostaphylus* and many others forming a low scrub which is highly inflammable. Fires are often started by lightning. If a fire occurs every

12 years or so, the chaparral can perpetuate itself. However, if fires occur too often, those species which rely on regeneration from seed, rather than on regrowth from roots are eliminated and the composition of the vegetation changes. If fires occur much less often than once every 12 years, fire-intolerant species such as *Prunus ilicifolia* and *Rhamnus crocea* invade.

Heathland occurs in Ireland, Scotland, England, Denmark and parts of northern Germany and southern Sweden. The vegetation is dominated by low shrubs in the heather family (*Ericaceae*). These heaths generally occur on very poor and shallow sandy podzols. The origins of heathland have been widely debated. One theory is that they are derived from forest by grazing and burning. Another theory is that they have never been forest; rather, they represent a 'sub-climax' community. Evidence from archaeological sites and from pollen analysis now favours the former theory. Further, the disappearance of tree pollen often coincides with the appearance of weed and crop pollen, strongly implicating the activities of early farmers. Much of the original clearance probably occurred in the Bronze Age or Iron Age.

Much of the heathland of northern Britain is still managed for grouse (*Lagopus scoticus*), a slow-flying bird which is easy for people to shoot. Regular burning of the heather at intervals of 10–15 years stimulates new growth on which the birds feed. The older, taller heather affords cover for breeding sites. Throughout much of Europe, heathland is being destroyed for conifer plantations, agricultural reclamation and housing developments. In Dorset, for instance, the Dartford warbler (*Sylvia undata*) is declining as the gorse (*Ulex* spp.) and tall heather in which it lives are destroyed (Fig. 21.2).

Fig. 21.2. *Dartford warbler* (*Sylvia undata*) nesting in gorse on Winfred Heath, Dorset.

Boreal Forest

Boreal forest is also known as taiga. It extends from north-eastern Europe across Russia to the Pacific Ocean and right across North America from Alaska to Newfoundland. To the north it merges into tundra; to the south it grades into deciduous forest or grassland. It can be very cold in the taiga: Eastern Siberia is the coldest area in the northern hemisphere with a mean January temperature of –50 to –60°C. Permafrost is widespread, so that the deepest layers of the soil remain frozen all year round and for much of the year snow lies on the ground. The soils are podzols.

The vegetation of the boreal forest is dominated by coniferous trees. With the exception of the larches (*Larix* spp.), these are evergreen. The important genera are pines (*Pinus* spp.), spruces (*Picea* spp.), firs (*Abies* spp.) and larches. Some broad-leaved species also occur, notably poplars (*Populus* spp.) and birches (*Betula* spp.). The ground flora includes some exquisite and delicate flowering plants such as twin flower (*Linnaea borealis*), one-flowered wintergreen (*Moneses uniflora*) and creeping lady's tresses (*Goodyera repens*).

With its harsh winters, the taiga poses severe problems for the animals found in it. Some of the birds and mammals migrate, some remain active during the long dark winter and some of the mammals hibernate. Most of the invertebrates survive as resting stages, typically eggs or pupae.

Many of the animals in the taiga have evolved specific adaptations to the conifers. The conifers produce large numbers of seeds, which provide a valuable food source for many birds. One group of finches, the crossbills, are particularly interesting because their mandibles, which cross at the tips, enable them to get at the seeds before the cones open (Fig. 21.3). Other seed-eating birds have to wait until the cones open before they can feed. Many ecological studies have been done on the animals in the taiga. Some of the best evidence for interspecific competition comes from studies on the birds there.

Fig. 21.3. A pair of crossbills (*Loxia curvirostra*) at their nest in a conifer tree in Scotland.

Savannah

Savannahs are tropical grasslands, often with scattered trees. They are most extensive in Africa, but are also found in Australia. South America and southern Asia, Savannah is subject to fire. Fires may result from lightning or be started by human. Much of the African savannah is burnt each year. Savannah trees have thick bark which insulates the living cambium from the heat of the fire.

The grasses that dominate the savannah are largely perennial. Most of them have coarse leaves that die back to form a protective layer around the meristems which give rise to new growth the following year. Most of the grasses have rhizomes and new plants usually arise vegetatively. Maximum growth rates are found at around 35°C, some 10°C warmer than the optimum for temperate grasses and a light intensity twice the optimum for temperate grasses. The reason for this is that most tropical grasses have a different photosynthetic mechanism to temperate plants. The initial compound formed when carbon dioxide is fixed has four carbon atoms. Because of this, these plants are known as C_4 plants. In temperate plants, the initial compound formed when carbon dioxide is fixed has three carbon atoms: such plants are called C_3 plants. It is this biochemical difference which allows C_4 plants to make better use of carbon dioxide at high temperatures and light intensities.

The savannah or grassland of East Africa is famed for its abundance of wildlife. Despite the activities of hunters, killing elephants for their ivory, rhinoceroses for their horns and cats for their skins, enormous numbers of large herbivores and carnivores still roam East Africa. More is known about the natural behaviour of the animals and their ecology in the Serengeti of northern Tanzania and southern Kenya than about any other region in the world.

The Serengeti still has vast herds of wildebeest (*Connochaetes taurinus*). These move around, following the same path that the rain does, as they seek the best grass on which to feed and calve (Fig. 21.4). Other large grazers found in great numbers include zebra (*Equus burchelli*), buffalo (*Syncerus caffer*) and Thomson's gazelle (*Gazella thomsoni*). Browsers include impala (*Aepyceros melampus*), giraffe (*Giraffa camelopardalis*), eland (*Tautotragus oryx*), black rhinoceros (*Diceros bicornis*) and gerenuk (*Litocranius walleri*). These herbivores support large numbers of mammalian carnivores including spotted hyena (*Crocuta crocuta*), lion (*panthem leo*), leopard (*P. pardus*), cheetah (*Acinonyx jubatus*), African wild dog (*Lycaan pictus*) and three species of jackal (*Canis* spp.).

Fig. 21.4. Migrating wildebeest (*Connochaetes taurinus*) in the Ngorongoro Crater, Tanzania.

Ecologists are only beginning to unravel the interspecific relationships in savannah ecosystems. In Africa, for example, the near extinction of elephants in certain areas may lead to an increase in woodland and accompanying decrease in grassland. This has obvious consequences for the relative numbers of grazers and browsers. The removal of one species may therefore greatly affect the population and species composition both of the flora and the fauna.

Temperate Grassland

Temperate grasslands are found across large areas of eastern Europe and Asia (steppe), central North America (prairie) and Argentina (pampas). The climate is moderately dry (between 200 and 750 mm of rain each year) and continental, so the summers are hot and the winters cold. The flora of these temperate grasslands is dominated by perennial grasses (Fig. 21.5). Associated with the grasses are broad-leaved perennials. The smaller broad-leaved plants flower early in the growing season, before the grasses reach their maximum height; the larger broad-leaved plants flower towards the end of the growing season, after the grasses have begun to die down.

Fig. 21.5. Pampas grass (*Cortaderia*) on the pampas of Uruguay, South America.

The Eurasian steppes lie between the forests to the north and the deserts to the south. The soils are chernozems and nowadays most of the steppe has been devoted to the production of wheat. Only in a few reserves and in parts of eastern Asia is natural steppe still found. Once the huge areas of wild grassland supported great herds of Przewalski's horse (*Equus przewalskii*). Przewalski's horse is now possibly extinct in the wild, though still found in a few zoos where it breeds well. It is a possible ancestor of the domestic horse which was first domesticated on the steppe about 6000 years ago. Przewalski's horse, with the wild ass (*E. hemionus*), saiga antelope (*Saiga tatarica*) and the now extinct tarpan (*E. gmelini*) are the ecological equivalents of the grazing antelopes and zebra found on the African savannah.

There are not very many species of birds on the steppe, probably because of the uniformity of the vegetation structure. Even some of the eagles (*Aquila* spp.) are forced to nest on the ground, given the absence of trees or cliffs. Nor is the steppe ideal for amphibians or reptiles. The few amphibians found are burrowers, while the short growing season gives little time for eggs to hatch and develop to adulthood. What reptiles there are tend to be ovoviviparous they produce eggs which hatch within the mother, so that the young are born live (without eggs) and can start to look for food straight away.

Alpine Shrubland

Alpine shrublands are sometimes found above the treeline on mountains. They occur in South America, Africa, the Himalayas and New Zealand. On mountains in Africa a characteristic vegetation is found in these circumstances dominated by large sparsely branched rosette plants which may be 5–8 m in height.

The genera *Dendrosenecio* and *Lobelia* are most prominent.

Alpine Grassland

The climate high up a mountain might be thought to be similar to the climate near the poles. However, day length remains constant up a mountain, while it changes as you approach the poles. Further, mountains often receive more rain and snow than regions near the poles. The thin atmosphere lets through more ultraviolet radiation and diurnal changes in temperature may be extreme. The soils may be very thin and large amounts of loose rock are often present. These differences combine to ensure that mountain vegetation differs significantly from the vegetation of the tundra.

Small alpine plants are often much loved by gardeners. They tend to grow slowly and have beautiful, delicate, brightly coloured flowers. Well known genera include *Pulsatilla*, *Hepatica*, *Saxifraga*, *Primula*, *Androsace*, *Gentiana*, *Campanula*, *Fritillaria* and *Crocus*.

The alpine marmot (*Marmota marmota*) occurs throughout the Alps of central Europe (Fig. 21.6) and is well adapted to life in the mountains. Marmots hibernate at least half the year. The entire family unit (up to 15 animals) retreats into an excavated den and sleeps huddled together throughout the winter. During hibernation, their heartbeats slow down and their body temperature falls to between 4 and 8°C. This helps the marmots to conserve energy. During the few months of the year when they are active they seem to spend much of their time standing on their hind legs checking whether there are any predators in the vicinity.

Fig. 21.6. Alpine marmot (*Marmota marmota*) feeding in front of its den.

Tundra

Huge areas of northern Asia and northern Canada are covered with low treeless vegetation known as tundra. The growing season is too short, the winters too cold and dry and the soil too unstable to support trees. Primary production is low, as might be expected. The soils are poor and the annual freezing and thawing gradually turns the soil over and sorts the soil particles out by size. On flat ground the result is often a characteristic polygonal pattern which results from the larger stones being forced to the surface.

The growing season is so short that few plants in the tundra are annuals. Sedges (*Carex* spp.), lichens and dwarf willows (*Salix* spp.) are found. When the short growing season does arrive, plant growth and flowering may be spectacular with enormous areas bathed in colour. Tremendous numbers of insects suddenly appear and for a few weeks productivity is high before winter sets in again for most of the year.

Warm Semi-Desert Scrub

Warm semi-desert scrubs occur throughout the world in dry warm-temperate and subtropical climates. The vegetation is characterised by scattered bushes, while the fauna is often rich in small mammals, birds, lizards and snakes.

Cool Semi-Desert

Cool semi-deserts are found in parts of North America, central Asia and Iran, Australia, Patagonia and the Andes. The vegetation often contains scrub and is found in places too dry for grassland.

Arctic-Alpine Semi-Desert

Arctic-alpine semi-deserts are found in arid regions either in the Arctic or on mountains above the treeline. The vegetation usually consists of low cushion plants.

Desert

Deserts are found throughout the world, mostly in the subtropical zone between 15° and 40° north and south of the equator. Among the great deserts of the world are the Namib and Kalahari deserts in the south of Africa, the Sahara in the north, the Arabian desert, the Gobi to the north of the Himalayas and perhaps the driest of them all, the coastal Atacama desert of Peru and Chile.

Some deserts receive up to 200 mm of rain a year, but most get less that 50 mm (Fig. 21. 7). Not only that, but the rain is unpredictable. For most desert organisms, the key to survival is being able to make use of the occasional heavy rainfalls. There are two extreme strategies open to plants. One is to survive as an ephemeral. Some of these ephemerals can complete their life cycle in a few weeks. They can take advantage of sudden rains by germinating, growing, flowering and seeding within the space of just 20–30 days. Other plants survive as perennials. Geophytes effectively avoid periods of drought by surviving as underground corms or bulbs. Above ground parts are produced only after heavy rains. Succulents, though, such as the cacti of America and the *Euphorbia* of Africa, have adaptations which enable them to survive above ground throughout the year. Succulents typically have thick cuticles, a very low surface area to volume ratio and sunken stomata which only open at night so as to minimise transpiration losses.

Animals in deserts face a formidable array of problems. Water is obviously scarce, but so is food. It may be very hot during the day, but cold at night. Sand makes locomotion difficult and there may be occasional blinding sandstorms. A wide variety of mechanisms have evolved to deal with these problems. Here we will concentrate on the camel (*Camelus dromedarius*).

Camels can tolerate a 30 per cent loss of their total water content. Most mammals die if 14 per cent of their water is lost. Even when severely dehydrated, camels continue to eat at almost their normal rate. In most other mammals, food is refused once the animal becomes dehydrated. Recovery from dehydration is spectacular. Camels can drink 20 per cent of their body weight in ten minutes. When a camel becomes short of water, its temperature fluctuates more than is otherwise the case (Fig. 21.8). By allowing its daily temperature to vary by as much as 7°C, the camel saves water which would otherwise have to evaporate to keep it cool. The thick fur of a camel insulates its body from the sun: in one experiment a sheared camel increased its water loss by about 50 per cent. Further adaptations to desert life are seen in the splayed hooves, ideal for walking on sand and in the hump which stores fat. This fat serves two functions, providing both energy and water on being respired.

Fig. 21.7. Sand dunes and creosote bushes from Death Valley, California, USA.

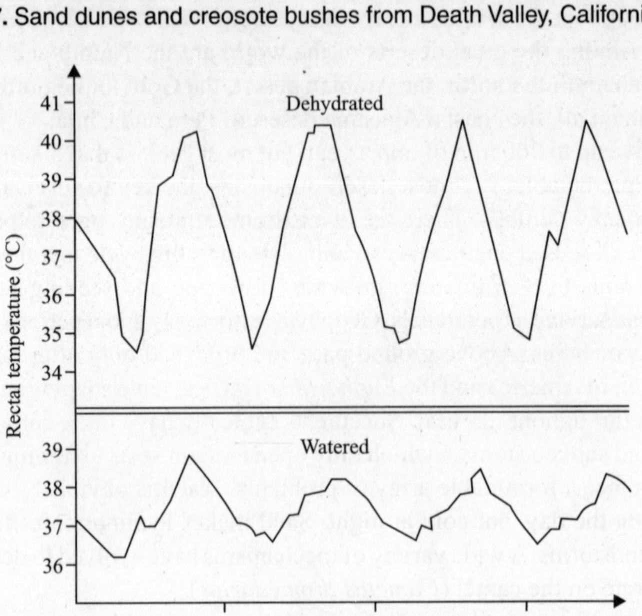

Fig. 21.8. The body temperature of a well-watered camel fluctuates on a hot day by only about 2°C. A dehydrated camel, though, must tolerate fluctuations of its body temperature of up to 7°C.

Arctic-Alpine Desert

Greenland, Antarctica and mountains well above the tree line support almost no plant life. The environment is dominated by rock, ice and snow. Arctic-alpine desert is the home for just a few wonderfully adapted animals. The Arctic is home to the polar bear (*Thalaretos maritimus*), while in Antarctica the emperor penguin (*Aptenodytes forsteri*) breeds in midwinter when it is appallingly cold and there is darkness 24 hours a day (Fig. 21.9).

Fig. 21.9. Emperor penguins (*Aptenodytes forsteri*).

WETLAND AND FRESHWATER BIOMES

Cool Temperate Bog

Bogs are wetlands that only receive water from rainfall. They tend to be nutrient poor and dominated by mosses in the genus *Sphagnum*.

Tropical Freshwater Swamp Forest

Swamps are wetlands in which trees are the dominant vegetation. The best developed tropical freshwater swamp forests are those of the Amazon basin. Minerals in the water and accompanying soils tend to be scarce, so that primary productivity is low. Consequently, there are few animals. In the blackwater tributaries of the Amazon, the water is dark brown due to the presence of humic acids which also lower the pH. Here the tamaqui fish (*Colosoma bidens*) has become dependent on the flowers and fruit which fall from the forest trees into the water.

Temperate Freshwater Swamp Forest

The vegetation of the temperate freshwater swamp forests of the Everglades. Animals are abundant. There are some 250 species of birds, 240 species of fish, 57 species of reptiles, 25 species of mammals and 17 species of amphibians. In the 1890s millions of feathers from egrets and other Florida birds were taken for hats and by 1930 over 100000 alligators were being taken each year for their hides. Stringent laws now protect these animals and the danger to the Everglade's ecosystem, as to so many wetlands, is from interference with the natural drainage.

Lakes and Ponds

As already discussed, crucial determinants of the ecology of standing freshwater are the extent to which the water is present all the year round (whether or not the pond is ephemeral) and whether nutrients are abundant (when the pond or lake is said to be eutrophic) or scarce (oligotrophic).

The North American Great Lakes are among the largest in the world. They drain a huge area which until 200 years ago consisted of deciduous forest to the south and conifer forest to the north. The history of the last 200 years has been one of progressive changes in the composition of the fish species. Fishing for salmon (*Salmo salar*) ceased by 1900. The salmon probably disappeared because of changes in the streams flowing into the lake. These streams were essential for spawning, but in the nineteenth century dams were built to operate sawmills and large quantities of sawdust polluted the streams. Repeated attempts to re-establish the salmon have failed because the stream flows are now insufficient to maintain the cool, oxygenated water and gravel bottoms which the salmon require.

By the end of the last century sea lampreys (*Petromyzon marinus*) had entered Lake Ontario. They may have come via the canal built in the early 1880s between the Hudson River and Lake Ontario. By 1946 the lamprey had spread to Lake Erie, Lake Huron, Lake Michigan and Lake Superior. The lamprey is a parasite. It attaches itself to other fish by the sucker which forms its jawless mouth. The advent of the lamprey caused many of the native fish in the Great Lakes to decline as they became parasitised.

The last 200 years has seen progressive eutrophication of the Great Lakes. Summer phosphorus concentrations in the epilimnion are now around 20–30 $\mu g/dm^3$, whereas 200 years ago there is little doubt that the lakes were clear and oligotrophic with phosphorus concentrations of less than 2 $\mu g/dm^3$. A number of measures are currently being tried to undo some of this ecological damage. There are restrictions on the use of detergents which contain phosphorus; legislation requires effluents to have phosphorus levels of less than 1 mg/dm^3; lampricides are being tried: restocking of some fish is being attempted. However, the huge size of the lakes means that it will probably be many years before the benefits, if any, are seen and a more natural balance is restored.

Streams and Rivers

Streams and rivers vary greatly. One only has to compare the huge mouth of the Amazon with a steep alpine stream (Fig. 21.10). However, as water winds its way along a stream or river from its source to its mouth, certain generalisations usually hold:

1. The speed at which the water moves (the current) decreases.
2. The volume of water passing down increases.
3. Oxygen levels fall.
4. The bed becomes composed of smaller particles (gravel gives way to silt).
5. The bed becomes less steep.
6. The summer temperature of the water often increases.
7. Eutrophication often occurs.
8. Invertebrate diversity often decreases.
9. Human influences increase.

The most obvious factors which distinguish a stream or river from a lake are the unidirectional and relatively rapid flow of the water. These mean that streams and rivers generally have communities quite different from adjacent lakes and probably account for the virtual absence of plankton in rivers or streams. The invertebrate families found in British streams and rivers and bird most characteristic of fast streams is the dipper (*Cinclus cinclus*). The dipper can walk along the stream bed and feed on the invertebrates found there.

Fig. 21.10. An alpine stream near Zermatt, Switzerland.

COASTAL AND MARINE BIOMES

Marine Rocky Shore

Often the border between the land and the sea (the littoral region) consists of sand or shingle. However, if waves or currents are particularly strong or if coastal cliffs descend to well below the low-water mark, the intertidal zone is rocky.

One of the most obvious biological features of a rocky shore is zonation. Low down on the shore organisms spend almost all their time submerged by water. High up on the shore they may be submerged only at certain times of the month; for the rest of the time they are essentially terrestrial, apart from the influence of sea spray. Because of this unidirectional gradient from the sea to the land, a characteristic zonation of organisms is found in littoral regions.

Figure 21.11 shows the regular zonation of five species of seaweed on the Isle of Cumbrae, Scotland. The ability to tolerate exposure to desiccating conditions increases in species from low to high shore levels. During the spring and summer, prolonged exposure during warm dry weather by neap tides (tides when the position of the Moon relative to the Sun causes the water level to be lower) may result in the death by desiccation of *Ascophyllum nodosum*, *Fucus spiralis* and *Pelvetia canaliculata* at the upper edges of their zones. Experiments have shown that if *Fucus spiralis* is moved up into the *Pelvetia* zone, it grows poorly and eventually dies, even though not crowded out by *Pelvetia*. It is probable that the seaweeds growing high up on the shore are prevented from growing further down the shore by competition from the seaweeds naturally found there. Certainly, *Pelvetia* and *Fucus spiralis* grow well if transplanted downshore.

It is not just the plants that are zoned on a rocky shore. Both sedentary (e.g., barnacles) and mobile (e.g., periwinkles) animals show characteristic zonations too. The factors which affect the distribution of the barnacles *Chthamalus montagui* and *Semibalanus balanoides*.

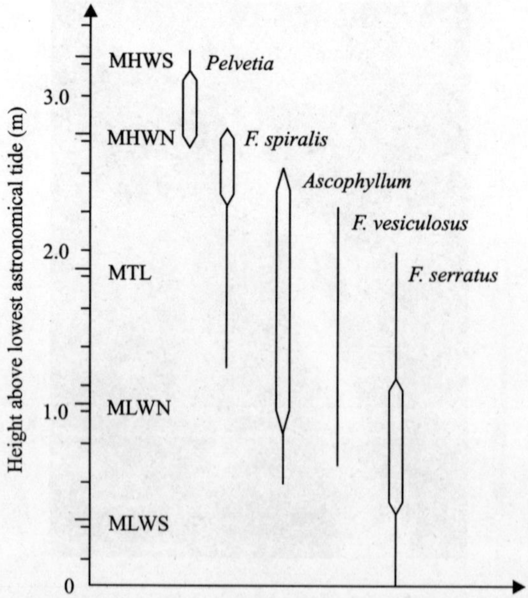

Fig. 21.11. Zonation of seaweeds on the Isle of Cumbrae, Scotland. MLWS = mean low water spring tide: MLWN = mean low water neap tide: MTL = mean tide level; MHWN = mean high water neap tide; MHWS = mean high water spring tide.

Marine Sandy Beach

Marine sandy beaches occur around the world in the littoral belt. To us they may look like an inviting environment. However, they pose considerable problems for the organisms living there. For a start, sand is abrasive and constantly on the move. That is sufficient to ensure that in the intertidal region no rooted plants can establish themselves. Further, sand is nutrient-poor, being composed almost entirely of silica (SiO_2), though there may be considerable amounts of calcareous matter from seashells. Diatoms and other algae occur in the surface layers of the sand where they can photosynthesise, but in the intertidal region the animals depend largely on organic matter such as plankton and detritus brought in by the waves.

Microscopic examination of sand grains in the intertidal zone reveals a world in miniature. Tiny creatures live between the grains. The mollusc *Caecum glabrum* and the sea cucumber *Leptosynapta minuta* are each only 2 mm long. Larger organisms are mainly worm-like or flattened. A large sieve and a long, narrow spade capable of digging down to about 25 cm or more reveals a variety of life on most British beaches. The most obvious are bivalve molluscs, including rounded cockles (*Cardium* spp.), flattened tellinids (*Tellina* spp.) and elongated razor shells (*Ensis* spp.). Worms may also be found, including lugworms (*Arenicola* spp.) which produce characteristic casts.

This rich invertebrate life often supports large numbers of wading birds. To some extent, birds with bills of different lengths specialise on different prey, as different invertebrates tend to occur at different depths in the sand. The ringed plover (*Charadrius hiaticula*) and sanderling (*Calidris alba*) have short bills, while the black-tailed godwit (*Limosa limosa*) and curlew (*Numenius arquata*) have much longer bills.

Marine Mud Flat

Marine mud flats are composed of fine silts. These tend to retain fine organic matter and the relative abundance of organic matter together with the fine particle size of the muds combine to produce anaerobic conditions only a few millimetres below the surface of the mud. This can be seen by scraping at the surface of the mud: the grey surface lies immediately above a thick black anoxic mud. Few species are adapted to survive in these conditions, but those that are may reach high densities because of the high productivity of the mud flats. Depending on where in the world the mud flat is situated, the fauna may include clams, crabs, gastropod molluscs, annelid worms and amphipod crustaceans. Marine mud flats often support even larger number of waders than are found on sandy beaches (Fig. 21.12).

Fig. 21.12. Bar-tailed godwits (*Limosa lapponica*) on the English Solway, West Cumbria.

Many of the mud flats around the British Isles have been colonised by dense stands of the grass *Spartina anglica*. In the early nineteenth century the cord-grass *S. alterniflora* was introduced from America to Southampton Water. There it hybridised with *S. maritima* giving rise to the sterile hybrid *S.* × *townsendii*. This subsequently doubled its chromosome number, becoming polyploid and formed the fertile *S. anglica* which vigorously out-competes its parents.

Temperate Salt Marsh

Salt marsh primary productivity is often high and the algae and detritus, in particular, support large numbers of insects and other invertebrates. These in turn are fed on by waders such as spotted redshank (*Tringa erythropus*) and oystercatcher (*Haematopus ostralegus*).

Coral Reef

In terms of their high productivity, high species diversity, complexities of co-evolution and sheer beauty, coral reefs are the tropical rainforests of the oceans. Corals are coelenterates, relatives of the sea anemones. Most corals are colonial and secrete a protective limestone skeleton from which they extend. At times of danger, the individual polyps can pull themselves down into this skeleton out of harm's way.

Corals are both autotrophic and heterotrophic. The polyps have tiny tentacles which catch food and stuff it into a central mouth. However, the polyps also contain symbiotic unicellular dinoflagellates apparently all belonging to one species, *Symbiodinium microadriaticum*. These exist in huge numbers and fix carbon dioxide in photosynthesis. It is the presence of these dinoflagellates which accounts for the high productivity of coral reefs.

Reef-building corals are only found in clear seas within about 50 m of the surface and where the temperature remains above 20°C throughout the year. It was Charles Darwin who recognised the three main categories of coral reef: fringing reef, barrier reef and atoll (Fig. 21.13). Darwin knew that a coral reef is made up of the limestone skeletons secreted by innumerable coral polyps, which are left behind, one on top of another. He imagined a coral reef developing around the top of a mountain subsiding into the sea slowly enough for the corals to be able to keep on secreting their skeletons and building a reef up. A critical test of Darwin's theory was provided in 1952 when engineers in the United States Navy sunk a shaft through layer after layer of increasingly ancient coral at Eniwetok atoll, 4222 feet down the drill hit volcanic rock—confirmation of Darwin's brilliant suggestion that beneath an atoll should be the sunken island on which coral polyps had begun building hundreds of thousands or millions of years ago.

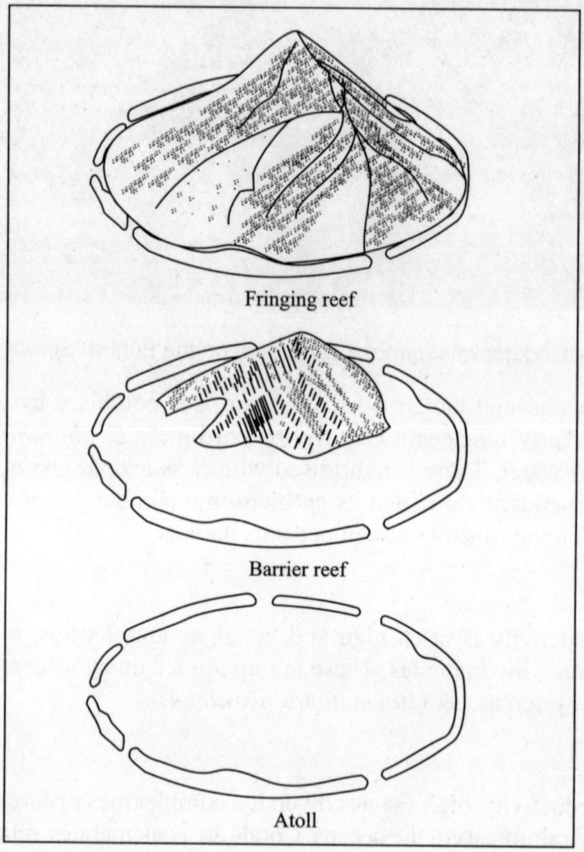

Fig. 21.13. Barrier reefs and atolls are formed from fringing reefs either by a rise in sea level or by the subsidence of the land they lie next to.

Many species of corals live together within a single reef. In turn these corals support many other animals. For a start, a number of species simply eat the corals. Many fish, including parrot fish, have teeth and jaws which enable them to attack the coral. Polyps are also eaten by a number of invertebrates of which perhaps the most notorious is the crown-of-thorns starfish (*Acanthaster planci*). These are sometimes found in enormous numbers, when they can cause immense damage to a reef. Recovery may take 40 years. It is an indication of how little we really know about coral reef ecology that it still is not known for sure whether these population explosions of *Acanthaster* are a natural feature of coral reef life or are the result of some human activity such as pollution.

Some of the most fascinating instances of coevolution in coral reefs are found in cleaning symbioses. In these, one organism, a fish or shrimp, cleans another organism, a fish, of ectoparasites. Numerous species of fish benefit as hosts in this way. About 50 species of fish and half-a-dozen species of shrimp, are known to be cleaners. Cleaning symbioses appear to have evolved independently many times. The cleaners are almost never eaten by their hosts, despite the fact that they sometimes enter into the gill-chambers and mouth of the host fish to clean. Trivers argued that cleaners are worth more to hosts alive than dead. If a host eats a cleaner, it may find it difficult to locate another cleaner when it next needs to be cleaned. The importance of cleaners to their hosts has been demonstrated in experiments where divers have removed all the cleaners from an area. Within a couple of weeks almost all the fish in the area leave. Those that remain are the territorial ones and many of these have white fuzzy infections, ulcerated sores and frayed fins.

Marine Surface Pelagic

Pelagic communities inhabit the ocean water, not the sea floor. The marine surface pelagic biome includes ocean water only down to depths penetrated by light. The maximum depth at which photosynthesis occurs is approximately 250 m and this is only in full sunlight in the most clear and tranquil open ocean water. Figures of less than 150 m are more typical for open ocean water and ones of 30 m for coastal water. The surface zone where photosynthesis occurs is known as the photic zone. It is here that phytoplankton occurs. Phytoplankton consists of photosynthetic organisms between 2 and 200 μm in diameter. Cyanobacteria, dinoflagellates, diatoms and green algae are all represented in the phytoplankton.

The phytoplankton of the marine surface pelagic biome supports large numbers of animals. As already discussed, the oceans, despite their huge size, are relatively unproductive. Further, there simply are not that many distinct niches in open water. This is probably the reason why 98 per cent of all marine species are benthic. Animals in the sea may belong to the plankton, when they are referred to as zooplankton or to the nekton. Phytoplankton and zooplankton are at the mercy of gravity, ocean currents and the wind. The nekton, though, includes larger organisms that can swim. Many animals have young that belong to the plankton before developing into larger more independent individuals.

A variety of animals comprise the nekton of the marine surface pelagic biome, including many species of fish, turtles, penguins and seals, though many of these animals also forage in the marine deep pelagic biome.

Marine Deep Pelagic

Marine deep pelagic communities are those found beneath the photic zone but still in open water: bottom-dwelling organisms are excluded. As no photosynthesis is possible, the creatures found here chiefly rely on the gravitational shower of food from the communities above. Carnivorous copepods and other crustaceans are frequent, along with Foraminifera and Radiolaria. Because so little light penetrates from above, the organisms here are either blind or produce their own light. These light-emitting organs serve a variety of functions. They can help in the detection of prey, confuse predators and help individuals to find each other for mating purposes. The further down one goes in the oceans, the less food there is. Organisms have responded to this lack of food in a variety of ways. Many deep-sea pelagic fish have hugely distensible stomachs and enormous mouths, so that fish twice as long as themselves may be taken as prey. Most deepsea pelagic fish minimise their metabolic rate: they move slowly and have poorly developed muscles. With so little food available, population densities are very low and this, coupled with the total darkness, may make finding a mate difficult. A number of species of ceratioid anglerfish have a remarkable way of solving this problem. Males are equipped with specialised teeth with which they grasp hold of a passing female. Attachment is followed by the uniting of the circulatory systems so that the male, whose sole function is to provide sperm, becomes totally dependent on the female for his nutrition. As a result, the female is far larger than the male, as you can see in Fig. 21.14.

Fig. 21.14. Female anglerfish (*Edriolychnus schmidti*) with two parasitic males.

Continental Shelf Benthos

Marine benthic organisms live on the bottom of the ocean. The considerable continental shelves that surround many of the continents lie, on average, 130 m below sea level, so that in their shallower regions benthic algae and plants can photosynthesise. Here are found the impressive kelp forests. Kelps are large brown algae, such as *Laminaria*. They attach themselves to the substratum by means of a holdfast and have one or more flat blades or fronds which can grow to more than 50 m in length. Kelps are often found in areas where ocean currents and the action of waves ensure a plentiful supply of nutrients. Their productivity may be very high, with individual fronds growing 25 cm a day. Surprisingly, few animals feed on them. Most of their production enters the food web in the form of detritus.

The continental shelf benthos supports large numbers of animals from a variety of phyla. Polychaete worms, nemertine worms, molluscs, sea squirts, bryozoans, sponges, sea spiders, crustaceans and echinoderms are all found along with a number of fish.

Deep Ocean Benthos

The deep ocean benthos extends from the edge of the continental shelf to the deepest ocean trenches. Until recently these communities were assumed to be wholly heterotrophic, but it is now known that whole communities depend on the sulphur spewed from deep volcanic vents. The other deep ocean benthic organisms depend on the slow rain of food from above that has escaped the attentions of the deep pelagic organisms. Indeed, there is evidence that decaying whale carcasses serve as 'stepping stones', allowing animals on the deep ocean floor to move from one volcanic vent to another. A surprising variety of organisms is found on the deep floor of the ocean despite the perpetual cold (nearly all the water is always at a steady 4°C), the continual darkness and the enormous pressures caused by the tremendous weight of water above.

Inevitably, our knowledge of the deep ocean benthos is rather sparse. The limited data available suggest that growth rates may be 50 times less than those typical in shallow water, so that small molluscs are thought to take 50 years to reach adulthood.

Recent research has shown that the deep ocean benthos has not remained unchanged for millions of years. The species composition of ostracods (a group of Crustacea) found on the North Atlantic seabed during the Pliocene (some 2.5 million years ago) at a depth of 2400–3400 m varies over a 10^3–10^4 year time scale. Benthic diversity seems to have been coupled with surface productivity, itself affected by the glacial-interglacial cycles at the time.

Currently, the deep ocean benthos must be one of the world's few biomes still relatively unaffected by human activity. Whether that continues to be the case remains to be seen. Every few years proposals are put forward for the deep sea mining of the nodules of manganese and other elements found on the ocean floor. Thankfully, such proposals remain uneconomic.

References

Adleman, K.W. and Lloyd, R., *Basic Concepts of Ecology*, Butterworths, London.
Allan, D.M., *Ecology and Ecosystems,* Marcel Dekker Inc., New York.
Amato, J.M., *Population Ecology*, Applied Science Publishers, London.
Birch, R.L., *Acidic Precipitation, Charles C. Thomas,* Spring field, New York.
Blackman, G.W., *Study of Plant Community*, Academic Press, London.
Brill, G.W., *Grassland Ecosystems,* Cambridge University Press, Cambridge.
Budyko, M.I., *Global Ecology,* Progress Publishers, Moscow.
Cattabeni, S., Cavallaro, A. and Galli, G., *Ecological Imperialism,* Pergamon Press, New York.
Connell, D.W. and Miller, G.J., *Ecological Applications,* John Wiley & Sons, New York.
Coolingwood, R.W., *Biological Aspects of Air Pollution*, John Wiley & Sons, New York.
Curds, C.R. and Hawkes, H.A., *Ecological Aspects of Air Pollution*, Academic Press, London.
Dix, H.M., *Environmental Pollution,* John Wiley & Sons, New York.
Duggins, P.R., *Animal Ecology,* Plenum Publishing Corporation, London.
Fruh, G.E., Gloyna, E.F. and Eckenfelder, W.W., *Global Environment,* University of Texas Press, Austin.
Gascon, Tyler, E.G. and Wayne, A.P., *Trends in Ecology and Evolution*, McGraw-Hill, New York.
Goldberg, E.D., *Atmosphere and Pollution,* Gordon and Breach, Science Publishers, New York.
Hardin, C.M., *Concepts of Ecology,* Castle Housing Publications Ltd., London.
James, A. and Evison L., *Air Pollution Technology,* John Wiley & Sons, New York.
Kathern, R.L., *Population Ecology, Distribution and Surveillance,* Harwood Academic Publishers, New York.
Lenihan, J. and Fletcher W.W., *Advances in Ecological Research*, Blackie, Glasgow and London.
Lowman, F.G., *Introduction to Atmosphere,* National Academy of Sciences, Washington DC.
Masters, G.M., *Introduction to Environmental Science and Technology,* John Wiley & Sons, New York.
McCaull, J. and Crossland, J., *Air Pollution,* Harcourt Brace Jovanovich, New York.
Michael, P., *Ecological Methods of Field and Laboratory Investigations,* Tata McGraw-Hill, New Delhi.
Neff, J.M., *Atmospheric Dispersion of Pollution,* Applied Science Publishers, London.
Odum, E.P., *Fundamentals of Ecology,* Saunders, Philadelphia.
Southwood, T.R.E., *Ecological Methods,* Chapman and Hall, London.
Stevens, G.C., *Ecology and Sustainable Development,* Applied Science Publishers, London.
Tien, H.Y., *Conservation Biology,* Van Nostrand Reinhold, New York.
Watts, Bela. G., *Ecology and Tropical Biology,* Chilton Book Company, Radnor, Pennsylvania.
Wells, N.L., *Food Webs*, Addison-Wesley Publishing Company, Philippines.

Index

U

V

W

X

Y